SpringerWienNewYork

Bernhard Hofmann-Wellenhof
Herbert Lichtenegger
Elmar Wasle

GNSS – Global Navigation Satellite Systems

GPS, GLONASS, Galileo,
and more

SpringerWienNewYork

Dr. Bernhard Hofmann-Wellenhof
Dr. Herbert Lichtenegger
Institut für Navigation und Satellitengeodäsie
Technische Universität Graz, Graz, Austria

Dr. Elmar Wasle
TeleConsult Austria GmbH, Graz, Austria

This work is subject to copyright.
All rights are reserved, whether the whole or part of the material is concerned, specifically those of translation, reprinting, re-use of illustrations, broadcasting, reproduction by photocopying machines or similar means, and storage in data banks.
Product Liability: The publisher can give no guarantee for the information contained in this book. This also refers to that on drug dosage and application thereof. In each individual case the respective user must check the accuracy of the information given by consulting other pharmaceutical literature.
The use of registered names, trademarks, etc. in this publication does not imply, even in the absence of a specific statement, that such names are exempt from the relevant protective laws and regulations and therefore free for general use.

© 2008 Springer-Verlag Wien
Printed in Austria
SpringerWienNewYork is part of Springer Science+Business Media,
springeronline.com

Typesetting: Composition by authors
Cover illustration: Elmar Wasle, Graz
Printing: Strauss GmbH, Mörlenbach, Germany
Printed on acid-free and chlorine-free bleached paper
SPIN 11524427

With 95 Figures

Library of Congress Control Number 2007938636

ISBN 978-3-211-73012-6 SpringerWienNewYork

Was du ererbt von deinen Eltern hast,
Erwirb es, um es zu besitzen.

Johann Wolfgang von Goethe
Faust, Der Tragödie Erster Teil, Nacht.
(slightly modified)

To our Parents!

Foreword

Some years ago, I discussed with the Springer publishing company the issue of a book on Galileo because the contours of the European development began to evolve and the puzzle of so many contributing pieces revealed some recognizable features. Springer, however, successfully convinced me to combine the planned Galileo book with the existing "GPS – theory and practice" book. Originally, I had declared the fifth edition of this book as the last one. However, in combination with Galileo, the generic parts could easily be used after an appropriate update. Since GLONASS showed clear indications of a soon renaissance after a very long period of insufficient maintenance with respect to the number of available satellites, a proper consideration in the book was also required.

"GNSS – GPS, GLONASS, Galileo & more" – is this title correct? This simple question is not that easily to be answered. Following a definition as given in the document A/CONF.184/BP/4 on satellite navigation and location systems published in 1998 by the United Nations as one contribution in the frame of the Third United Nations Conference on the Exploration and Peaceful Uses of Outer Space, "the Global Navigation Satellite System (GNSS) is a space-based radio positioning system that includes one or more satellite constellations, augmented as necessary to support the intended operation, and that provides 24-hour three-dimensional position, velocity, and time information to suitably equipped users anywhere on, or near, the surface of the earth (and sometimes off earth)". The definition continues with the two (current) core elements of satellite navigation systems, namely GPS and GLONASS.

The title "GNSS – GPS, GLONASS, Galileo & more" adequately fits into this definition. However, it is necessary to spend a few more sentences on this subject because the use of the acronym GNSS is not unique. By a large majority, the acronym GNSS is used for global navigation satellite systems, where the point is the plural of the word "system". Some authors like Glen Gibbons, the editor of the magazine Inside GNSS, even stress this by writing GNSSes. The plural of the word "system" is justified by the fact that there are more than one system, e.g., GPS and GLONASS, and each of these systems is a global navigation satellite system.

However, in the strict sense of the definition given above, considering these systems together and denoting them by a single term yields (now singular!) the global navigation satellite system.

There is one more item of the subtitle to be discussed. The ampersand "&" is a symbol standing for the word "and". Since there is no series comma between the word "Galileo" and the ampersand (because it does not look nicely), "Galileo & more" form one entity. This may be argued by the similarity of the current stage of development and deployment of Galileo and the other systems like the Chinese Beidou or the Indian IRNSS.

This book is a university-level introductory textbook. As long as possible, the book sticks to GNSS in the generic sense to describe various reference systems, satellite orbits, satellite signals, observables, mathematical models for positioning, data processing, and data transformation. With respect to the individual systems GPS, GLONASS, Galileo, and others, primarily the specific reference systems, the services, the space and the control segment, as well as the signal structure are described. Thus, it is really a book primarily on GNSS to cover also possibly evolving future systems.

The reader should be aware of the fact that the main scientific background of all authors is geodesy. This is narrowed even more by the fact that the Graz University of Technology is their common alma mater.

Herbert Lichtenegger and I are members of the Institute of Navigation and Satellite Geodesy of the Graz University of Technology. Elmar Wasle has been employed at the TeleConsult Austria GmbH since 2001, a company dealing with national and international research and development projects on GNSS. He is also linked to the same institute by teaching Galileo in a regular course.

This is important to stress because the geodetic background and geodetic perspectives may sometimes dominate.

Dr. Benjamin W. Remondi, retired from the US National Geodetic Survey, deserves credit and thanks. He has carefully read and corrected almost the full volume. His many suggestions and improvements, critical remarks and proposals are gratefully acknowledged.

Dipl.-Ing. Hans-Peter Ranner from the Institute of Navigation and Satellite Geodesy of the Graz University of Technology has ambitiously supported the genesis of the book. He has helped in many respects, e.g., by searching for proper references, by structuring some concepts for the derivation of formulas, or by recalculating some of the numerical examples.

Prof. Dr. Manfred Wieser from the Institute of Navigation and Satellite Geodesy of the Graz University of Technology has given us a special lecture on how to correctly interpret and fully understand rotation matrices.

The index of the book was produced using a computer program written by Elmar Wasle. This program also helped in the detection of spelling errors.

The book is compiled based on the text system LaTeX and the figures are drawn by using CorelDRAW.

We are also grateful to the Springer publishing company for supporting advice and cooperation.

April 2007

B. Hofmann-Wellenhof

Preface

The book is divided into 14 chapters. A list of acronyms, a section of references, and a detailed index, which should immediately help in finding certain topics of interest, complement the book.

The first chapter provides a brief historical review. It shows the origins of surveying and how global surveying techniques have been developed. In addition, the main aspects of positioning and navigating using satellites are described.

The second chapter deals with the reference systems, such as coordinate and time systems. The celestial and the terrestrial reference frames are explained in the section on coordinate systems, and the transformation between them is shown. The definition of different times is given in the section on time systems, together with appropriate conversion formulas.

The third chapter is dedicated to satellite orbits. This chapter specifically describes orbit representation, the determination of the Keplerian and the perturbed motion, as well as the dissemination of the orbital data.

The fourth chapter covers the satellite signal in a generic form. It shows the fundamentals of the signal structure with its various components and the principles of signal processing.

The fifth chapter deals with the observables. The data acquisition comprises code and phase pseudoranges and Doppler data. The chapter also contains the data combinations, both the phase combinations and the phase/code range combinations. Influences affecting the observables are described: the atmospheric and relativistic effects, the impact of the antenna phase center, and multipath.

The sixth chapter covers mathematical models for positioning. Models for observed data are investigated. Therefore, models for point positioning, differential positioning, and relative positioning, based on various data sets, are derived.

The seventh chapter comprises the data processing and deals with cycle slip detection and repair. This chapter also discusses phase ambiguity resolution. The method of least-squares adjustment is assumed to be known to the reader and, therefore, only a brief review (including the principle of Kalman filtering) is presented. Consequently, no details are given apart from the linearization of the mathematical models, which are the input for the adjustment procedure.

The eighth chapter links the GNSS results to a local datum. The necessary transformations are given. The combination of GNSS and terrestrial data is also considered.

The chapters nine through eleven focus on GPS, GLONASS, and Galileo. The respective reference systems for coordinates and time are explained and the space segment and the control segment are described. The signal structure is specified.

The twelfth chapter deals with additional system developments and investigations like Beidou, QZSS, and others. Also differential systems and system augmentations like WAAS, EGNOS, and others are treated.

The thirteenth chapter describes some applications of GNSS. Among some others, position determination, attitude determination, and time transfer are described in a general way. The combination of satellite-based systems per se and the integration with other systems, such as inertial navigation systems (INS), are mentioned.

The fourteenth chapter deals with the future of GNSS and how the user may benefit from the ongoing development. This future will substantially be affected by the international competition on the GNSS market.

In the list of abbreviations and acronyms, the first letter of the explanations is always a capital letter; otherwise, capital letters are generally used only if a distinct organization or a uniquely specified system is described. Within the text, the writing appears analogously. When the plural of an acronym is needed, no lowercase "s" is added. Articles before acronyms are frequently omitted even if they would be required when replacing the acronyms by their explanations.

Symbols representing a vector or a matrix are written in boldface. To indicate a transposition, the superscript "T" is used. The inner or scalar product of two vectors is indicated by a dot ".". The norm of a vector, i.e., its length, is indicated by two double-bars "||". Vectors not related to matrices are written either as column or as row vectors, whatever is more convenient.

Geodesists will not find the traditional "±" for accuracy or precision values. Implicitly, this double sign is certainly implied. Thus, if a measured distance of 100 m has a precision of 0.05 m, the geodetic writing (100 ± 0.05) m means that the solution may be in the range of 99.95 m and 100.05 m.

Internet citations within the text omit the part "http://" if the address contains "www"; therefore, "www.esa.int" means "http://www.esa.int". Also, there are no dates given to specify a guaranteed correctness of the address. Implicitly this means, all Internet addresses were tested to work properly before the manuscript was handed over to the publisher in April 2007.

Usually, Internet addresses given in the text are not repeated in the list of references. Therefore, the list of references does not yield a complete picture of the references used.

The use of the Internet sources caused some troubles for the following reason. When looking for a proper and concise explanation or definition, quite often identical descriptions were found at different locations. So the unsolvable problem arose to figure out the earlier and original source. In these cases, sometimes the decision was made, to avoid a possible conflict of interests, by omitting the citation of the source at all. This means that some phrases or sentences may have been adapted from Internet sources. On the other side, as soon as this book is released, it may and will also serve as an input source for several homepages.

The (American) spelling of a word is adopted from Webster's Dictionary of the English Language (third edition, unabridged), which may also be accessed electronically at www.merriam-webster.com. Apart from typical differences like the American "leveling" in contrast to the British "levelling", this may lead to other divergences when comparing dictionaries. Webster's Dictionary always combines the negation "non" and the following word without hyphen unless a capital letter follows. Therefore "nongravitational", "nonpropulsed", "nonsimultaneity" and "non-European" are corresponding spellings.

For the bibliographical references, the general guideline was to cite no source published before 1990. However, this rule needs an exception for some publications playing a fundamental role.

Finally, the authors do not endorse products or manufacturers. The inclusion by name of a commercial company or product does not constitute an endorsement by the authors. In principle, such inclusions were avoided whenever possible. Only those names which played a key role in the technological development are mentioned for historical purposes.

April 2007   B. Hofmann-Wellenhof   H. Lichtenegger   E. Wasle

Contents

Abbreviations		**xxi**
1	**Introduction**	**1**
1.1	The origins of surveying	1
1.2	Development of global surveying techniques	1
	1.2.1 Optical global triangulation	2
	1.2.2 Electromagnetic global trilateration	2
	1.2.3 Satellite-based positioning	3
1.3	Positioning and navigating with satellites	8
	1.3.1 Position determination	8
	1.3.2 Velocity determination	9
	1.3.3 Attitude determination	10
	1.3.4 Terminology	11
2	**Reference systems**	**13**
2.1	Introduction	13
2.2	Coordinate systems	15
	2.2.1 Definitions	15
	2.2.2 Transformation between celestial and terrestrial frames	17
	2.2.3 Transformation between terrestrial frames	21
2.3	Time systems	22
	2.3.1 Definitions	22
	2.3.2 Conversions	23
	2.3.3 Calendar	24
3	**Satellite orbits**	**27**
3.1	Introduction	27
3.2	Orbit description	27
	3.2.1 Keplerian motion	27
	3.2.2 Perturbed motion	33
	3.2.3 Disturbing accelerations	35
3.3	Orbit determination	39
	3.3.1 Keplerian orbit	40
	3.3.2 Perturbed orbit	43
3.4	Orbit dissemination	47
	3.4.1 Tracking networks	47

		3.4.2 Ephemerides	49
4	**Satellite signals**		**55**
	4.1	Introduction	55
		4.1.1 Physical fundamentals	56
		4.1.2 Propagation effects	61
		4.1.3 Frequency standards	67
	4.2	Generic signal structure	68
		4.2.1 Signal design parameter	68
		4.2.2 Carrier frequency	73
		4.2.3 Ranging code layer	75
		4.2.4 Data-link layer	84
		4.2.5 Satellite multiplexing	84
	4.3	Generic signal processing	84
		4.3.1 Receiver design	85
		4.3.2 Radio frequency front-end	87
		4.3.3 Digital signal processor	90
		4.3.4 Navigation processor	103
5	**Observables**		**105**
	5.1	Data acquisition	105
		5.1.1 Code pseudoranges	105
		5.1.2 Phase pseudoranges	106
		5.1.3 Doppler data	108
		5.1.4 Biases and noise	109
	5.2	Data combinations	111
		5.2.1 Linear phase pseudorange combinations	111
		5.2.2 Code pseudorange smoothing	113
	5.3	Atmospheric effects	116
		5.3.1 Phase and group velocity	116
		5.3.2 Ionospheric refraction	118
		5.3.3 Tropospheric refraction	128
		5.3.4 Atmospheric monitoring	138
	5.4	Relativistic effects	141
		5.4.1 Special relativity	141
		5.4.2 General relativity	144
		5.4.3 Relevant relativistic effects for GNSS	144
	5.5	Antenna phase center offset and variation	148
		5.5.1 General remarks	148
		5.5.2 Relative antenna calibration	150
		5.5.3 Absolute antenna calibration	150

Contents xv

		5.5.4	Numerical investigation	152
	5.6	Multipath		154
		5.6.1	General remarks	154
		5.6.2	Mathematical model	156
		5.6.3	Multipath reduction	158

6 Mathematical models for positioning 161

	6.1	Point positioning		161
		6.1.1	Point positioning with code ranges	161
		6.1.2	Point positioning with carrier phases	163
		6.1.3	Point positioning with Doppler data	165
		6.1.4	Precise point positioning	166
	6.2	Differential positioning		169
		6.2.1	Basic concept	169
		6.2.2	DGNSS with code ranges	170
		6.2.3	DGNSS with phase ranges	171
		6.2.4	Local-area DGNSS	172
	6.3	Relative positioning		173
		6.3.1	Basic concept	173
		6.3.2	Phase differences	174
		6.3.3	Correlations of the phase combinations	178
		6.3.4	Static relative positioning	183
		6.3.5	Kinematic relative positioning	185
		6.3.6	Pseudokinematic relative positioning	187
		6.3.7	Virtual reference stations	188

7 Data processing 193

	7.1	Data preprocessing		193
		7.1.1	Data handling	193
		7.1.2	Cycle slip detection and repair	194
	7.2	Ambiguity resolution		202
		7.2.1	General aspects	202
		7.2.2	Basic approaches	205
		7.2.3	Search techniques	214
		7.2.4	Ambiguity validation	236
	7.3	Adjustment, filtering, and quality measures		238
		7.3.1	Theoretical considerations	238
		7.3.2	Linearization of mathematical models	250
		7.3.3	Network adjustment	257
		7.3.4	Dilution of precision	262
		7.3.5	Quality parameters	266

| | | 7.3.6 | Accuracy measures . . . . | 272 |

8 Data transformation — 277
8.1 Introduction — 277
8.2 Coordinate transformations — 277
8.2.1 Cartesian coordinates and ellipsoidal coordinates — 277
8.2.2 Global coordinates and local-level coordinates — 280
8.2.3 Ellipsoidal coordinates and plane coordinates — 283
8.2.4 Height transformation — 290
8.3 Datum transformations — 293
8.3.1 Three-dimensional transformation — 293
8.3.2 Two-dimensional transformation — 297
8.3.3 One-dimensional transformation — 300
8.4 Combining GNSS and terrestrial data — 302
8.4.1 Common coordinate system — 302
8.4.2 Representation of measurement quantities — 303

9 GPS — 309
9.1 Introduction — 309
9.1.1 Historical review — 309
9.1.2 Project phases — 309
9.1.3 Management and operation — 310
9.2 Reference systems — 313
9.2.1 Coordinate system — 313
9.2.2 Time system — 315
9.3 GPS services — 315
9.3.1 Standard positioning service — 316
9.3.2 Precise positioning service — 318
9.3.3 Denial of accuracy and access — 319
9.4 GPS segments — 322
9.4.1 Space segment — 322
9.4.2 Control segment — 324
9.5 Signal structure — 327
9.5.1 Carrier frequencies — 329
9.5.2 PRN codes and modulation — 329
9.5.3 Navigation messages — 337
9.6 Outlook — 339
9.6.1 Modernization — 339
9.6.2 GPS III — 340

10 GLONASS — 341

- 10.1 Introduction — 341
  - 10.1.1 Historical review — 341
  - 10.1.2 Project phases — 342
  - 10.1.3 Management and operation — 343
- 10.2 Reference systems — 345
  - 10.2.1 Coordinate system — 345
  - 10.2.2 Time system — 346
- 10.3 GLONASS services — 347
  - 10.3.1 Standard positioning service — 347
  - 10.3.2 Precise positioning service — 348
- 10.4 GLONASS segments — 348
  - 10.4.1 Space segment — 348
  - 10.4.2 Control segment — 351
- 10.5 Signal structure — 354
  - 10.5.1 Carrier frequencies — 356
  - 10.5.2 PRN codes and modulation — 357
  - 10.5.3 Navigation messages — 360
- 10.6 Outlook — 362

11 Galileo — 365

- 11.1 Introduction — 365
  - 11.1.1 Historical review — 365
  - 11.1.2 Project phases — 367
  - 11.1.3 Management and operation — 368
- 11.2 Reference systems — 369
  - 11.2.1 Coordinate system — 369
  - 11.2.2 Time system — 369
- 11.3 Galileo services — 370
  - 11.3.1 Open service — 370
  - 11.3.2 Commercial service — 370
  - 11.3.3 Safety-of-life service — 371
  - 11.3.4 Public regulated service — 372
  - 11.3.5 Search and rescue service — 372
- 11.4 Galileo segments — 373
  - 11.4.1 Space segment — 374
  - 11.4.2 Ground segment — 378
- 11.5 Signal structure — 382
  - 11.5.1 Carrier frequencies — 383
  - 11.5.2 PRN codes and modulation — 384
  - 11.5.3 Navigation messages — 390

    11.6 Outlook 394

12 More on GNSS   397
    12.1 Global systems 397
        12.1.1 Comparison of GPS, GLONASS, and Galileo 397
        12.1.2 Beidou-2 / Compass 401
        12.1.3 Other global systems 403
    12.2 Regional systems 406
        12.2.1 Beidou-1 406
        12.2.2 QZSS 409
        12.2.3 Other regional systems 414
    12.3 Differential systems 415
        12.3.1 Principles 415
        12.3.2 Differential correction domains 416
        12.3.3 Examples of differential systems 417
    12.4 Augmentation systems 420
        12.4.1 Space-based augmentation systems 421
        12.4.2 Ground-based augmentation systems 426
    12.5 Assistance systems 429
    12.6 Outlook 430

13 Applications   431
    13.1 Products of GNSS measurements 431
        13.1.1 Satellite coordinates 431
        13.1.2 Position determination 431
        13.1.3 Velocity determination 440
        13.1.4 Attitude determination 441
        13.1.5 Time transfer 445
        13.1.6 Other products 445
    13.2 Data transfer and formats 447
        13.2.1 RTCM format 447
        13.2.2 RINEX format 449
        13.2.3 NMEA format 450
    13.3 System integration 450
        13.3.1 GNSS and inertial navigation systems 452
        13.3.2 Radionavigation plans 452
    13.4 User segment 453
        13.4.1 Receiver features 453
        13.4.2 Control networks 457
        13.4.3 Information services 458
    13.5 Selected applications 460

	13.5.1 Navigation	460
	13.5.2 Surveying and mapping	464
	13.5.3 Scientific applications	465

14 Conclusion and outlook   **467**

References   **471**

Subject index   **501**

Abbreviations

ACF	Autocorrelation function
A/D	Analog to digital
AFB	Air force base
AFS	Atomic frequency standard
AGC	Automatic gain control
AGNSS	Assisted (or aided) GNSS
AL	Alarm/alert limit
AltBOC	Alternative binary offset carrier
AOC	Auxiliary output chip
AOR	Atlantic ocean region
APOS	Austrian positioning service
ARGOS	Advanced research and global observation satellite
ARNS	Aeronautical radionavigation service
AROF	Ambiguity resolution on-the-fly
ARP	Antenna reference point
ARPL	Aeronomy and Radiopropagation Laboratory
A-S	Anti-spoofing
BCRS	Barycentric celestial reference system
BDT	Barycentric dynamic time
BER	Bit error rate
BIPM	Bureau International des Poids et Mesures (International Bureau of Weights and Measures)
BNTS	Beidou navigation test satellite
BOC	Binary offset carrier
BPF	Band-pass filter
BPSK	Binary phase-shifted key
C/A	Coarse/acquisition
CDMA	Code division multiple access
CEP	Celestial ephemeris pole
CEP	Circular error probable
CHAMP	Challenging minisatellite payload (mission)
CIGNET	Cooperative international GPS network
CIO	Conventional international origin
CIR	Cascade integer resolution
C/N_0	Carrier-to-noise power density ratio
C/NAV	Commercial navigation message

CNES	Centre National d'Etudes Spatiales (National Center for Space Research)
CODE	Center for Orbit Determination in Europe
CORS	Continuously operating reference station
COSPAS	Cosmicheskaya sistyema poiska avariynich sudov (Space system for the search of vessels in distress)
CRC	Cyclic redundancy check
CRF	Celestial reference frame
CRPA	Controlled reception pattern antenna
CRS	Celestial reference system
CS	Commercial service
CSOC	Consolidated Space Operations Center
DARPA	Defense Advanced Research Projects Agency
DASS	Distress alerting satellite system
DD	Double-difference
DEM	Digital elevation model
DGNSS	Differential GNSS
DGPS	Differential GPS
DLL	Delay lock loop
DMA	Defense Mapping Agency
DME	Distance measuring equipment
DoD	Department of Defense
DOP	Dilution of precision
DORIS	Doppler orbitography by radiopositioning integrated on satellite
DoT	Department of Transportation
DRMS	Distance root mean square
DSP	Digital signal processor
EC	European Community
ECAC	European Civil Aviation Conference
ECEF	Earth-centered, earth-fixed (coordinates)
EDAS	EGNOS data access system
EGM96	Earth Gravitational Model 1996
EGNOS	European geostationary navigation overlay service
EIRP	Equivalent isotropic radiated power
EKF	Extended Kalman filter
ENU	East, north, up
EOP	Earth orientation parameter
ERNP	EU radionavigation plan
ERTMS	European rail traffic management system
ESA	European Space Agency
ESTB	EGNOS system test bed

EU	European Union
EUPOS	European position (determination system)
EWAN	EGNOS wide-area network
FAA	Federal Aviation Administration
FARA	Fast ambiguity resolution approach
FASF	Fast ambiguity search filter
FCC	Federal Communications Commission
FDF	Flight dynamics facility
FDMA	Frequency division multiple access
FEC	Forward error correction
FGCS	Federal Geodetic Control Subcommittee
FLL	Frequency lock loop
F/NAV	Freely accessible navigation message
FOC	Full operational capability
FRP	Federal radionavigation plan
GACF	Ground asset control facility
GAGAN	GPS and geoaugmented navigation
GALA	Galileo overall architecture definition
GATE	Galileo test and development environment
GBAS	Ground-based augmentation system
GCC	Ground control center
GCRS	Geocentric celestial reference system
GCS	Ground control segment
GDGPS	Global DGPS
GDOP	Geometric dilution of precision
GEO	Geostationary (satellite)
GES	Ground earth station
GGSP	Galileo geodetic service provider
GGTO	GPS to Galileo time offset
GIM	Global ionosphere map
GIOVE	Galileo in-orbit validation element
GIS	Geographic information system
GIVD	Grid ionospheric vertical delay
GIVE	Grid ionospheric vertical error
GJU	Galileo Joint Undertaking
GLONASS	Global'naya Navigatsionnaya Sputnikovaya Sistema (Global Navigation Satellite System)
GMS	Ground mission segment
G/NAV	Governmental navigation message
GNSS	Global navigation satellite system
GOC	Galileo operating company

GOCE	Gravity field and steady-state ocean circulation explorer (mission)
GOTEX	Global orbit tracking experiment
GPS	Global Positioning System
GRACE	Gravity recovery and climate experiment
GRAS	Ground-based regional augmentation system
GRS-80	Geodetic Reference System 1980
GSA	GNSS Supervisory Authority
GSFC	Goddard Space Flight Center
GSS	Galileo sensor station
GST	Galileo system time
GSTB	Galileo system test bed
GTRF	Galileo terrestrial reference frame
HANDGPS	High-accuracy nationwide DGPS
HDOP	Horizontal dilution of precision
HEO	Highly inclined elliptical orbit (satellite)
HIRAN	High range navigation (system)
HLD	High level definition
HMI	Hazardously misleading information
HOW	Hand-over word
HPL	Horizontal protection level
HTTP	Hypertext transfer protocol
IAC	Information Analytical Center
IAG	International Association of Geodesy
IAU	International Astronomical Union
ICAO	International Civil Aviation Organization
ICD	Interface control document
ICRF	International celestial reference frame
IERS	International Earth Rotation Service
IF	Integrity flag; intermediate frequency
IGEB	Interagency GPS Executive Board
IGEX-98	International GLONASS Experiment 1998
IGP	Ionospheric grid point
IGS	International GNSS (formerly GPS) Service for Geodynamics
IMO	International Maritime Organization
I/NAV	Integrity navigation message
INS	Inertial navigation system
IOC	Initial operational capability
ION	Institute of Navigation
IOR	Indian ocean region
IOV	In-orbit validation
IPF	Integrity processing facility

Abbreviations

I/Q	In-phase/quadrature phase
IRM	IERS reference meridian
IRNSS	Indian Regional Navigation Satellite System
IRP	IERS reference pole
ITCAR	Integrated three-carrier ambiguity resolution
ITRF	International terrestrial reference frame
ITS	Intelligent transportation system
ITU	International Telecommunication Union
ITU-R	ITU, Radiocommunication (sector)
IVHS	Intelligent vehicle highway system
IWV	Integrated water vapor
JD	Julian date
JGS	Japanese geodetic system
JPL	Jet Propulsion Laboratory
JPO	Joint Program Office
LAAS	Local-area augmentation system
LAD	Local-area differential
LADGNSS	Local-area DGNSS
LAMBDA	Least-squares ambiguity decorrelation adjustment
LBS	Location-based service
LEO	Low earth orbit (satellite)
LEP	Linear error probable
LFSR	Linear feedback shift register
LHCP	Left-handed circular polarization
LLR	Lunar laser ranging
LNA	Low-noise amplifier
LO	Local oscillator
LOGIC	Loran GNSS interoperability channel
LOP	Line of position
LSAST	Least-squares ambiguity search technique
LUT	Local user terminal
MBOC	Multiplexed binary offset carrier
MCAR	Multiple carrier ambiguity resolution
MCC	Master control center
MCF	Mission control facility
MCS	Master control station
MDDN	Mission data dissemination network
MEDLL	Multipath estimating delay lock loop
MEO	Medium earth orbit (satellite)
MGF	Message generation facility
MJD	Modified Julian date

MOPS	Minimum operational performance standards
MRSE	Mean radial spherical error
MS	Monitoring station
MSAS	MTSAT space-based augmentation system
MSF	Mission support facility
MTSAT	Multifunctional transport satellite
NAD-27	North American Datum 1927
NAGU	Notice advisories to GLONASS users
NANU	Notice advisories to NAVSTAR users
NAP	NDS analysis package
NASA	National Aeronautics and Space Administration
NAVCEN	Navigation Center
NAVSTAR	Navigation system with timing and ranging
NCO	Numerically controlled oscillator
NDGPS	Nationwide DGPS
NDS	Nuclear detection system
NGA	National Geospatial-Intelligence Agency
NGS	National Geodetic Survey
NIMA	National Imagery and Mapping Agency
NIS	Navigation information service
NLES	Navigation land earth station
NMEA	National Marine Electronics Association
NNSS	Navy Navigation Satellite System
NOAA	National Oceanic and Atmospheric Administration
NSGU	Navigation signal generator unit
NTRIP	Networked transport of RTCM via Internet protocol
OCS	Operational control segment
OEM	Original equipment manufacturer
OMEGA	Optimal method for estimating GPS ambiguities
OS	Open service
OSPF	Orbit determination and time synchronization processing facility
OSU	Ohio State University
OTF	On-the-fly
OTR	On-the-run
PC	Personal computer
PCO	Phase center offset
PCV	Phase center variation
PDOP	Position dilution of precision
PE	Position error
PE-90	Parameter of the Earth 1990
PHM	Passive hydrogen maser

PL	Protection level
PLL	Phase lock loop
PNT	Positioning, navigation, and timing
POR	Pacific ocean region
PPP	Precise point positioning; public-private partnership
PPS	Precise positioning service
PPS-SM	PPS-security module
PRARE	Precise range and range rate equipment
PRC	Pseudorange correction
PRN	Pseudorandom noise
PRS	Public regulated service
PSD	Power spectral density
PSK	Phase-shifted key
PTF	Precise timing facility
PVS	Position and velocity of the satellite
PVT	Position, velocity, and time
PZ-90	Parametry Zemli 1990 (Parameter of the Earth 1990)
QASPR	Qualcomm automatic satellite position reporting
QPSK	Quadrature phase-shifted key
QZSS	Quasi-Zenith Satellite System
RAFS	Rubidium atomic frequency standard
RAIM	Receiver autonomous integrity monitoring
RF	Radio frequency
RHCP	Right-handed circular polarization
RIMS	Receiver integrity monitoring station
RINEX	Receiver independent exchange (format)
RIS	River information service
RMS	Root mean square
RNP	Required navigation performance
RNSS	Radionavigation satellite service
RRC	Range rate correction
RTCM	Radio Technical Commission for Maritime (Services)
RTK	Real-time kinematic
RX	Receiver/receive
SA	Selective availability
SAASM	Selective ability anti-spoofing module
SAIF	Submeter accuracy with integrity function
SAPOS	Satellite positioning service
SAR	Search and rescue
SARPS	Standards and recommended practices

SARSAT	SAR satellite-aided tracking
SBAS	Space-based augmentation system
SCCF	Spacecraft constellation control facility
SD	Selective denial
SDCM	System for differential correction and monitoring
SDDN	Satellite data distribution network
SDGPS	Satellite DGPS
SEP	Spherical error probable
SGS-85	Soviet Geodetic System 1985
SGS-90	Soviet Geodetic System 1990
SIL	Safety integrity level
SINEX	Software independent exchange (format)
SIS	Signal in space
SISA	SIS accuracy
SISE	SIS error
SISMA	SIS monitoring accuracy
SISNeT	SIS over Internet
SLR	Satellite laser ranging
S/N	Signal-to-noise ratio
SNAS	Satellite navigation augmentation system
SoL	Safety of life
SOP	Surface of position
SPF	Service products facility
SPS	Standard positioning service
SSR	Sum of squared residuals
SU	Soviet Union
SV	Space vehicle
TACAN	Tactical air navigation
TAI	Temps atomique international (International atomic time)
TCAR	Three-carrier ambiguity resolution
TCS	Tracking control station
TDMA	Time division multiple access
TDOP	Time dilution of precision
TDRSS	Tracking and data relay satellite system
TDT	Terrestrial dynamic time
TEC	Total electron content
TECU	TEC units
THR	Tolerable hazard rate
TLM	Telemetry word
TOA	Time of arrival

TOW	Time of week
TRF	Terrestrial reference frame
TT	Terrestrial time
TTA	Time to alarm/alert
TT&C	Telemetry, tracking and control (or command)
TTFF	Time to first fix
TV	Television
TVEC	Total vertical electron content
TX	Transmitter/transmit
UDRE	User differential range error
UERE	User equivalent range error
ULS	Uplink station
URE	User range error
US	United States (of America)
USA	Unites States of America
USNO	US Naval Observatory
USSR	Union of Soviet Socialist Republics
UT	Universal time
UTC	Coordinated universal time
UTM	Universal transverse Mercator
VBS	Virtual base station
VDB	VHF data broadcast
VDOP	Vertical dilution of precision
VHF	Very high frequency
VLBI	Very long baseline interferometry
VPL	Vertical protection level
VRS	Virtual reference station
WAAS	Wide-area augmentation system
WAD	Wide-area differential
WARTK	Wide-area real-time kinematic
WGS-84	World Geodetic System 1984
WMS	Wide-area master station
WRS	Wide-area reference station
XOR	Exclusive-or

1 Introduction

1.1 The origins of surveying

Since the dawn of civilization, man has looked to the heavens with awe searching for portentous signs. Some of these men became experts in deciphering the mystery of the stars and developed rules for governing life based upon their placement. The exact time to plant the crops was one of the events that was foretold by the early priest astronomers who in essence were the world's first surveyors. Today, it is known that the alignment of such structures as the pyramids and Stonehenge was accomplished by celestial observations and that the structures themselves were used to measure the time of celestial events such as the vernal equinox.

Some of the first known surveyors were Egyptian surveyors who used distant control points to replace property corners destroyed by the flooding Nile River. Later, the Greeks and Romans surveyed their settlements. The Dutch surveyor Snell van Royen was the first who measured the interior angles of a series of interconnecting triangles in combination with baselines to determine the coordinates of points long distances apart. Triangulations on a larger scale were conducted by the French surveyors Picard and Cassini to determine a baseline extending from Dunkirk to Collioure. The triangulation technique was subsequently used by surveyors as the main means of determining accurate coordinates over continental distances.

The chain of technical developments from the early astronomical surveyors to the present satellite geodesists reflects man's desire to be able to master time and space and to use science to foster society. The surveyor's role in society has remained unchanged from the earliest days; that is to determine land boundaries, provide maps of his environment, and control the construction of public and private works.

1.2 Development of global surveying techniques

The use of triangulation (later combined with trilateration and traversing) was limited by the line of sight. Surveyors climbed to mountain tops and developed special survey towers (e.g., Bilby towers) to extend this line of sight usually by small amounts. The series of triangles was generally oriented or fixed by astronomic points where selected stars had been observed to determine the position of that point on the surface of the earth. Since these astronomic positions could be in error by hundreds of meters, each continent was virtually (positionally) isolated and their interrelationship was imprecisely known.

1.2.1 Optical global triangulation

Some of the first attempts to determine the interrelationship between the continents were made using the occultation of certain stars by the moon. This method, known as the lunar-distance method, was cumbersome and had only limited success in the late 1600s. The launch of the first artificial satellite, i.e., the Russian Sputnik satellite on October 4, 1957, had tremendously advanced the connection of the various geodetic world datums. In the beginning of the era of artificial satellites, an optical method, based (in principle) on the stellar triangulation method developed in Finland as early as 1946, was applied very successfully. The worldwide satellite triangulation program, often called the BC-4 program (after the camera that was used), for the first time determined the interrelationships of the major world datums. This method involved photographing special reflective satellites against a star background with a metric camera that was fitted with a specially manufactured chopping shutter. The image that appeared on the photograph consisted of a series of dots depicting each star's path and a series of dots depicting the satellite's path. The coordinates of selected dots were precisely measured using a photogrammetric comparator, and the associated spatial directions from the observing site to the satellite were then processed using an analytical photogrammetric model. Photographing the same satellite from a neighboring site simultaneously and processing the data in an analogous way yields another set of spatial directions. Each pair of corresponding directions forms a plane containing the observing points and the satellite. The intersection of at least two planes results in the spatial direction between the observing sites. In the next step, these oriented directions were used to construct a global network with the scale being derived from several terrestrial traverses. An example is the European baseline running from Tromsø in Norway to Catania on Sicily. The main problem in using this optical technique was that clear sky was required simultaneously at a minimum of two observing sites separated by some 4 000 km, and the equipment was massive and expensive. Thus, optical direction measurement was soon supplanted by the electromagnetic ranging technique because of all-weather capability, greater accuracy, and lower cost of the newer technique.

1.2.2 Electromagnetic global trilateration

First attempts to (positionally) connect the continents by electromagnetic techniques was by the use of an electronic high-range navigation (HIRAN) system developed during World War II to position aircraft. Beginning in the late 1940s, HIRAN arcs of trilateration were measured between North America and Europe in an attempt to determine the difference in their respective datums. A significant technological breakthrough occurred in 1957 after the launch of Sputnik when scientists around the world (e.g., at the Johns Hopkins University Applied Physics Labora-

1.2 Development of global surveying techniques

tory) experienced that the Doppler shift in the signal broadcast by a satellite could be used as an observable to determine the exact time of closest approach of the satellite. This knowledge, together with the ability to compute satellite ephemerides according to Kepler's laws, led to the present capability of instantaneously determining precise position anywhere in the world.

1.2.3 Satellite-based positioning

Introduction

Satellite-based positioning is the determination of positions of observing sites on land or at sea, in the air and in space by means of artificial satellites. Thus, be aware not to misinterpret the frequently used shortened notation "satellite positioning".

Operational satellite-based positioning systems (as discussed in this textbook) assume that the satellite positions are known at every epoch.

The term global navigation satellite system (GNSS) covers throughout this textbook each individual global satellite-based positioning system as well as the combination or augmentation of these systems.

A historical review on the development of satellite-based positioning can be found, e.g., in Guier and Weiffenbach (1997) or Ashkenazi (2006).

Principle of satellite-based positioning

Operational satellites primarily provide the user with the capability of determining his position, expressed, for example, by latitude, longitude, and height. This task is accomplished by the simple resection process using ranges or range differences measured to satellites.

Imagine the satellites frozen in space at a given instant. The space vector ϱ^s relative to the center of the earth (geocenter) of each satellite (Fig. 1.1) can be computed from the ephemerides broadcast by the satellite by an algorithm presented in Chap. 3. If the receiver on ground defined by its geocentric position vector ϱ_r employed a clock that was set precisely to system time (Sect. 2.3), the geometric distance or range ϱ to each satellite could be accurately measured by recording the run time required for the (coded) satellite signal to reach the receiver. Each range defines a sphere (more precisely: surface of a sphere) with its center at the satellite position. Hence, using this technique, ranges to only three satellites would be needed since the intersection of three spheres yields the three unknowns (e.g., latitude, longitude, and height) which could be determined from the three range equations

$$\varrho = \|\varrho^s - \varrho_r\|. \tag{1.1}$$

Modern receivers apply a slightly different technique. They typically use an inexpensive crystal clock which is set approximately to system time. Thus, the clock of

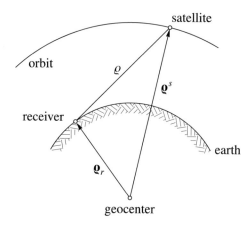

Fig. 1.1. Principle of satellite-based positioning

the receiver on ground is offset from true system time and, because of this offset, the distance measured to the satellite differs from the geometric range. Therefore, the measured quantities are called pseudoranges R since they represent the geometric range plus a range correction $\Delta\varrho$ resulting from the receiver clock error or clock bias δ. A simple model for the pseudorange is

$$R = \varrho + \Delta\varrho = \varrho + c\,\delta \tag{1.2}$$

with c being the speed of light.

Four simultaneously measured pseudoranges are needed to solve for the four unknowns; namely the three components of position plus the clock bias.

Satellite-based systems

Early systems

The immediate predecessor of today's modern positioning systems is the Navy Navigation Satellite System (NNSS), also called Transit system. This system was conceived in the late 1950s and developed in the 1960s by the US military, primarily, to determine the coordinates (and time) of vessels at sea and for military applications on land. Civilian use of this satellite system was eventually authorized, and the system became used worldwide both for navigation and surveying.

The system matured to six satellites in nearly circular polar low earth orbits (LEO) at altitudes of about 1 100 km. The satellites transmitted two carrier frequencies (150 and 400 MHz). Onto the carriers, time marks and orbital information were modulated.

1.2 Development of global surveying techniques

Receivers which could only track one of the two frequencies (denoted as single-frequency receivers) achieved position accuracies in the 100 m range. For dual-frequency receivers, the accuracy improved to about 20 m. Some of the early Transit experiments by the former US Defense Mapping Agency (DMA) and the US Coast & Geodetic Survey showed that accuracies of about one meter could be obtained by occupying a point for several days (or even weeks) or reducing the number of observations using the postprocessed precise ephemerides of the satellites. Later, groups of Doppler receivers in translocation mode (i.e., simultaneous observations) were used to determine the relative coordinates of points to submeter accuracy using the broadcast orbital information which are less accurate than the precise ephemerides. This system employed essentially the same Doppler observable used to track the Sputnik satellite; however, the orbits of the Transit satellites were precisely determined by tracking them at widely spaced fixed reference sites. The actual observation was the number of Doppler cycles (i.e., counts) between precise 2-minute timing marks from the onboard clock.

More details on Transit can be found in, e.g., Hofmann-Wellenhof et al. (2003: pp. 169–172). Note, however, that Transit is no longer operational since the end of 1996.

The Russian Tsikada (also written Cicada) system transmits the same two carrier frequencies as Transit and is similar to it with respect to the achievable accuracies. Ten LEO spacecraft were deployed in two complementary constellations, a military and a civilian network. The older (i.e., military) constellation consists of six satellites, where the first satellite was launched in 1974. The later (i.e., civilian) constellation has four satellites. Contrary to Transit, the Tsikada system is still operational.

The early systems had two major shortcomings. The main problem were the large time gaps between two satellite passes. In the case of early Transit, for example, nominally a satellite passed overhead every 90 minutes and users had to interpolate their position between "fixes" or passes. The second problem was the relatively low navigation accuracy. Particularly, the height determination was poor.

Present and future systems
The navigation system with timing and ranging (NAVSTAR) Global Positioning System (GPS) was developed by the US military to overcome the shortcomings of the early systems. In contrast to these systems, GPS answers the questions "What time, what position, and what velocity is it?" quickly, accurately, and inexpensively anywhere on the globe at any time. More details on GPS are given in Chap. 9.

The Global Navigation Satellite System (GLONASS) is the Russian counterpart to GPS and is operated by the Russian military. GLONASS differs from GPS in terms of the control segment, the space segment, and the signal structure. Details are given in Chap. 10.

Galileo is the European contribution to the future GNSS and will be discussed in detail in Chap. 11.

A Chinese system called Compass, which is the evolution of the first-generation regional system Beidou, is presently under development. Details on the system are provided in Sect. 12.1.2.

As mentioned, GNSS implies several existing systems like GPS, GLONASS, or Galileo. In addition, these systems are supplemented by space-based augmentation systems (SBAS) or ground-based augmentation systems (GBAS). Examples of SBAS are the US wide-area augmentation system (WAAS), the European geostationary navigation overlay service (EGNOS), or the Japanese multifunctional transport satellite (MTSAT) space-based augmentation system (MSAS). These systems augment the existing medium earth orbit (MEO) satellite constellations with geostationary (GEO) or geosynchronous satellites. For more details, the reader is referred to Sect. 12.4.

GNSS segments

Space segment

In order to provide a continuous global positioning capability, a constellation with a sufficient number of satellites must be developed for each GNSS to ensure that (at least) four satellites are simultaneously electronically visible at every site.

The selection of the satellite constellation has to follow various optimization routines. The design criteria are, without being exhaustive, the user position accuracy, the satellite availability, service coverage, and the satellite geometry. Furthermore, the size and weight of the satellites have to be taken into account, which are interrelated with the launch vehicle constraints and the costs of deployment, maintenance, and replenishment. The satellite orbit defines the degree of perturbing effects, which influence the maintenance maneuvers. With respect to the altitude, a distinction is made between LEO, MEO, and GEO satellites. Also, the satellite orbit influences the selection of the transmitted power. For MEO satellites for example, the effective earthward transmitted power is in the range of 25 watt. The transmission loss, however, attenuates the signal power to some 10^{-16} watt. Another design parameter is the eventuality of a satellite failure, which results in a diminished performance or even requires a reconstellation with new or spare satellites.

The GNSS satellites, essentially, provide a platform for atomic clocks, radio transceivers, computers, and various auxiliary equipment used to operate the system. The signals of each satellite allow the user to measure the pseudorange R to the satellite, and each satellite broadcasts a message which allows the user to determine the spatial position ϱ^s of the satellite for arbitrary instants. Given these capabilities, users are able to determine their position ϱ_r on or above the earth by resection. The auxiliary equipment of each satellite, among others, consists of solar

1.2 Development of global surveying techniques

panels for power supply and a propulsion system for orbit adjustments and stability control.

The satellites have various systems of identification. The launch sequence number, the orbital position number, the system specific name, and the international designation are mentioned to name a few.

Control segment

The control segment (also referred to as ground segment) is responsible for steering the whole system. The task includes the deployment and maintenance of the system, tracking of the satellites for the determination and prediction of orbital and clock parameters, monitoring of auxiliary data (e.g., ionosphere parameters), and upload of the data message to the satellites.

The control segment is also responsible for a possible encryption of data and the protection of services against unauthorized users.

Generally, the control segment comprises a master control station coordinating all activities, monitor stations forming the tracking network, and ground antennas being the communication link to the satellites.

User segment

The user segment can be classified into user categories, receiver types, and various information services.

User categories are subdivided into military and civilian users as well as authorized and unauthorized users. Civilian and unauthorized users do not have access to all signals or services of the GNSS.

A diversity of receiver types is on the market today. One characterization is based on the type of observables, i.e., the kind of pseudoranges. Another criterion is the ability to track one, two, or even more frequencies. Finally, one has to distinguish between receivers operating for one or more specific GNSS. An overview on receiver features is provided in Sect. 13.4.1.

Several governmental and private information services have been established to provide GNSS status information and data to the users. Generally, the information contains constellation status reports, scheduled outages, and orbital data. The latter are provided in the form of an almanac suitable for making satellite visibility predictions, and as precise ephemerides suitable for making the most precise positioning. Out of the variety of Internet sources only the International GNSS (formerly GPS) Service for Geodynamics (IGS) located at the US Jet Propulsion Laboratory (JPL) is mentioned here (http://igscb.jpl.nasa.gov).

1.3 Positioning and navigating with satellites

1.3.1 Position determination

Four simultaneously measured pseudoranges are needed to solve for the four unknowns at any time epoch; these are the three components of position plus the clock bias. Geometrically, the solution is accomplished by a sphere being tangent to the four spheres defined by the pseudoranges. The center of this sphere corresponds to the unknown position and its radius equals the range correction caused by the receiver clock error. In the two-dimensional case, the number of unknowns reduces to three and, thus, only three satellites are needed. This scenario is shown in Fig. 1.2 as adapted from Hofmann-Wellenhof et al. (2003: p. 37).

It is worth noting that the range error $\Delta\varrho$ could be eliminated in advance by differencing the pseudoranges measured from one site to two satellites or to two different positions of one satellite. In the second case, the resulting range difference or delta range corresponds to the observable as used in the Transit system. In both cases, the delta range defines a hyperboloid (or a hyperbola in the two-dimensional case) with its foci placed at the two satellites or the two different satellite positions for the geometric location of the receiver. Thus, pseudorange positioning by means of between-satellites differenced pseudoranges is also denoted hyperbolic positioning.

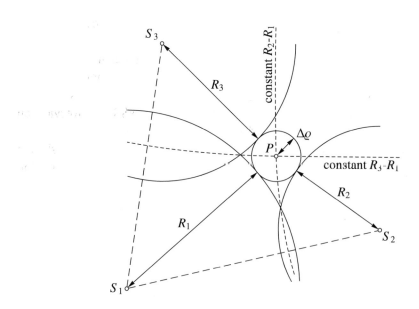

Fig. 1.2. Two-dimensional pseudorange positioning

1.3 Positioning and navigating with satellites

Differencing the pseudoranges measured at two sites (forming a baseline) to the satellite reduces or eliminates systematic errors in the satellite position and satellite clock biases. This (interferometric) approach has become fundamental for satellite-based surveying.

Considering the fundamental equation (1.1), one can conclude that the accuracy of the position determined using a single receiver is essentially affected by the following factors:

- accuracy of each satellite position,
- accuracy of pseudorange measurement,
- geometry.

As mentioned before, systematic errors or biases in the pseudoranges can be reduced or eliminated by differencing the measured pseudoranges either between satellites or between sites. However, no mode of differencing can overcome poor geometry.

A measure of satellite geometry with respect to the observing site is a factor known as geometric dilution of precision (GDOP). Assuming four satellites, in a geometric approach this factor is inversely proportional to the volume of a tetrahedron. This body is formed by points obtained from the intersection of a unit sphere with the vectors pointing from the observing site to the satellites. More details and an analytical approach on this subject are provided in Sect. 7.3.4.

1.3.2 Velocity determination

The determination of the instantaneous velocity of a moving vehicle is another goal of navigation. This can be achieved by using the Doppler principle of radio signals. Because of the relative motion of the satellites with respect to a moving vehicle, the frequency of a signal broadcast by the satellites is shifted when received at the vehicle. This measurable Doppler shift is proportional to the relative radial velocity or range rate.

Analytically, the Doppler observable D can be expressed by differentiating the pseudorange equation (1.2) with respect to time

$$D = \dot{R} = \dot{\varrho} + c\dot{\delta}, \tag{1.3}$$

where the time derivatives are indicated by a dot. The term $c\dot{\delta}$ considers the time derivation of the clock error which translates into a frequency bias. The range rate $\dot{\varrho}$ or radial velocity is obtained by differentiating (1.1) and is given by

$$\dot{\varrho} = \frac{(\varrho^s - \varrho_r)}{\varrho} \cdot (\dot{\varrho}^s - \dot{\varrho}_r) = \varrho_0 \cdot \Delta\dot{\varrho} \tag{1.4}$$

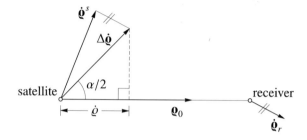

Fig. 1.3. Geometrical interpretation of a range rate

with ϱ_0 being the unit vector between the satellite and the receiver site and $\Delta\dot{\varrho}$ is the relative velocity vector describing the relative velocity between satellite and observing site.

Geometrically, Eq. (1.4) represents the projection of the relative velocity vector onto the line of sight, cf. Fig. 1.3.

If in Eq. (1.4), apart from position and velocity vector of the satellite, the position of the observing site is known, the velocity vector of the moving vehicle is the only remaining unknown and can thus be deduced from the Doppler observable. A minimum of four Doppler observables is required to solve for the three components of the vehicle's velocity vector and the frequency bias.

For the sake of completeness, also the inverse case is considered here. If the relative velocity vector $\Delta\dot{\varrho}$ is known, then Eq. (1.4) enables the computation of the direction vector ϱ_0. In the two-dimensional case, this vector defines a straight line as line of position (LOP) for the observing site. In the three-dimensional case, the corresponding surface of position (SOP) is a circular cone with the satellite as apex and its axis coinciding with the relative velocity vector (Hofmann-Wellenhof et al. 2003: p. 37). The aperture angle is given by $\alpha = 2\arccos(\dot{\varrho} / \|\Delta\dot{\varrho}\|)$.

1.3.3 Attitude determination

Attitude is defined as the orientation of a specific body frame attached to a land vehicle, ship, or aircraft with respect to a reference frame which is usually a local-level frame represented by north-, east-, and up-axis.

The parameters used to define three-dimensional attitude are r, p, y, the angles for roll, pitch, and yaw (or heading). In the case of an aircraft, the roll angle measures the rotation of the aircraft about the fuselage axis, the pitch angle measures the rotation about the wing axis, and the yaw angle measures the rotation about the vertical axis (Graas and Braasch 1992). Similar reference frames can be developed for other vehicles.

Traditionally, attitude parameters are derived from inertial navigation systems or other electronic devices. With the advent of low-cost high-performance sensors, multiantenna systems which integrate three or more GNSS antennas in a proper configuration provide an alternative and cost-effective means to obtain reliable and accurate attitude information.

For more details on attitude determination, the reader is referred to Sect. 13.1.4.

1.3.4 Terminology

Code pseudoranges versus phase pseudoranges

Typically, observables for satellite-based positioning are pseudoranges as derived from run-time observations of the coded satellite signal or from measurements of the phase of the carrier.

Generally speaking, the accuracy of code ranges is at the meter level, whereas the accuracy of carrier phases is in the millimeter range. The accuracy of code ranges can be improved, however, by the specific receiver technology or by smoothing techniques.

The disadvantage of phase ranges is the fact that they are ambiguous by an integer number of full wavelengths, whereas the code ranges are virtually unambiguous. The determination of the phase ambiguities is often a critical issue in high-accuracy satellite-based positioning.

Absolute versus relative positioning

The coordinates of a single point are determined by point positioning when using a single receiver which measures pseudoranges to four or more satellites. The terms "point positioning", "single-point positioning", and the term "absolute point positioning" are synonymously used. The term "absolute" reflects the opposite of "relative".

Instead of "relative positioning" the term "differential positioning" is often used. Note, however, that the two methods are (at least theoretically) different. Differential positioning is rather an improved single-point positioning technique and is based on applying (predicted) corrections to pseudoranges measured at an unknown site. The technique provides instantaneous solutions (usually denoted as real-time solutions) where improved accuracies with respect to a reference station are achieved.

Relative positioning is possible if (as in the case of differential positioning) two receivers are used and (code or carrier phase) measurements, to the same satellites, are simultaneously made at two sites. The measurements taken at both sites are (in contrast to differential positioning) directly combined. This direct combination further improves the position accuracy but prevents instantaneous solutions in the strict sense. Normally, the coordinates of one site are known and the position of the other site is to be determined relatively to the known site (i.e., the vector between the two sites is determined). In general, the receiver at the known site is stationary while observing.

In the past, point positioning was associated with navigation and relative positioning with surveying. Also, the term "relative" was used for carrier phase observations, whereas the term "differential" was used for code range observations. In practice, however, there is no universal agreement on these terms.

Static versus kinematic positioning

Static denotes a stationary observation location, while kinematic implies motion. A temporary loss of signal lock in static mode is not as critical as in kinematic mode.

Attention should be paid to the difference between the terms "kinematic" and "dynamic". The term "kinematic" describes the pure geometry of a motion, whereas "dynamic" considers the forces causing the motion. The following example may illustrate the difference. The type of modeling the satellite orbit is called dynamic. The positioning of a moving vehicle such as a plane or boat based on known satellite positions is regarded as kinematic procedure.

Real-time processing versus postprocessing

For real-time GNSS, the results must be available in the field immediately. The results are denoted as "instantaneous" if the observables of a single epoch are used for the position computation and the processing time is negligible. The concept of modern operational satellite techniques aims at instantaneous navigation of moving vehicles (i.e., cars, ships, aircraft) by unsmoothed code pseudoranges. A different and less stringent definition is "quasi (or near) real-time" which includes computing results with a slight delay. Today, radio data links allow the combination of measurements from different sites in (near) real time.

Postprocessing refers to applications when data are processed after the fact.

Surveying versus navigation

The fields of surveying and navigation are closely related. The goal of surveying, however, is mainly positioning, whereas navigation includes the determination of position, velocity, and attitude of moving objects. In the past, surveying was characterized by high positioning accuracies, static observations, and postprocessing procedures. In contrast, navigation requires lower accuracies but (near) real-time processing of kinematic observations. The differences between surveying and navigating modes have continued to diminish.

2 Reference systems

2.1 Introduction

The basic equation which relates the range ϱ with the instantaneous position vector $\boldsymbol{\varrho}^s$ of a satellite and the position vector $\boldsymbol{\varrho}_r$ of the observing site reads

$$\varrho = \|\boldsymbol{\varrho}^s - \boldsymbol{\varrho}_r\|. \tag{2.1}$$

In Eq. (2.1), both vectors must be expressed in a uniform coordinate system. The definition of a three-dimensional Cartesian system requires a convention for the orientation of the axes and for the location of the origin.

For global applications such as satellite geodesy, equatorial coordinate systems are appropriate. According to Fig. 2.1, a space-fixed or celestial system $\mathbf{X}_i^0$ and an earth-fixed or terrestrial system $\mathbf{X}_i$ must be distinguished, where $i = 1, 2, 3$. The earth's rotation vector $\boldsymbol{\omega}_e$ serves as $\mathbf{X}_3$-axis in both cases. The $\mathbf{X}_1^0$-axis for the space-fixed system points towards the vernal equinox and is, thus, the intersection line between the equatorial and the ecliptic plane. The $\mathbf{X}_1$-axis of the earth-fixed system is defined by the intersection line of the equatorial plane with the plane represented by the Greenwich meridian. The angle Θ_0 between the two systems is called Greenwich sidereal time. The $\mathbf{X}_2$-axis (in analogy to $\mathbf{X}_2^0$ not shown Fig. 2.1) is orthogonal to both the $\mathbf{X}_1$-axis and the $\mathbf{X}_3$-axis and completes a right-handed coordinate frame. A nonrotating coordinate system whose origin is located at the barycenter is at rest with respect to the solar system. It is, therefore, an inertial system which conforms to Newtonian mechanics. In a geocentric system, however, accelerations are present because the earth is orbiting the sun. Thus, in such a system the laws of general relativity must be taken into account. But, since the main relativistic effect is caused by the gravity field of the earth itself, the geocentric system is better suited for the description of the motion of a satellite close to the earth. Note that the axes of a geocentric coordinate system remain parallel because the motion of the earth around the sun is described by revolution without rotation.

The earth's rotation vector $\boldsymbol{\omega}_e$ oscillates due to several reasons. The basic differential equations describing the oscillations follow from classical mechanics and are given by

$$\mathbf{M} = \frac{d\mathbf{N}}{dt} \tag{2.2}$$

and

$$\mathbf{M} = \frac{\partial \mathbf{N}}{\partial t} + \boldsymbol{\omega}_e \times \mathbf{N}, \tag{2.3}$$

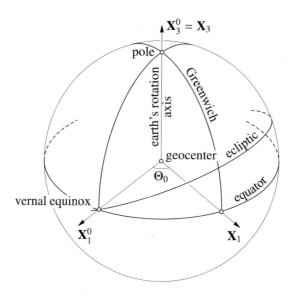

Fig. 2.1. Equatorial coordinate systems

where $\mathbf{M}$ denotes a torque vector, $\mathbf{N}$ is the angular momentum vector of the earth, and t indicates time (Moritz and Mueller 1988: Eqs. (2-54) and (2-59)). The symbol "×" in (2.3) indicates a vector (or cross) product. The torque $\mathbf{M}$ originates mainly from the gravitational forces of sun and moon; therefore it is closely related to the tidal potential. Equation (2.2) is valid in a (quasi-) inertial system such as $\mathbf{X}_i^0$ and Eq. (2.3) applies for the rotating system $\mathbf{X}_i$. The partial derivative expresses the temporal change of $\mathbf{N}$ with respect to the earth-fixed system, and the vector product considers the rotation of this system with respect to the inertial system.

The earth's rotation vector $\boldsymbol{\omega}_e$ is related to the angular momentum vector $\mathbf{N}$ by the inertia tensor $\mathbf{C}$ as

$$\mathbf{N} = \mathbf{C}\,\boldsymbol{\omega}_e\,. \tag{2.4}$$

Introducing for $\boldsymbol{\omega}_e$ its unit vector $\boldsymbol{\omega}$ and its norm $\omega_e = \|\boldsymbol{\omega}_e\|$, the relation

$$\boldsymbol{\omega}_e = \omega_e\,\boldsymbol{\omega} \tag{2.5}$$

can be formed.

The differential equations (2.2) and (2.3) can be separated into two parts. The oscillations of $\boldsymbol{\omega}$, i.e., the variations of the $\mathbf{X}_3$-axis, are considered in the subsequent section. The oscillations of the norm ω_e cause variations in the speed of rotation which are treated in the section on time systems.

Considering only the homogeneous part (i.e., $\mathbf{M} = \mathbf{0}$) of Eqs. (2.2) and (2.3) leads to free oscillations. The inhomogeneous solution gives the forced oscillation.

In both cases, the oscillations can be related to the inertial or to the terrestrial system. A further criterion for the solution concerns the inertia tensor. For a rigid earth and neglecting internal mass shifts, this tensor is constant; this is not the case for a deformable earth.

2.2 Coordinate systems

2.2.1 Definitions

Oscillations of axes

The oscillation of ω with respect to the inertial space is called nutation. For the sake of convenience, the effect is partitioned into the secular precession and the periodic nutation. The oscillation with respect to the terrestrial system is named polar motion. A simplified representation of polar motion is given in Fig. 2.2. The image of a mean position of ω is denoted by P in this polar perspective. The free oscillation results in a motion of the rotation axis along a circular cone, with its mean position as axis, and an aperture angle of about 0.4 arcseconds. On the earth, this motion is represented by a 6 m radius circle around P. The image of an instantaneous position of the free oscillating earth's rotation axis is denoted by R_0. The period of the free motion amounts to about 430 days and is known as the Chandler period. The forced motion can also be described by a cone. In Fig. 2.2, this cone is mapped by the circle around the free position R_0. The radius of this circle is related to the tidal deformation and amounts to approximately 0.5 m. The nearly diurnal period of the forced motion corresponds to the tesseral part of the tidal potential of second degree. The respective motions of the angular momentum axis, which is within one milliarcsecond of the rotation axis, are very similar. The free motion of the angular momentum axis deserves special attention because the forced motion

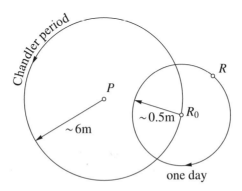

Fig. 2.2. Polar motion of the earth's rotation axis

can be removed by modeling the tidal attractions. The free polar motion is long-periodic and the free position in space is fixed since for $\mathbf{M} = \mathbf{0}$ the integration of Eq. (2.2) yields $\mathbf{N}$ = constant. By the way, this result implies the law of conservation of angular momentum as long as no external forces are applied. Because of the above mentioned properties, the angular momentum axis is appropriate to serve as a reference axis and the scientific community has named its free position in space celestial ephemeris pole (CEP). A candidate for serving as reference axis in the terrestrial system is the mean position of the rotation axis denoted by P (Fig. 2.2). This position is called conventional international origin (CIO).

Conventional celestial reference system

By convention, the $\mathbf{X}_3^0$-axis is identical to the position of the angular momentum axis at a standard epoch denoted by J2000.0 (Sect. 2.3.3). The $\mathbf{X}_1^0$-axis points to the associated vernal equinox. This equinox is defined, for example, by a set of fundamental stars, cf. European Space Agency (1997) or Wielen et al. (1999). Since this system is defined conventionally and the practical realization does not necessarily coincide with the theoretical system, it is called (conventional) celestial reference frame (CRF). Sometimes the term "quasi-inertial" is added to point out that a geocentric system is not rigorously inertial because of the accelerated motion of the earth around the sun. One example of such a celestial reference frame is that established by the International Earth Rotation Service (IERS) (McCarthy and Petit 2004). This frame is called ICRF where the first letter indicates the IERS origin. The ICRF is kinematically defined by a set of precise coordinates of extragalactic radio sources, mostly quasars and galactic nuclei.

Conventional terrestrial reference system

Again by convention, the $\mathbf{X}_3$-axis is identical to the mean position of the earth's rotation axis as defined by the CIO. The $\mathbf{X}_1$-axis is associated with the mean Greenwich meridian. The realization of this system is named the (conventional) terrestrial reference frame (TRF) and is defined by a set of terrestrial control stations serving as reference points. Most of the reference stations are equipped with very long baseline interferometry (VLBI), lunar laser ranging (LLR), satellite laser ranging (SLR), or GNSS capabilities.

An example for a terrestrial reference frame is the international terrestrial reference frame (ITRF) produced by the IERS (McCarthy and Petit 2004). The $\mathbf{X}_3$-axis is defined by the IERS reference pole (IRP) and the $\mathbf{X}_1$-axis lies in the IERS reference meridian (IRM). The ITRF is realized by a number of terrestrial sites where temporal effects (plate tectonics, tidal effects) are also taken into account. Thus, ITRF is regularly updated and the acronym is supplemented by digits to mark the last year whose data were used in the formation of the frame. Since October 2006, the ITRF2005 (http://itrf.ensg.ign.fr) has been the operative version.

Another terrestrial reference frame is the World Geodetic System 1984 (WGS-84), which is applied for GPS. After some modifications, the present version of WGS-84 is almost identical with the ITRF2005. A detailed description of WGS-84 is given in Sect. 9.2.1.

The coordinates in GLONASS are based on the Parameter of the Earth 1990 (PE-90) frame. Details of the PE-90 frame are given in Sect. 10.2.1 or Feairheller and Clark (2006: p. 605).

Also, the Galileo terrestrial reference frame (GTRF) is theoretically identical with ITRF2005. For more details on GTRF, the reader is referred to Sect. 11.2.1.

2.2.2 Transformation between celestial and terrestrial frames

General remarks

Following the resolutions of the International Astronomical Union (IAU) adopted in 2000, the "new" concept of the transformation has to consider relativistic effects since the celestial reference system (CRS) is split into the barycentric celestial reference system (BCRS) and the so far considered geocentric celestial reference system (GCRS). In the following section, however, the "old" concept is retained to keep the mathematical apparatus of the transformation as simple as possible. The reader interested in more details is referred to Capitaine et al. (2002).

The transformation between the (geocentric) celestial reference frame and the terrestrial reference frame is performed by means of rotations. For an arbitrary vector $\mathbf{x}$, the transformation is given by

$$\mathbf{x}_{\text{TRF}} = \mathbf{R}^M \mathbf{R}^S \mathbf{R}^N \mathbf{R}^P \mathbf{x}_{\text{CRF}}, \tag{2.6}$$

where

$\mathbf{R}^M$ ... rotation matrix for polar motion,
$\mathbf{R}^S$ ... rotation matrix for Greenwich sidereal time,
$\mathbf{R}^N$ ... rotation matrix for nutation,
$\mathbf{R}^P$ ... rotation matrix for precession.

The CRF, defined at the standard epoch J2000.0, is transformed into the instantaneous or true system at observation epoch by applying the corrections due to precession and nutation. The $\mathbf{X}_3^0$-axis of the true CRF represents the free position of the angular momentum axis and, thus, points to the CEP at observation epoch. Rotating this system about the $\mathbf{X}_3^0$-axis and through the sidereal time by the matrix $\mathbf{R}^S$ does not change the position of the CEP. Finally, the CEP is rotated into the CIO by $\mathbf{R}^M$ which completes the transformation.

The rotation matrices in Eq. (2.6) are composed of the elementary matrices $\mathbf{R}_i\{\alpha\}$ describing a rotation of the coordinate system about the $\mathbf{X}_i$-axis and through the angle α. The rotation matrices are given by

$$\mathbf{R}_1\{\alpha\} = \begin{bmatrix} 1 & 0 & 0 \\ 0 & \cos\alpha & \sin\alpha \\ 0 & -\sin\alpha & \cos\alpha \end{bmatrix},$$

$$\mathbf{R}_2\{\alpha\} = \begin{bmatrix} \cos\alpha & 0 & -\sin\alpha \\ 0 & 1 & 0 \\ \sin\alpha & 0 & \cos\alpha \end{bmatrix}, \qquad (2.7)$$

$$\mathbf{R}_3\{\alpha\} = \begin{bmatrix} \cos\alpha & \sin\alpha & 0 \\ -\sin\alpha & \cos\alpha & 0 \\ 0 & 0 & 1 \end{bmatrix}.$$

Note that these matrices are consistent with right-handed coordinate systems. The rotation angle α has a positive sign for clockwise rotation as viewed from the origin to the positive $\mathbf{X}_i$-axis.

Precession

A graphic representation of precession is given in Fig. 2.3. The position of the mean vernal equinox at the standard epoch t_0 is denoted by E_0 and the position at the observation epoch t is denoted by E. The precession matrix

$$\mathbf{R}^P = \mathbf{R}_3\{-z\}\,\mathbf{R}_2\{\vartheta\}\,\mathbf{R}_3\{-\zeta\} \qquad (2.8)$$

is composed of three successive rotation matrices, where z, ϑ, ζ are the precession parameters. Explicitly, performing the multiplication,

$$\mathbf{R}^P = \begin{bmatrix} \cos z \cos\vartheta \cos\zeta & -\cos z \cos\vartheta \sin\zeta & -\cos z \sin\vartheta \\ -\sin z \sin\zeta & -\sin z \cos\zeta & \\ \sin z \cos\vartheta \cos\zeta & -\sin z \cos\vartheta \sin\zeta & -\sin z \sin\vartheta \\ +\cos z \sin\zeta & +\cos z \cos\zeta & \\ \sin\vartheta \cos\zeta & -\sin\vartheta \sin\zeta & \cos\vartheta \end{bmatrix} \qquad (2.9)$$

is obtained. The precession parameters are computed from the time series

$$\zeta = 2306.2181''\,T + 0.30188''\,T^2 + 0.017998''\,T^3,$$
$$z = 2306.2181''\,T + 1.09468''\,T^2 + 0.018203''\,T^3, \qquad (2.10)$$
$$\vartheta = 2004.3109''\,T - 0.42665''\,T^2 - 0.041833''\,T^3$$

as given in Seidelmann (1992: Table 3.211.1). The parameter T represents the timespan expressed in Julian centuries of 36 525 mean solar days between the standard

2.2 Coordinate systems

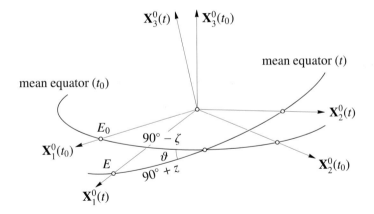

Fig. 2.3. Precession

epoch J2000.0 and the epoch of observation. For a numerical example, consider an observation epoch J1990.5 which corresponds to $T = -0.095$. Substituting T into Eq. (2.10), the numerical values $\zeta = -219.0880''$, $z = -219.0809''$, and $\vartheta = -190.4134''$ are obtained. The substitution of these values into Eq. (2.9) gives the following numerical precession matrix:

$$\mathbf{R}^P = \begin{bmatrix} 0.999997318 & 0.002124301 & 0.000923149 \\ -0.002124301 & 0.999997744 & -0.000000981 \\ -0.000923149 & -0.000000981 & 0.999999574 \end{bmatrix}.$$

Nutation

A graphic representation of nutation is given in Fig. 2.4. The mean vernal equinox at the observation epoch is denoted by E and the true equinox by E_t. The nutation matrix $\mathbf{R}^N$ is composed of three successive rotation matrices where both the nutation in longitude $\Delta\psi$ and the nutation in obliquity $\Delta\varepsilon$ can be treated as differential quantities:

$$\mathbf{R}^N = \mathbf{R}_1\{-(\varepsilon + \Delta\varepsilon)\}\, \mathbf{R}_3\{-\Delta\psi\}\, \mathbf{R}_1\{\varepsilon\}. \tag{2.11}$$

Explicitly,

$$\mathbf{R}^N = \begin{bmatrix} 1 & -\Delta\psi \cos\varepsilon & -\Delta\psi \sin\varepsilon \\ \Delta\psi \cos\varepsilon & 1 & -\Delta\varepsilon \\ \Delta\psi \sin\varepsilon & \Delta\varepsilon & 1 \end{bmatrix} \tag{2.12}$$

is obtained. The mean obliquity of the ecliptic ε has been determined (Seidelmann 1992: p. 114) as

$$\varepsilon = 23°26'21.448'' - 46.8150''\, T - 0.00059''\, T^2 + 0.001813''\, T^3, \tag{2.13}$$

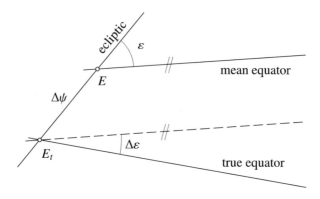

Fig. 2.4. Nutation

where T is the same time factor as in Eq. (2.10). The nutation parameters $\Delta\psi$ and $\Delta\varepsilon$ are computed from the harmonic series:

$$\Delta\psi = \sum_{i=1}^{106} a_i \sin\left(\sum_{j=1}^{5} e_j E_j\right) = -17.2'' \sin\Omega_m + \dots ,$$

$$\Delta\varepsilon = \sum_{i=1}^{64} b_i \cos\left(\sum_{j=1}^{5} e_j E_j\right) = 9.2'' \cos\Omega_m + \dots ,$$

(2.14)

where the amplitudes a_i, b_i as well as the integer coefficients e_j are tabulated, for example, in Seidelmann (1992: Table 3.222.1). The five fundamental arguments E_j describe mean motions in the sun-earth-moon system. The mean longitude Ω_m of moon's ascending node is one of the arguments. The moon's node retrogrades with a period of about 18.6 years and this period appears in the principal terms of the nutation series.

Sidereal time

The rotation matrix for sidereal time $\mathbf{R}^S$ is

$$\mathbf{R}^S = \mathbf{R}_3\{\Theta_0\} .$$

(2.15)

The computation of the apparent Greenwich sidereal time Θ_0 is shown in the section on time systems (Sect. 2.3.2).

Polar motion

The previous computations yield the instantaneous CEP. The CEP must still be rotated into the CIO. This is achieved by means of the pole coordinates x_P, y_P, which

2.2 Coordinate systems

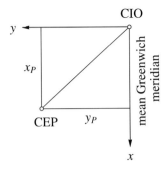

Fig. 2.5. Pole coordinates

define the position of the CEP with respect to the CIO (Fig. 2.5). The pole coordinates are determined by the IERS and are published on its Web site www.iers.org. The rotation matrix for polar motion $\mathbf{R}^M$ is given by

$$\mathbf{R}^M = \mathbf{R}_2\{-x_P\}\,\mathbf{R}_1\{-y_P\} = \begin{bmatrix} 1 & 0 & x_P \\ 0 & 1 & -y_P \\ -x_P & y_P & 1 \end{bmatrix}. \tag{2.16}$$

The rotation matrices $\mathbf{R}^S$ and $\mathbf{R}^M$ are often combined to a single matrix $\mathbf{R}^R$ for earth rotation:

$$\mathbf{R}^R = \mathbf{R}^M \mathbf{R}^S. \tag{2.17}$$

In the case of operational satellite-based positioning systems, the space-fixed coordinate system is already related to the CEP; hence, $\mathbf{R}^R$ is the only rotation matrix which must be applied for the transformation into the terrestrial system. For most practical purposes, the effect of polar motion is negligible.

2.2.3 Transformation between terrestrial frames

The transformation between the various (static) terrestrial reference frames (i.e., datum transformation) is generally performed by three-dimensional similarity transformations. This conformal transformation contains seven parameters and is given by

$$\mathbf{X}_{\mathrm{TRF}_1} = \mathbf{c} + \mu\,\mathbf{R}\,\mathbf{X}_{\mathrm{TRF}_2}, \tag{2.18}$$

where $\mathbf{X}_{\mathrm{TRF}_1}$ denotes the three-dimensional position vector of a site in one and $\mathbf{X}_{\mathrm{TRF}_2}$ denotes the corresponding vector represented in another terrestrial coordinate reference frame. The vector $\mathbf{c} = [c_1, c_2, c_3]$ is the translation vector between the two coordinate frames, μ is a scale factor, and **R** is an orthonormal matrix. The latter is composed of three successive rotations α_i about the coordinate frame axes.

If the respective terrestrial reference frames are kinematically defined (e.g., the various ITRF), Eq. (2.18) must be expanded to a fourteen-parameter transformation by taking into account time derivatives (i.e., rates) of the seven aforementioned parameters. Details on this kind of datum transformation as well as numerical results for the transformation parameters from ITRF2000 to past ITRF are provided in McCarthy and Petit (2004: Sect. 4.1 and Table 4.1).

A vector **X** in the terrestrial reference frame can be represented by Cartesian coordinates X, Y, Z as well as by ellipsoidal coordinates φ, λ, h. The rectangular coordinates are often called earth-centered, earth-fixed (ECEF) coordinates. Details on the transformation of Cartesian and ellipsoidal coordinates are provided in Sect. 8.2.

2.3 Time systems

2.3.1 Definitions

Several time systems are in current use. They are based on various periodic processes such as earth rotation and are listed in Table 2.1.

Solar and sidereal times

A measure of earth rotation is the hour angle which is the angle between the meridian of a celestial body and a reference meridian (preferably the Greenwich meridian). Universal time (UT) is defined by the Greenwich hour angle augmented by 12 hours of a fictitious sun uniformly orbiting in the equatorial plane. Sidereal time is defined by the hour angle of the vernal equinox. Taking the mean equinox as the reference leads to mean sidereal time and using the true equinox as a reference

Table 2.1. Time systems

Periodic process	Representative time systems
Earth rotation	Universal time (UT)
	Greenwich sidereal time (Θ_0)
Earth revolution	Terrestrial dynamic time (TDT)
	Barycentric dynamic time (BDT)
Atomic oscillations	International atomic time (TAI)
	Coordinated UT (UTC)
	Reference time in satellite-based positioning systems

2.3 Time systems

yields true or apparent sidereal time. Both, solar and sidereal time are not uniform since the angular velocity (i.e., the earth's rotation rate) ω_e is not constant. The fluctuations are partly due to changes in the polar moment of inertia exerted by tidal deformation as well as other mass transports. Another factor is due to the oscillations of the earth's rotation axis itself. In this case, the universal time corrected for polar motion is denoted by UT1.

Dynamic times

The time systems derived from planetary motions in the solar system are called dynamic times. The barycentric dynamic time (BDT) is an inertial time system in the Newtonian sense and provides the time variable in the equations of motion. The quasi-inertial terrestrial dynamic time (TDT) was formerly called ephemeris time and serves for the integration of the differential equations for the orbital motion of satellites around the earth. In 1991, the IAU introduced the term terrestrial time (TT) to replace TDT. Furthermore, the terminology of coordinate times according to the theory of general relativity was introduced. More details on this subject are given in Seidelmann and Fukushima (1992).

Atomic times

In practice, the dynamic time system is achieved by the use of atomic time scales. The coordinated UT (UTC) system is a compromise. The unit of the system is the atomic second, but to keep the system close to UT1 and approximate civil time, integer leap seconds are inserted at distinct epochs. Thus, UTC is not a continuous time scale.

2.3.2 Conversions

The times derived from earth rotation (i.e., UT1, the mean solar time corrected for polar motion, and Θ_0, the apparent sidereal time) are related by the formula

$$\Theta_0 = 1.002\,737\,9093 \text{ UT1} + \vartheta_0 + \Delta\psi \cos\varepsilon. \tag{2.19}$$

The first term on the right side of Eq. (2.19) accounts for the different scales of solar and sidereal time. The quantity ϑ_0 represents the actual sidereal time at Greenwich midnight (i.e., 0^h UT). The third term describes the projection of $\Delta\psi$ onto the equator and considers the effect of nutation. The mean sidereal time follows from Eq. (2.19) by neglecting the nutation term.

A time series has been determined for ϑ_0 as

$$\begin{aligned}\vartheta_0 = {} & 24\,110.548\,41^S + 8\,640\,184.812\,866^S\,T_0 \\ & + 0.093\,104^S\,T_0^2 - 6.2^S \cdot 10^{-6}\,T_0^3,\end{aligned} \tag{2.20}$$

where T_0 represents the timespan expressed in Julian centuries of 36 525 mean solar days between the standard epoch J2000.0 and the day of observation at 0^h UT (Seidelmann 1992: p. 50).

The time UT1 is related to UTC by the quantity dUT1 which is time-dependent and is reported by the IERS:

$$UT1 = UTC + dUT1. \tag{2.21}$$

When the absolute value of dUT1 becomes larger than 0.9^S, a leap second is inserted into the UTC system.

For the conversion between the atomic and the dynamic time system, the following relations are defined:

$$\begin{aligned} TAI &= TDT - 32.184^S & \text{constant offset,} \\ TAI &= UTC + 1.000^S\, n & \text{variable offset as leap seconds are substituted.} \end{aligned} \tag{2.22}$$

The actual integer n is reported by the IERS. In January 2007, for example, the integer value was $n = 33$.

2.3.3 Calendar

Definitions

The Julian date (JD) defines the number of mean solar days elapsed since the epoch January 1.5^d, 4713 before Christ.

The modified Julian date (MJD) is obtained by subtracting 2 400 000.5 days from JD. This convention saves digits and MJD commences at civil midnight instead of noon. For the sake of completeness, Table 2.2 with the Julian date for two standard epochs is given. This table enables, for example, the calculation of the parameter T for the GPS standard epoch. Subtracting the respective Julian dates and dividing by 36 525 (i.e., the number of days in a Julian century) yields $T = -0.199\,876\,7967$.

Table 2.2. Standard epochs

Civil date	Julian date	Explanation
January 1.5^d, 2000	2 451 545.0	Current standard epoch (J2000.0)
January 6.0^d, 1980	2 444 244.5	GPS standard epoch

2.3 Time systems

Date conversions

The relations for date conversions are taken from Montenbruck (1984) and are slightly modified so that they are only valid for an epoch between March 1900 and February 2100.

Let the civil date be expressed by integer values for the year Y, month M, day D, and a real value for the time in hours UT. Then

$$JD = INT[365.25\,y] + INT[30.6001\,(m+1)] \\ + D + UT/24 + 1\,720\,981.5 \qquad (2.23)$$

is the conversion into Julian date, where INT denotes the integer part of a real number and y, m are given by

$$y = Y - 1 \quad \text{and} \quad m = M + 12 \quad \text{if} \quad M \leq 2,$$
$$y = Y \quad \text{and} \quad m = M \quad \text{if} \quad M > 2.$$

The inverse transformation, that is, the conversion from Julian date to civil date, is carried out stepwise. First, the auxiliary numbers

$$a = INT[JD + 0.5],$$
$$b = a + 1537,$$
$$c = INT[(b - 122.1)/365.25],$$
$$d = INT[365.25\,c],$$
$$e = INT[(b - d)/30.6001]$$

are calculated. Afterwards, the civil date parameters are obtained from the relations

$$D = b - d - INT[30.6001\,e] + FRAC[JD + 0.5],$$
$$M = e - 1 - 12\,INT[e/14], \qquad (2.24)$$
$$Y = c - 4715 - INT[(7 + M)/10],$$

where FRAC denotes the fractional part of a number. As a by-product of date conversion, the day of week can be evaluated by the formula

$$N = \text{modulo}\{INT[JD + 0.5], 7\}, \qquad (2.25)$$

where $N = 0$ denotes Monday, $N = 1$ means Tuesday, and so on.

A further task is the calculation of the number of weeks since a reference epoch:

$$WEEK = INT[(JD_{\text{observation epoch}} - JD_{\text{reference epoch}})/7]. \qquad (2.26)$$

The formulas given here can be used to prove the different dates in Table 2.2 or to verify the fact that the epoch J2000.0 corresponds to Saturday in the 1042nd GPS week.

3 Satellite orbits

3.1 Introduction

The applications of operational satellite methods depend substantially on knowing the satellite orbits. For single receiver positioning, an orbital error is highly correlated with the position error. In the case of baselines, relative orbital errors are approximately equal to relative baseline errors.

Orbital information is either transmitted by the satellite as part of the broadcast message or can be obtained in the form of precise ephemerides from several sources. While good precise orbits are available in near real time, the final precise ephemerides are available after several days.

This chapter provides a review of orbital theory to introduce the reader into the methods of computing ephemerides.

3.2 Orbit description

3.2.1 Keplerian motion

Orbital parameters

Assume two point masses m_1 and m_2 separated by the distance r. Considering for the moment only the attractive force between the masses and applying Newtonian mechanics, the movement of mass m_2 relative to m_1 is defined by the homogeneous differential equation of second order

$$\ddot{\mathbf{r}} + \frac{G(m_1 + m_2)}{r^3} \mathbf{r} = \mathbf{0}, \tag{3.1}$$

where

$\mathbf{r}$ ... relative position vector with $\|\mathbf{r}\| = r$,

$\ddot{\mathbf{r}} = \dfrac{d^2 \mathbf{r}}{dt^2}$ ... relative acceleration vector,

G ... universal gravitational constant,

and the time parameter t being an inertial (i.e., dynamic) time. In fact, the inertial time is provided by the system time of the respective GNSS.

In the case of an artificial satellite orbiting the earth, both bodies can be considered in a first approximation as point masses and the mass of the satellite can be neglected. The product of G and the earth's mass M_e is denoted as the geocentric

Table 3.1. Keplerian orbital parameters

Parameter	Notation
Ω	Right ascension of ascending node
i	Inclination of orbital plane
ω	Argument of perigee
a	Semimajor axis of orbital ellipse
e	Numerical eccentricity of ellipse
T_0	Epoch of perigee passage

gravitational constant μ. According to the current IERS conventions, the numerical value for μ is

$$\mu = G M_e = 3\,986\,004.418 \cdot 10^8 \text{ m}^3 \text{ s}^{-2}.$$

The analytical solution of differential equation (3.1) can be found in textbooks on celestial mechanics (e.g., Brouwer and Clemence 1961; Beutler 1991, 1992) and leads to the Keplerian motion defined by six orbital parameters which correspond to the six integration constants of the second-order vector equation (3.1). Satellite orbits are elliptical, and the six associated parameters are listed in Table 3.1. The point of closest approach of the satellite with respect to the earth's center of mass is called perigee and the most distant position is the apogee. The intersection between the equatorial and the orbital plane with the unit sphere is termed the nodes, where the ascending node defines the northward crossing of the equator. A graphical representation of the Keplerian orbit is given in Fig. 3.1.

The mean angular satellite velocity n (also known as the mean motion) with revolution period P follows from Kepler's third law given by

$$n = \frac{2\pi}{P} = \sqrt{\frac{\mu}{a^3}}. \tag{3.2}$$

Assume an orbit with the semimajor axis $a = 26\,560$ km. The substitution of a into Eq. (3.2) yields an orbital period of roughly 11.97 hours in the dynamic time system or, equivalently, 12 sidereal hours. The ground track of the satellite, thus, repeats every sidereal day.

The instantaneous position of the satellite within its orbit is described by angular quantities known as anomalies. The term anomaly has been retained for historical reasons. Table 3.2 lists anomalies commonly used. The mean anomaly $M(t)$ is a mathematical abstraction relating to mean angular motion, while both the eccentric anomaly $E(t)$ and the true anomaly $v(t)$ are geometrically producible (Fig. 3.1). The

3.2 Orbit description

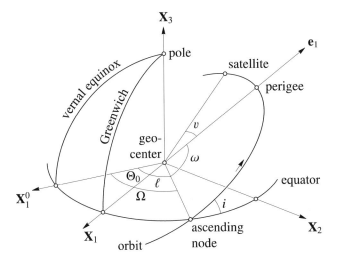

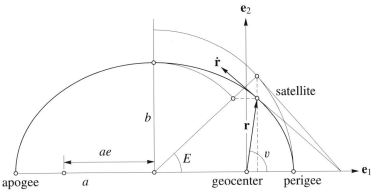

Fig. 3.1. Keplerian orbit

three anomalies are related by the formulas

$$M(t) = n(t - T_0), \tag{3.3}$$

$$E(t) = M(t) + e \sin E(t), \tag{3.4}$$

$$v(t) = 2 \arctan\left[\sqrt{\frac{1+e}{1-e}} \tan \frac{E(t)}{2}\right], \tag{3.5}$$

where e denotes the first numerical eccentricity. Equation (3.3) is valid by definition and shows that the mean anomaly can be used instead of T_0 as a defining parameter for the orbit. Equation (3.4) is known as Kepler's equation and is obtained in the course of the analytical integration of Eq. (3.1). Finally, Eq. (3.5) follows purely from geometric relations. The proof is left to the reader.

Table 3.2. Anomalies of the Keplerian orbit

Notation	Anomaly
$M(t)$	Mean anomaly
$E(t)$	Eccentric anomaly
$v(t)$	True anomaly

To become more familiar with the various anomalies, assume an orbit with a semidiurnal orbital period and an eccentricity of $e = 0.1$. At an epoch 3 hours after perigee passage, the mean anomaly is $M = 90.0000°$. The calculation of the eccentric anomaly requires iteration and gives the value $E = 95.7012°$. The true anomaly is obtained as $v = 101.3838°$.

Orbit representation

The coordinate system e_1, e_2 defining the orbital plane is shown in Fig. 3.1. The position vector $\mathbf{r}$ and the velocity vector $\dot{\mathbf{r}} = d\mathbf{r}/dt$ of the satellite can be represented by means of the eccentric as well as the true anomaly:

$$\mathbf{r} = a \begin{bmatrix} \cos E - e \\ \sqrt{1 - e^2} \sin E \end{bmatrix} = r \begin{bmatrix} \cos v \\ \sin v \end{bmatrix}, \quad (3.6)$$

$$r = a(1 - e \cos E) = \frac{a(1 - e^2)}{1 + e \cos v}, \quad (3.7)$$

$$\dot{\mathbf{r}} = \frac{n a^2}{r} \begin{bmatrix} -\sin E \\ \sqrt{1 - e^2} \cos E \end{bmatrix} = \sqrt{\frac{\mu}{a(1 - e^2)}} \begin{bmatrix} -\sin v \\ \cos v + e \end{bmatrix}, \quad (3.8)$$

$$\dot{r} = \frac{n a^2}{r} \sqrt{1 - (e \cos E)^2} = \sqrt{\mu \left(\frac{2}{r} - \frac{1}{a}\right)}. \quad (3.9)$$

The components of the vector $\mathbf{r}$ are evident from the geometry in Fig. 3.1 where the semiminor axis b of the orbital ellipse is replaced by $a\sqrt{1 - e^2}$. The geocentric distance $r = r(E)$ corresponds to the norm $\|\mathbf{r}(E)\|$ and follows from simple algebra. The representation $r = r(v)$ is known as polar equation of the ellipse (Bronstein et al. 2005: p. 213).

The derivation of the velocity vector $\dot{\mathbf{r}}$ and its norm $\dot{r}$ is laborious and the result is given without proof. It is worth noting that Eq. (3.9) when squared and divided by 2 relates kinetic energy on the left side with potential energy on the right side,

3.2 Orbit description

where a is constant by definition. Hence, Eq. (3.9) can be recognized as the law of energy conservation in the earth-satellite system!

The transformation of $\mathbf{r}$ and $\dot{\mathbf{r}}$ into the equatorial system $\mathbf{X}_i^0$ is performed by a rotation matrix $\mathbf{R}$ and results in vectors denoted by ϱ and $\dot{\varrho}$. The superscript "s" generally indicating a satellite is omitted here for simplicity. The vectors expressed in the orbital system must be considered as three-dimensional vectors for the transformation. Therefore, the axes $\mathbf{e}_1$, $\mathbf{e}_2$ are supplemented with an $\mathbf{e}_3$-axis which is orthogonal to the orbital plane. Since $\mathbf{r}$ and $\dot{\mathbf{r}}$ are vectors in the orbital plane (represented by $\mathbf{e}_1$, $\mathbf{e}_2$), their $\mathbf{e}_3$-component is zero.

The transformation is defined by

$$\varrho = \mathbf{R}\,\mathbf{r},$$
$$\dot{\varrho} = \mathbf{R}\,\dot{\mathbf{r}}, \tag{3.10}$$

where the matrix $\mathbf{R}$ is composed of three successive rotation matrices and is given by

$$\mathbf{R} = \mathbf{R}_3\{-\Omega\}\,\mathbf{R}_1\{-i\}\,\mathbf{R}_3\{-\omega\}$$

$$= \begin{bmatrix} \cos\Omega\cos\omega \\ -\sin\Omega\sin\omega\cos i & -\cos\Omega\sin\omega \\ -\sin\Omega\cos\omega\cos i & \sin\Omega\sin i \\ \sin\Omega\cos\omega \\ +\cos\Omega\sin\omega\cos i & -\sin\Omega\sin\omega \\ +\cos\Omega\cos\omega\cos i & -\cos\Omega\sin i \\ \sin\omega\sin i & \cos\omega\sin i & \cos i \end{bmatrix} \tag{3.11}$$

$$= \begin{bmatrix} \mathbf{e}_1 & \mathbf{e}_2 & \mathbf{e}_3 \end{bmatrix}.$$

The column vectors of the orthonormal matrix $\mathbf{R}$ are the axes of the orbital coordinate system represented in the equatorial system $\mathbf{X}_i^0$.

In order to rotate the system $\mathbf{X}_i^0$ into the terrestrial system $\mathbf{X}_i$, an additional rotation through the angle Θ_0, the Greenwich sidereal time, is required. The transformation matrix, therefore, becomes

$$\mathbf{R}' = \mathbf{R}_3\{\Theta_0\}\,\mathbf{R}_3\{-\Omega\}\,\mathbf{R}_1\{-i\}\,\mathbf{R}_3\{-\omega\}. \tag{3.12}$$

The product $\mathbf{R}_3\{\Theta_0\}\,\mathbf{R}_3\{-\Omega\}$ can be expressed by a single matrix $\mathbf{R}_3\{-\ell\}$, where $\ell = \Omega - \Theta_0$ is the longitude of the ascending node. Hence, Eq. (3.12) can be written in the form

$$\mathbf{R}' = \mathbf{R}_3\{-\ell\}\,\mathbf{R}_1\{-i\}\,\mathbf{R}_3\{-\omega\} \tag{3.13}$$

and the matrix $\mathbf{R}'$ corresponds to the matrix $\mathbf{R}$ if in Eq. (3.11) the parameter Ω is replaced by ℓ.

For a numerical example, assume a satellite orbiting in a Kepler ellipse with the following parameters: $a = 26\,000$ km, $e = 0.1$, $\omega = -140°$, $i = 60°$, $\ell = 110°$. To calculate the position and velocity vector in the earth-fixed equatorial system at an epoch where the eccentric anomaly is $E = 45°$, the vectors are first calculated in the orbital plane using Eqs. (3.6) through (3.9). Then, the transformation into the equatorial system is performed by means of Eq. (3.10), but using the rotation matrix $\mathbf{R}'$. The final result is

$$\boldsymbol{\varrho} = [11\,465,\ 3\,818,\ -20\,922] \quad [\text{km}],$$

$$\dot{\boldsymbol{\varrho}} = [-1.2651,\ 3.9960,\ -0.3081] \quad [\text{km s}^{-1}].$$

In addition to the fixed orbital system $\mathbf{e}_i$, another orthonormal system $\mathbf{e}_i^*$ may be defined. This system rotates about the $\mathbf{e}_3$-axis because the $\mathbf{e}_1^*$-axis always points towards the instantaneous satellite position. Hence, the unit vectors $\mathbf{e}_i^*$ can be derived from the position and velocity vectors by

$$\mathbf{e}_1^* = \frac{\boldsymbol{\varrho}}{\|\boldsymbol{\varrho}\|}, \qquad \mathbf{e}_3^* = \frac{\boldsymbol{\varrho} \times \dot{\boldsymbol{\varrho}}}{\|\boldsymbol{\varrho} \times \dot{\boldsymbol{\varrho}}\|} = \mathbf{e}_3, \qquad \mathbf{e}_2^* = \mathbf{e}_3^* \times \mathbf{e}_1^*. \qquad (3.14)$$

Note that the base vectors $\mathbf{e}_i^*$ correspond to the column vectors of a modified rotation matrix $\mathbf{R}^*$ if the parameter ω is replaced by $(\omega + v)$ in Eq. (3.11).

The transformation of a change $\Delta\boldsymbol{\varrho}$ in the position vector into the orbital system $\mathbf{e}_i^*$ results in a vector $\Delta\mathbf{r} = [\Delta r_1, \Delta r_2, \Delta r_3]$ with its components computed along the respective axes $\mathbf{e}_i^*$:

$$\begin{aligned}
\Delta r_1 &= \mathbf{e}_1^* \cdot \Delta\boldsymbol{\varrho} && \text{radial component}, \\
\Delta r_2 &= \mathbf{e}_2^* \cdot \Delta\boldsymbol{\varrho} && \text{along-track component}, \\
\Delta r_3 &= \mathbf{e}_3^* \cdot \Delta\boldsymbol{\varrho} && \text{across-track component}.
\end{aligned} \qquad (3.15)$$

Inversely, $\Delta\boldsymbol{\varrho}$ is calculated if the vector $\Delta\mathbf{r}$ is given. The solution follows from the inversion of Eq. (3.15) and leads to

$$\Delta\boldsymbol{\varrho} = \mathbf{R}^* \Delta\mathbf{r}. \qquad (3.16)$$

For a numerical examination assume a change $\Delta\boldsymbol{\varrho} = [0.1, 1.0, -0.5]$ [km] in the satellite position of the previous example. Applying Eqs. (3.14) and (3.15) gives $\Delta\mathbf{r} = [0.638, 0.914, 0.128]$ [km].

Differential relations

The derivatives of $\boldsymbol{\varrho}$ and $\dot{\boldsymbol{\varrho}}$ with respect to the six Keplerian parameters are required in the subsequent section. The differentiation can be separated into two groups

3.2 Orbit description

because in Eq. (3.10) the vectors $\mathbf{r}$ and $\dot{\mathbf{r}}$ depend only on the parameters a, e, T_0, whereas the matrix $\mathbf{R}$ is only a function of the remaining parameters ω, i, Ω.

The differentiation of $\mathbf{r}$ and $\dot{\mathbf{r}}$ leads to time-dependent vectors which are given in Hofmann-Wellenhof et al. (2001: Eqs. (4.20) through (4.25)), where $dm = -n\,dT_0$ is substituted. Considering Eq. (3.11), the differentiation of the matrix $\mathbf{R}$ with respect to the parameters ω, i, Ω is simple and does not pose any problem.

Hence, the differential relations

$$d\boldsymbol{\varrho} = \mathbf{R}\frac{\partial \mathbf{r}}{\partial a}da + \mathbf{R}\frac{\partial \mathbf{r}}{\partial e}de + \mathbf{R}\frac{\partial \mathbf{r}}{\partial m}dm + \frac{\partial \mathbf{R}}{\partial \omega}\mathbf{r}\,d\omega + \frac{\partial \mathbf{R}}{\partial i}\mathbf{r}\,di + \frac{\partial \mathbf{R}}{\partial \Omega}\mathbf{r}\,d\Omega,$$

$$d\dot{\boldsymbol{\varrho}} = \mathbf{R}\frac{\partial \dot{\mathbf{r}}}{\partial a}da + \mathbf{R}\frac{\partial \dot{\mathbf{r}}}{\partial e}de + \mathbf{R}\frac{\partial \dot{\mathbf{r}}}{\partial m}dm + \frac{\partial \mathbf{R}}{\partial \omega}\dot{\mathbf{r}}\,d\omega + \frac{\partial \mathbf{R}}{\partial i}\dot{\mathbf{r}}\,di + \frac{\partial \mathbf{R}}{\partial \Omega}\dot{\mathbf{r}}\,d\Omega$$

(3.17)

are obtained. Note that all terms in Eq. (3.17) are time-dependent, although the derivatives of the matrix $\mathbf{R}$ with respect to the parameters ω, i, Ω are constant.

3.2.2 Perturbed motion

The Keplerian orbit is a theoretical orbit and does not include actual perturbations. Consequently, disturbing accelerations $d\ddot{\boldsymbol{\varrho}}$ must be added to Eq. (3.1), which is now expressed in the equatorial system. The perturbed motion, thus, is based on an inhomogeneous differential equation of second order:

$$\ddot{\boldsymbol{\varrho}} + \frac{\mu}{\varrho^3}\boldsymbol{\varrho} = d\ddot{\boldsymbol{\varrho}}.$$ (3.18)

One should note that, for MEO satellites, the acceleration $\|\ddot{\boldsymbol{\varrho}}\|$ due to the central attractive force μ/ϱ^2 is at least 10^4 times larger than the disturbing accelerations. Hence, for the analytical solution of Eq. (3.18), perturbation theory may be applied, where, initially, only the homogeneous part of the equation is considered. This leads to a Keplerian orbit defined by the six parameters p_{i0}, $i = 1, 2, \ldots, 6$ at the reference epoch t_0. Each disturbing acceleration $d\ddot{\boldsymbol{\varrho}}$ causes temporal variations $\dot{p}_{i0} = dp_{i0}/dt$ in the orbital parameters. Therefore, at an arbitrary epoch t, the parameters p_i describing the osculating ellipse are given by

$$p_i = p_{i0} + \dot{p}_{i0}\,(t - t_0).$$ (3.19)

In order to obtain time derivatives $\dot{p}_{i0}$, the Keplerian motion is compared to the perturbed motion. In the first case, the parameters p_i are constant, whereas in the second case they are time-dependent. Thus, for the position and velocity vector of the perturbed motion one may write

$$\boldsymbol{\varrho} = \boldsymbol{\varrho}\{t, p_i(t)\},$$
$$\dot{\boldsymbol{\varrho}} = \dot{\boldsymbol{\varrho}}\{t, p_i(t)\}.$$
(3.20)

Differentiating the above equations with respect to time and taking into account Eq. (3.18) leads to

$$\dot{\boldsymbol{\varrho}} = \frac{\partial \boldsymbol{\varrho}}{\partial t} + \sum_{i=1}^{6} \left(\frac{\partial \boldsymbol{\varrho}}{\partial p_i} \frac{dp_i}{dt} \right), \tag{3.21}$$

$$\ddot{\boldsymbol{\varrho}} = \frac{\partial \dot{\boldsymbol{\varrho}}}{\partial t} + \sum_{i=1}^{6} \left(\frac{\partial \dot{\boldsymbol{\varrho}}}{\partial p_i} \frac{dp_i}{dt} \right) = -\frac{\mu}{\varrho^3} \boldsymbol{\varrho} + d\ddot{\boldsymbol{\varrho}}. \tag{3.22}$$

Since an (osculating) ellipse is defined for any epoch t, the Eqs. (3.21) and (3.22) must also be valid for Keplerian motion. Evidently, equivalence is obtained with the following conditions

$$\sum_{i=1}^{6} \left(\frac{\partial \boldsymbol{\varrho}}{\partial p_i} \frac{dp_i}{dt} \right) = \mathbf{0},$$

$$\sum_{i=1}^{6} \left(\frac{\partial \dot{\boldsymbol{\varrho}}}{\partial p_i} \frac{dp_i}{dt} \right) = d\ddot{\boldsymbol{\varrho}}. \tag{3.23}$$

In the following, for simplicity, only one disturbing acceleration is considered. The two vector equations (3.23) correspond to six linear equations which, in vector notation, are given by

$$\mathbf{A}\mathbf{x} = \boldsymbol{\ell}, \tag{3.24}$$

where

$$\mathbf{A} = \begin{bmatrix} \frac{\partial \boldsymbol{\varrho}}{\partial a} & \frac{\partial \boldsymbol{\varrho}}{\partial e} & \frac{\partial \boldsymbol{\varrho}}{\partial m} & \frac{\partial \boldsymbol{\varrho}}{\partial \omega} & \frac{\partial \boldsymbol{\varrho}}{\partial i} & \frac{\partial \boldsymbol{\varrho}}{\partial \Omega} \\ \frac{\partial \dot{\boldsymbol{\varrho}}}{\partial a} & \frac{\partial \dot{\boldsymbol{\varrho}}}{\partial e} & \frac{\partial \dot{\boldsymbol{\varrho}}}{\partial m} & \frac{\partial \dot{\boldsymbol{\varrho}}}{\partial \omega} & \frac{\partial \dot{\boldsymbol{\varrho}}}{\partial i} & \frac{\partial \dot{\boldsymbol{\varrho}}}{\partial \Omega} \end{bmatrix}$$

$$= \begin{bmatrix} \mathbf{R}\frac{\partial \mathbf{r}}{\partial a} & \mathbf{R}\frac{\partial \mathbf{r}}{\partial e} & \mathbf{R}\frac{\partial \mathbf{r}}{\partial m} & \mathbf{r}\frac{\partial \mathbf{R}}{\partial \omega} & \mathbf{r}\frac{\partial \mathbf{R}}{\partial i} & \mathbf{r}\frac{\partial \mathbf{R}}{\partial \Omega} \\ \mathbf{R}\frac{\partial \dot{\mathbf{r}}}{\partial a} & \mathbf{R}\frac{\partial \dot{\mathbf{r}}}{\partial e} & \mathbf{R}\frac{\partial \dot{\mathbf{r}}}{\partial m} & \dot{\mathbf{r}}\frac{\partial \mathbf{R}}{\partial \omega} & \dot{\mathbf{r}}\frac{\partial \mathbf{R}}{\partial i} & \dot{\mathbf{r}}\frac{\partial \mathbf{R}}{\partial \Omega} \end{bmatrix},$$

$$\mathbf{x} = \begin{bmatrix} \frac{da}{dt} & \frac{de}{dt} & \frac{dm}{dt} & \frac{d\omega}{dt} & \frac{di}{dt} & \frac{d\Omega}{dt} \end{bmatrix}^T$$

$$= \begin{bmatrix} \dot{a} & \dot{e} & \dot{m} & \dot{\omega} & \dot{i} & \dot{\Omega} \end{bmatrix}^T,$$

$$\boldsymbol{\ell} = \begin{bmatrix} \mathbf{0} \\ d\ddot{\boldsymbol{\varrho}} \end{bmatrix}.$$

3.2 Orbit description

The [6×6] matrix **A** requires the derivatives of $\boldsymbol{\varrho}$ and $\dot{\boldsymbol{\varrho}}$ with respect to the Keplerian parameters which have been developed in the preceding section, cf. Eq. (3.17). The [6 × 1] vector $\boldsymbol{\ell}$ contains the disturbing acceleration. Finally, the six unknown time derivatives appear in the [6 × 1] vector **x**.

The inversion of the system (3.24) leads to Lagrange's equations, which are given in Hofmann-Wellenhof et al. (2001: Eq. (4.34)), where the disturbing potential R, associated with the disturbing acceleration by $d\ddot{\boldsymbol{\varrho}} = \operatorname{grad} R$, has been introduced. Note that the system fails for $e = 0$ or $i = 0$. This singularity can be avoided by the substitution of auxiliary parameters (Arnold 1970: p. 28).

The Lagrange equations presuppose that the disturbing potential R is expressed in function of the Keplerian parameters. When the acceleration $d\ddot{\boldsymbol{\varrho}}$ is represented by components K_i along the axes $\mathbf{e}_i^*$, the Lagrange equations can be transformed using the identity

$$\frac{\partial R}{\partial p_i} = \operatorname{grad} R \cdot \frac{\partial \boldsymbol{\varrho}}{\partial p_i} = \left[K_1 \mathbf{e}_1^* + K_2 \mathbf{e}_2^* + K_3 \mathbf{e}_3^* \right] \cdot \frac{\partial \boldsymbol{\varrho}}{\partial p_i}. \quad (3.25)$$

The simple but cumbersome algebra leads to the Gaussian equations. The result is given in Hofmann-Wellenhof et al. (2001: Eq. (4.36)). Note that in the temporal variations $\dot{i}$ and $\dot{\Omega}$ only the component orthogonal to the orbital plane, K_3, appears, whereas the variations $\dot{a}$, $\dot{e}$, $\dot{m}$ are affected by both components in the orbital plane, K_1 and K_2. The variation $\dot{\omega}$ contains all the components K_i.

3.2.3 Disturbing accelerations

In reality, many disturbing accelerations act on a satellite and are responsible for the temporal variations of the Keplerian elements. Roughly speaking, they can be divided into two groups, namely those of gravitational and those of nongravitational origin (Table 3.3). Because the MEO satellites are orbiting at an altitude of approximately 20 000 km, the indirect effect of solar radiation pressure as well as air drag may be neglected.

Table 3.3. Sources for disturbing accelerations

Gravitational	Nonsphericity of the earth
	Tidal attraction (direct and indirect)
Nongravitational	Solar radiation pressure (direct and indirect)
	Air drag
	Relativistic effects
	Others (e.g., solar wind, magnetic field forces)

On the other hand, the shape (and, thus, the cross section) of the satellites is irregular, which renders the modeling of direct solar radiation pressure more difficult. The variety of materials used for the satellites, each has a different heat absorption which results in additional and complicated perturbing accelerations. Also, accelerations may arise from leaks in the container of the gas propellant as mentioned by Lichten and Neilan (1990).

To demonstrate the effect of disturbing accelerations, an example is computed by assuming a constant disturbance $d\ddot{\varrho} = 10^{-9}$ m s^{-2} acting on a MEO satellite. The associated shift in the position of the satellite results from double integration over time t and yields $d\varrho = (t^2/2) \, d\ddot{\varrho}$. Substituting the numerical value $t = 12$ hours gives the shift after one revolution which is $d\varrho \approx 1$ m. This can be considered as typical value.

Nonsphericity of the earth

The earth's potential V can be represented by a spherical harmonic expansion (Hofmann-Wellenhof and Moritz 2006: Eq. (7-1)) by

$$V = \frac{\mu}{r}\left[1 - \sum_{n=2}^{\infty}\left(\frac{a_e}{r}\right)^n J_n P_n(\sin\varphi) \right.$$
$$\left. + \sum_{n=2}^{\infty}\sum_{m=1}^{n}\left(\frac{a_e}{r}\right)^n [C_{nm}\cos m\lambda + S_{nm}\sin m\lambda] P_{nm}(\sin\varphi)\right],$$

(3.26)

where a_e is the semimajor axis of the earth, r is the geocentric distance of the satellite, and φ, λ are its latitude and longitude. The parameters J_n, C_{nm}, S_{nm} denote the zonal and tesseral coefficients of the harmonic development known from an earth model. Finally, P_n are the Legendre polynomials and P_{nm} are the associated Legendre functions.

The term μ/r on the right side of Eq. (3.26) represents the potential V_0 for a spherical earth, and its gradient, $\mathrm{grad}(\mu/r) = (\mu/r^3)\,\mathbf{r}$, is regarded as the central force for the Keplerian motion. Hence, the disturbing potential R is given by the difference

$$R = V - V_0.$$
(3.27)

It is shown below that the disturbing acceleration due to J_2, the term representing the oblateness, is smaller by a factor of 10^4 than the acceleration due to V_0. On the other hand, the oblateness term is approximately three orders of magnitude larger than any other coefficient.

3.2 Orbit description

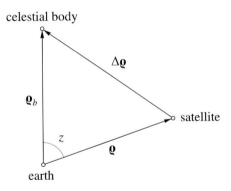

Fig. 3.2. Three-body problem

A numerical assessment of the central acceleration of MEO satellites gives $\|\ddot{\mathbf{r}}\| = \mu/r^2 \approx 0.57 \, \mathrm{m\,s^{-2}}$. The acceleration corresponding to the oblateness term in the disturbing potential R is given by $\|d\ddot{\mathbf{r}}\| \approx \|\partial R/\partial r\| = 3\mu \, (a_e/r^2)^2 \, J_2 \, P_2(\sin\varphi)$. The latitude of a satellite can only reach the value of its orbital inclination. Assuming $i = 55°$, the maximum of the function $P_2(\sin\varphi) = (3\sin\varphi^2 - 1)/2$, therefore, becomes 0.5. Finally, with $J_2 \approx 1.1 \cdot 10^{-3}$, the numerical value $\|d\ddot{\mathbf{r}}\| \approx 5 \cdot 10^{-5} \, \mathrm{m\,s^{-2}}$ is obtained.

In the early days of operational satellite-based positioning systems a subset of earth model coefficients complete up to degree and order eight was considered sufficient for satellite arcs of a few revolutions. Today, subsets of coefficients from models such as the Earth Gravitational Model 1996 (EGM96) up to degree and order 70 are recommended for high-accuracy orbit determination.

Tidal effects

Consider a celestial body with mass m_b and the geocentric position vector ϱ_b (Fig. 3.2). Note that the geocentric angle z between the celestial body and the satellite can be expressed as a function of ϱ_b and ϱ, the latter denoting the geocentric position vector of the satellite, by

$$\cos z = \frac{\varrho_b}{\|\varrho_b\|} \cdot \frac{\varrho}{\|\varrho\|}. \tag{3.28}$$

The additional mass exerts an acceleration with respect to the earth as well as with respect to the satellite. For the perturbed motion of the satellite around the earth, only the difference of the two accelerations is relevant; consequently, the disturbing acceleration is given by

$$d\ddot{\varrho} = G\,m_b \left[\frac{\varrho_b - \varrho}{\|\varrho_b - \varrho\|^3} - \frac{\varrho_b}{\|\varrho_b\|^3} \right]. \tag{3.29}$$

Among all the celestial bodies in the solar system, only the sun and the moon must be considered because the effects of the planets are negligible. The geocentric position vector of the sun and the moon are obtained by evaluating known analytical expressions for their motion.

The maximum of the perturbing acceleration is reached when the three bodies in Fig. 3.2 are in collinear position. In this case, Eq. (3.29) reduces to $\|d\ddot{\varrho}\| = Gm_b [1/\|\varrho_b - \varrho\|^2 - 1/\|\varrho_b\|^2]$. For a numerical assessment, the corresponding numerical values for the sun ($Gm_b \approx 1.3 \cdot 10^{20}$ m^3 s^{-2}, $\varrho_b \approx 1.5 \cdot 10^{11}$ m) and of the moon ($Gm_b \approx 4.9 \cdot 10^{12}$ m^3 s^{-2}, $\varrho_b \approx 3.8 \cdot 10^8$ m) are substituted. The resulting numerical values for the perturbing acceleration acting on MEO satellites are $2 \cdot 10^{-6}$ m s^{-2} for the sun and $5 \cdot 10^{-6}$ m s^{-2} for the moon.

Apart from the direct effect of the tide-generating bodies, indirect effects due to the tidal deformation of the solid earth and the oceanic tides must be taken into account. Considering only the tidal potential W_2 of second degree, the disturbing potential R due to the tidal deformation of the solid earth (Melchior 1978) is given by

$$R = k \left(\frac{a_e}{\varrho}\right)^3 W_2 = \frac{1}{2} k G m_b \frac{a_e^5}{(\varrho \varrho_b)^3} (3\cos^2 z - 1) \tag{3.30}$$

with $k \approx 0.3$ being one of the Love numbers. The associated acceleration of a MEO satellite is in the order of 10^{-9} m s^{-2} as the reader may verify.

The model for the indirect effect due to the oceanic tides is more complicated. Tidal charts with the distribution of the oceanic tides are required. In addition, loading coefficients are needed. These coefficients describe the response of the solid earth to the load of the oceanic water masses. The perturbing acceleration is again in the order of 10^{-9} m s^{-2}.

As a consequence of the tidal deformation and the oceanic loading, the geocentric position vector ϱ_r of an observing site varies with time. This variation must be taken into account when modeling receiver-dependent biases in the observation equations.

Solar radiation pressure

Following Fliegel et al. (1985), the perturbing acceleration due to the direct solar radiation pressure has two components. The principal component $d\ddot{\varrho}_1$ is directed away from the sun and the smaller component $d\ddot{\varrho}_2$ acts along the satellite's y-axis. This is an axis orthogonal to both the vector pointing to the sun and the antenna which is nominally directed towards the center of the earth.

The principal component is usually modeled by

$$d\ddot{\varrho}_1 = \nu K \varrho_\odot^2 \frac{\varrho - \varrho_\odot}{\|\varrho - \varrho_\odot\|^3}, \tag{3.31}$$

where $\varrho_\odot$ denotes the geocentric position vector of the sun. The factor K depends linearly on the solar radiation term, a factor defining the reflective properties of the satellite, and the area-to-mass ratio of the satellite. The quantity ν is an eclipse factor. An eclipse occurs twice a year for each satellite when the sun is in or near the orbital plane and lasts one hour at the maximum. The eclipse factor is zero when the satellite is completely in the earth's shadow, equals one when the satellite is in sunlight, and for the penumbra regions the relation $0 < \nu < 1$ applies.

The magnitude of $d\ddot{\varrho}_1$ is in the order of 10^{-7} m s^{-2}. Hence, an accurate model for the factors K and ν is required even for short arcs. The modeling is extremely difficult since the solar radiation term varies unpredictably over the year and a single factor for the reflective properties is not adequate for the satellite. Although the mass in orbit is usually known well, the irregular shape of the satellites does not allow for an exact determination of the area-to-mass ratio. A further problem is the modeling of the earth's penumbra and the assignment of an eclipse factor, particularly in the transition zone between illumination and shadow.

The component $d\ddot{\varrho}_2$ is often called y-bias and is believed to be caused by a combination of misalignments of the solar panels and thermal radiation along the y-axis. Since the magnitude of this bias can remain constant for several weeks, it is usually introduced as an unknown parameter which is determined in the course of the orbit determination. Note that this bias is two orders of magnitudes smaller than the principal term.

That portion of the solar radiation pressure which is reflected back from the earth's surface causes an effect called albedo. In the case of MEO satellites, the associated perturbing accelerations are smaller than the y-bias and can be neglected.

Relativistic effect

The relativistic effect on the satellite orbit is caused by the gravity field of the earth and gives rise to a perturbing acceleration which is (simplified) given (Beutler 1991: Eq. (2.5)) by

$$d\ddot{\varrho} = -\frac{3\mu^2 a (1-e^2)}{c^2} \frac{\varrho}{\varrho^5}, \tag{3.32}$$

where c denotes the speed of light. Numerically assessed, the perturbing acceleration results in an order of $3 \cdot 10^{-10}$ m s^{-2} (Zhu and Groten 1988). This effect is smaller than the indirect effects by one order of magnitude and is mentioned for the sake of completeness.

3.3 Orbit determination

In this section, orbit determination essentially means the determination of orbital parameters and satellite clock biases. In principle, the problem is inverse to the navigational or surveying task. In the fundamental equation for the range ϱ and the range rate $\dot{\varrho}$ between the satellite s and the observing site r,

$$\varrho = \|\boldsymbol{\varrho}^s - \boldsymbol{\varrho}_r\|, \tag{3.33}$$

$$\dot{\varrho} = \frac{(\boldsymbol{\varrho}^s - \boldsymbol{\varrho}_r)}{\|\boldsymbol{\varrho}^s - \boldsymbol{\varrho}_r\|} \cdot \dot{\boldsymbol{\varrho}}^s, \tag{3.34}$$

the position vector $\boldsymbol{\varrho}^s$ and the velocity vector $\dot{\boldsymbol{\varrho}}^s$ of the satellite are considered unknown, whereas the position vector $\boldsymbol{\varrho}_r$ of the (stationary) observing site is assumed to be known in a geocentric system.

The ranges in (3.33) are obtained with high precision as outlined in Sect. 5.1. This is particularly true for delta range data since biases are eliminated by differencing the ranges. The range rates in Eq. (3.34) are less accurate and are derived from frequency shifts due to the Doppler effect. The observations for the orbit determination are in most cases performed at terrestrial sites, but data could also be obtained from orbiting receivers.

In the following, the satellite clock biases and other parameters are neglected to emphasize the actual orbit determination which is performed in two steps. First, a Kepler ellipse is fitted to the observations. In the second step, this ellipse serves as reference for the subsequent improvement of the orbit by taking into account perturbing accelerations.

3.3.1 Keplerian orbit

For the moment it is assumed that both the position and the velocity vector of the satellite have been derived from observations. Now the question arises how to use these data for the derivation of the Keplerian parameters.

The position and velocity vector given at the same epoch t define an initial value problem; and two position vectors at different epochs t_1 and t_2 define a (first) boundary value problem. In principle, a second and a third boundary value problem could also be defined; however, these problems are of no practical importance in the context of this textbook and are not treated here.

Initial value problem

As stated previously, the derivation of the Keplerian parameters from position and velocity vectors, both given at the same epoch and expressed in an equatorial system such as $\mathbf{X}_i$, is an initial value problem for solving the differential equation (3.1). Recall that the two given vectors contain six components which allow for the calculation of the six Keplerian parameters. Since both vectors refer to the same epoch, the time parameter is omitted.

3.3 Orbit determination

The solution corresponds to a transformation inverse to Eq. (3.10) and makes use of the fact that quantities like distances or angles are invariant with respect to rotation. Hence, the following equations are obtained:

$$\|\varrho^s\| = \|\mathbf{r}\|,$$
$$\|\dot{\varrho}^s\| = \|\dot{\mathbf{r}}\|,$$
$$\varrho^s \cdot \dot{\varrho}^s = \mathbf{r} \cdot \dot{\mathbf{r}},$$
$$\|\varrho^s \times \dot{\varrho}^s\| = \|\mathbf{r} \times \dot{\mathbf{r}}\|.$$
(3.35)

In addition, by substituting Eqs. (3.6) and (3.8), the relations

$$\varrho^s \cdot \dot{\varrho}^s = \sqrt{\mu a}\,(e \sin E),\tag{3.36}$$

$$\|\varrho^s \times \dot{\varrho}^s\| = \sqrt{\mu a (1 - e^2)}\tag{3.37}$$

are derived.

Now the inverse transformation is solved as follows. First, the geocentric distance r and the velocity $\dot{r}$ are calculated from the given vectors ϱ^s and $\dot{\varrho}^s$. Based on these two quantities, the semimajor axis a follows from Eq. (3.9). With a and r determined, $e \cos E$ is calculated using Eq. (3.7) and $e \sin E$ is derived using Eq. (3.36). Hence, the eccentricity e and the eccentric anomaly E, and consequently the mean and true anomalies M and v, are calculated. The cross product of ϱ^s and $\dot{\varrho}^s$ is equivalent to the vector of angular momentum and is orthogonal to the orbital plane. Therefore, this vector is, after normalization, identical to the vector $\mathbf{e}_3$ in Eq. (3.11) from which the parameters i and $\ell = \Omega - \Theta_0$ are deduced. According to Eq. (3.37), the norm of the vector product allows for a check of the calculated parameters a and e. For the determination of ω, the unit vector $\mathbf{k} = [\cos \ell, \sin \ell, 0]$ pointing from the geocenter to the ascending node is defined. From Fig. 3.1 one can obtain the relations $\varrho^s \cdot \mathbf{k} = r \cos(\omega + v)$ and $\varrho^s \cdot \mathbf{X}_3 = r \sin i \sin(\omega + v)$. The two equations can be uniquely solved for ω, since r, v, i are known.

For a numerical example, start with a position vector ϱ^s and the corresponding velocity vector $\dot{\varrho}^s$ given by

$$\varrho^s = [11\,465,\ 3\,818,\ -20\,923] \quad [\text{km}],$$
$$\dot{\varrho}^s = [-1.2651,\ 3.9960,\ -0.3081] \quad [\text{km s}^{-1}]$$

to determine the Keplerian parameters. This is the task inverse to the numerical example in Sect. 3.2.1 and the result can be checked from there.

Boundary value problem

Now it is assumed that two position vectors $\varrho^s(t_1)$ and $\varrho^s(t_2)$ at epochs t_1 and t_2 are available. Note that position vectors are preferred for orbit determination since they are more accurate than velocity vectors. The given data correspond to boundary values in the solution of the basic second-order differential equation, cf. Eq. (3.1).

An approximate method for the derivation of the Keplerian parameters makes use of initial values defined for an average epoch $t = (t_1 + t_2)/2$:

$$\varrho^s(t) = \frac{\varrho^s(t_2) + \varrho^s(t_1)}{2},$$
$$\dot{\varrho}^s(t) = \frac{\varrho^s(t_2) - \varrho^s(t_1)}{t_2 - t_1}. \tag{3.38}$$

The rigorous solution starts with the computation of the geocentric distances

$$r_1 = r(t_1) = \|\varrho^s(t_1)\|,$$
$$r_2 = r(t_2) = \|\varrho^s(t_2)\|. \tag{3.39}$$

The unit vector $\mathbf{e}_3$, orthogonal to the orbital plane, is obtained from a vector product by

$$\mathbf{e}_3 = \frac{\varrho^s(t_1) \times \varrho^s(t_2)}{\|\varrho^s(t_1) \times \varrho^s(t_2)\|} \tag{3.40}$$

and produces the longitude ℓ and the inclination angle i, cf. Eqs. (3.11) and (3.13). As demonstrated earlier, the argument of the latitude $u = \omega + v$ is defined as the angle between the satellite position and the ascending node vector $\mathbf{k} = [\cos \ell, \sin \ell, 0]$. Hence, the relation

$$r_i \cos u_i = \mathbf{k} \cdot \varrho^s(t_i), \qquad i = 1, 2 \tag{3.41}$$

is obtained from which the u_i with $u_2 > u_1$ can be deduced uniquely. There are now two equations, cf. Eq. (3.7),

$$r_i = \frac{a(1 - e^2)}{1 + e \cos(u_i - \omega)}, \qquad i = 1, 2, \tag{3.42}$$

where the parameters a, e, ω are unknown. The system can be solved for a and e after assigning a preliminary value such as the nominal one to ω, the argument of the perigee. Based on the assumed ω and the u_i, the true anomalies v_i and, subsequently, the mean anomalies M_i are obtained. Therefore, the mean angular velocity n can be calculated twice by the formulas

$$n = \sqrt{\frac{\mu}{a^3}} = \frac{M_2 - M_1}{t_2 - t_1}, \tag{3.43}$$

3.3 Orbit determination

cf. Eqs. (3.2) and (3.3). The equivalence is achieved by varying ω. This iterative procedure is typical for boundary value problems. Finally, the epoch of perigee passage T_0 follows from the relation

$$T_0 = t_i - \frac{M_i}{n}. \tag{3.44}$$

For a numerical solution of the boundary value problem assume two position vectors $\varrho^s(t_1)$ and $\varrho^s(t_2)$, both represented in the earth-fixed equatorial system $\mathbf{X}_i$ and with $\Delta t = t_2 - t_1 = 1$ hour:

$$\varrho^s(t_1) = [11\,465, \ 3\,818, \ -20\,923] \ [\text{km}],$$

$$\varrho^s(t_2) = [\ 5\,220, \ 16\,754, \ -18\,421] \ [\text{km}].$$

The application of Eqs. (3.39) through (3.44) results, apart from rounding errors, in the following set of parameters for the associated Kepler ellipse: $a = 26\,000$ km, $e = 0.1$, $\omega = -140°$, $i = 60°$, $\ell = 110°$, and $T_0 = t_1 - 1.3183^{\text{h}}$.

Orbit improvement

If there are redundant observations, the parameters of an instantaneous Kepler ellipse can be improved. The position vector ϱ_0^s associated with the reference ellipse can be computed. Each observed range, for example, gives rise to the relation

$$\varrho = \varrho_0 + d\varrho = \|\varrho_0^s - \varrho_r\| + \frac{\varrho_0^s - \varrho_r}{\|\varrho_0^s - \varrho_r\|} \cdot d\varrho^s. \tag{3.45}$$

The vector $d\varrho^s$ can be expressed as a function of the six Keplerian parameters, cf. Eq. (3.17). Thus, Eq. (3.45) actually contains the differential increments for the six orbital parameters.

3.3.2 Perturbed orbit

Analytical solution

As known from previous sections, the perturbed motion is characterized by temporal variations of the orbital parameters. The analytical expressions for these variations are given by the Lagrange equations or Gaussian equations, cf. Sect. 3.2.2.

In order to be suitable for Lagrange's equations, the disturbing potential must be expressed as a function of the Keplerian parameters. Kaula (1966) was the first who has performed this transformation for the earth potential. The resulting relation for the disturbing potential is

$$R = \sum_{n=2}^{\infty} A_n(a) \sum_{m=0}^{n} \sum_{p=0}^{n} F_{nmp}(i)$$

$$\cdot \sum_{q=-\infty}^{\infty} G_{npq}(e)\, S_{nmpq}(\omega, \Omega, M; \Theta_0, C_{nm}, S_{nm}), \tag{3.46}$$

Table 3.4. Perturbations due to the earth's gravity field

Parameter	Secular	Long-periodic	Short-periodic
a	no	no	yes
e	no	yes	yes
i	no	yes	yes
Ω	yes	yes	yes
ω	yes	yes	yes
M	yes	yes	yes

where the original formulation has been slightly changed since the former notations $J_{nm} = -C_{nm}$ and $K_{nm} = -S_{nm}$ are not used any more (Hofmann-Wellenhof and Moritz 2006: p. 257). Recall that n denotes the degree and m the order of the spherical harmonics in the disturbing potential development. Each of the functions A_n, F_{nmp}, G_{npq} contains only one parameter of a Kepler ellipse; however, the function S_{nmpq} is composed of several parameters and can be expressed as function of the frequency

$$\dot{\psi} = (n-2p)\dot{\omega} + (n-2p+q)\dot{M} + m(\dot{\Omega} - \dot{\Theta}_0), \tag{3.47}$$

which is a measure of the spectrum of the perturbations.

The conditions $(n-2p) = (n-2p+q) = m = 0$ lead to $\dot{\psi} = 0$ and, thus, to secular variations. Because of $m = 0$, they are caused by zonal harmonics. If $(n-2p) \neq 0$, then the variations depend on $\dot{\omega}$ and are, therefore, generally long-periodic. Finally, the conditions $(n-2p+q) \neq 0$ and/or $m \neq 0$ result in short-periodic variations. The integer value $(n-2p+q)$ gives the frequency in cycles per revolution and m the frequency in cycles per (sidereal) day.

A rough overview of the frequency spectrum of the Keplerian parameters due to the gravity field of the earth is given in Table 3.4. Summarizing, one can state that the even-degree zonal coefficients produce primarily secular variations and the odd-degree zonal coefficients give rise to long-periodic perturbations. The tesseral coefficients are responsible for short-periodic terms. From Table 3.4 one can see that short-periodic variations occur in each parameter. With the exception of the semimajor axis, the parameters are also affected by long-periodic perturbations. Secular effects are only contained in Ω, ω, M. The analytical expression for the secular variations of these parameters due to the oblateness term J_2 is given as an

3.3 Orbit determination

example:

$$\dot{\Omega} = -\frac{3}{2} n a_e^2 \frac{\cos i}{a^2 (1 - e^2)^2} J_2,$$

$$\dot{\omega} = \frac{3}{4} n a_e^2 \frac{5 \cos^2 i - 1}{a^2 (1 - e^2)^2} J_2, \qquad (3.48)$$

$$\dot{m} = \frac{3}{4} n a_e^2 \frac{3 \cos^2 i - 1}{a^2 \sqrt{(1 - e^2)^3}} J_2.$$

The first equation describes the regression of the node in the equatorial plane, the second equation expresses the rotation of the perigee, and the third equation contributes to the variation of the mean anomaly by $\dot{M} = n + \dot{m}$. Assuming an orbital inclination $i = 55°$ of a MEO satellite, the numerical values $\dot{\Omega} \approx -0.03°$ per day, $\dot{\omega} \approx 0.01°$ per day, and $\dot{m} \approx 0$ are obtained. The result for $\dot{m}$ is verified immediately since the term $3 \cos^2 i - 1$ becomes approximately zero for the chosen inclination. Special attention must be paid to resonance effects which occur when the period of revolution corresponds to a harmonic in the gravity potential. Thus, MEO satellites are placed in such orbits where an orbital period very close to half a sidereal day is avoided.

The tidal potential also has a harmonic representation and, thus, the tidal perturbations can be analytically modeled. This was performed first by Kozai (1959) and, analogous to the earth's potential effect, has led to analytical expressions for the secular variations of the node's right ascension Ω and of the perigee's argument ω. The reader may find the respective formulas in Kozai (1959).

Numerical solution

If the disturbing acceleration cannot be expressed in analytical form, one has to apply numerical methods for the solution. Therefore, in principle, with initial values such as the position and velocity vectors $\varrho(t_0)$ and $\dot{\varrho}(t_0)$ at a reference epoch t_0, a numerical integration of Eq. (3.18) could be performed. This simple concept can be improved by the introduction of a Kepler ellipse as a reference. By this means, only the smaller difference between the total and the central acceleration must be integrated. The integration results in an increment $\Delta\varrho$ which, when added to the position vector computed for the reference ellipse, gives the actual position vector.

The second-order differential equation is usually transformed to a system of two differential equations of the first order. The velocity vector is defined by

$$\dot{\varrho}(t) = \dot{\varrho}(t_0) + \int_{t_0}^{t} \ddot{\varrho}(t_0) \, dt = \dot{\varrho}(t_0) + \int_{t_0}^{t} \left[d\ddot{\varrho}(t_0) - \frac{\mu}{\varrho^3(t_0)} \varrho(t_0) \right] dt \qquad (3.49)$$

and the position vector reads

$$\varrho(t) = \varrho(t_0) + \int_{t_0}^{t} \dot{\varrho}(t_0) \, dt \, . \tag{3.50}$$

The numerical integration of this coupled system can be performed by applying the standard Runge–Kutta algorithm (e.g., Kreyszig 2006: p. 892).

Numerical methods can also be applied to the integration of the disturbing equations of Lagrange or Gauss. These equations have the advantage of being differentials of the first order and, therefore, must only be integrated once over time.

This section concludes with a short outline of the first-order Runge–Kutta algorithm. Let $y(x)$ be a function defined in the interval $x_1 \le x \le x_2$ and denote by $y' = dy/dx$ its first derivative with respect to the argument x. The general solution of the ordinary differential equation of the first order

$$y' = \frac{dy}{dx} = y'(y, x) \tag{3.51}$$

follows from integration. The particular solution is found after assigning the given numerical initial value $y_1 = y(x_1)$ to the integration constant. For the application of numerical integration, first the integration interval is partitioned into n equal and sufficiently small $\Delta x = (x_2 - x_1)/n$, where n is an arbitrary integer number greater than 0. Then the difference between successive functional values is obtained by the weighted mean

$$\Delta y = y(x + \Delta x) - y(x) = \tfrac{1}{6} \left[\Delta y^{(1)} + 2 \left(\Delta y^{(2)} + \Delta y^{(3)} \right) + \Delta y^{(4)} \right] , \tag{3.52}$$

where

$$\Delta y^{(1)} = y'(y, x) \, \Delta x \, ,$$

$$\Delta y^{(2)} = y' \left(y + \Delta y^{(1)}/2, \, x + \Delta x/2 \right) \Delta x \, ,$$

$$\Delta y^{(3)} = y' \left(y + \Delta y^{(2)}/2, \, x + \Delta x/2 \right) \Delta x \, ,$$

$$\Delta y^{(4)} = y' \left(y + \Delta y^{(3)}, \, x + \Delta x \right) \Delta x \, .$$

Hence, starting with the initial value y_1 for the argument x_1, the function can be calculated for the successive argument $x_1 + \Delta x$ and so on.

For a numerical example, assume the ordinary first-order differential equation $y' = y - x + 1$ with the initial value $y_1 = 1$ for $x_1 = 0$. The differential equation should be solved for the argument $x_2 = 1$ with increments $\Delta x = 0.5$. Starting with the initial values for the first interval, successively the values $\Delta y^{(1)} = 1.000$,

$\Delta y^{(2)} = 1.125$, $\Delta y^{(3)} = 1.156$, and $\Delta y^{(4)} = 1.328$ are obtained. The corresponding weighted mean is $\Delta y = 1.148$. Replacing the initial values in the second interval by $x = x_1 + \Delta x = 0.5$ and $y = y_1 + \Delta y = 2.148$ and proceeding as before yields $\Delta y = 1.569$ and, thus, the final result $y_2 = y_1 + \Sigma \Delta y = 3.717$. Note that the function $y = e^x + x$, satisfying the differential equation, gives the true value $y_2 = 3.718$. With an increment $\Delta x = 0.1$, the numerical integration would provide an accuracy in the order of 10^{-6}.

3.4 Orbit dissemination

3.4.1 Tracking networks

Objectives and strategies

The orbit determination for GNSS satellites is based on observations at monitor stations of the respective control segment.

Global networks lead to higher accuracy and reliability of the orbits compared to those determined from regional networks. The tie of the orbital system to terrestrial reference frames is achieved by the colocation of GNSS receivers with, e.g., VLBI and SLR trackers. The distribution of the sites is essential to achieve the highest accuracy. Two different approaches may be compared. In the first case, the sites are regularly distributed around the globe; in the second case, each network site is surrounded by a cluster of additional points to facilitate ambiguity resolution and, thus, strengthen the solution of the orbital parameters by a factor of three to five.

Examples for global networks

Several global networks have been established for orbit determination. Some networks are of regional or even continental size such as the Australian GPS orbit determination network. Subsequently, some examples of global networks are given.

The first civil large-scale global orbit tracking experiment (GOTEX) was performed in the fall of 1988 and aimed at the colocation of GPS equipment at existing VLBI and SLR sites. The data were collected to permit an accurate tie between the WGS-84 and the VLBI/SLR systems. About 25 sites, distributed worldwide, were occupied during the three weeks of the campaign.

The cooperative international GPS network (CIGNET) was operated by the US National Geodetic Survey (NGS) with tracking stations located at VLBI sites. The service started in 1988 with 8 stations in North America, Europe, and Japan. By 1991, already 20 globally distributed stations were participating (Chin 1991). In the following three years, 30 more stations were incorporated into the network. The charter of CIGNET was limited to the collection and distribution of the raw tracking data and there was no intention to provide global orbits.

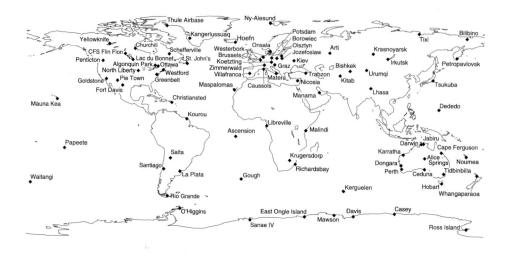

Fig. 3.3. IGS tracking network (reference frame stations only) as per 2006

In 1990, the International Association of Geodesy (IAG) decided to install the International GPS Service for Geodynamics (IGS) (Mueller 1991). After a test campaign, routine activities started on January 1, 1994. The main purpose of this service (where meanwhile the term GPS has been replaced by GNSS) is orbit determination for geodynamic applications which require highest accuracy.

The IGS is headed by the Central Bureau located at the US Jet Propulsion Laboratory (JPL). In August 2006, the tracking network (Fig. 3.3) was based on more than 330 globally distributed active tracking sites with coordinates (and velocities) related to the ITRF. The IGS stations collect code ranges and carrier phases from all GNSS satellites in view using dual-frequency receivers. The data are analyzed independently by seven agencies and are archived daily in receiver- and software-independent exchange formats (RINEX and SINEX) by global and regional data centers. Today, the IGS routinely provides high-quality orbits for all GNSS satellites. Predicted orbits with an accuracy of about 10 cm are available in near real time, whereas postprocessed orbits have a delay of about 1 day (rapid solution) or two weeks (final solution). The accuracy of the final solution is estimated to be better than 5 cm. Also, the raw tracking data, the satellite clock parameters, the earth orientation parameters (EOP), and other data like ionospheric and tropospheric information are available through the service. Associated with the IGS is an information service which is accessible at http://igscb.jpl.nasa.gov. Further details on the IGS are given in Gurtner (1995), Beutler (1996), Neilan and Moore (1999).

3.4 Orbit dissemination

3.4.2 Ephemerides

Three sets of data are available to determine position and velocity vectors of the satellites in a terrestrial reference frame at any instant: almanac data, broadcast ephemerides, and precise ephemerides. The data differ in accuracy (Table 3.5) and are available either in real time or with some delay (i.e., latency) after the fact.

Almanac data

The purpose of the almanac data is to provide the user with adequate data to facilitate receiver satellite acquisition and for planning tasks such as the computation of visibility charts. The almanac data are regularly updated and are broadcast as part of the satellite message. The almanac message essentially contains parameters for the orbit and satellite clock correction terms for all satellites of the respective GNSS.

Table 3.6 serves as an example for GPS satellites, but similar data are available for GLONASS and Galileo. All angles are expressed in semicircles (i.e., multiples of 180°). The parameter ℓ_0 denotes the difference between the node's right ascension at epoch t_a and the Greenwich sidereal time at t_0, the beginning of the current GPS week. The reduction of the Keplerian parameters to the observation epoch t is obtained by the formulas

$$\begin{aligned} M &= M_0 + n\,(t - t_a)\,, \\ i &= 54° + \delta i\,, \\ \ell &= \ell_0 + \dot{\Omega}\,(t - t_a) - \omega_e\,(t - t_0)\,, \end{aligned} \qquad (3.53)$$

where $\omega_e = 7\,292\,115.1467 \cdot 10^{-11}$ rad s^{-1} is the (untruncated) angular velocity of the earth. The other three Keplerian parameters a, e, ω remain unchanged. Note that in the formula for ℓ in Eq. (3.53) the second term on the right side of the equation considers the node's regression and the third term expresses the uniform change in the sidereal time since epoch t_0. An estimate for the satellite clock bias

Table 3.5. Uncertainties of ephemerides

Ephemerides	Uncertainty	Remark
Almanac	Some kilometers	Depending on the age of data
Broadcast ephemerides	~ 1 m	Or even better
Precise ephemerides	0.05–0.20 m	Depending on the latency

Table 3.6. Almanac data (GPS)

Parameter	Explanation
ID	Satellite identification number
WEEK	Current GPS week
t_a	Reference epoch in seconds within the current week
$\sqrt{a}$	Square root of semimajor axis in $\sqrt{\text{meter}}$
e	Eccentricity
M_0	Mean anomaly at reference epoch
ω	Argument of perigee
δi	Inclination offset from 0.3 semicircles ($\hat{=} 54°$)
ℓ_0	Longitude of the node at weekly epoch t_0
$\dot{\Omega}$	Drift of node's right ascension per second
a_0	Satellite clock offset in seconds
a_1	Satellite clock drift coefficient

is given by

$$\delta^s = a_0 + a_1 (t - t_a). \tag{3.54}$$

Almanac data are also available from a variety of information services.

Broadcast ephemerides

The broadcast ephemerides are based on observations at the monitor stations of the respective control segment. The most recent of these data are used to compute a reference orbit for the satellites. Additional tracking data are entered into a Kalman filter and the improved orbits are used for extrapolation. The master station of the control segment is responsible for the computation of the ephemerides and the subsequent upload to the satellites.

The broadcast ephemerides are part of the satellite message. Essentially, the ephemerides contain records with general information, records with orbital information, and records with information on the satellite clock. The orbital information is provided in the form of Keplerian parameters together with their temporal variations (e.g., for GPS) or position and velocity vectors at equidistant epochs (e.g., for GLONASS). The information on the satellite clock is in most cases given in the form of coefficients to model the clock offset from system time by polynomials.

Table 3.7, serves as an example for GPS satellites. The parameters in the block of orbital information are the reference epoch, six parameters to describe a Kepler ellipse at the reference epoch, three secular correction terms, and six periodic correction terms. The nine correction terms consider the perturbation effects due to the

3.4 Orbit dissemination

Table 3.7. Broadcast ephemerides (GPS)

Parameter	Explanation
ID	Satellite identification number
WEEK	Current GPS week
t_e	Ephemerides reference epoch
$\sqrt{a}$	Square root of semimajor axis in $\sqrt{\text{meter}}$
e	Eccentricity
M_0	Mean anomaly at reference epoch
ω_0	Argument of perigee
i_0	Inclination
ℓ_0	Longitude of the node at weekly epoch t_0
Δn	Mean motion difference
$\dot{i}$	Rate of inclination angle
$\dot{\Omega}$	Rate of node's right ascension
C_{uc}, C_{us}	Correction coefficients (argument of perigee)
C_{rc}, C_{rs}	Correction coefficients (geocentric distance)
C_{ic}, C_{is}	Correction coefficients (inclination)
t_c	Satellite clock reference epoch
a_0	Satellite clock offset
a_1	Satellite clock drift coefficient
a_2	Satellite clock frequency drift coefficient

nonsphericity of the earth, the direct tidal effect, and the solar radiation pressure. The ephemerides are regularly updated and should only be used during a specified time interval.

In order to compute the satellite position at the observation epoch t, the following quantities, apart from the parameters a and e, are needed:

$$M = M_0 + \left[\sqrt{\frac{\mu}{a^3}} + \Delta n\right](t - t_e),$$

$$\ell = \ell_0 + \dot{\Omega}(t - t_e) - \omega_e(t - t_0),$$

$$\omega = \omega_0 + C_{uc}\cos(2u) + C_{us}\sin(2u), \qquad (3.55)$$

$$r = r_0 + C_{rc}\cos(2u) + C_{rs}\sin(2u),$$

$$i = i_0 + C_{ic}\cos(2u) + C_{is}\sin(2u) + \dot{i}(t - t_e),$$

where $u = \omega_0 + v$ is the argument of latitude. The geocentric distance r_0 is calculated by Eq. (3.7) using a, e, E at the observation epoch. The vector **r** in the orbital plane follows from the second representation in (3.6). Based on the reference epoch t_e, the computation of ℓ is analogous to that in Eq. (3.53). Note that again the untruncated value for the earth rotation rate must be used.

The block with clock parameters enables the computation of the satellite clock error at an observation epoch t by

$$\delta^s = a_0 + a_1 (t - t_c) + a_2 (t - t_c)^2 . \tag{3.56}$$

Precise ephemerides

The most accurate orbital information is provided by the IGS in the form of various data sets for precise ephemerides. An overview on these products, their accuracy and latency can be found at http://igscb.jpl.nasa.gov/components/prods.html. Currently, IGS data and products are free of charge for all users.

The precise ephemerides consist of satellite positions and velocities at equidistant epochs. Typical spacing of the data is 15 minutes. Since 1985, NGS has been involved in the distribution of precise GPS orbital data. At that time, the data were distributed in the specific ASCII formats SP1 and SP2 and their binary counterparts ECF1 and ECF2. Later, ECF2 was modified to EF13 format. The formats SP1 and ECF1 contain position and velocity data, whereas SP2 and ECF2 contain just position data. This almost halves the storage amount since the velocity data can be computed from the position data by numerical differentiation. In 1989, NGS decided to add the GPS satellite clock offset data to the orbital formats. Furthermore, the second-generation formats can handle up to 85 satellites (GPS and others) instead of 35 GPS satellites included in the first-generation formats. Apart from the clock corrections, the files often contain only position data; however, a header character on the first line allows for the inclusion of velocity data. The corresponding ASCII format is denoted SP3 and the binary counterpart is ECF3 or (in a modified version) EF18. The format SP3 is widely used and was also adopted by the IGS.

Each NGS format consists of a header containing general information (start time, epoch interval, orbit type, etc.) followed by the data section for successive epochs. These data are repeated for each satellite. The positions are given in kilometers and the velocities are given in kilometers per second. The NGS formats are described in Remondi (1991). Also, NGS provides software to translate orbital files from one format to another.

The position and velocity vectors between the given epochs are obtained by interpolation where the Lagrange interpolation, based on polynomial base functions, is used. Note that Lagrange interpolation is also applicable to variable epoch series and that the coefficients determined can be applied to considerably longer series without updating the coefficients. This interpolation method is a fast procedure and

3.4 Orbit dissemination

can easily be programmed. Extensive studies by B.W. Remondi concluded that for GPS satellites a 30-minute epoch interval and a 9th-order interpolator suffices for an accuracy of some decimeters (i.e, about 10^{-8}). Another study by Remondi (1991) using a 17th-order interpolator demonstrates that millimeter-level accuracies can be achieved based on a 40-minute epoch interval.

For those not familiar with Lagrange interpolation, the principle of this method and a numerical example are given. Assume functional values $f(t_j)$ are given at epochs t_j, $j = 0, 1, \ldots, n$. Then,

$$\ell_j(t) = \frac{(t - t_0)(t - t_1) \cdots (t - t_{j-1})(t - t_{j+1}) \cdots (t - t_n)}{(t_j - t_0)(t_j - t_1) \cdots (t_j - t_{j-1})(t_j - t_{j+1}) \cdots (t_j - t_n)} \quad (3.57)$$

is the definition of the corresponding base functions $\ell_j(t)$ of degree n related to an arbitrary epoch t. The interpolated functional value at epoch t follows from the summation

$$f(t) = \sum_{j=0}^{n} f(t_j) \, \ell_j(t) \, . \quad (3.58)$$

The following numerical example assumes the functional values $f(t_j)$ given at the epochs t_j:

$$f(t_0) = f(-3) = 13 \, ,$$
$$f(t_1) = f(+1) = 17 \, ,$$
$$f(t_2) = f(+5) = 85 \, .$$

The base functions are polynomials of second degree

$$\ell_0(t) = \frac{(t - t_1)(t - t_2)}{(t_0 - t_1)(t_0 - t_2)} = \frac{1}{32}(t^2 - 6t + 5) \, ,$$

$$\ell_1(t) = \frac{(t - t_0)(t - t_2)}{(t_1 - t_0)(t_1 - t_2)} = -\frac{1}{16}(t^2 - 2t - 15) \, ,$$

$$\ell_2(t) = \frac{(t - t_0)(t - t_1)}{(t_2 - t_0)(t_2 - t_1)} = \frac{1}{32}(t^2 + 2t - 3) \, ,$$

and, according to Eq. (3.58), the interpolated value for $t = 4$ is $f(t) = 62$. The result is immediately verified since the given functional values were generated by the polynomial $f(t) = 2t^2 + 5t + 10$.

4 Satellite signals

4.1 Introduction

The different methods of satellite navigation are classified into passive and active as well as into one-way (uplink = earth-to-space; downlink = space-to-earth) and two-way ranging systems. Active systems require the user to emit signals. The three major GNSS (GPS, GLONASS, Galileo) are passive one-way downlink ranging systems. The satellites emit modulated signals that include the time of transmission to derive ranges as well as the modeling parameters to compute satellite positions. A three-layer model describes the emitted satellite signals best (Fig. 4.1).

The physical layer characterizes the physical properties of the transmitted signals. The ranging code layer describes the method of measuring the propagation time. Rather than using a time pulse for propagation time observation, the ranging code layer is based on a continuous but periodic modulated signal exploited within correlation techniques. The periodicity is strictly synchronized to the time system of the satellites and to the data message. Finally, the data-link layer commonly contains the time of transmission, satellite ephemerides, etc. The data-link layer may also be accomplished by other satellite or terrestrial communication links. This, however, still requires a high level of synchronization between ranging code and data-link layer.

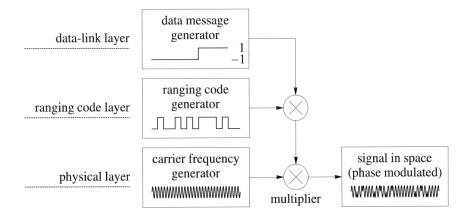

Fig. 4.1. Composition of the navigation satellite signal

4.1.1 Physical fundamentals

General remarks

Satellite navigation relies on electromagnetic waves, whose characteristics are described by Maxwell's equations. Oscillating electric or magnetic forces generate the waves. The duality of electromagnetic waves manifests itself in the combined propagation of an electric field and a magnetic field. An oscillating electric field causes a magnetic field through magnetic flux. Electric induction is the affiliated inverse effect.

The electric field vector $\mathbf{E} = [E_x, E_y]$ determines the polarization of the electromagnetic waves. A constant direction of $\mathbf{E}$ characterizes linear polarization. In homogeneous, isotropic, and stationary media, which are media that do not show any spatial or temporal variations of their physical properties, the electric field vector $\mathbf{E}$ and the magnetic field vector $\mathbf{B}$ are orthogonal to each other and also orthogonal to the propagation direction of the electromagnetic wave (Fig. 4.2). A varying ratio between the elements of the field vector $\mathbf{E}$ indicates an elliptically or circularly (i.e., constant amplitude of the field vector) polarized wave (Fig. 4.2). If the electric field vector rotates clockwise, when looking into the direction of propagation, the electromagnetic wave is right-handed polarized, otherwise left-handed. Electromagnetic waves traveling through ionized gases or through the earth magnetic field undergo a change in their polarization. This effect, known as Faraday rotation, causes linearly polarized signals to become elliptically or circularly polarized. The magnitude of rotation fluctuates. Satellite navigation treats this effect by using circularly polarized signals by definition. Under certain circumstances, specular reflecting surfaces change the right-handed circular polarization (RHCP) to left-handed circular polarization (LHCP) and vice versa.

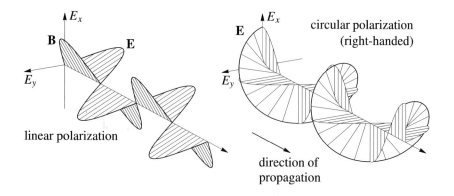

Fig. 4.2. Linear and circular polarization

4.1 Introduction

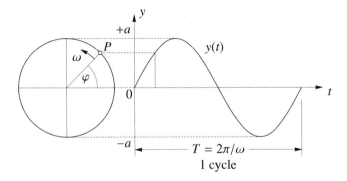

Fig. 4.3. Electromagnetic wave representation

Electromagnetic wave representation

Electromagnetic waves are represented best by sinusoidal waves applying the theory of harmonic motion (Fig. 4.3):

$$y(t) = a \sin(2\pi f t). \tag{4.1}$$

The characteristic parameters describing the sinusoidal wave are the amplitude a, the linear frequency f, and the time parameter t. The amplitude a is defined as the maximum or peak value respectively of a periodically varying quantity. The linear frequency f describes the number of identical cycles per second (cps) and is usually given in hertz (Hz) or decimal multiples of hertz (e.g., kHz = 10^3 Hz, MHz = 10^6 Hz, GHz = 10^9 Hz).

The period T denotes the time needed to complete one cycle, thus,

$$T = \frac{1}{f}. \tag{4.2}$$

The circular frequency ω, also denoted as angular velocity, is defined by

$$\omega = 2\pi f. \tag{4.3}$$

A short numerical example may clarify the above relations. Assume a linear frequency $f = 0.5$ Hz, which corresponds to 0.5 cps. The associated circular frequency is $\omega = \pi$ radians per second or, equivalently, $180°$ per second. Finally, the period T equals 2 seconds.

Referring to Fig. 4.3, the point P moves along the perimeter of the circle with the angular velocity ω. The current state for a specific t is defined by the phase angle or briefly phase φ, which is given by

$$\varphi = \omega t, \tag{4.4}$$

when measured in radians or by

$$\varphi = f t, \tag{4.5}$$

when measured in cycles. The latter form will be used throughout this textbook.

Differentiation of (4.5) with respect to time yields

$$f = \frac{d\varphi}{dt}, \tag{4.6}$$

and integrating the frequency f between the initial epoch t_0 and epoch t leads to the equation

$$\varphi(t) - \varphi(t_0) = \int_{t_0}^{t} f \, dt, \tag{4.7}$$

where $\varphi(t_0)$ is denoted as initial phase.

Equations (4.4) through (4.7) and Fig. 4.3 describe the phase as a function of time. However, considering a single epoch, the phase also varies with increasing distance between emitter and receiver. In this spatial context, the phase is expressed by range and wavelength. The latter is the distance between two recurrent states of the wave, i.e., after one cycle or after the variation of φ by 2π. The wavelength is commonly denoted by λ and given in meters. Thus, the remarkable relation

$$\varphi = \frac{t}{T} \bigg|_{\substack{\varrho = \text{constant} \\ \varphi = \varphi(t)}} = \frac{\varrho}{\lambda} \bigg|_{\substack{t = \text{constant} \\ \varphi = \varphi(\varrho)}} \tag{4.8}$$

describes the proportionality between the temporal and spatial variation of the phase, where ϱ is the range equivalent to the phase proportional to t.

The complete phase equation combines both variations. Thus, the phase expressed in cycles is given by

$$\varphi = f t - \frac{\varrho}{\lambda}. \tag{4.9}$$

Using the relation

$$c = \lambda f, \tag{4.10}$$

with $c = 299\,792\,458 \text{ m s}^{-1}$ being the speed of light, yields

$$\varphi = f(t - \frac{\varrho}{c}) \tag{4.11}$$

or, finally,

$$\varphi = f(t - t_\varrho), \tag{4.12}$$

4.1 Introduction

where t_ϱ is the time interval the wave needs to propagate the distance ϱ.

For a numerical example consider an electromagnetic wave with a frequency $f = 1.5\,\text{GHz}$ and calculate the phase 2 seconds after emission at a location 20 000 km distant from the emitter. According to (4.11), the continuous phase corresponds to 2 899 930 771.44... cycles. The fractional part of the continuous phase is denoted as observable phase and corresponds in this example to 0.44... cycles.

Doppler frequency shift

The Austrian physicist C. Doppler postulated in 1842, that a relative motion between transmitter and receiver will cause a frequency shift, today commonly known as Doppler shift. An approaching transmitter seems to squeeze the waves, whereas a receding transmitter lengthens the waves. In a first approximation, the Doppler shift Δf reads

$$\Delta f = f_r - f^s = -\frac{1}{c} v_\varrho f^s = -\frac{1}{\lambda^s} v_\varrho, \tag{4.13}$$

where f^s denotes the emitted frequency (the superscript s is chosen to associate a satellite), f_r the received frequency, and v_ϱ is the radial relative velocity (line-of-sight velocity) of the emitter with respect to the receiver. Denoting the distance between emitter and receiver by ϱ leads to

$$v_\varrho = \frac{d\varrho}{dt} = \dot\varrho. \tag{4.14}$$

The Doppler shift is, thus, a measure for velocity or, after integration over time, proportional to range differences

$$\Delta\varrho = \int_{t_0}^{t} \dot\varrho\, dt = -\lambda^s \int_{t_0}^{t} \Delta f\, dt = -\lambda^s\, \Delta\varphi. \tag{4.15}$$

Assume as an example a satellite orbiting in 20 000 km height which corresponds to a mean velocity of about $3.9\,\text{km s}^{-1}$ according to (3.9). Neglecting the earth's rotation, a stationary terrestrial receiver will not measure any Doppler shift at the epoch of closest approach of the satellite since the relative radial velocity equals zero at this moment. The maximum radial velocity, $0.9\,\text{km s}^{-1}$, occurs when the satellite crosses the horizon. Assume a transmitted frequency $f^s = 1.5\,\text{GHz}$, then the Doppler shift corresponds to $\Delta f \cong 4.7\,\text{kHz}$. This frequency shift results in a phase change of 4.7 cycles after 1 millisecond, which corresponds to a change in range of about 0.9 m. The result can be verified by multiplying the radial velocity with the time span of 1 millisecond.

Electromagnetic spectrum

The electromagnetic spectrum, also referred to as frequency spectrum, is represented in Fig. 4.4. Different systems and services use different bands of the electromagnetic spectrum, but its usage is strictly regulated by the International Telecommunication Union (ITU). The ITU allocates different parts of the frequency spectrum to different services, assigns frequencies to providers and users, and allots frequencies to countries (International Telecommunication Union 2004).

Satellite navigation has been allocated to the L-, S-, and C-band. In the near future, the latter is planned to be used as an uplink band; and in the more distant future, it is considered as an option for further navigation signals. L-, S-, and C-band belong to the microwave frequency window.

The assignment of frequencies follows the principle of first-come, first-served. A primary service is not allowed to interfere with services in neighboring frequency bands. Secondary services shall additionally neither interfere with services in the same frequency band, nor claim protection from harmful interference of primary services (International Telecommunication Union 2004). More specifically, the ITU specifies the maximum level of interference between different services, thus, a signal may place a certain amount of energy in neighboring frequency bands.

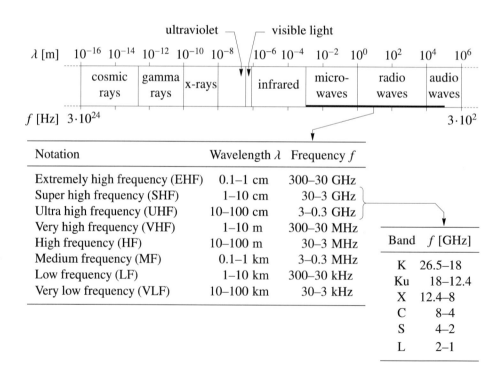

Fig. 4.4. Electromagnetic spectrum

4.1 Introduction

ITU classified GNSS as radionavigation satellite service (RNSS) and aeronautical radionavigation service (ARNS). Especially the frequency bands allocated to ARNS are strictly regulated, and thus particularly useful for safety-critical operations. GNSS is a primary service in the ARNS/RNSS frequency band 1 559–1 610 MHz. The other frequency bands used by GNSS are coprimarily allocated to satellite navigation and other services.

4.1.2 Propagation effects

Different physical phenomena affect electromagnetic wave propagation. Most of these effects are frequency-dependent. Subsequently, selected phenomena are described following the definitions of terms for radio wave propagation of the Institute of Electrical and Electronics Engineers (1997).

Geometry of wave propagation

Reflection
Electromagnetic waves meeting a medium surface are partly reflected (cf. Fig. 4.5). The incident and the reflected waves are symmetric to the normal of the surface and together with it span a plane. A more general form of reflection is scattering, where the energy of a wave is dispersed in various directions due to the interaction with inhomogeneities of the medium. Maximal scattering occurs if the inhomogeneities equal the wavelength λ. A smooth surface can be regarded as a speculum so that incident electromagnetic waves are specularly reflected. Specular reflection causes fluctuation of phase and amplitude and is to a great part deterministic. A rough surface creates diffuse reflection. The differentiation between smooth and rough surface is given by the Rayleigh criterion (Institute of Electrical and Electronics Engineers 1997: p. 28).

Diffraction expresses the deviation of the direction of energy flow of a wave when touching lightly an obstacle in passing. According to Huygens–Fresnel's principle, a wavefront is made up of an infinite number of isotropic radiators. A wave, thus, does not reflect from a single point but from an entire surface. The superposition of all different wave crests results in the wave propagating in the di-

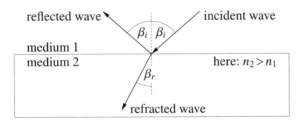

Fig. 4.5. Reflection and refraction of an electromagnetic wave

rection under consideration. In this way, waves seem to be bent around an obstacle. Jong et al. (2002: p. 98) conclude that in this way signals can be received when no line-of-sight conditions occur.

Multipath signals are electromagnetic waves propagating not along the line of sight between transmitter and receiver but reflected or scattered by any kind of object (cf. Sect. 5.6).

Refraction
Refraction describes the change of propagation direction when a wavefront passes from one to the other medium (cf. Fig. 4.5). According to Snell's law, the sine of the angle of incidence, β_i, and the sine of the angle of refraction, β_r, define a constant ratio. This ratio is equivalent to the ratio of the refractive indices n_i of both media:

$$\frac{\sin \beta_i}{\sin \beta_r} = \frac{n_2}{n_1} = \text{constant}. \tag{4.16}$$

The refractive index n_i is defined as the ratio between the velocity v_i of the wave in a medium and the speed of light c in vacuum, which is a universal constant,

$$n_i = \frac{c}{v_i}. \tag{4.17}$$

Snell's law, thus, can be written in the form

$$n_1 v_1 = n_2 v_2 = c. \tag{4.18}$$

Equation (4.18) implies that the refractive index for vacuum equals unity. A refractive index greater than 1 implies that the electromagnetic waves are delayed compared to the time needed to travel the same distance in vacuum with the speed of light. The refractive indices are functions of water vapor, temperature, pressure, frequency of the electromagnetic signals, and the amount of free electrons. Details on this subject are given in Sect. 5.3.

Dispersion is the dependency of the phase velocity and, thus, of refraction on the frequency. If dispersion occurs, a medium is called dispersive, e.g., the ionosphere is a dispersive medium at 1.5 GHz, while the troposphere is not.

Type of waves
An electromagnetic wave propagates in homogeneous media according to Fermat's principle along the path of shortest time. Taking into account Eqs. (4.10), (4.17), and Snell's law, the geometry of electromagnetic wave propagation is a function of the frequency f. According to Fig. 4.6, three wave types, i.e., ground waves, sky waves, line-of-sight waves, can be distinguished. Ground waves ($f \lesssim 1.6$ MHz) follow the curvature of the earth. Sky waves ($1.6 \lesssim f \lesssim 30$ MHz) are reflected by

4.1 Introduction

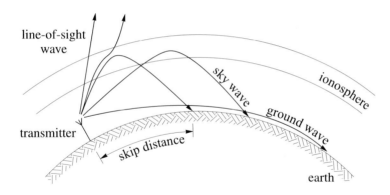

Fig. 4.6. Electromagnetic wave propagation (Hofmann-Wellenhof et al. 2003: Sect. 4.2.4)

the ionosphere. As Uttam et al. (1997: p. 106) state, electromagnetic waves are reflected from the ionized layer depending on the degree of ionization, the frequency, and the angle of incidence. One specific angle defines the critical or skip distance, which defines the maximum elevation angle, where waves will be reflected by the ionosphere. The critical distance changes during the day as a function of the ionization. Line-of-sight waves ($f \gtrsim 30$ MHz) propagate through the ionosphere, although the ray path is still influenced by it.

Energy change during wave propagation

Definitions

Absorption is defined as the conversion of the energy of an electromagnetic wave into heat. Absorption occurs when electromagnetic waves propagate through a medium. Generally speaking, the higher the frequency, the larger the absorption in the atmosphere.

Attenuation describes the relation between decreasing power with increasing distance from the emitter. Attenuation, also denoted as transmission loss, is a function of absorption, the refractive index, and geometrical spreading. The latter effect also occurs in free space, i.e., an idealized medium without any magnetic or electric field and missing any obstruction. Gain, the opposite of attenuation, describes an increasing ratio between received and emitted power.

Fading and scintillation, respectively, describe the temporal variation of the signal power due to alternating physical characteristics along the transmission path. Scintillation in particular describes irregular effects causing variability of phase and amplitude. Ionospheric scintillation, affected by solar activity for example, is of particular concern in auroral and polar regions.

Interference is the effect of energy change due to the superposition of electromagnetic waves.

Measures

Power is the amount of energy transferred per unit time. Let P^s be the power emitted from the satellite and P_r be the power measured at the receiver. A power ratio $P_r/P^s < 1$ describes a transmission loss and $P_r/P^s > 1$ defines a gain. A measure for the power ratio is given in decibel units defined by

$$10 \log_{10} \frac{P_r}{P^s} = n \quad [\text{dB}]. \tag{4.19}$$

Accordingly, $n < 0$ defines loss, whereas $n > 0$ describes gain. For example, $n = -3$ dB means that the received power is half the emitted power. A variation of Eq. (4.19) is used to express an absolute power in decibel units. Therefore, the power is related to 1 W, thus, $n = 10 \log_{10}(P/1)$, where n expresses the power of P and is given in decibel watt (dBW). Assume, for example, an emitted power of 25 W ≅ 14 dBW and a received power of 10^{-16} W ≅ -160 dBW, hence, the transmission loss amounts to -174 dB.

In free space, the transmitted power is geometrical-homogeneously spread over the surface of a sphere $4\pi \varrho^2$, where ϱ describes the distance between transmitter and receiver. Antennas may concentrate the radiated power into certain directions. The antenna directivity is described by variations of the antenna gain G. The principal interrelation between the antenna gain and the effective antenna area (antenna aperture) A is given by

$$G = 4\pi A \frac{f^2}{c^2}. \tag{4.20}$$

With G^s and G_r denoting the antenna gains of the transmitting and receiving antenna, the received power P_r follows, according to Betz (2006), from

$$P_r = P^s G^s G_r L_0, \tag{4.21}$$

where L_0 denotes the free-space transmission loss. Therefore, increasing the effective antenna area will also increase the received power. The factor $P^s G^s$ is called the equivalent isotropic radiated power (EIRP). The quantity L_0 follows from the Friis transmission formula (Institute of Electrical and Electronics Engineers 1997: p. 13)

$$L_0 = \left(\frac{c}{4\pi \varrho f}\right)^2, \tag{4.22}$$

where the original formula has been inverted to define L_0 as a transmission loss. For a numerical example, consider a distance of $\varrho = 20\,000$ km and a frequency of $f = 1.5$ GHz. This results in a free-space loss of -182 dB. To account for any form of attenuation, especially due to the atmosphere, foliage penetration (particularly

4.1 Introduction

under wet conditions), or building penetration, the factor k is introduced and the actual transmission loss L reads

$$L = k L_0, \qquad (4.23)$$

where $k=0$ means total signal blockage, whereas $k=1$ reveals no additional signal loss except for L_0. Using the actual transmission loss, (4.21) reads

$$P_r = P^s G^s G_r L. \qquad (4.24)$$

Since the transmission loss is a function of the distance to the satellite and satellite elevation, it changes as a function of time. The minimum received power level of GNSS signals on the earth's surface has typically a magnitude in the order of about -160 dBW, depending on the emitted power, the free-space loss, and the antenna gains (e.g., ARINC Engineering Services 2006a).

Earth's atmosphere

The atmosphere of the earth is categorized into different layers according to their physical properties and influences onto the electromagnetic waves. With respect to the electromagnetic structure, the atmosphere is divided into the neutral atmosphere and the ionosphere. While the neutral atmosphere comprises the troposphere and the stratosphere, the GNSS community abbreviates this to the troposphere and calls the delay due to the neutral atmosphere "tropospheric delay".

The troposphere extends from the earth's surface to about 50 km height. The troposphere is nondispersive for frequencies up to 30 GHz. The tropospheric refractive index is a function of temperature, pressure, and partial water vapor pressure. The latter consists of a dry and a wet component. About 90% of the tropospheric delay is caused by the dry or hydrostatic part, which is again mainly a function of pressure. The wet part depends on the water vapor and is, due to its high variability, difficult to model (Rothacher 2001a). The magnitude of rain, fog, and cloud attenuation is in contrast to the tropospheric delay negligible for L-band frequencies (Mansfeld 2004: p. 64). As a consequence, GNSS is specified all-weather operable. However, the rain, fog, and cloud attenuation must be considered, e.g., for C-band transmissions (Irsigler et al. 2002).

The ionosphere is the electrically charged component of the higher atmosphere. It is characterized by its free, neutral, and charged particles, where diversity varies as a function of the time of day. The ionosphere is categorized into several layers, in particular into D, E, and F in ascending height order. The ionization of the D-layer (50–90 km) varies with sunlight. The low electron density in the D-layer and the high particle density causes nearly complete deionization during night times. The ionization of the E-layer (90–150 km), denoted as the Kennelly–Heaviside layer (Arbesser-Rastburg 2001), is caused by the ultraviolet and x-rays during day, and

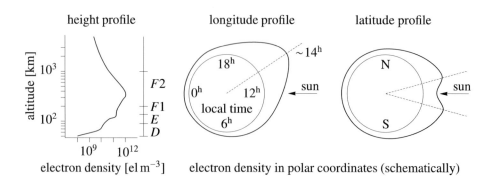

Fig. 4.7. Total electron content profiles (Issler et al. 2001)

cosmic rails and meteors during night. In the F-layer (150–1000 km), also called the Appleton layer, the ionization is maximal around noon and decreases toward sunset. The F-layer splits into the region $F1$ (150–200 km) and $F2$ (200–1000 km) during day. The maximal electron density can be found in the $F2$-layer. The ionospheric delay typically ramps up rapidly around 10^h local time and peaks around 14^h local time (Fig. 4.7). Typically it is low in the early morning hours.

The ionospheric refraction is modeled as a function of the electron density represented by the total electron content (TEC). The TEC is influenced by the solar activity, diurnal and seasonal variations, and the earth's magnetic field. Schematic profiles of the electron density are given in Fig. 4.7. TEC can be modeled on a global and continental level. Small-scale variations of TEC inhomogeneities are not yet possible to predict.

Coronal mass ejections and extreme ultraviolet solar radiation cause large geomagnetic storms in the earth's magnetic field (Volpe National Transportation Systems Center 2001). These storms may lead to a loss of lock in satellite tracking and may cause acquisition problems. The phenomena, due to the characteristics of the magnetic field of the earth, are especially critical in the polar regions. In general, they increase the spatial and temporal variation of the electron content and cause additional ionospheric scintillations in phase and amplitude, which is critical for weak signal tracking or codeless receivers.

The solar activity is generally quantified by the sunspot number. Sunspots are strong magnetic regions that appear as dark areas on the surface of the sun. In 1858, the Swiss astronomer R. Wolf defined the sunspot number by combining groups of sunspots with individual sunspots. The sunspot number shows an 11-year variation cycle of the solar activity (Fig. 4.8).

In a dispersive medium, the phase velocity differs from the group velocity (cf. Sect. 5.3.1). The group velocity describes the velocity of the envelope of a group of electromagnetic waves. The ionized gases in the ionosphere cause the phases

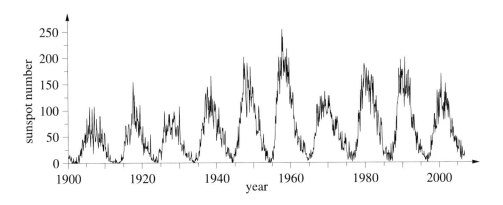

Fig. 4.8. Sunspot number (Solar Influences Data Analysis Center 2007)

of the electromagnetic waves to shift. The phase advance leads to a phase velocity greater than the speed of light. This does not contradict Einstein's postulate of the universal characteristic of the speed of light, since no information is transmitted by a single electromagnetic wave. The phase advance and the group delay are equal in size but different in sign. Practically speaking, code pseudoranges become longer and phase pseudoranges get shorter.

The gradient of the refractive index with height along the path through the atmosphere bends the electromagnetic wave. Brunner and Gu (1991) computed that, depending on the frequency, the ionospheric activity, and the other assumptions they made, the ray path is separated from direct line of sight at a satellite elevation angle of 15° by about 55–300 m.

4.1.3 Frequency standards

The key to the accuracy of satellite navigation is the fact that all signal components are precisely controlled by atomic clocks. These clocks are based on atomic frequency standards (AFS) which produce the reference frequency by stimulated radiation. Today's frequency standards show a stability over a day in the level of $\Delta f/f = 10^{-12}$ to 10^{-15} (Mansfeld 2004: p. 43). Quartz and rubidium frequency standards show good short-time stability, whereas cesium and hydrogen maser have a better long-time performance. The emitted frequency can be modeled as function of time t by

$$f(t) = f_n + \Delta f + (t - t_0)\dot{f} + \tilde{f}(t), \tag{4.25}$$

where f_n denotes the nominal frequency, Δf is the frequency offset, $\dot{f}$ corresponds to the frequency drift, $\tilde{f}(t)$ is the random frequency component, and t_0 is the reference epoch (Misra and Enge 2006: p. 111). Drift and offset can be modeled or calibrated, whereas the random error is of primary concern. The oscillator stability is statistically estimated using the Allan variance or, if a frequency drift exists, the Hadamard variance. A mathematical formulation can be found, for example, in Wiederholt (2006).

The AFS in GNSS satellites are required to have long-time as well as short-time stability with low failure rate. For this reason, AFS are used in the satellites complementarily and redundantly. In this way it is guaranteed that the modeling parameters of the oscillator errors, as provided by the control station and transmitted to the users, are valid for a long time. The AFS in the satellites are affected by the relative motion of the satellites with respect to the user and the variation of the gravitational potential. Both effects are described by Einstein's theory of the special and general relativity (cf. Sect. 5.4).

4.2 Generic signal structure

The satellite signals should enable real-time range measurements as well as data transmission capabilities. Any satellite signal has to serve an unlimited number of users without interfering other systems, satellites, or services. The methodology of range measurement is based on correlating two signals, namely the received satellite signal with a locally generated replica.

4.2.1 Signal design parameter

The GNSS signal design (Fig. 4.1) depends primarily on the availability of the carrier frequency bands according to ITU assignments. The selection of the frequency band is a function of the service requirements, the propagation effects, and the technical requirements regarding transmitter and receiver specifications. The ranging code layer design is driven by the acquisition and tracking characteristics, the correlation properties, the interoperability with other systems, as well as on the implementation complexity. The data-link layer, finally, has to be constructed carefully to avoid a negative influence on the tracking performance of the receiver and to ensure low bit error rates.

Power spectral density

The French mathematician J. Fourier postulated in 1807 that any arbitrary function of time, i.e., signal $s(t)$, can be represented by a superposition of trigonometric functions of different phase, amplitude, and frequency. Thus, any arbitrary signal may be represented in the time as well as in the frequency domain. Refer to Brigham (1988) or Oppenheim et al. (1999) for a detailed introduction into Fourier transformation and digital signal processing in general. The Fourier transformation

4.2 Generic signal structure

for aperiodic continuous signals, which is the most general case, reads

$$S(f) = \int_{-\infty}^{\infty} s(t) e^{-i 2\pi f t} \, dt, \tag{4.26}$$

where i is the imaginary unit (i.e., $i^2 = -1$). There exists a close relation between continuity and periodicity of the time and frequency domain representation as discussed in Brigham (1988).

The intentional alternation of the frequency spectrum is denoted as filtering. Filters can only be designed to process a limited time interval. Real-time filters are additionally limited by causality, which requires that signal values out of the future cannot be processed. Thus, it is not possible to design ideal low-pass, high-pass, band-pass, or band-stop filters for real-time processes.

Parceval's theorem postulates that the energy in a signal is equal to the result of integrating the signal squared over time. Taking into account the interrelations between time and frequency domain, the energy E reads

$$E = \int_{-\infty}^{\infty} s^2(t) \, dt = \int_{-\infty}^{\infty} |S(f)|^2 \, df, \tag{4.27}$$

where $|S(f)|$ denotes the amplitude of the Fourier transform $S(f)$ at frequency f. Due to the orthogonality of the base functions, $|S(f)|^2$ represents the energy contributed by every single base function component $S(f)$ to the total power of the signal. Thus, $|S(f)|^2$ is an indicator of the power spectral density (PSD), and commonly expressed in dBW. White noise leads to a constant PSD over the frequency spectrum.

Correlation property

The (cross-) correlation function is defined by

$$R(\tau) = \int_{-\infty}^{\infty} s_1(t) \, s_2(t + \tau) \, dt \tag{4.28}$$

and describes the degree of correspondence of two signals $s_1(t)$, $s_2(t)$ as a function of the time shift τ between them. The correlation function of two periodic signals, both with period T, is defined by

$$R(\tau) = \int_0^T s_1(t) \, s_2(t + \tau) \, dt. \tag{4.29}$$

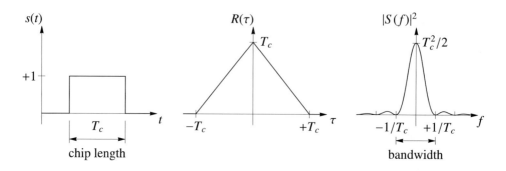

Fig. 4.9. Rectangular signal (left), autocorrelation function (center), and power spectral density (right)

A decreasing correlation coefficient $R(\tau)$ indicates increasing orthogonality of the signals, thus, $R(\tau) = 0$ specifies perfect orthogonality of signals.

If $s_1(t) = s_2(t)$, Eq. (4.28) expresses the autocorrelation function (ACF). The ACF is an even function of τ, i.e., $R(-\tau) = R(\tau)$. At zero lag, $\tau = 0$, the ACF corresponds to the computation of the energy of the signal (cf. Eq. (4.27)). The Fourier transform of the ACF of a signal $s(t)$ corresponds to the PSD of the signal.

Figure 4.9 shows a rectangular pulse in the time domain, its autocorrelation function, and the PSD of the rectangular pulse; positive and negative frequencies are shown, as usually applied by convention, and the PSD is halved accordingly (Brigham 1988). The Fourier transform of the rectangular pulse corresponds to the sine-cardinal (sinc) function

$$S(f) = T_c \frac{\sin(\pi f T_c)}{\pi f T_c} = T_c \operatorname{sinc}(\pi f T_c), \qquad (4.30)$$

where T_c denotes the width of the rectangular pulse, also called chip length. Note that the sine-cardinal function is infinite in length, i.e., from $-\infty$ to $+\infty$.

The distribution of the energy of a signal in the frequency spectrum is characterized by the bandwidth parameter. A generally valid definition for the bandwidth is the distance between the nulls of the mainlobe, which corresponds to a two-sided bandwidth (Fig. 4.9). Bandwidth B and T_c are inversely proportional:

$$B = \frac{2}{T_c}. \qquad (4.31)$$

Although most of the energy of a signal is within the frequency band allocated by ITU, there are also out-of-band frequency components which interact with other signals. Remember the infinite length of the sinc function. In the same way other signals interact with the in-band frequencies of a navigation satellite signal. The level of interference, in-band as well as out-of-band, is strictly regulated by ITU.

4.2 Generic signal structure

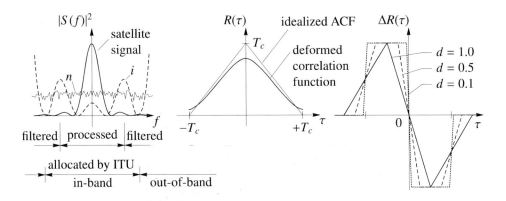

Fig. 4.10. Power spectral density (left), deformed correlation function (center), and idealized discriminator functions (right)

Discriminator function

In terms of GNSS, the satellite signal is correlated with a locally generated signal to derive, e.g., run time information. Subsequently, for simplification the satellite signal is reduced to one rectangular pulse, without losing generality. The in-band frequency components of the satellite signal (cf. Fig. 4.10) will be affected by noise n and in-band interferences i. Out-of-band interference effects are greatly avoided by filtering all high frequencies. Filtering, furthermore, is necessary to avoid aliasing effects during analog to digital conversion. These high frequencies, however, cause the sharp edges of the impulse and of the correlation function. As a consequence, the correlation function will be deformed by noise and in-band interference and rounded at the edges due to filter processes, as schematically shown in Fig. 4.10.

Note that notation has changed from the ACF to a general correlation function. Theoretically, the satellite signal and the locally generated one should be identical, but in terms of practical implementation they are not, because of, e.g., noise, interference, filtering.

Detection of the correlation maximum despite of the deformed correlation function is based on a discriminator function, e.g.,

$$\frac{dR(\tau)}{d\tau} \approx \Delta R(\tau) = \frac{R\left(\tau + \frac{d}{2}T_c\right) - R\left(\tau - \frac{d}{2}T_c\right)}{dT_c}, \tag{4.32}$$

which is based upon two successive correlation coefficients, also denoted as early and late, while d on the right sight of Eq. (4.32) denotes the correlation spacing. Figure 4.10 shows idealized discriminator functions. To find the maximum of the correlation function is achieved by searching for a null in the discriminator function.

Time-delayed indirect signals

If the direct (line-of-sight) signal $s_d(t)$ of the satellite is overlaid by a time-delayed indirect signal $s_m(t)$, i.e., multipath signal,

$$s_r(t) = s_d(t) + \beta\, s_m(t + \tau_m), \qquad (4.33)$$

and correlated with the locally generated signal $s_\ell(t)$, then the correlation function reads

$$\begin{aligned}
R_r(\tau) &= \int_0^T s_r(t)\, s_\ell(t + \tau)\, dt \\
&= \int_0^T s_d(t)\, s_\ell(t + \tau)\, dt + \int_0^T \beta\, s_m(t + \tau_m)\, s_\ell(t + \tau)\, dt \\
&= R_d(\tau) + R_m(\tau).
\end{aligned} \qquad (4.34)$$

The separation of $R_d(\tau)$ and $R_m(\tau)$ is not that simple if possible at all; thus, in general, $R_r(\tau)$ has to be used to search for the null in the discriminator function (Fig. 4.11). The search result τ_0 is affected by the time delay τ_m, the damping factor of the power of the delayed signal β, and the correlation spacing d. The sharper the autocorrelation peak and the smaller the correlation spacing, the smaller will be the influence of the time-delayed signal, and, consequently, the better the time measurement of the satellite signal propagation. The multipath error is generally presented as an error envelope by plotting the time delay τ_m of one multipath signal against the signal tracking error τ_0.

Narrow correlation spacing is one method to decrease the multipath influence.

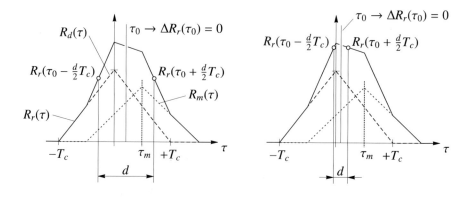

Fig. 4.11. Correlation function affected by time-delayed signals

4.2 Generic signal structure

Another one is the double-delta correlator method, which uses a discriminator function based on four successive correlation values, thus, decreasing the multipath influence but increasing at the same time the processing load.

Figure 4.11 shows a multipath signal which is in phase with the direct line-of-sight signal, therefore denoted as constructive multipath, since it increases the general correlation result. A destructive multipath signal is out of phase, i.e., the correlation function is negative compared to the line-of-sight correlation function, and decreases the correlation result.

4.2.2 Carrier frequency

A number of different frequency band options have been discussed for GNSS signals (e.g., Spilker 1996a: Sect. II.B). None of the frequency bands is optimal with respect to all design criteria; however, the L-band was chosen as the best compromise between frequency availability, propagation effects, and system design. As mentioned by Hammesfahr et al. (2001) or Irsigler et al. (2002), the C-band provides better performances with respect to other criteria and might be an option for future navigation signals. Generally speaking, the higher the frequency, the lower the ionospheric delays and free space loss, and the higher the antenna gain. Nevertheless, higher frequencies will result in higher atmospheric attenuation, increased Doppler uncertainty, and more technological constraints.

Modulation methods

Amplitude, frequency, or phase modulations describe a temporal variation of the respective electromagnetic wave parameter to carry information (Fig. 4.12). Simple carrier modulation schemes distinguish only between two states of the parameter, e.g., in case of phase modulation the phase changes between $+\pi$ and $-\pi$ per step. More complex carrier modulation schemes distinguish between several states, thus transmitting more than one bit per step. Increasing complexity and information density increases at the same time the interferences susceptibility and bit error rate.

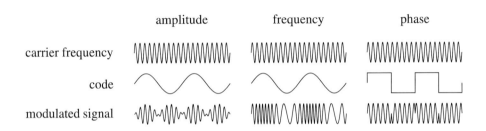

Fig. 4.12. Modulation methods

The phase modulation shifts the phase of the carrier by π whenever the chip sequence changes its state from $+1$ to -1 and vice versa. This modulation method knows only two conditions of phase shift and is, consequently, also referred to as binary phase-shifted key (BPSK). The frequency spectrum of the code-modulated carrier is a simple shift of the code spectrum to the carrier frequency. Apart from the code, the data bits of the data-link layer are modulated on the carrier. The data $d(t)$ and code $c(t)$ modulated on the carrier frequency f finally define the satellite signal $s(t)$,

$$s(t) = \sqrt{2P}\, d(t)\, c(t)\, \cos(2\pi f t), \tag{4.35}$$

where P corresponds to the power of the signal component. Introducing $\sqrt{2P}$ as amplitude is a direct consequence of Eq. (4.27) and the fact that power is the amount of energy transferred per unit time, thus,

$$P = \lim_{T \to \infty} \frac{1}{2T} \int_{-T}^{T} s^2(t)\, dt. \tag{4.36}$$

Signal multiplexing

Satellite navigation systems commonly transmit more than one ranging code and data message on every carrier frequency. The different code sequences are modulated on the in-phase I and $\pi/2$ shifted quadrature phase channel Q of the carrier frequency. Assume two ranging codes modulated on the I and Q channel

$$s(t) = \sqrt{2P_1}\, c_1(t)\, d_1(t)\, \cos(2\pi f t) + \sqrt{2P_2}\, c_2(t)\, d_2(t)\, \sin(2\pi f t), \tag{4.37}$$

where $c_1(t)$, $c_2(t)$ denote two different ranging codes and $d_1(t)$, $d_2(t)$ two different data messages. Both ranging codes change the phase of the signal in two ways, therefore a total of four different phase shifts has to be considered, thus, this particular modulation method is denoted quadrature phase-shifted key (QPSK).

Further assume that the ranging codes $c_1(t)$ and $c_2(t)$ themselves are composed by three other ranging codes $c_3(t)$, $c_4(t)$, and $c_5(t)$ in the form

$$c_1(t) = \alpha_1\, c_3(t) + \alpha_2\, c_4(t) + \alpha_3\, c_5(t), \tag{4.38}$$

$$c_2(t) = \beta_1\, c_3(t) + \beta_2\, c_4(t) + \beta_3\, c_5(t). \tag{4.39}$$

In this way it is possible to modulate several different spreading codes onto the carrier, depending on the factors α_i and β_i.

Another method for signal multiplexing is to modulate two ranging codes onto the carrier frequency following the time division multiplexing method. Thus, one or several chips of ranging code c_1 will be followed by one or several chips of ranging code c_2. Consequently, the multiplexed code will either have a higher code rate or a longer code length.

4.2 Generic signal structure

4.2.3 Ranging code layer

Spectral spreading

A GNSS satellite signal is based on direct sequence spread spectrum modulation. A carrier frequency, thus, is modulated by a chip sequence, which is composed of the data message modulated by the ranging code. The ranging code itself consists of a periodic sequence of rectangular pulses, where the amplitude changes quasi randomly between the logic levels 0 and 1. These codes are denoted pseudorandom noise (PRN) codes. The operation of data modulation corresponds to an exclusive-or (XOR) algorithm, i.e., the binary sum of two states, where a binary 0 results if the two states are equal at logic level. The XOR operation corresponds to a multiplication if signal levels 1 and -1 are used instead of the logic levels.

Following Fig. 4.9, one chip has a period T_c. The length of the ranging code sequence T_p with N_c chips is

$$T_p = N_c T_c \tag{4.40}$$

and is sometimes also referred to as code epoch or simply epoch. The reciprocal of the chip length is denoted as chipping rate or code rate $R_c = 1/T_c$ which is proportional to the bandwidth (cf. Eq. (4.31)). The frequency spectrum of a ranging code looks similar to the one of a rectangular pulse with two major differences. First, the periodicity of the ranging code causes the frequency spectrum to be discrete. The interval between the frequency lines is inversely proportional to the period of the code sequence (Fig. 4.13)

$$f_p = \frac{1}{T_p}. \tag{4.41}$$

Second, the finite number of changes between the signal levels causes the spectral lines to slightly deviate from a sine-cardinal envelope given by Eq. (4.30). The

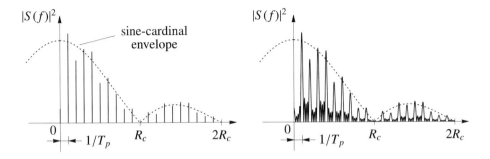

Fig. 4.13. Power spectral density of a PRN code (left) and a PRN code modulated by data message (right) (schematically)

longer the code sequence and the shorter the period T_c, the more will the frequency spectrum assimilate to the sinc function. Deviations from the envelope are undesired since the interference susceptibility increases and also the requirements for the linearity of power amplifier increases. For simplification, the PSD of all signals will be displayed by the envelope only throughout this textbook.

The bit length of the data message T_d is generally greater than the period of the code sequence T_p. The frequency spectrum of the data message is (due to its aperiodicity) continuous and the bandwidth smaller than the one of the ranging code. Modulating the data message with the ranging code will create a signal with the bandwidth of the ranging code but with continuous frequency spectrum while the spectrum will keep the peaklike shape (Fig. 4.13). In other words, the ranging code spreads the power needed for data transmission over a wider bandwidth. Thus, the ranging codes are also referred to as spreading codes, the code chips as spreading chips.

In terms of data transmission performances, spectral spreading is inefficient. The transmission of the navigation message, generally requiring frequency bandwidths of 10 to 250 Hz (Misra and Enge 2006: p. 346) to transmit 50 bits per second, is spread onto a frequency band covering 20 to 50 MHz. The spreading technique, however, fulfills four tasks, making it indispensable for satellite navigation. Periodic spread spectrum signals enable run-time measurements. Further, orthogonal spread spectrum signals used by different satellites enable code division multiple access (CDMA) as briefly described in Sect. 4.2.5. Third, the spreading and despreading procedure reduces the influence of interferences. Finally, demodulating the data message becomes possible despite of low signal power.

The spectral spread data message $d(t)$ will be despread in the receiver by multiplication of the received satellite signal $s(t)$ with the locally generated ranging code $c(t)$. Equation (4.42) describes the spreading operation, whereas Eq. (4.43) shows the appropriate despreading (bandwidth collapsing) operation in a simplified formulation by neglecting the time dependency:

$$s = c\,d, \tag{4.42}$$

$$(s + i + n)\,c = c^2 d + i\,c + n\,c = d + i\,c + n\,c, \tag{4.43}$$

where n denotes noise and i interference. This operation relies on the fact that a squared code sequence will result in a constant signal level

$$c^2(t) = 1. \tag{4.44}$$

The data message $d(t)$ collapses to its original bandwidth. In contrast, any interference signal, in particular its power, will be spread by the same process as shown in Fig. 4.14 (Issler et al. 2001).

4.2 Generic signal structure

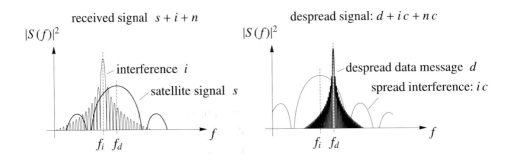

Fig. 4.14. Spectral spreading (schematically)

Ranging codes

Definition

In mathematical terminology two ranging codes c_i, c_j with noiselike characteristics have to meet the following ideal requirements (assume signal level $+1, -1$):

$$M[c_i(t)] = M[c_j(t)] = 0, \tag{4.45}$$

$$M[c_i^2(t)] = M[c_j^2(t)] = T_p, \tag{4.46}$$

$$M[c_i(t+\tau)c_j(t)] = 0 \quad \forall\, i \neq j, \tag{4.47}$$

$$M[c_i(t+\tau)c_i(t)] = 0 \quad \forall\, \tau \text{ modulo } T_p \neq 0, \tag{4.48}$$

where M[ ] denotes a mathematical operator defined as the integral over the period T_p. Equation (4.45) characterizes the mean value (balance) of the code sequence. Equation (4.46) describes the autocorrelation at zero lag. Equation (4.47) defines the crosscorrelation property, and Eq. (4.48) specifies the values of the autocorrelation function for (τ modulo T_p) $\neq 0$, accounting for the periodicity of the signals.

Assume two satellite code sequences meeting these requirements are correlated with a locally generated code sequence in the receiver

$$M\left[c_i(t+\tau_1)\left(c_i(t+\tau_1) + c_j(t+\tau_2)\right)\right] = T_p + 0. \tag{4.49}$$

The orthogonality of the different signals enables to isolate one satellite from the other and to perform time measurements on selected satellites. The second term in Eq. (4.49), $c_i(t+\tau_1)c_j(t+\tau_2)$, is the smaller the lower the crosscorrelation between the two signals is. In an ideal environment the crosscorrelation equals zero. Orthogonality and good autocorrelation characteristics are fundamental requirements for high-accuracy time measurements and good interference mitigation.

Periodic code sequences meeting the requirements of Eqs. (4.45) through (4.48) are called pseudorandom noise (PRN) codes. These sequences have noiselike behavior with a maximum autocorrelation at zero lag ($\tau = 0$).

Code generation

PRN codes for navigation signals are commonly generated using linear feedback shift registers (LFSR). An LFSR is characterized by the number of register cells n and the characteristic polynomial $p(x)$, which defines the feedback cells. The states of the feedback cells of the register are XOR-added and fed back as new input into the LFSR. The XOR-adders thereby characterize the linearity of LFSR (Holmes 1982: p. 306). An increasing number of register cells results in a longer PRN code and in a better correlation property. The maximum length, N_m, of the PRN code is defined by

$$N_m = 2^n - 1. \tag{4.50}$$

Therefore, after a maximum of N_m states, the PRN code repeats itself. The only state of an n-bit LFSR which will not be generated is the all-zero state. Not all LFSR have maximal length, thus, the sequence is already repeated before N_m code states. Holmes (1982: p. 309), e.g., states that LFSR with an odd number of feedback cells do not have maximal length. A simple example may illustrate the function of an LFSR consisting of $n = 3$ register cells and the characteristic polynomial $p(x) = 1 + x^1 + x^3$, which defines the cells R_1 and R_3 as feedback cells (Fig. 4.15). Starting with the initial state 1 1 1, the bits are shifted to the right at any clock pulse where the content of the rightmost cell is read as output. The new value of the leftmost cell is determined by the XOR operation of the two feedback cells. Following this procedure results in a code sequence 1 1 1 0 1 0 0 1 1 1 0 1 0 0 1 1 ..., where the code repeats after $N_m = 7$ states. Starting with a different initial state, e.g., 1 0 1, will generate the same code sequence but time shifted.

Maximal-length LFSR show excellent autocorrelation property; however, not all of these sequences have good crosscorrelation characteristics. R. Gold proposed, as described in Dixon (1984: p. 79), to combine two maximal-length LFSR sequences of the same length with good autocorrelation and crosscorrelation property to generate a great family of nearly balanced PRN codes with excellent correlation characteristics. Following Gold's recommendation, codes of length $N_m = 2^n - 1$,

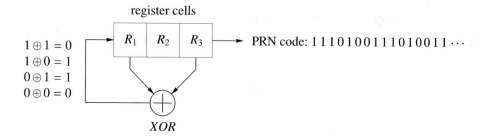

Fig. 4.15. Linear feedback shift register

4.2 Generic signal structure

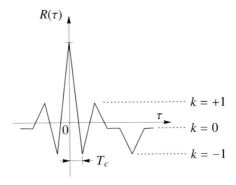

Fig. 4.16. Autocorrelation of a PRN Gold code

where n = even and (n modulo 4) $\neq 0$, show good correlation property. The level of crosscorrelation of two Gold codes is given by Holmes (1982: p. 553) as

$$R_k(\tau) = 2^{(n+2-a)/2} k - 1, \tag{4.51}$$

where $k = \{-1, 0, 1\}$ (Fig. 4.16), $a = (n \text{ modulo } 2)$, and the chip length has been set equal to 1.

Beside Gold codes, GNSS signals further apply truncated Gold codes, or codes that are built by combining codes of different lengths. For example, tiered codes consist of a long high-frequent primary code XOR-added by a short low-frequent secondary code. In this way it is possible to combine the advantages of short high-frequency codes with those of long low-frequency ones. Apart from the LFSR PRN codes there is a great number of optimized pseudorandom noise sequences (Hein et al. 2006b), whereas codes that do not underlie a specific rule of generation are called memory codes.

PRN sequences are evaluated by computing the maximum absolute crosscorrelation value, which characterizes the orthogonality of two code sequences. Commonly, this value is referenced to the maximum autocorrelation value, whereby the respective power levels are compared. Assume, for example, a code sequence $c_1(t)$ periodically repeated, i.e., a satellite signal, correlated with one period of a locally generated code $c_2(t)$, both codes having a period $T_p = N_c T_c$. The output of the correlation is a signal having constant signal level $R(\tau)$, where (τ modulo T_p) = constant. The power of this signal (= constant value $R^2(\tau)$) in relation to the maximum power possible (= result of the autocorrelation of $c_2(t)$ squared) defines the level of crosscorrelation, thus,

$$M = 10 \log_{10} \left(\frac{R(\tau)}{N_c T_c} \right)^2 \quad [\text{dB}], \tag{4.52}$$

or considering the maximum crosscorrelation of Gold codes

$$M = 20 \log_{10} \left(\frac{2^{(n+2-a)/2} + 1}{2^n - 1} \right) \quad [\text{dB}], \quad (4.53)$$

where $N_c = N_m$ and $T_c = 1$. The better the orthogonality of two code sequences the smaller is the level of crosscorrelation.

The ranging code sequences determine important performance characteristics. The design parameters are above all the code length, the code rate, and the autocorrelation and crosscorrelation properties. For example, short ranging codes allow fast acquisition, whereas long ranging codes enhance weak signal tracking performance. A high code rate will positively influence the correlation result and, thus, the tracking accuracy and the interference rejection capability. Short chip lengths furthermore decrease the influence of time-delayed signals. In contrast, high code rates cause wide bandwidths, which require a wideband radio frequency circuitry, fast sampling rates, and a higher processing load. Typical chipping rates are in the range of megachips per second (Mcps). A careful selection of the random noise codes is necessary to avoid interferences with other signals using the same and adjacent frequency bands, whilst maximizing the interoperability and compatibility with other satellite navigation systems.

Code modulation / submodulation

The objective of code modulation (submodulation) is to shape the frequency spectrum of the ranging code in order to assign signal energy to dedicated frequency parts. PRN codes with no additional code modulation are denoted as BPSK signals. Binary offset carrier (BOC) modulation uses a rectangular subcarrier with frequency f_s to modulate the PRN spreading code of frequency f_c (Betz 2002). The BOC rectangular subcarrier resembles a meander sequence. BOC modulation splits the power spectral density mainlobe of a BPSK modulation into two sidelobes and

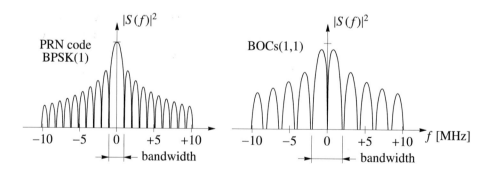

Fig. 4.17. Power spectral density of code-modulated ranging codes

4.2 Generic signal structure

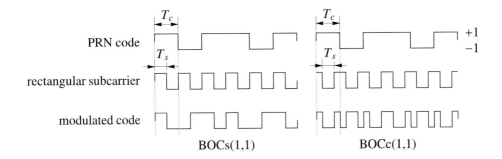

Fig. 4.18. Binary offset carrier modulation

places them symmetrically about the center frequency (Fig. 4.17), thus, also denoted as split-spectrum signal. GNSS BOC modulation is commonly referenced to $f_0 = 1.023$ MHz and

$$f_s = n f_0 = \frac{1}{2T_s}, \quad (4.54)$$

$$f_c = m f_0 = \frac{1}{T_c} = \frac{1}{kT_s} = \frac{2}{k} f_s, \quad (4.55)$$

and the modulation is denoted as BOC(n, m). Imitating this notation, signals with no extra code (sub-) modulation are denoted as BPSK(m). The parameters n and m define the spectral separation and the shape of the mainlobe about the center frequency. The factor $k = T_c/T_s$ is a positive integer. If $f_s = f_c$, the modulation is referred to as Manchester coding. Additionally, sine-phased BOC (BOCs) and cosine-phased BOC (BOCc) modulated signals are distinguished depending on the phase of the rectangular subcarrier (Fig. 4.18). As Ward et al. (2006) mention, the factor k is a measure of the number of positive and negative peaks in the autocorrelation function, e.g., BOCs has $2k - 1$ peaks.

The normalized power spectral density of a signal with sine-phased BOC modulation and $k =$ even is given in Betz (2000) as

$$|S_{\text{BOCs}(n,m)}(f)|^2 = f_c \left(\frac{\sin\left(\frac{\pi f}{2f_s}\right) \sin\left(\frac{\pi f}{f_c}\right)}{\pi f \cos\left(\frac{\pi f}{2f_s}\right)} \right)^2. \quad (4.56)$$

Normalization is achieved by dividing the power spectral density by chip length T_c. BOC-modulated signals cover different frequency bands with a higher energy level than BPSK signals. The larger bandwidth of BOC-modulated signals results in a better tracking performance. Bandwidth is defined here as the bandwidth between the outer nulls of the two spectral mainlobes. The spectral separation of two mainlobes additionally favors interference mitigation. Furthermore, single sideband processing allows to take advantage of the energy within one mainlobe, even when the other is disturbed by interference, since both sidebands redundantly contain the ranging and data information (Betz 2002). The major disadvantage of the BOC modulation is that the autocorrelation function (Fig. 4.19) has several local maxima, which means searching for a maximum of the ACF is ambiguous. Note that BOCc signals in combination with narrow correlators have a higher tracking accuracy due to the narrower ACF (Fig. 4.19).

A BOC-like modulated signal can be generated by two BPSK modulated signals. The BPSK mainlobes are shifted from the center frequency to a higher and lower part, respectively. The two BPSK-modulated signals do not necessarily have to use the same PRN code. Signals with different PRN codes and generated in this way are called alternative BOC (AltBOC) modulated signals (Hein et al. 2002).

Different modulation schemes in conjunction with signal multiplexing maximize the benefits of spectral separation and spectral shaping to improve the signal tracking performances. Hein et al. (2006a), therefore, presented the method of multiplexed binary offset carrier (MBOC). The power spectral density of the MBOC(6,1,1/11), as shown in Fig. 4.20, is given by

$$\overline{|S(f)|}^2 = \frac{10}{11} \overline{|S_{BOCs(1,1)}(f)|}^2 + \frac{1}{11} \overline{|S_{BOCs(6,1)}(f)|}^2, \qquad (4.57)$$

thus the notation MBOC(6,1,1/11) specifies the combination of a BOCs(1,1) with a BOCs(6,1) power spectral density, whereby the BOCs(6,1) component holds 1/11 of the overall energy. Note that simple shaping steps in the frequency domain may lead to more difficult design operations in the time domain. The advantages of the MBOC design have been highlighted in the discussions in Stansell et al. (2006) and Gibbons et al. (2006).

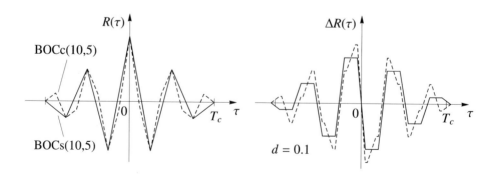

Fig. 4.19. BOCs(10,5) and BOCc(10,5) autocorrelation functions (left) and discriminator functions (right)

4.2 Generic signal structure

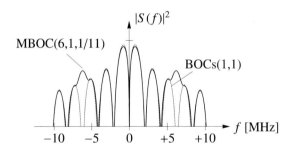

Fig. 4.20. Power spectral density of MBOC(6,1,1/11)

Pilot signals

Not all signals are composed of ranging code and data message. Pilot signals (pilot tones) reject the data message. In this way coherent integration (cf. Sect. 4.3.3) becomes even possible for very long integration times, thus increasing the sensitivity of receivers. Very long pilot codes commonly use the principle of tiered codes. The short primary codes allow a fast acquisition, whereas the long tiered code increases the sensitivity accordingly. Pilot tones are commonly signal multiplexed with data signals with similar ranging codes. Consequently, once tracking the tones allows to more easily acquire the data signals.

Interference – jamming, spoofing, meaconing

The spread spectrum GNSS signal design allows to mitigate a great deal of interference signals. However, as Spilker and Natali (1996: p. 756) emphasize, even spread spectrum does not provide sufficient protection against high-power interferences. Interference is differentiated into intentional and unintentional interference. To the latter belong out-of-band emissions mainly from other services or in-band emissions in particular from other systems, both regulated by ITU. Intrasystem interference denotes interaction of signals from the same system. Intersystem interference, furthermore, specifies the interference of signals of different satellite navigation systems. The Volpe National Transportation Systems Center (2001) categorizes intentional interference into jamming, spoofing, and meaconing. Jamming denotes the operation of drowning the navigation signals in high-power signals to cause loss of lock and to avoid reacquisition. Spoofing is the operation of emitting legitimate-appearing false signals to shift the computed position solution of a user. Meaconing is similar to spoofing; however, the signal is not generated but rebroadcast using received and delayed signals.

Interferences are mitigated using optimized antenna and filter designs to filter the signal in the spatial, time, and frequency domain. Adaptive mitigation methods estimate and mitigate time-, frequency-, or space-varying interferences.

Any intentional interference is unlawful and, as Corrigan et al. (1999) request, has to be prosecuted.

4.2.4 Data-link layer

The interrelation between code chip duration T_c, the code length T_p, and the data bit duration T_d is given by

$$T_d = N_p T_p = N_p N_c T_c, \tag{4.58}$$

where N_c defines the number of the ranging code chips and N_p describes the number of code epochs per data bit. In combination with information transmission, the chips of the data message are commonly denoted as bits. The period of the carrier wave is commonly several times smaller than the chip length of the ranging code. The bit length of one data bit is commonly again several times longer than the period of the spreading code. Decreasing data bit duration increases at the same time the data transmission frequency. Increasing T_d, in contrast, enhances low bit error rates and allows signal acquisition in weak signal environments (cf. Sect. 4.3.3).

4.2.5 Satellite multiplexing

The signal design must avoid interferences between signals of different satellites; in this way the receiver is able to differentiate the signals from various space vehicles. The satellite multiplexing methods exploit the one or the other orthogonality between signals. Code division multiple access (CDMA) guarantees access to different satellites by using orthogonal code sequences. Due to the characteristics of the code sequences, CDMA is also denoted as spread spectrum multiple access. The signals of the different satellites overlap in the frequency and time spectrum. Frequency division multiple access (FDMA) exploits the spectral separation of different satellite signals. The signals of different satellites overlap in the time and code domain. Time division multiple access (TDMA) avoids intrasystem interference by emitting signals from different satellites at different instances of time.

4.3 Generic signal processing

The satellites generate a signal by modulating a ranging code, according to the spectral shaping scheme, and the data message onto the carrier frequency (cf. Fig. 4.1). The different signals are then multiplexed and right-handed circularly polarized emitted from the satellite antennas. The signals emitted are bound by the regulations of ITU. Susceptible out-of-band interferences are either avoided by the signal design or filtered (Dobrosavljevic and Spicer 2004). The loss of correlation

4.3 Generic signal processing

thereby is specified in the interface specification documents of the navigation systems. Misra and Enge (2006: p. 431) describe the transmitted signal to have an EIRP of 500 W or an effective earthward power of about 25 W. The transmission loss attenuates the signal power to about 10^{-16} W. These signals are finally processed in the receiver to derive position information.

4.3.1 Receiver design

Basic concept

The generic GNSS receiver is composed of three functional blocks (Fig. 4.21): the radio frequency (RF) front-end, the digital signal processor (DSP), and the navigation processor. The differentiation into the three blocks is based on their functions and not on the hardware technology. MacGougan et al. (2005) state that an increasing share of the software part up to a pure software-based receiver provides a high level of flexibility and cost-effectiveness. Nevertheless, a specialized hardware-software combination highly integrated on a chip will have a higher performance in terms of throughput. Refer to Tsui (2005) or Borre et al. (2007) for a detailed discussion of software-based receivers.

The RF front-end receives and conditions the incoming satellite signals, downconverts them to an intermediate frequency (IF), and an analog to digital (A/D) converter samples the signals. The RF front-end further implements the frequency standard, which provides the reference frequencies and timing information.

The DSP correlates the locally generated signals with the satellite signals and provides the observables, i.e., code ranges, carrier phases, and Doppler frequencies, as well as the navigation data streams. The observables are in principle by-products of the tracking loops.

The navigation processor decodes the navigation message to gain time, ephemerides, and almanac data and finally computes position, velocity, and time (PVT) information. All three functional blocks of the receiver exchange information to increase the performance.

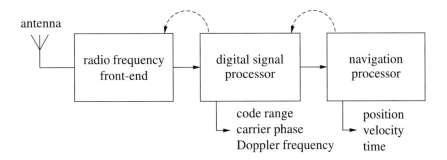

Fig. 4.21. GNSS receiver functional blocks

Signal-to-noise ratio

The noise is generally described using a temperature equivalent parameter, the thermal noise. Thermal noise, which, as Langley (1997) emphasizes, is not the actual physical temperature of the receiver, indicates the motion of electrons in the receiver circuitry. Thermal noise is commonly assumed to be white and Gaussian distributed. The noise power density N_0 reads

$$N_0 = kT \quad [\text{W Hz}^{-1}], \quad (4.59)$$

where $k = -228.6\,\text{dBW K}^{-1}\,\text{Hz}^{-1}$ defines the Boltzmann constant and T is the temperature equivalent given in kelvin (K). The (thermal) noise power N is the product of noise power density N_0 and bandwidth B_r processed by the receiver (Butsch 2002):

$$N = N_0 B_r = kTB_r \quad [\text{W}]. \quad (4.60)$$

The noise power density of a typical receiver is in the order of $N_0 = -201$ to $-204\,\text{dBW Hz}^{-1}$ (cf. Ward et al. 2006: p. 263; Langley 1997).

The signal-to-noise ratio (S/N) describes the performance of a functional block by relating the signal power P to the noise power N:

$$\text{S/N} = 10\log_{10}\frac{P}{N} \quad [\text{dB}]. \quad (4.61)$$

The carrier-to-noise power density ratio C/N_0 is a bandwidth-independent index number that relates the (carrier) power to noise per 1 Hz bandwidth

$$C/N_0 = 10\log_{10}\frac{P_r}{N_0} \quad [\text{dBHz}]. \quad (4.62)$$

Whereas S/N is generally used in conjunction with signals at baseband after despreading operations, C/N_0 is more commonly used to quantify the signal power P_r of the received signal (Langley 1997). Carrier-to-noise power density ratios below 34 dBHz characterize weak signals.

The minimum received signal strength is defined to be in the order of -160 dBW (e.g., ARINC Engineering Services 2006a). Inserting this power level into Eq. (4.61) and taking the noise power N for a signal bandwidth of 2 MHz according to (4.60), the S/N becomes negative, i.e., the signal is drowned by noise, thus, not detectable by spectrum analyzers. Only the overlay of several signals from different satellites and higher power levels than the minimum specified will increase the combined signal power above the noise level (Borre et al. 2007: p. 65).

The correlation of the satellite signal with the locally generated signal despreads the data signal and thereby increases the power level (Fig. 4.14). This

4.3 Generic signal processing

spread signal design is one reason why GNSS receivers can use small, omnidirectional antennas rather than large dish antennas.

The receiver correlates the incoming signal with a signal replica. The correlation result, taking into account the power of the replica, is proportional to the incoming signal power (4.24), thus it can be used to estimate the signal power. Butsch (2002) further emphasizes that an estimate of the noise power can be derived from the mean squared deviation of the correlation result over time. Dixon (1984: p. 10) emphasizes that the difference in output and input S/N of any operation is denoted as processing gain and processing loss, respectively.

4.3.2 Radio frequency front-end

Antenna design

Antennas receive the satellite signals, transform the energy of the electromagnetic waves into electric currents, and forward them to the RF front-end, while rejecting multipath and interference signals as far as possible. In general, the antenna gain is a function of azimuth and elevation. Omnidirectional antennas, however, have a uniform antenna gain pattern in all directions. Such antennas are generally used in GNSS applications. For static applications, the gain is limited as far as possible to the upper hemisphere by using, e.g., ground plane or choke ring design, rejecting signals coming from below the horizon, therefore rejecting signals reflected from the ground. Other applications, e.g., marine applications, require a uniform gain pattern also below the horizon to compensate for rolling and pitching of the ship. As mentioned in Sect. 4.1.1, some reflected signals may change the polarization from RHCP to LHCP. Antennas are commonly designed to have low gain for LHCP signals. Figure 4.22 shows schematically the antenna gain in polar coordinates.

The controlled reception pattern antenna (CRPA) technology combines several antenna elements to an array to exploit beam forming, beam steering, and null steer-

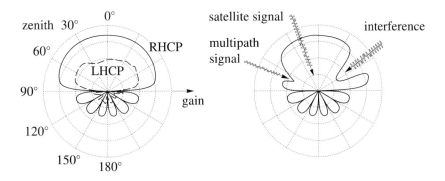

Fig. 4.22. Antenna gain pattern (left) and the influence of beam forming, beam steering, and null steering (right) (schematically)

ing techniques (Fu et al. 2001). In this way the gain is adaptively maximized toward satellite signals, and minimized to any other interfering signal source (Fig. 4.22).

Active antennas implement preamplifiers which integrate burnout protection elements, frequency filters, and low-noise amplifier (LNA) components. Filtering is accomplished based upon a trade-off of out-of-band rejection and signal distortion (Fig. 4.10).

Reference oscillator

The frequency standards of the receivers, commonly based on quartz crystal local oscillators (LO), do not provide the same high level of stability as the atomic clocks in the satellites. Although the time offset to GNSS time is introduced in the observation equations as a fourth unknown, the oscillators must meet short-time stability requirements to enable high satellite acquisition and tracking performance. Furthermore, the reference oscillators should have low sensitivity to vibration or high dynamics and have low phase noise.

Radio section

Incoming satellite signals are amplified by an LNA, filtered by a band-pass filter (BPF), and downconverted to IF using a frequency mixer. The signal is filtered again and high-power signal parts normalized by the automatic gain control (AGC) before the analog to digital (A/D) converter samples the signal. Although this generic design (Fig. 4.23) is common to a vast majority of receivers, all of them employ different strategies and different frequency plans to optimize the receiver design on user requirements in trade-off of the complexity and cost. Every additional filtering and downconversion step reduces out-of-band interferences without in-band distortions, thereby minimizing aliasing effects in the subsequent A/D conversions. However, every intermediate step adds additional noise power.

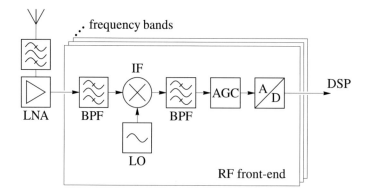

Fig. 4.23. Generic radio frequency front-end functional block diagram

4.3 Generic signal processing

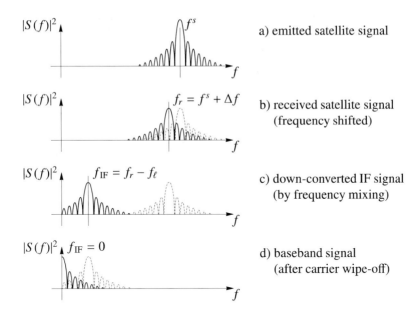

Fig. 4.24. Signal processing steps displayed in the power spectrum

Navigation signals of two separated frequency bands are processed using either one very wideband RF front-end or two completely separated, but in principle identical, signal paths.

Intermediate frequency
Downconversion is a simple shift of the frequencies in the frequency spectrum. This is accomplished by mixing the incoming signal centered on the frequency f_r with a locally generated pure harmonic signal f_ℓ (Wells et al. 1987: p. 7.07). The mathematical formulation corresponds to

$$\cos(2\pi f_r t)\,\cos(2\pi f_\ell t) = \tfrac{1}{2}\left[\cos\left(2\pi(f_r+f_\ell)\,t\right) + \cos\left(2\pi(f_r-f_\ell)\,t\right)\right]. \tag{4.63}$$

The term $f_{\mathrm{IF}} = f_r - f_\ell$, referred to as intermediate frequency (IF), corresponds to the downconverted part of the signal. The upconverted (high-frequency) part, $f_r + f_\ell$, is filtered using, e.g., a band-pass filter as illustrated in Fig. 4.23. Frequency mixing shifts the frequencies in the electromagnetic spectrum; however, it does not influence the phase shift and the frequency dependent Doppler shifts in the code sequence (cf. Fig. 4.24).

Not all receiver designs use the concept of intermediate frequency. Some downconvert directly to zero or near zero frequency, respectively, i.e., down to baseband ($f_\ell = f_r \rightarrow f_{IF} = 0$). Baseband denotes the frequency band occupied by a demodulated signal (Langley 1997). Other IF concepts do not use any downconversion at all but discretize the signals directly.

Signal amplification
The power levels of the received signal are normalized to a reasonable level to facilitate the signal processing. Signal amplification to a normalized level is performed by the AGC. Holmes (1982: p. 105) mentions that the AGC avoids circuitry saturation and damage. The AGC receives information about the power level by the correlators in the DSP. Blanking elements may suppress any pulse-type high-power interference before the signal is amplified (Giraud et al. 2005).

Analog to digital conversion
The A/D converter discretizes the incoming analog signals in time (sampling) and in magnitude (quantization). The minimum sampling rate is defined by the Nyquist (Shannon) theorem. The out-of-band frequencies that are not filtered in advance are aliased into the in-band frequencies (Oppenheim et al. 1999: p. 86), consequently increasing the noise level.

Typically, sampling rates of 2 to 20 times the PRN code chipping rate are used (Dorsey et al. 2006: p. 108). One-bit quantization would generally be sufficient for signal acquisition and tracking, higher quantization levels (e.g., 2-bit or 4-bit), however, show a better signal-to-noise ratio and furthermore decrease the susceptibility to interference. The statistical distribution of signal and noise and the averaging process during the correlation algorithms counteract the signal distortions introduced by the quantization. Borre et al. (2007: p. 62) estimate the signal distortion of a 1-bit quantization to be less than 2 dB.

Although the satellite signals after the A/D conversion are discrete $s[n]$, the signals are still represented using the continuous form $s(t)$ for all following operations to express their general validity.

4.3.3 Digital signal processor

In the first step of the digital signal processor (DSP) functional block, the signal is split into a number of channels. Every channel outputs code ranges, range rates, time tag information, the navigation data message, and affiliated information like the S/N. High-end receivers also perform phase measurements in the channels. The generic signal processor functional block diagram (Fig. 4.25) shows one possible realization.

4.3 Generic signal processing

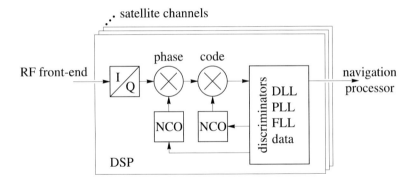

Fig. 4.25. Generic signal processor functional block diagram

Channel multiplexing

An increasing number of channels increases the requirements of the hardware components and circuitry. Older receivers used a limited number of channels to save power and cost. The satellites have been tracked by sequencing the satellites on the same channel. Parallel or continuous receivers increase the performance by assigning every satellite signal to a dedicated channel. Meanwhile the number of channels corresponds to the maximum number of visible satellites multiplied by the number of frequencies and ranging codes processed. Multichannel receivers however have to calibrate interchannel timing biases, which would otherwise deteriorate the position solution.

Acquisition and tracking

The very first time a receiver is switched on, it neither has information about its position nor about the approximate time, nor about the position of the satellites. The receiver has no information which satellites are in view. The receiver starts a sky search by analyzing the input signal with respect to all known satellite ranging codes. If the receiver has been initialized before, it generally uses almanac and ephemerides, the approximate user position, and approximate time estimate to provide aiding information in form of estimated Doppler shift and estimated time shift to the tracking loops. Depending on the availability of the information, four different acquisition modes are distinguished: cold start, warm start, hot start, and reacquisition. No information is available in cold-start conditions (sky search). Warm start relies on almanac, user position, and time estimation. Hot start, in contrast, relies not only on almanac but also on ephemerides data. Reacquisition is the process when satellite signals have just been lost and are acquired again, therefore the receiver has good knowledge of time and Doppler shift.

The core element of the receiver is the correlation between the received signal and a locally generated replica. The input signal $s(t)$ to the DSP channels consists of a superposition of signals from different satellites, multipath signals, interferences, and noise. In a simplified form the signal reads:

$$s(t) = \sum_{r=1}^{k} s_r(t + \tau_r) + n(t) + i(t) =$$

$$= \sum_{r=1}^{k} \sqrt{2P_r}\, c_r(t + \tau_r)\, d_r(t + \tau_r) \cos\left(2\pi(f^s - f_\ell + \Delta f_r)(t + \tau_r)\right)$$

$$+ n(t) + i(t),$$

(4.64)

where n denotes white Gaussian noise, i represents interference, Δf_r the frequency offset, and τ_r the code delay. The identification of one dedicated satellite signal is performed by searching for the maximum of the autocorrelation function (cf. Eq. (4.29)). The crosscorrelation, consequently, works like a filter for all other satellite signal components (cf. Eq. (4.49)). At the same time the data message is despread (cf. Eq. (4.43)).

The process of searching the maximum correlation is denoted as acquisition. The subsequent process of following the maximum over time defines tracking. As soon as the search process has been successful and the receiver tracks a specific satellite, its signal is called to be locked. The acquisition and tracking processes are closely related. For acquiring satellites, the C/N_0 has to be a few dB higher than for tracking (Hein and Issler 2001).

Acquisition consists of a two-dimensional search of all possible combinations between frequency offset Δf and code delay τ. Concentrate for the moment onto one particular signal neglecting interference and any noise figure. The code delay and frequency offset of this signal are denoted by $\tau_{\max}$ and $\Delta f_{\max}$, whereby both are a function of time and not known a priori. The correlation result between this signal and a locally generated one depends on the code delay τ and frequency offset Δf of the locally generated signal. The code delay is a function of satellite signal run time. A signal emitted from a satellite in zenith with an altitude of 20 000 km will have a run time of about 70 ms. The ranging code period should be longer than the maximum run time in order to unambiguously resolve the range. If the ranging code period is shorter, the ambiguity can be solved by an approximate user position, by Doppler measurements, by considering the data message reception time of different satellites, or by other means.

The intermediate frequency $f_{\text{IF}} = f^s - f_\ell + \Delta f$ is a function of the locally generated harmonic, the relative user–satellite motion (Doppler shift), any sort of frequency instability of the receiver and satellite frequency standard, and some other effects, like relativity or atmospheric scintillation. Since $f^s - f_\ell$ is known a priori,

4.3 Generic signal processing 93

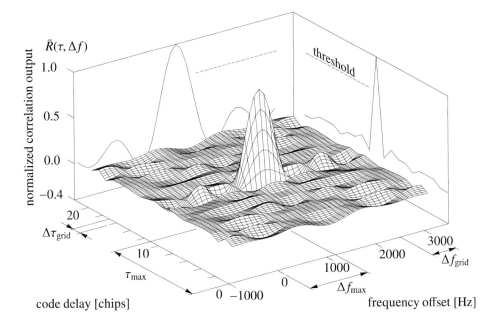

Fig. 4.26. Correlation results as a function of the time delay and frequency offset

the remaining frequency offset corresponds mainly to the Doppler shift. Thus, the frequency offset is sometimes referred to as Doppler offset.

Figure 4.26 shows an excerpt of the normalized autocorrelation results $\bar{R}(\tau, \Delta f)$ as a function of Δf and τ. Consider that the receiver does not compute $\bar{R}(\tau, \Delta f)$ at one epoch for all combinations of Δf and τ as shown in Fig. 4.26, and then searches for the maximum. But for reasons of practical implementation, the receiver computes one $\bar{R}(\tau, \Delta f)$ value and decides whether a maximum has been found or not, i.e., whether the threshold has been met.

The receiver does not necessarily need to search for the frequency offset, i.e., track the carrier to resolve code ranges (code delays). However the carrier phase adds increasing accuracy. The combined code and carrier tracking technique is called carrier-aided or coherent tracking. Note that signal waves and waveforms with constant phase difference are denoted coherent. Aiding information from an external source is necessary if the carrier is not tracked. As Ward et al. (2006: p. 154) state, in the case of noncoherent tracking, the frequency response roll-off characteristics will attenuate the code correlation result. Ward et al. (2006: p. 154) further emphasize that a receiver that loses phase lock, subsequently loses code lock. Note that carrier measurements need a higher technical expenditure than it is generally needed for carrier tracking.

The frequency offset and code delay of the locally generated signal are varied according to a defined search pattern until the correlation coefficient meets a predefined threshold. In the next step, the coarse global search is replaced by a fine local search algorithm. The practical implementation of a correlation function corresponds to a multiplication of the two signals followed by an integrate and dump function. Dwell time denotes the time needed to compute one integration result. Considering Fig. 4.26 and Eq. (4.64), it should be evident that any nonorthogonality between the signal replica and any other satellite signal or interferer will cause several correlation peaks. In addition, very strong signals generate high-level crosscorrelation peaks, thus influencing the tracking capabilities of weak signals. A second peak above the threshold in the frequency-code-shift domain may lead to false lock and, therefore, to wrong range measurements.

The search space is commonly snooped for a correlation result using a regular search grid. The number of grid cells is defined by the increments in frequency Δf_{grid} and in time $\Delta \tau_{grid}$. The increments in the frequency and the time space have to be fine enough to detect any correlation peak. An increasing density will increase the probability also to identify weak signals; consequently, the sensitivity of the receiver is increased. However, the acquisition time increases with the density. The increments in the time space depend on the code modulation, e.g., for BPSK modulated signals, typically 1/2 chip length search intervals are used. Wilde et al. (2006) show that for BOCs(1,1)-modulated signals three times as many search intervals have to be used compared to BPSK modulation, thus, 1/6 chip length, to guarantee a similar level of average crosscorrelation. The increments in the frequency space are, for example, $\Delta f_{grid} = 1/(2T_{dwell})$ kHz, where T_{dwell} corresponds to the dwell time in milliseconds.

Frequency offsets of ± 6 to ± 12 kHz have to be accounted for, where ± 5 kHz are induced by satellite motion, the rest by user motion and local oscillator drift.

Take as an example a short Gold ranging code of one millisecond length and composed of 1023 chips. The dwell time for one correlation result, thus, corresponds to 1 ms. Searching in 1/2 code intervals will result in 2046 correlation results. Therefore 2046 ms of search time are needed in the code delay range. In addition, assume $\Delta f_{grid} = 500$ Hz and a frequency offset range of 12 kHz resulting in 24 correlation results (= 12 000/500) in the frequency offset range, finally ending up with 49 104 correlation coefficients. In this example, acquisition would take a maximum of 49 seconds. It should be evident now that every channel relies not on a single correlator but rather on numerous to decrease acquisition time.

An increase of search speed can also be achieved by applying other strategies, e.g., exploiting the duality of time and frequency domain. Borre et al. (2007: Chap. 6), e.g., discuss the frequency search space concept or the parallel code phase approach.

Lock detectors compare the correlation result with the threshold to take out the decision whether a maximum has been found. Short code sequences or, more gen-

4.3 Generic signal processing

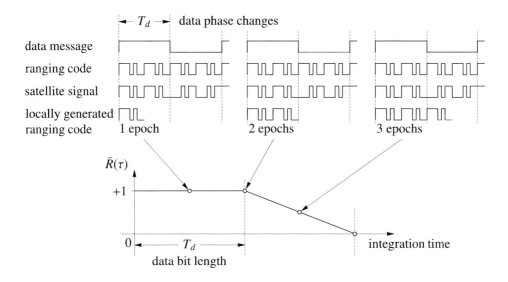

Fig. 4.27. Normalized correlation result roll-off as a function of integration time

erally, short integration times will result in high local maxima. Remember that the level of maximum crosscorrelation is a function of code length. Additionally, short integration times will be more affected by noise. The effect of Gaussian distributed noise can be reduced by using several code periods for computing the integration result; however, this will not decrease the level of crosscorrelation of the same ranging codes of different satellites.

Short integration times together with a low threshold increase the probability of false acquisition, i.e., identifying a local maximum as global maximum. The threshold can be increased; however, this will at the same time decrease the sensitivity of the receiver. Recall that the correlation result (4.29) and (4.64) is also a function of signal power; thus, weak signals will cause low correlation results. Long code sequences accumulate a great deal of signal energy, thus, positively influencing the correlation results. However, within this long integration time, the Doppler and the code delay will change. The integration result, hence, corresponds to a "smeared" average.

As already mentioned, the ranging code length and therefore the integration time can be lengthened by using more than one code epoch at a time (Fig. 4.27). Coherent integration time denotes any integration time where no data phase transition occurs. Assume a ranging code length of 1 ms and a data rate of 50 Hz, then a maximum of 20 ranging code epochs can be used for coherent integration (in Fig. 4.27 two have been used for simplification). This requires, however, the integration interval to be synchronized with the data phase transition. Noncoherent integration methods are used to lengthen the integration time while avoiding effects of the data phase transition onto the integration result. The concept of noncoherent integration is to integrate over coherent time epochs, square the results, and sum them. The squaring process eliminates the data phase transitions but, as a tradeoff, doubles the noise power. Nevertheless, the performance of long noncoherent integration time is generally better than the one of short coherent integration time intervals. Other methods to increase the correlation time without being affected by the data phase transition rely on wiping off the data bits in advance. This could be done either when the data chips are known or estimated (navigation messages do not change rapidly), or received from an external source.

Tracking loops

The receiver, actually, generates two realizations of the reference carrier frequency and three realizations of the ranging code that are correlated with the satellite signal. The two realizations of the carrier are 90° phase shifted and are called in-phase and quadrature phase signals. The respective steps in the receiver are the I/Q operations. These operations are necessary since the phase of the incoming signal is unknown. Even if the incoming signal and the locally generated carrier have the same frequency but are out of phase, the integration result will decrease as a function of phase shift. This is also called amplitude fading or energy loss. Counteracting this effect, the received signal is correlated with two 90° phase shifted locally generated signals. The correlation result is summed up accordingly. This preserves the energy in the correlation result.

Receivers will generally align the generated carrier in the I channel to the phase and frequency of the satellite signal. This allows in the follow-up data demodulation to sense data phase transitions by only considering the output of the I channel.

The locally generated (harmonic) carrier that is mixed with the satellite signal shifts the frequency spectrum of the signal to zero frequency (Fig. 4.23d). This is a similar operation as during the IF downconversion. If the frequency of the harmonic and the carrier frequency of the IF signal match, then the carrier is called to be wiped off and the signal is in baseband. This operation and the following operations, thus, are commonly denoted as baseband processing.

The three realizations of the ranging codes are time shifted by the correlation spacing to provide an early (E), prompt (P), and late (L) correlation result. The early and late correlation coefficients in the in-phase and quadrature phase branches (I_E, I_L, Q_E, Q_L) are only used for the code tracking loop. The prompt correlation result (I_P, Q_P) in contrast is also processed in the carrier tracking loop.

Note that the prompt values have been used before for the acquisition algorithm. The sum of the squares of the integration results in the I and Q path ($I_P^2 + Q_P^2$) are used by the lock detector to determine the correlation between incoming signal and replica. The sum expresses at the same time the average signal power which is used for S/N estimation and which is used by the AGC. As soon as the lock detector

4.3 Generic signal processing

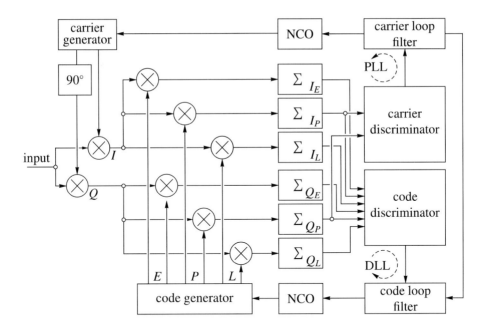

Fig. 4.28. Tracking loops

confirms acquisition, the feedback control loops track the carrier and code.

A phase lock loop (PLL) tracks the carrier phase while providing at the same time the navigation data bit transitions. The delay lock loop (DLL) continuously adjusts the code replica to the satellite signal. The tracking loops must accommodate any satellite and vehicle dynamics which change the Doppler shift and the code delay and any other frequency offset. The alignment of the generated signal to the incoming one is done in the DLL and PLL in parallel, therefore the two loops are closely interlocked.

Carrier tracking loop
Neglect for the moment the ranging code and data message. The satellite signal is split into the I and Q branches, one multiplied with the carrier phase as output by the carrier generator, the other multiplied by a 90° phase shifted carrier. If the frequency of the generated signal and the incoming one coincide, the carrier will be completely wiped off, and the center frequency f_{IF} of the resulting frequency corresponds to zero (Fig. 4.24d). Any deviation between generated frequency and incoming one is a measure of the remaining Doppler shift and will result in a beat frequency and a beat phase

$$\Delta\varphi_{\ell r} = \varphi_r - \varphi_\ell, \tag{4.65}$$

where φ_ℓ corresponds to the phase of the generated carrier frequency, φ_r to the one

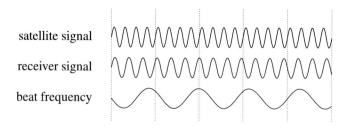

Fig. 4.29. Beat frequency

of the satellite signal as it is output by the RF front-end (Fig. 4.29).

The process of carrier wipe-off is based on the principle given in Eq. (4.63). The high-frequency component is filtered using a low-pass filter. The low-pass filter is a simple integrate and dump operation resulting in the variables I_P and Q_P. Finally, the carrier loop discriminator function computes the phase shift. The output of the discriminator function is filtered in the carrier loop noise filter, which in particular characterizes the phase lock loop behavior and tracking accuracy. The carrier loop filter analyzes the beat phase and extracts the remaining frequency offset as well as the remaining phase shift between satellite signal and I channel harmonic. A wide filter bandwidth will be useful for high dynamics; thus, fast frequency offset changes can be accommodated. A high bandwidth, however, increases the noise level. Different loop designs have been constructed to optimize the performance for user requirements. The numerically controlled oscillator (NCO), which is driven by the output of the carrier loop noise filter, finally controls the carrier generator. This procedure closes the loop. Carrier wipe-off is an iterative process which, due to changing Doppler shift, has to be performed continuously. The carrier generator setting is a measure of the Doppler shift.

Any data phase transition will cause a 180° phase shift within the PLL. The Costas loop is one realization of the PLL, with the additional feature of being insensitive to any 180° phase shift. In principle, the discriminator function of the Costas loop takes the output of the integrate and dump filters squared, what causes the frequency to double, the noise to increase, but also to eliminate the data phase transitions.

Commonly used carrier phase discriminator functions, without being exhaustive, are the decision-directed discriminator, the dot-product discriminator, or the arctangent discriminator (Ward et al. 2006). The latter is defined by

$$D_\varphi = \arctan\left(\frac{Q_P}{I_P}\right). \tag{4.66}$$

If the discriminator function outputs phase changes over time (= frequency) instead of phase, the PLL transfers into a frequency lock loop (FLL). Misra and

4.3 Generic signal processing

Enge (2006: p. 483) mention that the FLL may be applied for coarse frequency tracking before applying the more accurate phase lock loop.

The error (1σ) of the phase lock loop is given by

$$\sigma_{\text{PLL}} = \frac{\lambda}{2\pi} \sqrt{\frac{B_{\text{PLL}} N_0}{P_r} \left(1 + \frac{N_0}{2 P_r T_I}\right)} \quad [\text{m}], \quad (4.67)$$

where λ is the wavelength of the satellite signal carrier, B_{PLL} is the single-sided carrier loop noise bandwidth of the carrier loop noise filter given in hertz, T_I denotes the coherent (predetection) integration time (Langley 1997). The Radio Technical Commission for Maritime Services (2001: Appendix B) emphasizes that the error induced by the phase lock loop is driven by the signal-to-noise ratio, not by other errors such as multipath, ionospheric errors, or even receiver-specific characteristics like quantization level. A decreasing carrier loop noise bandwidth decreases the error level; however, at the same time the loop is too inert to follow any high dynamics of the receiver or other means of phase scintillations. Stationary receivers use a bandwidth of 2 Hz or even less (Langley 1998: p. 182). Borre et al. (2007: p. 93) emphasize that for land applications, bandwidths of 20 Hz are common. One method to use a small bandwidth but to have a good performance in high dynamic environments is to use aiding information from, e.g., an inertial measurement unit.

The second part in the parentheses of Eq. (4.67) expresses the squaring loss and can be neglected if no data is on the satellite signal (pilot signal) or the data can be wiped off by other means.

Langley (1998: p. 182) assumes, for example, a carrier-to-noise density ratio of 45 dBHz, a carrier loop noise bandwidth of 2 Hz, a wavelength of 0.2 m, and an integration time of 1 ms, then the carrier loop tracking error is 0.2 mm.

Code tracking loop

Likewise to the PLL also the DLL differentiates between in-phase and quadrature phase components in order to preserve the energy in the signal.

Already mentioned in Sect. 4.2.1, the exact shape of the correlation function is, due to noise, restricted filter design, etc., unknown, therefore the concept of a discriminator function using a correlator spacing d is applied. Standard correlators use a correlation spacing of one chip. Narrow correlators reduce the correlation spacing to, e.g., 0.1 chips. In this way the noise figures are reduced due to their higher temporal correlation and the influence of multipath effects as shown in Fig. 4.11 is reduced as well.

The code discriminator functions are designed to avoid influences of phase shift induced by data phase transitions. A simple coherent DLL discriminator is

$$D_c = (I_E - I_L) \operatorname{sign}(I_P), \quad (4.68)$$

where sign(I_P) corresponds in particular to the data message bit modulation, as long as the carrier phase is coherently tracking. Noncoherent discriminators eliminate the data phase transitions using squaring techniques, e.g., early-minus-late power discriminator or the dot-product discriminator that reads

$$D_{nc} = (I_E - I_L)I_P + (Q_E - Q_L)Q_P. \tag{4.69}$$

The output of the discriminator function is filtered in the tracking loop filter and fed back into the NCO. The relative user–satellite motion-induced Doppler influences the code sequence according to (4.13) in dependence of the code rate. For example, a code rate of 1.023 MHz will cause a Doppler shift of 3.4 Hz ($v_\varrho \sim 1000$ m s^{-1}). This influence creates a dynamics-driven discriminator output. The output of the carrier loop filter is scaled and used in the code tracking filter to eliminate the user–satellite dynamics (Langley 1998). The carrier loop filter estimate of the dynamics is more accurate and, therefore, preferably used in the code loop filter. The rate-aided code loop (Misra and Enge 2006: p. 479), also denoted as carrier-aided code loop (Ward et al. 2006: p. 162), allows to use low code loop bandwidth and, thus, low noise in the code measurement. The remaining dynamics in the code loop is ionospheric and noise induced.

The DLL tracking becomes more fragile when considering BOC-modulated ranging codes and their ambiguous discriminator function. There are different concepts to avoid false tracking. One of them is to track only one sidelobe of the BOC frequency spectrum, consequently neglecting half of the energy in the signal. A method which takes full advantage of the BOC energy spectral spreading is the bump and jump technique, relying on two more correlators, a very early and a very late correlator. Refer to, e.g., Julien et al. (2004b) or Julien (2005) for further discussions of unambiguous BOC tracking.

The DLL tracking error, assuming no remaining Doppler offset, is estimated using the received power P_r, the noise density N_0, the equivalent single-sided tracking loop bandwidth B_{DLL} of the tracking loop filter, the coherent (predetection) integration time T_I, and the correlation spacing d. The mathematical model presented here follows Dierendonck (1996: p. 373) and the Radio Technical Commission for Maritime Services (2001: Appendix B), respectively:

$$\sigma_{DLL} = c\,T_c \sqrt{\frac{B_{DLL}\,N_0\,d}{2P_r}\left(1 + \frac{2N_0}{(2-d)P_r T_I}\right)} \quad [m], \tag{4.70}$$

where σ_{DLL} corresponds to the estimated tracking error (1σ) of a BPSK-modulated signal using a noncoherent early-minus-late power discriminator. The speed of light is denoted by c, T_c defines the chip length in seconds. The higher the tracking loop bandwidth B_{DLL} the faster the DLL synchronizes to the satellite signal. However, the increasing bandwidth increases the noise level. If an additional smoothing filter

4.3 Generic signal processing

is placed after the tracking loop filter, B_{DLL} has to be replaced by this filter bandwidth (Radio Technical Commission for Maritime Services 2001: Appendix B). Sleewaegen et al. (2004) present a more general formulation for BPSK, BOC, or AltBOC modulation assuming solely pilot tracking and a dot-product power discriminator. Minimum tracking errors are commonly described by the Cramer–Rao lower bound as given in Spilker (1996a: p. 111).

Codeless receivers

For security reasons, the navigation service providers do not publish all ranging codes. In some cases only the original ranging code is known, but the encryption method is defined only in classified interface control documents. Methods have been developed to use signals with encrypted ranging codes. Without losing generality, denote the known ranging code by P, the encryption regulation W, and the encrypted ranging code as Y = P W. The frequency of W is assumed to be lower than the one of the P-code. Further assume that the ranging code may be modulated onto two carriers with frequencies $f_1 > f_2$.

Four methods are distinguished: the codeless and the quasi-codeless methods, which themselves are differentiated into squaring and crosscorrelation techniques (Lachapelle 1998). All four approaches suffer from a substantial degradation in the S/N. Figure 4.30 summarizes the characteristics of the four techniques. The graphical diagrams are adapted from Ashjaee and Lorenz (1992) and Eissfeller (1993).

The squaring technique was first presented in 1981 by C. Counselman. The received signal is mixed with itself and, hence, all modulations are removed. The result is the unmodulated carrier with twice the frequency and, thus, half the wavelength. The S/N is substantially reduced in the squaring process (Ashjaee 1993).

The crosscorrelation technique is another codeless technique which was first described in 1985 by P. MacDoran. The technique is based on the fact that the unknown Y-code is identical on two carriers which enable crosscorrelation of the f_1 and f_2 signal. Due to the frequency-dependent propagation of an electromagnetic wave through ionosphere, the delay between f_1 and f_2 has to be taken into account. The observables resulting from the correlation process are the time delay which corresponds to the range difference between the two signals, and a phase difference $\Phi_2 - \Phi_1$.

The code correlation plus squaring technique was patented by Keegan (1990). The method is also denoted code-aided squaring and involves correlating the received Y-code on the f_2 signal with a locally generated replica of the P-code. This correlation is possible because the Y-code originates from an XOR sum of the P-code and the encrypted W-code. Since the chipping rate of the W-code is less than the frequency of the Y-code, there always exist Y-code portions which are identical to the original P-code portions (Eissfeller 1993). After the correlation, a low-pass filter is applied to limit the bandwidth and, subsequently, the signal is squared to

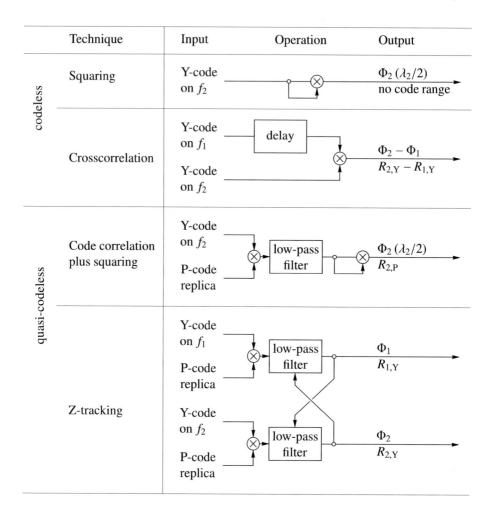

Fig. 4.30. Codeless techniques

get rid of the code similar to the procedures in the tracking loops to free the data bits. The technique provides code range and, because of squaring, half wavelength carrier phase.

An improved quasi-codeless technique is denoted Z-tracking^{TM} and has been reported in Ashjaee and Lorenz (1992). The Y-code on both the f_1 and f_2 signal is separately correlated with a receiver-generated replica of the P-code. Since there is a separate correlation on f_1 and f_2, the W-code on each frequency is obtained. The subsequent low-pass filter allows to estimate phase-bit of the remaining W-code. This estimate is used in the f_1, f_2 path mutually to eliminate the W-code. There is no need to know the W-code because it is only used for synchronization purposes (Breuer et al. 1993). The removal of the encrypting code leads to the same signals

4.3.4 Navigation processor

The navigation processor fulfills three main tasks. The result of the data demodulation process is used to decode the navigation message and compute satellite positions. Secondly, the code, phase, and Doppler measurements are used to compute position, velocity, and time information. The third task of the navigation processor is to provide aiding information to the tracking loops and to the filters. The time needed for a GNSS receiver between power up and providing the first position information is denoted as time to first fix (TTFF).

Data demodulation involves three operations: symbol synchronization, frame synchronization, and message decoding. Carrier-aided tracking systems wipe-off the code and carrier from the signal, therefore the remaining absolute phase changes between $+\pi$ and $-\pi$ correspond to the data symbol transition. These phase changes are detectable by analyzing sign(I_P), if the generated I-channel carrier has been aligned to the incoming signal (coherent tracking). The bit synchronization operation, thus, senses any sign changes of the prompt correlation values. Frame synchronization is accomplished using a well-known data bit sequence (preamble) of the message, which is periodically transmitted. The receiver additionally checks the frame length, the parity, and any other deterministic or predictable information. The navigation message decoding that follows, optionally performs block deinterleaving, forward error correction, deciphering, and finally recovers the data bit train of the original message. The exact process varies with the different messages of the various systems and services. The different error correction encoding methods are implemented to increase the reliability of data transmission and reduce the bit error rate.

One concept applied for nearly all modernized and future GNSS signals is the half-rate convolutional encoding and decoding. Therefore, the symbol rate in symbols per second (sps) is twice the original data rate in bits per second (bps). The convolutional coding is characterized by the characteristic polynomials $G_1 = 171$ and $G_2 = 133$, both in octal notation (cf. Fig. 4.31). The octal number of G_1 transforms into the binary number 001 111 001, which, neglecting the first two zeros, defines the polynomial $p(x) = 1 + x^1 + x^2 + x^3 + x^4 + x^7$. The polynomial of G_2 is coded similarly. Holmes (1982: p. 264) denotes the resulting code as optimal nonsystematic convolutional code. At every clock impulse one bit is fed into the convolutional encoder and two symbols G_1, G_2 are output (Holmes 1982: p. 251). The output symbols have twice the rate as the input bits. Thus, this convolutional encoding scheme is denoted half-rate.

In the block interleaving process, the bits and symbols respectively fill up a

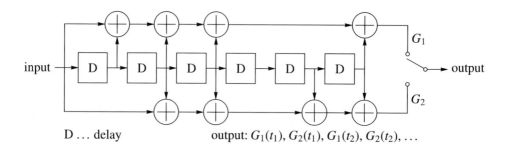

Fig. 4.31. Convolutional encoder

matrix column by column. The symbols, in the sequel, are transmitted row by row. Assume that a sequence of more than one symbol is corrupted while transmitted due to, e.g., pulse-type jamming. Deinterleaving the received symbols will spread the corrupted bits over the whole message. In this way a reconstruction of the original symbols and bits, respectively, becomes possible using the other error correction methods.

The content of the data message differs from system to system. Furthermore, different services use different data messages. Common to all systems is that there is at least one service to transmit ephemerides data as well as time information. In this way the receiver can autonomously compute satellite position information. If the satellite parameters are not detectable on the satellite signal, e.g., due to low S/N, the receiver may acquire them by other means, i.e., communication links.

Some GNSS also apply satellite and frequency diversity concepts to decrease the time needed for decoding the complete navigation message. Satellites transmitting the same information but at a different time epoch is called satellite diversity. For example, satellite 1 transmits the almanac of satellites 5 and 6, whereas satellite 2 transmits at the same period of time the almanac of satellites 7 and 8. Thus the navigation message is alternated as a function of time and satellite. In this way, when tracking more than one satellite, demodulation of almanac information from several satellites becomes possible in a short period of time. This furthermore decreases the TTFF and increases the number of tracked satellites despite of short tracking periods. Frequency diversity applies the same principle; however, the navigation message is not alternated as a function of the satellite but as a function of the carrier frequency. The different sequencing allows for a fast demodulation of the navigation message when tracking more than one frequency.

Once the data message has been decoded and the receiver synchronized to the GNSS time, it is not necessary to continuously receive or decode the navigation data, as long as the receiver continuously tracks the code or carrier. This is advantageous since, as Spilker and Natali (1996: p. 748) emphasize, the S/N threshold for tracking is generally significantly lower than the threshold for data detection.

5 Observables

5.1 Data acquisition

In concept, the satellite navigation observables are ranges which are deduced from measured time or phase differences based on a comparison between received signals and receiver-generated signals. Unlike the terrestrial electronic distance measurements, satellite navigation uses the "one-way concept" where two clocks are involved, namely one in the satellite and the other in the receiver. Thus, the ranges are biased by satellite and receiver clock errors and, consequently, they are denoted as pseudoranges.

5.1.1 Code pseudoranges

Let us denote by $t^s(\text{sat})$ the signal emission time referred to the reading of the satellite clock and by $t_r(\text{rec})$ the signal reception time referred to the reading of the receiver clock. Recall that the satellite clock reading $t^s(\text{sat})$ is transmitted in the navigation message via the PRN code. The errors (or biases) of the clocks with respect to a common time system (i.e., the respective system time) are termed δ^s and δ_r.

The difference between the clock readings is equivalent to the time shift Δt which aligns the satellite and reference signal during the code correlation procedure in the receiver. Thus,

$$t_r(\text{rec}) - t^s(\text{sat}) = [t_r + \delta_r] - [t^s + \delta^s] = \Delta t + \Delta \delta, \tag{5.1}$$

indicating that in $t_r(\text{rec})$ and $t^s(\text{sat})$ two different time systems are involved but that now on the right-hand side t_r and t^s refer to the common system time and where $\Delta t = t_r - t^s$ and $\Delta \delta = \delta_r - \delta^s$. The bias δ^s of the satellite clock can be modeled if the respective information is transmitted accordingly, e.g., by a polynomial with the coefficients being transmitted in the navigation message. Assuming the correction δ^s has been applied, $\Delta \delta$ equals the receiver clock bias.

When multiplying the time interval $t_r(\text{rec}) - t^s(\text{sat})$ of Eq. (5.1), which is affected by the clock errors, by the speed of light c, the code pseudorange

$$R = c\,[t_r(\text{rec}) - t^s(\text{sat})] = c\,\Delta t + c\,\Delta \delta = \varrho + c\,\Delta \delta \tag{5.2}$$

is obtained, where $\varrho = c\,\Delta t$ has been introduced. The range ϱ is calculated from the true signal travel time. In other words, ϱ corresponds to the distance between the position of the satellite at epoch t^s and the position of the antenna of the receiver at epoch t_r. Remember that both epochs refer to the common system time. Since ϱ is a function of two different epochs, it is often expanded into a Taylor series with respect to, e.g., the emission time

$$\varrho = \varrho(t^s, t_r) = \varrho(t^s, (t^s + \Delta t)) = \varrho(t^s) + \dot{\varrho}(t^s) \Delta t, \qquad (5.3)$$

where $\dot{\varrho}$ denotes the time derivative of ϱ or the radial velocity of the satellite relative to the receiving antenna. All epochs in Eq. (5.3) are expressed in a common system time.

The maximum radial velocity for GNSS satellites in the case of a stationary receiver is $\dot{\varrho} \approx 1.0 \,\mathrm{km\,s^{-1}}$, and the travel time of the satellite signal is from about 0.06 s to 0.10 s. The amount of correction term in Eq. (5.3), thus, is greater than 60 m.

The precision of a pseudorange derived from code measurements has been traditionally about 1% of the chip length. Therefore, a chip length of 300 m for a coarse code would yield a precision of roughly 3 m and an assumed chip length of 30 m for a precise code would yield a precision of 0.3 m. However, more recent developments demonstrate that a precision of about 0.1% of the chip length is possible.

5.1.2 Phase pseudoranges

Let us denote by $\varphi^s(t)$ the phase of the received and reconstructed carrier with frequency f^s and by $\varphi_r(t)$ the phase of a reference carrier generated in the receiver with frequency f_r. Here, the parameter t is an epoch in a common time system reckoned from an initial epoch $t_0 = 0$. According to Eq. (4.11), the following phase equations are obtained

$$\varphi^s(t) = f^s t - f^s \frac{\varrho}{c} - \varphi_0^s,$$
$$\varphi_r(t) = f_r t - \varphi_{0r}, \qquad (5.4)$$

where the phases are expressed in cycles. The initial phases φ_0^s, φ_{0r} are caused by clock errors and are equal to

$$\varphi_0^s = -f^s \delta^s,$$
$$\varphi_{0r} = -f_r \delta_r. \qquad (5.5)$$

Hence, the beat phase $\varphi_r^s(t)$ is given by

$$\varphi_r^s(t) = \varphi^s(t) - \varphi_r(t),$$
$$= -f^s \frac{\varrho}{c} + f^s \delta^s - f_r \delta_r + (f^s - f_r) t. \qquad (5.6)$$

5.1 Data acquisition

The deviation of the frequencies f^s, f_r from the nominal frequency f is in the order of only some fractional parts of hertz. This may be verified by considering, e.g., a short-time stability in the frequencies of $df/f = 10^{-12}$. With the nominal carrier frequency $f \approx 1.5$ GHz, the frequency error, thus, becomes $df = 1.5 \cdot 10^{-3}$ Hz. Such a frequency error may be neglected because during signal propagation (i.e., $t = 0.07$ s) a maximum error of 10^{-4} cycles in the beat phase is generated which is below the noise level. The clock errors are in the range of milliseconds and are, thus, less effective. Summarizing, Eq. (5.6) may be written in the simplified form

$$\varphi_r^s(t) = -f\frac{\varrho}{c} - f\Delta\delta, \qquad (5.7)$$

where again $\Delta\delta = \delta_r - \delta^s$ has been used. If the assumption of frequency stability is incorrect and the oscillators are unstable, then their behavior has to be modeled by, for example, polynomials where clock and frequency offsets and a frequency drift are determined. Historically, a complete carrier phase model which includes the solution of large (e.g., 1 second) receiver clock errors was developed by Remondi (1984). In practice, eventual residual errors can be eliminated by differencing the measurements.

Switching on a receiver at epoch t_0, the instantaneous fractional beat phase is measured. The initial integer number N of cycles between satellite and receiver is unknown. However, when tracking is continued without loss of lock, the number N, also called integer ambiguity, remains the same and the beat phase at epoch t is given by

$$\varphi_r^s(t) = \Delta\varphi_r^s\Big|_{t_0}^{t} + N, \qquad (5.8)$$

where $\Delta\varphi_r^s$ denotes the (measurable) fractional phase at epoch t augmented by the number of integer cycles since the initial epoch t_0. A geometrical interpretation of Eq. (5.8) is provided in Fig. 5.1, where $\Delta\varphi_i$ is a shortened notation for $\Delta\varphi_r^s|_{t_0}^{t_i}$ and, for simplicity, the initial fractional beat phase $\Delta\varphi_0$ is assumed to be zero. Substituting Eq. (5.8) into Eq. (5.7) and denoting the negative observation quantity by $\Phi = -\Delta\varphi_r^s$ yields the equation for the phase pseudoranges

$$\Phi = \frac{1}{\lambda}\varrho + \frac{c}{\lambda}\Delta\delta + N, \qquad (5.9)$$

where the wavelength λ has been introduced according to Eq. (4.10). Multiplying the above equation by λ scales the phase expressed in cycles to a range (given in meter):

$$\lambda\Phi = \varrho + c\Delta\delta + \lambda N. \qquad (5.10)$$

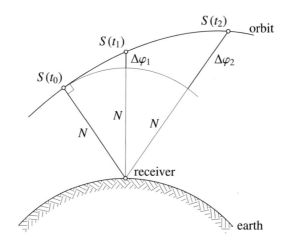

Fig. 5.1. Geometrical interpretation of phase range

This equation differs from the code pseudorange only by the integer multiples of λ. Again, the range ϱ represents the distance between the satellite at satellite transmission time epoch t and the receiver at reception time epoch $t + \Delta t$ (considering both epochs in a common system time). The phase of an electromagnetic wave can be measured to better than 0.01 cycles which corresponds to millimeter precision for a frequency in the gigahertz range.

It should be noted that a plus sign convention has been chosen for Eq. (5.9). This choice is somewhat arbitrary since quite often the phase Φ and the distance ϱ have different signs. Actually, the sign is receiver-dependent because the beat phase is generated in the receiver and the combination of the satellite and the receiver signal may differ for various receiver types.

5.1.3 Doppler data

Historically, the Transit system used the integrated Doppler shifts (i.e., phase differences) which were scaled to delta ranges. Today, the raw Doppler shift, being linearly dependent on the radial velocity, cf. Eq. (4.13), and, thus, allowing for velocity determination in real time is important for navigation. Considering Eq. (5.9), the equation for the observed Doppler shift scaled to range rate is given by

$$D = \lambda \dot{\Phi} = \dot{\varrho} + c \Delta \dot{\delta}, \tag{5.11}$$

where the derivatives with respect to time are indicated by a dot. The raw Doppler shift measurement is less accurate than the integrated Doppler shift. An estimate of the achievable accuracy is 0.001 Hz. Using Eq. (4.13), this corresponds to $3 \cdot 10^{-4}$ m s^{-1} if the Doppler shift is based on an emitted frequency of 1 GHz.

5.1 Data acquisition

Remember when the satellite is moving toward the GNSS receiver, the Doppler shift is positive; so one gets more Doppler counts when the range is diminishing.

Apart from its meaning in navigation, it is worth noting here that the raw Doppler shift is also applied to determine integer ambiguities in kinematic surveying or may be used as an additional independent observable for point positioning.

5.1.4 Biases and noise

The code pseudoranges, cf. Eq. (5.2), and phase pseudoranges, cf. Eq. (5.9), are affected by systematic errors or biases and random noise as well. Note that Doppler measurements are affected by the bias rates only. The error sources can be classified into three groups, namely satellite-related errors, propagation-medium-related errors, and receiver-related errors. Some range biases are listed in Table 5.1.

Some of the systematic errors can be modeled and give rise to additional terms in the observation equations which will be explained in detail in later sections. As mentioned earlier, systematic effects can also be eliminated (or at least strongly reduced) by appropriate combinations of the observables. Differencing measurements of two receivers to the same satellite eliminates satellite-specific biases; differencing between two satellites and one receiver eliminates receiver-specific biases. As a consequence, double-difference pseudoranges are, to a high degree, free of systematic errors originating from the satellites and from the receivers. With respect to refraction, this is only true for short baselines where the measured ranges at both endpoints are affected equally. In addition, ionospheric refraction can virtually be eliminated by an adequate combination of dual-frequency data. Antenna phase center variations are treated in Sect. 5.5. Multipath is caused by multiple reflections of the signal (which can also occur at the satellite during signal emission). The interference between the direct and the reflected signal is largely not random; however, it may also appear as a noise. Wells et al. (1987) report a similar effect called imaging where a reflecting obstacle generates an image of the real antenna

Table 5.1. Range biases

Source	Effect
Satellite	Clock bias
	Orbital errors
Signal propagation	Ionospheric refraction
	Tropospheric refraction
Receiver	Antenna phase center variation
	Clock bias
	Multipath

Table 5.2. Typical magnitude of range biases

Range	Bias
Code range (coarse code)	300 cm
Code range (precise code)	30 cm
Phase range	5 mm

which distorts the antenna pattern. Both effects, multipath and imaging, can be considerably reduced by selecting sites protected from reflections (buildings, vehicles, trees, etc.) and by an appropriate antenna design. It should be noted that multipath is frequency (i.e., wavelength) dependent. Therefore, carrier phases are less affected than code ranges (Lachapelle 1990). More details on the multipath problems are given in Sect. 5.6.

The random noise mainly contains the actual observation noise plus multipath. Assuming a typical chip length of 300 m for a coarse code and of 30 m for a precise code, the pseudorange noise is summarized in Table 5.2.

The signal-in-space (SIS) user range error (URE) is an estimate comprising errors of ephemerides data, satellite clock, and the ionospheric and tropospheric delay. It does not consider errors caused by the environment (e.g., multipath) or by the user equipment (e.g., receiver noise including antenna offset and variation). Extending the URE by the user equipment and environmental errors, the user equivalent range error (UERE) is obtained. Even if there are some correlations, the individual error contributions are considered to be independent. Therefore, the UERE is computed as square root of the summed squares of the six error constituents ephemerides data, satellite clock, ionosphere, troposphere, multipath, and receiver measurement. The receiver measurement component can be further split into receiver clock error and white noise. In Table 5.3 adapted from Parkinson (1996: p. 481), the UERE is calculated based on typical values (all given in meters) for the individual quantities. The column headed by "Total" results from the square root of the sum of the squared bias and the squared random quantity, e.g., for the satellite clock the total is obtained by $\sqrt{2.0^2 + 0.7^2} = 2.1$. Note that linked to this UERE computation is the 1σ probability level which amounts to 68.3%, see Sect. 7.3.6.

In combination with a dilution of precision (DOP) factor, which will be explained in Sect. 7.3.4, UERE allows for an estimation of the achievable point positioning accuracy (Sect. 7.3.1).

As Kuusniemi (2005) indicates, the value of the UERE as computed in Table 5.3 is limited because in real situations too many variables must be taken into account, e.g., the elevation angle of the satellite which influences the signal path length, the strength of the received signal, and the changing multipath environment.

5.2 Data combinations

Table 5.3. UERE computation

Error source	Bias [m]	Random [m]	Total [m]
Ephemerides data	2.1	0.0	2.1
Satellite clock	2.0	0.7	2.1
Ionosphere	4.0	0.5	4.0
Troposphere	0.5	0.5	0.7
Multipath	1.0	1.0	1.4
Receiver measurement	0.5	0.2	0.5
UERE [m]	5.1	1.4	5.3

The signal strength is described by the carrier-to-noise power density ratio and the signal-to-noise (S/N) ratio. Details on these parameters are given in Sect. 4.3.1. The carrier-to-noise power density ratio is, according to Lachapelle (2003), the fundamental navigation signal quality parameter.

5.2 Data combinations

GNSS observables are obtained from the ranging code information or the carrier wave in the broadcast satellite signal. Assuming two carriers based on the respective frequencies f_1 and f_2 and one code modulated on each of the two carriers, one could measure the code ranges R_1, R_2, the carrier phases Φ_1, Φ_2, and the corresponding Doppler shifts D_1, D_2 for a single epoch, where the subscript indicates the respective frequency. Subsequently, the Doppler observables are not considered. In general, the number of observables of a GNSS receiver can differ, for example, a single-frequency receiver delivers only data from one frequency; considering a receiver that may manage three carriers and two codes, the number of observables increases accordingly.

The objective of this section is to show how linear combinations are developed for dual-frequency data, and how code range smoothing by means of carrier phases is performed.

5.2.1 Linear phase pseudorange combinations

General remarks

Suppose two frequencies f_1 and f_2 and denote the respective phase pseudoranges by Φ_1 and Φ_2. The linear combination of two phase pseudoranges is defined by

$$\Phi = n_1 \Phi_1 + n_2 \Phi_2, \tag{5.12}$$

where n_1 and n_2 are arbitrary numbers. The substitution of the relations $\Phi_i = f_i t$ for the corresponding frequencies f_1 and f_2 gives

$$\Phi = n_1 f_1 t + n_2 f_2 t = f t. \tag{5.13}$$

Therefore,

$$f = n_1 f_1 + n_2 f_2 \tag{5.14}$$

is the frequency and

$$\lambda = \frac{c}{f} \tag{5.15}$$

is the wavelength of the linear combination.

Compared to the noise of a single phase, the noise level for the linear combination differs by the factor $\sqrt{n_1^2 + n_2^2}$ which follows from the application of the error propagation law and assuming the same noise level for both phases.

Linear combinations with integer numbers

The simplest nontrivial linear combinations of the two phase pseudoranges Φ_1 and Φ_2 in Eq. (5.12) are $n_1 = n_2 = 1$, yielding the sum

$$\Phi_1 + \Phi_2, \tag{5.16}$$

and $n_1 = 1$, $n_2 = -1$, leading to the difference

$$\Phi_1 - \Phi_2. \tag{5.17}$$

According to (5.15), increasing the frequency reduces (or narrows) the wavelength and decreasing the frequency increases (or widens) the wavelength. Accordingly, the combination $\Phi_1 + \Phi_2$ is denoted as narrow lane and $\Phi_1 - \Phi_2$ as wide lane. The lane signals are used for ambiguity resolution (Sect. 7.2).

The advantage of a linear combination with integer numbers is that the integer nature of the ambiguities is preserved.

Linear combinations with real numbers

A slightly more complicated linear combination results from the choice

$$n_1 = 1, \quad n_2 = -\frac{f_2}{f_1} \tag{5.18}$$

leading to the combination

$$\Phi_1 - \frac{f_2}{f_1} \Phi_2, \tag{5.19}$$

5.2 Data combinations

which is often denoted as geometric residual. This quantity is the kernel in a combination used to reduce ionospheric effects (Sect. 5.3.2).

Another linear combination follows from the reciprocal values of (5.18)

$$n_1 = 1, \qquad n_2 = -\frac{f_1}{f_2} \tag{5.20}$$

leading to the combination

$$\Phi_1 - \frac{f_1}{f_2}\Phi_2, \tag{5.21}$$

which is often denoted as ionospheric residual. This quantity is used, e.g., in the context of cycle slip detection (Sect. 7.1.2).

The drawback of a linear combination with real numbers is that the integer nature of the ambiguity is generally lost.

5.2.2 Code pseudorange smoothing

The principle of code pseudorange smoothing by means of phase pseudoranges is an important issue in accurate real-time positioning.

Assuming dual-frequency measurements for epoch t_1, the code pseudoranges $R_1(t_1)$, $R_2(t_1)$ and the carrier phase pseudoranges $\Phi_1(t_1)$, $\Phi_2(t_1)$ are obtained. Further assume the code pseudoranges are scaled to cycles (but still being denoted as R) by dividing them by the corresponding carrier wavelength. Note that pseudoranges scaled to cycles are sometimes denoted code phases. Using the two frequencies f_1, f_2, the combination

$$R(t_1) = \frac{f_1 R_1(t_1) - f_2 R_2(t_1)}{f_1 + f_2} \tag{5.22}$$

is formed for the code pseudoranges and the wide-lane signal

$$\Phi(t_1) = \Phi_1(t_1) - \Phi_2(t_1) \tag{5.23}$$

for the carrier phase pseudoranges. From Eq. (5.22) it can be verified by applying the error propagation law that the noise of the combined code pseudorange $R(t_1)$ is reduced by the factor $\sqrt{f_1^2 + f_2^2}/(f_1 + f_2)$ which amounts to 0.7 for present GNSS compared to the noise of the single code measurement. The increase of the noise in the wide-lane signal by a factor of $\sqrt{2}$ has no effect because the noise of the carrier phase pseudoranges is much lower than the noise of the code pseudoranges. Note that both signals $R(t_1)$ and $\Phi(t_1)$ have the same frequency and, thus, the same wavelength as may be verified by applying Eq. (5.14).

Combinations (5.22) and (5.23) are formed for each epoch. Additionally, for all epochs t_i after t_1, extrapolated values $R(t_i)_{ex}$ of the code pseudoranges can be calculated from

$$R(t_i)_{ex} = R(t_1) + (\Phi(t_i) - \Phi(t_1)). \tag{5.24}$$

The smoothed value $R(t_i)_{sm}$ is finally obtained by the arithmetic mean

$$R(t_i)_{sm} = \tfrac{1}{2}\left(R(t_i) + R(t_i)_{ex}\right). \tag{5.25}$$

Generalizing the above formulas for an arbitrary epoch t_i (with the preceding epoch t_{i-1}), a recursive algorithm is given by

$$\begin{aligned}
R(t_i) &= \frac{f_1\, R_1(t_i) - f_2\, R_2(t_i)}{f_1 + f_2}, \\
\Phi(t_i) &= \Phi_1(t_i) - \Phi_2(t_i), \\
R(t_i)_{ex} &= R(t_{i-1})_{sm} + (\Phi(t_i) - \Phi(t_{i-1})), \\
R(t_i)_{sm} &= \tfrac{1}{2}\left(R(t_i) + R(t_i)_{ex}\right),
\end{aligned} \tag{5.26}$$

which works under the initial condition $R(t_1) = R(t_1)_{ex} = R(t_1)_{sm}$ for all $i > 1$.

The above algorithm assumes data free of gross errors. However, carrier phase data are sensitive to changes in the integer ambiguity (i.e., cycle slips). To circumvent this problem, a variation of the algorithm is given subsequently. Using the same notations as before for an epoch t_i, the smoothed code pseudorange is obtained by

$$R(t_i)_{sm} = w\, R(t_i) + (1 - w)\, R(t_i)_{ex}, \tag{5.27}$$

where w is a time-dependent weight factor. Note that from the previous algorithm $R(t_i)_{ex} = R(t_{i-1})_{sm} + \Phi(t_i) - \Phi(t_{i-1})$ could be substituted into (5.27).

For the first epoch $i = 1$, the weight is set $w = 1$; thus, putting the full weight on the measured code pseudorange. For consecutive epochs, the weight of the code pseudoranges is continuously reduced and, thus, emphasizes the influence of the carrier phases. A reduction of the weight by 0.01 from epoch to epoch was tested in a kinematic experiment with a data sampling rate of 1 Hz. After 100 seconds, only the extrapolated value is taken into account. Again, in the case of cycle slips, the algorithm would fail. A simple check of the carrier phase difference for two consecutive epochs by the Doppler shift multiplied by the time interval may detect data irregularities such as cycle slips. After the occurrence of a cycle slip, the weight is reset to $w = 1$, which fully eliminates the influence of the erroneous carrier phase data. The clue of this approach is that cycle slips must be detected but do not have

5.2 Data combinations

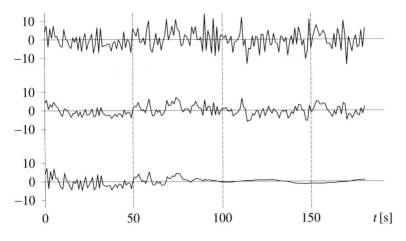

Fig. 5.2. Code pseudoranges in [m]: unsmoothed (top), smoothed (middle) by algorithm (5.26), and weighted smoothing (bottom) using Eq. (5.27)

to be repaired. Note, however, repair is possible if there is enough redundancy in the measurements.

To demonstrate the effect of the smoothing algorithm, real data are presented in Fig. 5.2. The code pseudoranges for a data sample of 170 epochs measured with a 1 Hz data rate are shown in the top graph (after eliminating the trend due to the satellite motion). In the middle graph, $R(t_i)_{sm}$ of the smoothing algorithm (5.26) is given. Finally, the bottom graph of Fig. 5.2 shows the weighting effect of (5.27). As described above, the weight reduction by 0.01 from epoch to epoch shows the decreasing influence of the code and the emphasized impact of the carrier phases. Another smoothing algorithm for code pseudoranges uses phase differences $\Delta\Phi(t_i, t_1)$ obtained by the integrated Doppler shift between the current epoch t_i and the starting epoch t_1. Note that the integrated Doppler shifts are insensitive to cycle slips. From each code pseudorange $R(t_i)$ at epoch t_i, an estimate of the code pseudorange at epoch t_1 can be given by

$$R(t_1)_i = R(t_i) - \Delta\Phi(t_i, t_1), \qquad (5.28)$$

where the subscript i on the left side of the equation indicates the epoch that the code pseudorange $R(t_1)$ is computed from. Obtaining an estimate consecutively for each epoch, the arithmetic mean $R(t_1)_m$ of the code pseudorange for n epochs is calculated by

$$R(t_1)_m = \frac{1}{n} \sum_{i=1}^{n} R(t_1)_i, \qquad (5.29)$$

and the smoothed code pseudorange for an arbitrary epoch results from

$$R(t_i)_{sm} = R(t_1)_m + \Delta\Phi(t_i, t_1). \tag{5.30}$$

The advantage of this procedure lies in the reduction of the noise in the initial code pseudorange by averaging an arbitrary number n of measured code pseudoranges. Note from the three formulas (5.28) through (5.30) that the algorithm may also be applied successively epoch by epoch where the arithmetic mean must be updated from epoch to epoch. Using the above notations, formula (5.30) also works for epoch t_1, where, of course, $\Delta\Phi(t_1, t_1)$ is zero and there is no smoothing effect.

All the smoothing algorithms are also applicable if only single-frequency data are available. In this case, $R(t_i)$, $\Phi(t_i)$, and $\Delta\Phi(t_i, t_1)$ denote the single-frequency code pseudorange, carrier phase pseudorange, and phase difference, respectively.

5.3 Atmospheric effects

5.3.1 Phase and group velocity

Consider a single electromagnetic wave propagating in space with wavelength λ and frequency f. The velocity of its phase

$$v_{ph} = \lambda f \tag{5.31}$$

is denoted phase velocity. For GNSS, the carrier waves are propagating with this velocity.

For a group of waves with slightly different frequencies, the propagation of the resultant energy is defined by the group velocity

$$v_{gr} = -\frac{df}{d\lambda} \lambda^2 \tag{5.32}$$

according to Bauer (2003: p. 106). This velocity has to be considered for GNSS code measurements.

A relation between phase and group velocity is derived by forming the total differential of Eq. (5.31) resulting in

$$dv_{ph} = f\, d\lambda + \lambda\, df, \tag{5.33}$$

which is rearranged to

$$\frac{df}{d\lambda} = \frac{1}{\lambda} \frac{dv_{ph}}{d\lambda} - \frac{f}{\lambda}. \tag{5.34}$$

The substitution of (5.34) into (5.32) yields

$$v_{gr} = -\lambda \frac{dv_{ph}}{d\lambda} + f\lambda \tag{5.35}$$

5.3 Atmospheric effects

or finally the Rayleigh equation

$$v_{gr} = v_{ph} - \lambda \frac{dv_{ph}}{d\lambda}. \tag{5.36}$$

The differentiation (5.33) implicitly contains the dispersion (Joos 1956: p. 57), which is defined as the dependence of the phase velocity on the wavelength or the frequency. Phase and group velocity are equal in nondispersive media and correspond to the speed of light in vacuum.

The wave propagation in a medium depends on the refractive index n. Generally, the propagation velocity is obtained from

$$v = \frac{c}{n}. \tag{5.37}$$

Applying this expression to the phase and group velocity, appropriate formulas for the corresponding refractive index n_{ph},

$$v_{ph} = \frac{c}{n_{ph}}, \tag{5.38}$$

and the refractive index n_{gr},

$$v_{gr} = \frac{c}{n_{gr}}, \tag{5.39}$$

are achieved. Differentiation of the phase velocity with respect to λ, that is,

$$\frac{dv_{ph}}{d\lambda} = -\frac{c}{n_{ph}^2} \frac{dn_{ph}}{d\lambda}, \tag{5.40}$$

and substitution of the last three equations into (5.36) yields

$$\frac{c}{n_{gr}} = \frac{c}{n_{ph}} + \lambda \frac{c}{n_{ph}^2} \frac{dn_{ph}}{d\lambda} \tag{5.41}$$

or

$$\frac{1}{n_{gr}} = \frac{1}{n_{ph}} \left(1 + \lambda \frac{1}{n_{ph}} \frac{dn_{ph}}{d\lambda}\right). \tag{5.42}$$

This equation may be inverted to

$$n_{gr} = n_{ph} \left(1 - \lambda \frac{1}{n_{ph}} \frac{dn_{ph}}{d\lambda}\right), \tag{5.43}$$

where the approximation $(1+\varepsilon)^{-1} \doteq 1 - \varepsilon$ has been applied accordingly. Thus,

$$n_{gr} = n_{ph} - \lambda \frac{dn_{ph}}{d\lambda} \tag{5.44}$$

is the modified Rayleigh equation. A slightly different form is obtained by differentiating the relation $c = \lambda f$ with respect to λ and f, that is,

$$\frac{d\lambda}{\lambda} = -\frac{df}{f}, \tag{5.45}$$

and by substituting the result into (5.44):

$$n_{gr} = n_{ph} + f \frac{dn_{ph}}{df}. \tag{5.46}$$

5.3.2 Ionospheric refraction

The ionosphere extends in various layers from about 50 km to 1 000 km above earth and is described in more detail in Sect. 4.1.2. It is a dispersive medium with respect to the GNSS radio signal. Following Seeber (2003: p. 54), the series

$$n_{ph} = 1 + \frac{c_2}{f^2} + \frac{c_3}{f^3} + \frac{c_4}{f^4} + \ldots \tag{5.47}$$

approximates the phase refractive index. The coefficients c_2, c_3, c_4 do not depend on frequency but on the quantity N_e denoting the number of electrons per cubic meter (i.e., the electron density) along the propagation path. Using an approximation by cutting off the series expansion after the quadratic term, that is

$$n_{ph} = 1 + \frac{c_2}{f^2}, \tag{5.48}$$

differentiating this equation leading to

$$dn_{ph} = -\frac{2c_2}{f^3} df, \tag{5.49}$$

and substituting (5.48) and (5.49) into (5.46) yields

$$n_{gr} = 1 + \frac{c_2}{f^2} - f \frac{2c_2}{f^3} \tag{5.50}$$

or

$$n_{gr} = 1 - \frac{c_2}{f^2}. \tag{5.51}$$

It can be seen from (5.48) and (5.51) that the group and the phase refractive indices deviate from unity with opposite sign. With an estimate for c_2 (Seeber 2003: p. 54),

$$c_2 = -40.3 \, N_e \quad [\text{Hz}^2], \tag{5.52}$$

5.3 Atmospheric effects

the relation $n_{gr} > n_{ph}$ and, thus, $v_{gr} < v_{ph}$ follows because the electron density N_e is always positive. As a consequence of the different velocities, a group delay and a phase advance occur. In other words, GNSS ranging codes are delayed and the carrier phases are advanced. Therefore, the measured code pseudoranges are too long and the measured carrier phase pseudoranges are too short compared to the geometric range between the satellite and the receiver. The amount of the difference is the same in both cases.

According to Fermat's principle, the measured range s is defined by

$$s = \int n \, ds, \tag{5.53}$$

where the integral must be extended along the path of the signal. The geometric range s_0 along the straight line between the satellite and the receiver may be obtained analogously by setting $n = 1$:

$$s_0 = \int ds_0. \tag{5.54}$$

The difference Δ^{Iono} between measured and geometric range is called ionospheric refraction and follows from

$$\Delta^{Iono} = \int n \, ds - \int ds_0, \tag{5.55}$$

which may be written for a phase refractive index n_{ph} from (5.48) as

$$\Delta^{Iono}_{ph} = \int \left(1 + \frac{c_2}{f^2}\right) ds - \int ds_0 \tag{5.56}$$

and for a group refractive index n_{gr} from (5.51) as

$$\Delta^{Iono}_{gr} = \int \left(1 - \frac{c_2}{f^2}\right) ds - \int ds_0. \tag{5.57}$$

A simplification is obtained when approximating the integration for the first term in (5.56) and (5.57) along the geometric range. In this case, ds becomes ds_0 and the formulas

$$\Delta^{Iono}_{ph} = \int \frac{c_2}{f^2} ds_0, \qquad \Delta^{Iono}_{gr} = -\int \frac{c_2}{f^2} ds_0 \tag{5.58}$$

result, which can also be written as

$$\Delta^{Iono}_{ph} = -\frac{40.3}{f^2} \int N_e \, ds_0, \qquad \Delta^{Iono}_{gr} = \frac{40.3}{f^2} \int N_e \, ds_0, \tag{5.59}$$

where (5.52) has been substituted. Defining the total electron content (TEC) by

$$\text{TEC} = \int N_e \, ds_0 \tag{5.60}$$

and substituting TEC into (5.59) yields

$$\Delta_{\text{ph}}^{\text{Iono}} = -\frac{40.3}{f^2} \text{TEC}, \qquad \Delta_{\text{gr}}^{\text{Iono}} = \frac{40.3}{f^2} \text{TEC} \tag{5.61}$$

as the final result (in meter). Usually, the TEC is given in TEC units (TECU), where

$$1 \text{ TECU} = 10^{16} \text{ electrons per m}^2. \tag{5.62}$$

For a numerical example, the delay $\Delta_{\text{ph}}^{\text{Iono}} = -0.18$ m is obtained if a frequency of say 1.5 GHz and one TECU is substituted.

Note that TEC as introduced in (5.60) is the total electron content along the straight signal path between the satellite and the receiver. The integral is assumed to include the electrons in a column with a cross section of 1 m² and extending from the receiver to the satellite. Usually, the total vertical electron content (TVEC) is modeled. More figuratively, this quantity is sometimes denoted as total overhead electron content. If TVEC is introduced in (5.61), the quantities apply only for satellites at zenith. For arbitrary lines of sight (Fig. 5.3), the zenith angle of the satellite must be taken into account by

$$\Delta_{\text{ph}}^{\text{Iono}} = -\frac{1}{\cos z'} \frac{40.3}{f^2} \text{TVEC}, \qquad \Delta_{\text{gr}}^{\text{Iono}} = \frac{1}{\cos z'} \frac{40.3}{f^2} \text{TVEC} \tag{5.63}$$

since the path length in the ionosphere varies with a changing zenith angle. These two quantities differ only with respect to the sign. Introducing the notation

$$\Delta^{\text{Iono}} = \frac{1}{\cos z'} \frac{40.3}{f^2} \text{TVEC} \tag{5.64}$$

for the (positive) amount of the ionospheric influence on a measured pseudorange allows the omission of the subscripts "ph" or "gr" but requires the consideration of the correct sign for the appropriate models. This means that the ionospheric influence for the code pseudorange is modeled by $+\Delta^{\text{Iono}}$ and for the phase pseudorange by $-\Delta^{\text{Iono}}$.

Figure 5.3 represents a single-layer model with the assumption that all free electrons are concentrated in an infinitesimally thin spherical shell at the height h_m and containing the ionospheric point IP. From Fig. 5.3, the relation

$$\sin z' = \frac{R_e}{R_e + h_m} \sin z_0 \tag{5.65}$$

5.3 Atmospheric effects

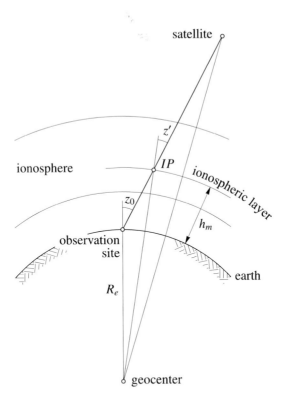

Fig. 5.3. Geometry for the ionospheric path delay

is derived, where R_e is the mean radius of the earth, h_m is a mean value for the height of the ionosphere, and z' and z_0 are the zenith angles at the ionospheric point and at the observing site. The zenith angle z_0 can be calculated for a known satellite position and approximate coordinates of the observing site. For h_m a value in the range between 300 km and 400 km is typical. The height is only sensitive for low satellite elevations.

As shown by (5.61), the change of range caused by the ionospheric refraction may be restricted to the determination of the TEC. However, the TEC itself is a fairly complicated quantity. As mentioned in Sect. 4.1.2, it depends on sunspot activities (approximately 11-year cycle), seasonal and diurnal variations, the line of sight, which includes elevation and azimuth of the satellite, and the position of the observing site. The TEC may be measured, estimated, its effect computed by models, or eliminated.

Measuring the TEC
Considering a nationwide example among many others, Japan has for a long time been very ambitious in measuring TEC. One of the experiments is based on the cor-

relation between the TEC and the critical plasma frequency provided by Japanese ionospheric observatories on an hourly basis. By interpolation, any arbitrary location in Japan can be covered.

Another Japanese experiment essentially uses the ionospheric residual as described in (5.21). More than 1 000 GNSS receivers installed in Japan establish a dense network with an average distance of 25 km between two receivers (Otsuka et al. 2002). In the three-step procedure, first an hourly average of TEC is estimated by applying a weighted least-squares adjustment individually to the data from a single receiver. Some biases are eliminated in the second step. Finally, a two-dimensional map of TVEC with a time resolution of 30 s and a spatial resolution of $0.15° \times 0.15°$ in latitude and longitude is generated. Similar experiments have been running around the world since "the ionosphere is increasingly used as a laboratory in which active plasma experiments are performed" (Stubbe 1996).

In the global sense, the measuring of TEC is covered by the more general topic of atmospheric monitoring as described in Sect. 5.3.4.

Computing the effect of the TEC

Klobuchar model

Here, the entire vertical ionospheric refraction is approximated by the model of Klobuchar (1986) and yields the vertical time delay for the code measurements. Although the model is an approximation, it is nevertheless of importance because it uses the ionospheric coefficients broadcast within the navigation message. The Klobuchar model is

$$\Delta T_v^{\text{Iono}} = A_1 + A_2 \cos\left(\frac{2\pi(t - A_3)}{A_4}\right), \quad (5.66)$$

where

$$A_1 = 5 \cdot 10^{-9} \text{ s} = 5 \text{ ns},$$
$$A_2 = \alpha_1 + \alpha_2\, \varphi_{IP}^m + \alpha_3\, \varphi_{IP}^{m\,2} + \alpha_4\, \varphi_{IP}^{m\,3},$$
$$A_3 = 14^h \text{ local time},$$
$$A_4 = \beta_1 + \beta_2\, \varphi_{IP}^m + \beta_3\, \varphi_{IP}^{m\,2} + \beta_4\, \varphi_{IP}^{m\,3}. \quad (5.67)$$

The values for A_1 and A_3 are constant, the coefficients $\alpha_i, \beta_i, i = 1,\ldots,4$, are uploaded to the satellites and broadcast to the user. The parameter t in (5.66) is the local time of the ionospheric point IP (Fig. 5.3) and may be derived from

$$t = \frac{\lambda_{IP}}{15} + t_{\text{UT}}, \quad (5.68)$$

5.3 Atmospheric effects

where λ_{IP} is the geomagnetic longitude positive to east for the ionospheric point in degrees and t_{UT} is the observation epoch in universal time (UT). Finally, φ_{IP}^m in Eq. (5.67) is the spherical distance between the geomagnetic pole and the ionospheric point. Denoting the coordinates of the geomagnetic pole by φ_P, λ_P and those of the ionospheric point by φ_{IP}, λ_{IP}, then φ_{IP}^m is obtained by

$$\cos \varphi_{IP}^m = \sin \varphi_{IP} \sin \varphi_P + \cos \varphi_{IP} \cos \varphi_P \cos(\lambda_{IP} - \lambda_P), \tag{5.69}$$

where the coordinates of the geomagnetic pole are

$$\begin{aligned} \varphi_P &= 78.3° \text{ N}, \\ \lambda_P &= 291.0° \text{ E}. \end{aligned} \tag{5.70}$$

Summarizing, the evaluation of the Klobuchar model may be performed by the following steps:

- Compute the azimuth a and the zenith angle z_0 of the satellite for epoch t_{UT}.
- Choose a mean height of the ionosphere and compute the distance s between the observing site and the ionospheric point obtained from the triangle formed by geocenter – observing site – ionospheric point (Fig. 5.3).
- Compute the coordinates φ_{IP}, λ_{IP} of the ionospheric point by means of the quantities a, z_0, s.
- Calculate φ_{IP}^m from (5.69).
- Calculate A_2 and A_4 from (5.67), where the coefficients α_i, β_i, $i = 1, \ldots, 4$, are received via the satellite navigation message.
- Use (5.67) and (5.68) and compute the vertical (or zenith) delay ΔT_v^{Iono} by (5.66).
- By calculating z' from (5.65) and applying $\Delta T^{\text{Iono}} = \Delta T_v^{\text{Iono}} / \cos z'$, the transition from the vertical delay to the delay along the wave path is achieved. The result is obtained as a time delay in seconds which must be multiplied by the speed of light to convert it to a measure of change in range.

By taking into account the Klobuchar model, the influence of the ionospheric refraction is reduced by at least 50% (ARINC Engineering Services 2006a).

NeQuick model
The Aeronomy and Radiopropagation Laboratory (ARPL) of the Abdus Salam International Centre for Theoretical Physics in Trieste (Italy) and the Institute for Geophysics, Astrophysics and Meteorology of the University of Graz (Austria) have developed NeQuick, a three-dimensional, time-dependent ionospheric electron density model.

Since the model enables the calculation of the electron concentration at any given location in the ionosphere, the vertical or slant electronic density profile and TEC may easily be computed along any satellite-to-receiver path as required for GNSS.

The International Telecommunication Union, Radiocommunication (ITU-R) sector has adopted in 2001 the NeQuick model as suitable method for TEC modeling. NeQuick is an ionospheric electron concentration model able to give the electron density distribution on both the bottomside and topside of the ionosphere. The model input parameters are position, time, and solar flux; the output is the electron concentration at the given location and time.

The NeQuick model uses monthly average values of solar activity either expressed by the 12-month running mean sunspot number R_{12} or by the average 10.7 cm solar radio flux $F_{10.7}$. The two quantities are interrelated by $R_{12} = (F_{10.7} - 57)/0.93$.

Supplementing explanations according to R. Leitinger and S. Radicella are given in the software documentation for the NeQuick model, which is freely available from ITU-R.

When running the software, the NeQuick model has to be used once for each new set of season (input quantity is month), time (input quantity is UT), and solar activity (input data is solar radio flux $F_{10.7}$). After this, TEC may be computed for any latitude φ, longitude λ, and height h.

The available software also allows for the electronic density profile calculation for a path between a satellite and a ground station. At http://arpl.ictp.it/nq-online/index.html, the Web site of ARPL, an online calculation may be performed resulting in the electronic density profile as a function of altitude.

The solar flux (or the sunspot number, respectively) may be replaced by the effective ionization parameter Az. Three coefficients a_0, a_1, a_2 determine the effective ionization parameter by

$$Az = a_0 + a_1\mu + a_2\mu^2, \tag{5.71}$$

where μ is the modified magnetic dip obtained from the true magnetic dip I and the latitude φ of the site of interest by

$$\tan\mu = \frac{I}{\sqrt{\cos\varphi}}. \tag{5.72}$$

The magnetic dip I is also denoted as magnetic inclination (which is 0° at the magnetic equator and 90° at each of the magnetic poles).

In the framework of the European Galileo project, the NeQuick model has been proposed to be used for single-frequency positioning. Arbesser-Rastburg (2006) structures the possible procedure:

5.3 Atmospheric effects

- Sensor stations: the primary task of the globally distributed network of sensor stations is to continuously measure slant TEC and to optimize the effective ionization parameters required for the NeQuick model.
- Satellites: at least once a day, the satellites get an upload of the effective ionization parameters. This information is sent to the users in the navigation message.
- Users: from the navigation message, the receiver extracts the coefficients a_0, a_1, a_2 necessary to calculate the effective ionization parameter Az. This input is required for the NeQuick model to calculate the slant TEC (so that the respective pseudorange may be corrected).

Arbesser-Rastburg (2006) and Arbesser-Rastburg and Jakowski (2007) give even more detailed information with respect to the GNSS receiver inputs and the algorithm to be performed.

The receiver retrieves from the navigation message as input data

- the coefficients a_0, a_1, a_2 for the effective ionization parameter Az,
- the ionospheric disturbance flags (alert the user that the ionospheric correction coming from the navigation message might not meet the specified performance),
- the actual time (UT and month of the year),
- the satellite position (calculable from the Keplerian elements),
- the receiver estimated position (before ionospheric correction).

The receiver retrieves from its internal firmware as additional input data

- information on the earth's magnetic field (which should be updated every five years to account for the variation of the earth's magnetic field),
- the ITU-R maps in twelve files, one for each calendar month.

The algorithm carried out in the receiver consists of the following steps:

1. The receiver position φ, λ, h is estimated using pseudoranges (without ionospheric correction).
2. Based on the latitude φ and on the longitude λ of the receiver position, the magnetic dip I is computed with the internal firmware information on the earth's magnetic field using a third-order interpolation procedure.
3. The modified dip μ is computed using (5.72).
4. The effective ionization parameter Az is computed using (5.71).
5. Based on Az and using the NeQuick model, the electron density is calculated for a point along the satellite–receiver path.

6. All previous steps are repeated for many discrete points along the satellite–receiver path. The number and spacing of the points will be a trade-off between integration error and computational time.

7. All electron density values along the satellite–receiver ray are integrated numerically in order to obtain the slant TEC.

8. The slant TEC is converted to Δ^{Iono} using (5.61) to get a slant delay being used to correct pseudoranges.

Leitinger et al. (2005) and Nava et al. (2005) describe a modification of the NeQuick model: the topside formulation is now based on an empirical parameter that does not depend on the month of the year; furthermore, a new mapping procedure is introduced so that simplified ITU-R maps may be used, and some other improvements are implied.

Eliminating the effect of the TEC

It is difficult to find a satisfying model for the TEC because of the various time-dependent influences. The most efficient method is to eliminate the ionospheric refraction by using two signals with different frequencies. This dual-frequency method is the main reason why the GNSS satellites emit (at least) two carrier waves.

Starting with the phase pseudorange model (5.10) and taking into account the frequency-dependent ionospheric refraction (5.64) gives

$$\lambda_1 \Phi_1 = \varrho + c \Delta\delta + \lambda_1 N_1 - \Delta_1^{\text{Iono}},$$
$$\lambda_2 \Phi_2 = \varrho + c \Delta\delta + \lambda_2 N_2 - \Delta_2^{\text{Iono}},$$
(5.73)

where the subscripts 1 and 2 indicate the dependence on the respective frequency of the two carriers. After dividing by the corresponding wavelengths,

$$\Phi_1 = \frac{1}{\lambda_1} \varrho + \frac{c}{\lambda_1} \Delta\delta + N_1 - \frac{1}{\lambda_1} \Delta_1^{\text{Iono}},$$
$$\Phi_2 = \frac{1}{\lambda_2} \varrho + \frac{c}{\lambda_2} \Delta\delta + N_2 - \frac{1}{\lambda_2} \Delta_2^{\text{Iono}}$$
(5.74)

are obtained. Using the relation $c = f \lambda$ yields

$$\Phi_1 = \frac{f_1}{c} \varrho + f_1 \Delta\delta + N_1 - \frac{f_1}{c} \Delta_1^{\text{Iono}},$$
$$\Phi_2 = \frac{f_2}{c} \varrho + f_2 \Delta\delta + N_2 - \frac{f_2}{c} \Delta_2^{\text{Iono}},$$
(5.75)

5.3 Atmospheric effects

which can be written in the form

$$\Phi_1 = af_1 + N_1 - \frac{b}{f_1},$$
$$\Phi_2 = af_2 + N_2 - \frac{b}{f_2} \tag{5.76}$$

by introducing

$$a = \frac{\varrho}{c} + \Delta\delta \qquad \text{geometry term},$$
$$b = \frac{f_i^2}{c} \Delta^{\text{Iono}} = \frac{1}{c} \frac{40.3}{\cos z'} \text{TVEC} \qquad \text{ionosphere term}, \tag{5.77}$$

where the second expression for b may be verified by substituting Eq. (5.64). Note that the auxiliary quantities a and b are frequency independent. Therefore no subscript is needed to specify the frequency.

The ionosphere term can be eliminated by the following linear combination. Multiplying the first equation of (5.76) by f_1 and the second by f_2 and forming the difference yields

$$\Phi_1 f_1 - \Phi_2 f_2 = a(f_1^2 - f_2^2) + N_1 f_1 - N_2 f_2 \tag{5.78}$$

and, after multiplying the equation by $f_1/(f_1^2 - f_2^2)$ and a slight rearrangement, the ionosphere-free combination

$$\left[\Phi_1 - \frac{f_2}{f_1}\Phi_2\right] \frac{f_1^2}{f_1^2 - f_2^2} = af_1 + \left[N_1 - \frac{f_2}{f_1}N_2\right] \frac{f_1^2}{f_1^2 - f_2^2} \tag{5.79}$$

is obtained. Resubstituting for the geometry term a according to (5.77),

$$\left[\Phi_1 - \frac{f_2}{f_1}\Phi_2\right] \frac{f_1^2}{f_1^2 - f_2^2} = \frac{f_1}{c}\varrho + f_1 \Delta\delta + \left[N_1 - \frac{f_2}{f_1}N_2\right] \frac{f_1^2}{f_1^2 - f_2^2} \tag{5.80}$$

results for the ionosphere-free combination. The significant drawback of the combination is that the integer nature of the ambiguities is lost since f_2/f_1 is not an integer for current GNSS. Note that on the left side of the equation the geometric residual reappears, cf. Eq. (5.19). Thus, this quantity could also be denoted as reduced ionosphere-free signal.

The derivation of the ionosphere-free combination for code pseudoranges starts with the model equations

$$R_1 = \varrho + c\Delta\delta + \Delta_1^{\text{Iono}},$$
$$R_2 = \varrho + c\Delta\delta + \Delta_2^{\text{Iono}}, \tag{5.81}$$

where Δ^{Iono} is inversely proportional to the squared respective carrier frequency, cf. Eq. (5.64). Thus, multiplying the first equation of (5.81) by f_1^2 and the second by f_2^2 and then forming the difference yields

$$R_1 f_1^2 - R_2 f_2^2 = (f_1^2 - f_2^2)(\varrho + c\,\Delta\delta)\,, \tag{5.82}$$

where the ionosphere term is eliminated. After dividing the equation by $(f_1^2 - f_2^2)$ and a slight rearrangement, the ionosphere-free combination

$$\left[R_1 - \frac{f_2^2}{f_1^2} R_2\right] \frac{f_1^2}{f_1^2 - f_2^2} = \varrho + c\,\Delta\delta \tag{5.83}$$

is obtained.

The advantage of the ionosphere-free combination is the elimination (or more precisely, the reduction) of ionospheric effects. Remembering the derivation, it should be clear that the term "ionosphere-free" is not fully correct because there are some approximations involved, for instance, Eq. (5.48) or the integration is not carried out along the true signal path in (5.58). Brunner and Gu (1991) propose an improved model to account for the higher-order terms arising from the series expansion of the refractive index, the geomagnetic field effect, and the bending effects of the ray paths.

5.3.3 Tropospheric refraction

The effect of the neutral atmosphere (i.e., the nonionized part) is denoted as tropospheric refraction, tropospheric path delay, or simply tropospheric delay. The naming is slightly incorrect because the name excludes the stratosphere, which is another constituent of the neutral atmosphere. However, the dominant contribution of the troposphere explains the choice of the name.

The neutral atmosphere is a nondispersive medium with respect to radio waves up to frequencies of 15 GHz. Thus, the propagation is frequency independent. Consequently, a distinction between carrier phases and code ranges derived from different carriers is not necessary. The disadvantage is that an elimination of the tropospheric refraction by dual-frequency methods is not possible.

The tropospheric path delay is defined by

$$\Delta^{\text{Trop}} = \int (n - 1)\,ds_0\,, \tag{5.84}$$

which is analogous to the ionospheric formula (5.55), where again an approximation is introduced so that the integration is performed along the geometric range. Usually, instead of the refractive index n the refractivity

$$N^{\text{Trop}} = 10^6\,(n - 1) \tag{5.85}$$

5.3 Atmospheric effects

is used so that Eq. (5.84) becomes

$$\Delta^{\text{Trop}} = 10^{-6} \int N^{\text{Trop}} \, ds_0 \,. \tag{5.86}$$

Hopfield (1969) shows the possibility of separating N^{Trop} into a dry and a wet component,

$$N^{\text{Trop}} = N_d^{\text{Trop}} + N_w^{\text{Trop}} \,, \tag{5.87}$$

where the dry part results from the dry (hydrostatic) atmosphere and the wet part from the water vapor. Correspondingly, the relations

$$\Delta_d^{\text{Trop}} = 10^{-6} \int N_d^{\text{Trop}} \, ds_0 \,, \tag{5.88}$$

$$\Delta_w^{\text{Trop}} = 10^{-6} \int N_w^{\text{Trop}} \, ds_0 \,, \tag{5.89}$$

and

$$\Delta^{\text{Trop}} = \Delta_d^{\text{Trop}} + \Delta_w^{\text{Trop}}$$
$$= 10^{-6} \int N_d^{\text{Trop}} \, ds_0 + 10^{-6} \int N_w^{\text{Trop}} \, ds_0 \tag{5.90}$$

are obtained. As mentioned in Sect. 4.1.2, about 90% of the tropospheric refraction arises from the dry and about 10% from the wet component. In practice, models for the refractivities are introduced in Eq. (5.90) and the integration is performed by numerical methods or analytically by series expansions of the integrand. Models for the dry and wet refractivity at the surface of the earth have been known for some time (e.g., Essen and Froome 1951). The corresponding dry component on the surface (indicated by the subscript 0) is

$$N_{d,0}^{\text{Trop}} = \bar{c}_1 \frac{p}{T} \,, \qquad \bar{c}_1 = 77.64 \ \text{K mb}^{-1} \,, \tag{5.91}$$

where p is the atmospheric pressure in units of millibar (mb) and T is the temperature in kelvin (K). The wet component on the surface was found to be

$$N_{w,0}^{\text{Trop}} = \bar{c}_2 \frac{e}{T} + \bar{c}_3 \frac{e}{T^2} \,, \qquad \bar{c}_2 = -12.96 \ \text{K mb}^{-1} \,,$$
$$\bar{c}_3 = 3.718 \cdot 10^5 \ \text{K}^2 \text{mb}^{-1} \,, \tag{5.92}$$

where e is the partial pressure of water vapor in mb and T again the temperature in K. The overbar in the coefficients only stresses that there is absolutely no relationship to the coefficients for the ionosphere in, e.g., (5.47).

The values for $\bar{c}_1$, $\bar{c}_2$, and $\bar{c}_3$ are empirically determined and, certainly, cannot fully describe the local situation. An improvement is obtained by measuring meteorological data at the observation site. The following paragraphs present several models where meteorological surface data are taken into account.

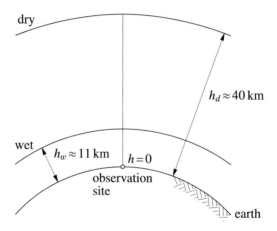

Fig. 5.4. Thickness of polytropic layers for the troposphere

Hopfield model

Using real data covering the whole earth, Hopfield (1969) has empirically found a representation of the dry refractivity as a function of the height h above the surface by

$$N_d^{\text{Trop}}(h) = N_{d,0}^{\text{Trop}} \left[\frac{h_d - h}{h_d} \right]^4 \tag{5.93}$$

under the assumption of a polytropic layer with thickness

$$h_d = 40\,136 + 148.72\,(T - 273.16) \quad [\text{m}], \tag{5.94}$$

as shown in Fig. 5.4. Substitution of (5.93) into (5.88) yields (for the dry part) the tropospheric path delay

$$\Delta_d^{\text{Trop}} = 10^{-6}\, N_{d,0}^{\text{Trop}} \int \left[\frac{h_d - h}{h_d} \right]^4 ds_0. \tag{5.95}$$

The integral can be solved if the delay is calculated along the vertical direction and if the curvature of the signal path is neglected. Thus, Eq. (5.95) becomes

$$\Delta_d^{\text{Trop}} = 10^{-6}\, N_{d,0}^{\text{Trop}} \frac{1}{h_d^4} \int_0^{h_d} (h_d - h)^4\, dh, \tag{5.96}$$

where the lower limit $h = 0$ corresponds to an observation site on the surface of the earth and where the constant denominator has been extracted. After integration,

$$\Delta_d^{\text{Trop}} = 10^{-6}\, N_{d,0}^{\text{Trop}} \frac{1}{h_d^4} \left[-\frac{1}{5}(h_d - h)^5 \Big|_{h=0}^{h=h_d} \right] \tag{5.97}$$

5.3 Atmospheric effects

is obtained. The evaluation of the expression between the brackets gives $h_d^5/5$ so that

$$\Delta_d^{\text{Trop}} = \frac{10^{-6}}{5} N_{d,0}^{\text{Trop}} h_d \tag{5.98}$$

is the dry portion of the tropospheric zenith delay.

The wet portion is more difficult to model because of the strong variations of the water vapor with respect to time and space. Nevertheless, due to lack of an appropriate alternative, the Hopfield model assumes the same functional model for both the wet and dry components. Thus,

$$N_w^{\text{Trop}}(h) = N_{w,0}^{\text{Trop}} \left[\frac{h_w - h}{h_w} \right]^4, \tag{5.99}$$

where the mean value

$$h_w = 11\,000 \text{ m} \tag{5.100}$$

is used. Sometimes other values such as $h_w = 12\,000$ m have been proposed. Unique values for h_d and h_w cannot be given because of their dependence on location and temperature. In Germany, a local model for estimating the tropospheric path delay at microwave frequencies using radiosonde data over 4.5 years yielded for the region of the observation site $h_d = 41.6$ km and $h_w = 11.5$ km. The effective troposphere heights are given as $40 \text{ km} \le h_d \le 45 \text{ km}$ and $10 \text{ km} \le h_w \le 13 \text{ km}$.

The integration of (5.99) is completely analogous to (5.95) and results in

$$\Delta_w^{\text{Trop}} = \frac{10^{-6}}{5} N_{w,0}^{\text{Trop}} h_w . \tag{5.101}$$

Therefore, the total tropospheric zenith delay is

$$\Delta^{\text{Trop}} = \frac{10^{-6}}{5} \left[N_{d,0}^{\text{Trop}} h_d + N_{w,0}^{\text{Trop}} h_w \right] \tag{5.102}$$

in units of meters. The model in its present form does not account for an arbitrary zenith angle of the signal. Considering the line of sight, an obliquity factor must be applied which, in its simplest form, is the projection from the zenith onto the line of sight. Frequently, the transition of the zenith delay to a delay with arbitrary zenith angle is denoted as the application of a mapping function.

Introducing the mapping function, Eq. (5.102) becomes

$$\Delta^{\text{Trop}} = \frac{10^{-6}}{5} \left[N_{d,0}^{\text{Trop}} h_d\, m_d(E) + N_{w,0}^{\text{Trop}} h_w\, m_w(E) \right], \tag{5.103}$$

where $m_d(E)$ and $m_w(E)$ are the mapping functions for the dry and the wet part and E (expressed in degrees) indicates the elevation angle at the observing site (where the line of sight is simplified as straight line). Explicitly,

$$m_d(E) = \frac{1}{\sin\sqrt{E^2 + 6.25}},$$

$$m_w(E) = \frac{1}{\sin\sqrt{E^2 + 2.25}}$$

(5.104)

are the mapping functions for the Hopfield model. In more compact form, (5.103) is represented as

$$\Delta^{\text{Trop}}(E) = \Delta_d^{\text{Trop}}(E) + \Delta_w^{\text{Trop}}(E),$$

(5.105)

where the terms on the right side of the equation are given by

$$\Delta_d^{\text{Trop}}(E) = \frac{10^{-6}}{5} \frac{N_{d,0}^{\text{Trop}} h_d}{\sin\sqrt{E^2 + 6.25}},$$

$$\Delta_w^{\text{Trop}}(E) = \frac{10^{-6}}{5} \frac{N_{w,0}^{\text{Trop}} h_w}{\sin\sqrt{E^2 + 2.25}}$$

(5.106)

or, after substituting (5.91), (5.94) and (5.92), (5.100) respectively, by

$$\Delta_d^{\text{Trop}}(E) = \frac{10^{-6}}{5} \frac{77.64}{\sin\sqrt{E^2 + 6.25}} \frac{p}{T} [40\,136 + 148.72\,(T - 273.16)],$$

$$\Delta_w^{\text{Trop}}(E) = \frac{10^{-6}}{5} \frac{(-12.96\,T + 3.718 \cdot 10^5)}{\sin\sqrt{E^2 + 2.25}} \frac{e}{T^2} 11\,000.$$

(5.107)

Measuring p, T, e at the observation location and calculating the elevation angle E, the total tropospheric path delay is obtained in meters by (5.105) after evaluating (5.107).

Modified Hopfield models

The empirical function (5.93) is now rewritten by introducing lengths of position vectors instead of heights. Denoting the radius of the earth by R_e, the corresponding

5.3 Atmospheric effects

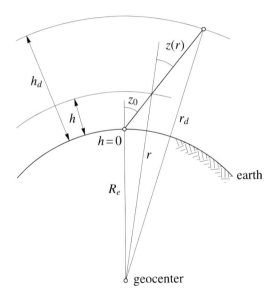

Fig. 5.5. Geometry for the tropospheric path delay

lengths are $r_d = R_e + h_d$ and $r = R_e + h$ (Fig. 5.5). Thus, the dry refractivity in the form

$$N_d^{\text{Trop}}(r) = N_{d,0}^{\text{Trop}} \left[\frac{r_d - r}{r_d - R_e} \right]^4 \tag{5.108}$$

is equivalent to (5.93). Applying Eq. (5.88) and introducing the mapping function $1/\cos z(r)$ gives the dry path delay

$$\Delta_d^{\text{Trop}}(z) = 10^{-6} \int_{R_e}^{r_d} N_d^{\text{Trop}}(r) \frac{1}{\cos z(r)} \, dr \tag{5.109}$$

for an observation site on the surface of the earth. Note that the zenith angle $z(r)$ is variable. Denoting the zenith angle at the observation site by z_0, the sine-law

$$\sin z(r) = \frac{R_e}{r} \sin z_0 \tag{5.110}$$

can be applied (Fig. 5.5). From Eq. (5.110) follows

$$\cos z(r) = \sqrt{1 - \frac{R_e^2}{r^2} \sin^2 z_0}, \tag{5.111}$$

which is equivalent to

$$\cos z(r) = \frac{1}{r}\sqrt{r^2 - R_e^2 \sin^2 z_0}. \tag{5.112}$$

Substituting (5.112) and (5.108) into (5.109) yields

$$\Delta_d^{\text{Trop}}(z) = \frac{10^{-6} N_{d,0}^{\text{Trop}}}{(r_d - R_e)^4} \int_{R_e}^{r_d} \frac{r(r_d - r)^4}{\sqrt{r^2 - R_e^2 \sin^2 z_0}} \, dr, \tag{5.113}$$

where the terms being constant with respect to the integration variable r have been extracted from the integral. Assuming the same model for the wet portion, the corresponding formula is given by

$$\Delta_w^{\text{Trop}}(z) = \frac{10^{-6} N_{w,0}^{\text{Trop}}}{(r_w - R_e)^4} \int_{R_e}^{r_w} \frac{r(r_w - r)^4}{\sqrt{r^2 - R_e^2 \sin^2 z_0}} \, dr. \tag{5.114}$$

Instead of the zenith angle z the elevation angle $E = 90° - z$ can also be used. Many modified Hopfield models have been derived, depending solely on the method to solve the integral. Here, one model is presented based on a series expansion of the integrand. The resulting formulas can be found, e.g., in Remondi (1984), where a subscript i is introduced which reflects either the dry component (replace i by d) or the wet component (replace i by w). With

$$r_i = \sqrt{(R_e + h_i)^2 - (R_e \cos E)^2} - R_e \sin E \tag{5.115}$$

the tropospheric delay in meters is

$$\Delta_i^{\text{Trop}}(E) = 10^{-12} N_{i,0}^{\text{Trop}} \left[\sum_{k=1}^{9} \frac{\alpha_{k,i}}{k} r_i^k \right], \tag{5.116}$$

where

$$\alpha_{1,i} = 1, \qquad \alpha_{6,i} = 4a_i b_i (a_i^2 + 3b_i),$$

$$\alpha_{2,i} = 4a_i, \qquad \alpha_{7,i} = b_i^2 (6a_i^2 + 4b_i),$$

$$\alpha_{3,i} = 6a_i^2 + 4b_i, \qquad \alpha_{8,i} = 4a_i b_i^3, \tag{5.117}$$

$$\alpha_{4,i} = 4a_i (a_i^2 + 3b_i), \qquad \alpha_{9,i} = b_i^4,$$

$$\alpha_{5,i} = a_i^4 + 12a_i^2 b_i + 6b_i^2,$$

5.3 Atmospheric effects

and

$$a_i = -\frac{\sin E}{h_i}, \qquad b_i = -\frac{\cos^2 E}{2 h_i R_e}. \tag{5.118}$$

Substituting i by d, the dry part results, for which in (5.116) for $N_{d,0}^{\text{Trop}}$ Eq. (5.91) and for h_d Eq. (5.94) must be introduced. Analogously, Eqs. (5.92) and (5.100) must be used for $N_{w,0}^{\text{Trop}}$ and for h_w.

Saastamoinen model

The refractivity can alternatively be deduced from gas laws. The Saastamoinen model is based on this approach where again some approximations have been employed (Saastamoinen 1973). Here, any theoretical derivation is omitted. Saastamoinen models the tropospheric delay, expressed in meters,

$$\Delta^{\text{Trop}} = \frac{0.002277}{\cos z} \left[p + \left(\frac{1255}{T} + 0.05 \right) e - \tan^2 z \right], \tag{5.119}$$

as a function of z, p, T and e. As before, z denotes the zenith angle of the satellite, p the atmospheric pressure in millibar, T the temperature in kelvin, and e the partial pressure of water vapor in millibar. A numerical assessment using parameters of a standard atmosphere at sea level ($p = 1013.25$ millibar, $T = 273.16$ kelvin, and $e = 0$ millibar) results in a tropospheric zenith delay of about 2.3 m.

Saastamoinen has refined this model by taking into account two correction terms, one being dependent on the height of the observing site and the other one on both the height and the zenith angle. The refined formula is

$$\Delta^{\text{Trop}} = \frac{0.002277}{\cos z} \left[p + \left(\frac{1255}{T} + 0.05 \right) e - B \tan^2 z \right] + \delta R, \tag{5.120}$$

where the correction terms B, δR are interpolated from Tables 5.4 and 5.5.

Models using the mapping function of Marini

In 1972, Marini developed a continued fraction of the mapping function. Herring (1992) specified this function with three constants and normalized to unity at the zenith. For the dry component, the mapping function

$$m_d(E) = \frac{1 + \dfrac{a_d}{1 + \dfrac{b_d}{1 + c_d}}}{\sin E + \dfrac{a_d}{\sin E + \dfrac{b_d}{\sin E + c_d}}} \tag{5.121}$$

Table 5.4. Correction term B for the refined Saastamoinen model

Height [km]	B [mb]
0.0	1.156
0.5	1.079
1.0	1.006
1.5	0.938
2.0	0.874
2.5	0.813
3.0	0.757
4.0	0.654
5.0	0.563

is used, where the coefficients are defined as

$$a_d = [1.2320 + 0.0139 \cos \varphi - 0.0209\, h + 0.00215\, (T - 283)] \cdot 10^{-3},$$
$$b_d = [3.1612 - 0.1600 \cos \varphi - 0.0331\, h + 0.00206\, (T - 283)] \cdot 10^{-3},$$
$$c_d = [71.244 - 4.293 \cos \varphi - 0.149\, h - 0.0021\, (T - 283)] \cdot 10^{-3}$$

(5.122)

depending on the latitude φ and height h in kilometer of the observing site and on the temperature T in kelvin.

For the wet part, the mapping function is the same as in (5.121) but the subscript d must be replaced by w. The corresponding coefficients are obtained as

$$a_w = [0.583 - 0.011 \cos \varphi - 0.052\, h + 0.0014\, (T - 283)] \cdot 10^{-3},$$
$$b_w = [1.402 - 0.102 \cos \varphi - 0.101\, h + 0.0020\, (T - 283)] \cdot 10^{-3},$$
$$c_w = [45.85 - 1.91 \cos \varphi - 1.29\, h + 0.015\, (T - 283)] \cdot 10^{-3}.$$

(5.123)

Niell (1996) uses the same type of mapping function as Herring, i.e., the continued fraction of the Marini mapping function restricted to three coefficients. The coefficients for the dry part depend on the latitude and the height at the observing site and on the day of the year, whereas the coefficients for the wet part depend only on the site latitude. Numerical values of the coefficients are given for some specific latitudes in Niell (1996). Interpolation must be used to obtain the coefficients for arbitrary latitudes and days.

5.3 Atmospheric effects

Table 5.5. Correction term δR [m] for the refined Saastamoinen model

Zenith angle	Station height above sea level [km]							
	0	0.5	1.0	1.5	2.0	3.0	4.0	5.0
60°00′	0.003	0.003	0.002	0.002	0.002	0.002	0.001	0.001
66°00′	0.006	0.006	0.005	0.005	0.004	0.003	0.003	0.002
70°00′	0.012	0.011	0.010	0.009	0.008	0.006	0.005	0.004
73°00′	0.020	0.018	0.017	0.015	0.013	0.011	0.009	0.007
75°00′	0.031	0.028	0.025	0.023	0.021	0.017	0.014	0.011
76°00′	0.039	0.035	0.032	0.029	0.026	0.021	0.017	0.014
77°00′	0.050	0.045	0.041	0.037	0.033	0.027	0.022	0.018
78°00′	0.065	0.059	0.054	0.049	0.044	0.036	0.030	0.024
78°30′	0.075	0.068	0.062	0.056	0.051	0.042	0.034	0.028
79°00′	0.087	0.079	0.072	0.065	0.059	0.049	0.040	0.033
79°30′	0.102	0.093	0.085	0.077	0.070	0.058	0.047	0.039
79°45′	0.111	0.101	0.092	0.083	0.076	0.063	0.052	0.043
80°00′	0.121	0.110	0.100	0.091	0.083	0.068	0.056	0.047

The transition to tropospheric models is achieved by substituting the mapping functions given in this paragraph into models for the zenith delay, e.g., Eq. (5.103).

Tropospheric problems

There are many other tropospheric models which are similar to the models given here. Janes et al. (1991) and Spilker (1996b) analyze several other tropospheric models. The question arises why there are so many different approaches. One reason is the difficulty in modeling the water vapor. The simple use of surface measurements cannot give the utmost accuracy so that water vapor radiometers have been developed. These instruments measure the sky brightness temperature by radiometric microwave observations along the signal path enabling the calculation of the wet path delay. Accurate water vapor radiometers are expensive and experience problems at low elevation angles since the tropospheric zenith delay is amplified by the mapping function.

The difficulty in modeling the tropospheric effect will require continuation of research and development for many years. One solution is to combine surface and radiosonde meteorological data, water vapor radiometer measurements and statistics. This is a major task and an appropriate model has not yet been found.

Any standard model suffers from the estimation of the zenith delay from measured ground parameters. Another approach is to estimate the zenith delay in the least-squares adjustment of the phase observations in a network. Some processing software programs offer this option. Usually one zenith delay for each site and session is estimated; however, it is good practice to estimate more than one zenith delay per session (Brunner and Welsch 1993).

For local networks, the strong correlation between the tropospheric delay and the station height may be exploited if, apart from the reference station, a calibration station is introduced. This is especially meaningful for landslide monitoring where the reference and the calibration station must be situated in stable bedrock. Rührnößl et al. (1998) describe a correction model yielding a height correction term being calculated from the reference and the calibration station and assuming a linear behavior of the tropospheric delay between the two stations. More details are found in Gassner and Brunner (2003) and Schön et al. (2005).

5.3.4 Atmospheric monitoring

Ionospheric tomography

Tomography has developed from a medical diagnostics tool, commonly denoted as computer tomography, to become an imaging technique for many applications, including geodesy and geophysics (Leitinger 1996). Referring to ionospheric tomography, the line integral of electron density, i.e., the TEC, is measured over a large number of ray paths transitioning the ionosphere. This dataset is inverted to produce an image of electron density in ionosphere maps.

More GNSS-specifically, TEC monitoring is possible using satellite-based positioning systems (Jakowski 1996). As one representative example, the Center for Orbit Determination in Europe (CODE) estimates global ionosphere maps (GIM) as an additional product since January 1, 1996. The main idea is to analyze the ionospheric residuals of dual-frequency phases, cf. (5.21), which contain the information on ionospheric refraction. Following closely Schaer (1997, 1999), the TEC is developed into a series of spherical harmonics adopting a single-layer model in a sun-fixed reference frame. For each day, a set of TEC coefficients is determined which approximates an average distribution of the vertical TEC on a global scale. The GIM produced may contribute to improve the ambiguity resolution, as demonstrated in the CODE processing. Also spaceborne applications, e.g., satellite altimetry, may benefit from the TEC maps. For ionosphere physicists, these maps are an alternative source of information about the deterministic behavior of the ionosphere that may be correlated with solar and geomagnetic parameters and compared to theoretical ionosphere models. All details on how to use the maps are given under www.cx.unibe.ch/aiub/igs.html.

The GIM are based on the single-layer model in Fig. 5.3. The electron density

5.3 Atmospheric effects

of the surface is modeled by

$$\text{TVEC}(\beta, \Delta\lambda) = \sum_{n=0}^{n_{max}} \sum_{m=0}^{n} [a_{nm} \cos m\Delta\lambda + b_{nm} \sin m\Delta\lambda] \bar{P}_{nm}(\sin\beta) \quad (5.124)$$

yielding the total vertical electron content as a function of the ionospheric point expressed by the geocentric latitude β and the longitude difference $\Delta\lambda = \lambda - \lambda_0$ between the earth-fixed longitude λ and the longitude of the sun λ_0. The coefficients a_{nm} and b_{nm} are the coefficients to be determined representing the parameters of the GIM. Finally, $\bar{P}_{nm}(\sin\beta)$ are the fully normalized associated Legendre functions of degree n and order m (Hofmann-Wellenhof and Moritz 2006: Sects. 1.7 and 1.10).

If the solar-geographical reference frame refers to the mean sun (i.e., a fictitious sun uniformly rotating in the equator), the geographic longitude of the sun may be written in function of the universal time (UT)

$$\lambda_0 = 12^h - \text{UT} \quad (5.125)$$

and the latitude of the sun is set to zero. Note that $\text{TVEC}(\beta, \Delta\lambda)$ may equivalently be expressed in the solar-geomagnetic frame.

The global ionosphere maps are generated on a daily basis by CODE. The TEC (more precisely, TVEC) is modeled with a spherical harmonic expansion up to degree $n = 12$ and order $m = 8$ referring to a solar-geomagnetic reference frame. Each day, twelve 2-hour sets are derived from data of the global IGS network (Sec. 3.4.1). From Schaer (1997) some statistical values are given: the maximum and minimum values in TEC units (TECU) for day 73 of 1996 are

$$\text{TVEC}_{max}(\beta, \Delta\lambda) = \text{TVEC}(-7.60°, 45.37°) = 35.79 \text{ TECU},$$

$$\text{TVEC}_{min}(\beta, \Delta\lambda) = \text{TVEC}(60.91°, -106.64°) = 0.34 \text{ TECU},$$

and the mean TECU, averaged from the one-day GIM, roughly describes the evolution of the ionospheric activity in a global sense and varies for a 28-month time span starting with January 1, 1995 between about 6 and 18 TECU.

Apart from the global ionosphere maps, also regional ionosphere maps based on some 30 European IGS stations are provided for Europe. The application is restricted to the corresponding definition area.

Troposphere sounding

Reliable information on global climate change processes over future decades and better weather forecasting on near- and medium-term time scales are only possible on the basis of global and regional data records. These data are used to accurately model atmospheric state parameters with high spatial and temporal resolution.

Water vapor is one of the most significant constituents of the troposphere. This parameter plays a fundamental role with regard to weather and climate since it has the capability to transport moisture and heat through the atmosphere. Meteorologists have started to use GNSS as a low-cost tool for measuring the water vapor. Thereby, tropospheric refraction (in the past considered as nuisance parameter) has become a well appreciated signal.

The tropospheric zenith delay is estimated during data processing. The dry component can be computed with high accuracy based on surface meteorological data; the remaining wet component is a function of the water vapor in the atmosphere. Short-periodic variations of the integrated water vapor (IWV) improve numerical weather prediction, whereas long-periodic variations have impact on climate research. More details on the subject are found in Bevis et al. (1992) and Gendt et al. (1999). An operational ground-based water vapor observing system using zenith delay measurements is described in Wolfe and Gutman (2000).

CHAMP mission

The challenging minisatellite payload (CHAMP) mission is used for geophysical research and application. The mission started in 2000 and was scheduled to last five years in order to provide a sufficiently long observation time to resolve long-term temporal variations in the magnetic field, in the gravity field, and within the atmosphere. Some specifications of the satellite of this mission: altitude 300–470 km, inclination 87.3 degree, eccentricity 0.001.

The measurable refractional effects on GNSS signals propagating through the atmospheric limb may be used to derive profiles for a variety of atmospheric parameters. Ionospheric refraction is used for the derivation of electron density in profiles between 60 km and the CHAMP orbital height. In conjunction with TEC measurements from a network of terrestrial stations, a comprehensive model of the ionosphere is possible with high resolution in space and time. Refractional effects in the atmosphere ranging from the earth surface up to about 60 km altitude (bending of signal path, tropospheric path delay) give rise to profiles for meteorological parameters (density, pressure, temperature, water vapor). The final objective of the CHAMP mission has been the determination of all these parameters in (near) real time.

More on CHAMP may be found under http://op.gfz-potsdam.de/champ of the GeoForschungsZentrum Potsdam, Germany, where the information of this section has been extracted from.

5.4 Relativistic effects

5.4.1 Special relativity

Lorentz transformation

Consider two four-dimensional systems $S(x,y,z,t)$ and $S'(x',y',z',t')$, where the union of space coordinates x, y, z and the time coordinate t characterizes space-time coordinates. The system S is at rest and, relative to S, the system S' is uniformly translating with velocity v. For simplicity, it is assumed that both systems coincide at an initial epoch $t = 0$ and that the translation occurs along the x-axis.

The transformation of the space-time coordinates is given by

$$
\begin{aligned}
x' &= \frac{x - vt}{\sqrt{1 - \frac{v^2}{c^2}}}, \\
y' &= y, \\
z' &= z, \\
t' &= \frac{t - \frac{v}{c^2} x}{\sqrt{1 - \frac{v^2}{c^2}}},
\end{aligned}
\quad (5.126)
$$

where c is the speed of light. Note that the equations above describe the moving system S' with respect to the system S at rest (more figuratively: as viewed from the moving system). Equivalently, the system S at rest may be described with respect to the moving system S' (more figuratively: as viewed from the system at rest). The corresponding formulas follow by solving Eq. (5.126) for the space-time coordinates in the system S at rest or simply by interchanging the role of the primed and the unprimed coordinates and reversing the sign of the velocity v. Thus, the relations

$$
\begin{aligned}
x &= \frac{x' + vt'}{\sqrt{1 - \frac{v^2}{c^2}}}, \\
y &= y', \\
z &= z', \\
t &= \frac{t' + \frac{v}{c^2} x'}{\sqrt{1 - \frac{v^2}{c^2}}},
\end{aligned}
\quad (5.127)
$$

are obtained. Equations (5.126) and (5.127) are known as Lorentz transformation. An elegant and simple derivation of these formulas can be found in Joos (1956:

p. 217) or in Moritz and Hofmann-Wellenhof (1993: Sect. 4.1). Using Eqs. (5.126) or (5.127), the relation

$$x^2 + y^2 + z^2 - c^2 t^2 = x'^2 + y'^2 + z'^2 - c^2 t'^2 \tag{5.128}$$

may be verified. This means that the norm of a vector in space-time coordinates is invariant with respect to the choice of its reference system. Note that in the case of $c \to \infty$, the Lorentz transformation (5.126) converts to the Galilei transformation

$$\begin{aligned} x' &= x - vt, \\ y' &= y, \\ z' &= z, \\ t' &= t, \end{aligned} \tag{5.129}$$

which is fundamental in classical Newtonian mechanics.

The theory of special relativity is, by definition, restricted to inertial systems. The application of the Lorentz transformation reveals some features of that theory.

Time dilation

Consider an observer moving with the system S'. At a specific location x' the time events t'_1 and t'_2 are recorded. The corresponding time events t_1 and t_2 in the system S at rest, according to Lorentz transformation (5.127), are

$$t_1 = \frac{t'_1 + \frac{v}{c^2} x'}{\sqrt{1 - \frac{v^2}{c^2}}}, \qquad t_2 = \frac{t'_2 + \frac{v}{c^2} x'}{\sqrt{1 - \frac{v^2}{c^2}}}. \tag{5.130}$$

The time interval $\Delta t' = t'_2 - t'_1$ in the moving system is called proper time and the time interval $\Delta t = t_2 - t_1$ in the system at rest is called coordinate time. The relation between proper and coordinate time is found by the difference of the two expressions in (5.130) yielding

$$\Delta t = \frac{\Delta t'}{\sqrt{1 - \frac{v^2}{c^2}}}, \tag{5.131}$$

which means that as viewed at the system at rest, the time interval recorded by the moving observer is lengthened or dilated. The same holds for the inverse situation: a time interval recorded by an observer in the system at rest is dilated for an observer in the moving system. The result $\Delta t' = \Delta t / \sqrt{1 - v^2/c^2}$ may be verified by the reader by using Eqs. (5.126) or is simply obtained by interchanging the role of the primed and the unprimed coordinates in (5.131) (reversing the sign of v has here no effect because this quantity is squared). The time dilation is the reason why moving clocks run slower than clocks at rest.

5.4 Relativistic effects

Lorentz contraction

The derivation of the Lorentz contraction is similar to that of time dilation. Consider now two locations x'_1 and x'_2 in the moving system S' at a specific epoch t'. The corresponding locations x_1 and x_2 in the system S at rest, according to the Lorentz transformation (5.127), are

$$x_1 = \frac{x'_1 + vt'}{\sqrt{1 - \frac{v^2}{c^2}}}, \qquad x_2 = \frac{x'_2 + vt'}{\sqrt{1 - \frac{v^2}{c^2}}}. \qquad (5.132)$$

Using the abbreviations $\Delta x = x_2 - x_1$ and $\Delta x' = x'_2 - x'_1$, the difference of the two expressions in (5.132) gives

$$\Delta x = \frac{\Delta x'}{\sqrt{1 - \frac{v^2}{c^2}}}, \qquad (5.133)$$

which means that as viewed from the system S at rest, $\Delta x'$ is lengthened to Δx. Expressing it in another way, the dimension of a body moving with the observer in the system S' seems to be contracted.

Second-order Doppler effect

Since frequency is inversely proportional to time, one can deduce immediately from the considerations on time dilation the formula

$$f = f' \sqrt{1 - \frac{v^2}{c^2}}, \qquad (5.134)$$

which means that the frequency f' of a moving emitter would be reduced to f. This is the second-order Doppler effect.

Mass relation

Special relativity also affects masses. Denoting the masses in the two reference frames S and S' by m and m', respectively, then

$$m = \frac{m'}{\sqrt{1 - \frac{v^2}{c^2}}} \qquad (5.135)$$

is the corresponding mass relation.

The previous formulas include the same square root. Expansion into binomial series yields

$$\frac{1}{\sqrt{1-\frac{v^2}{c^2}}} = 1 + \frac{1}{2}\left(\frac{v}{c}\right)^2 + \ldots,$$

$$\sqrt{1-\frac{v^2}{c^2}} = 1 - \frac{1}{2}\left(\frac{v}{c}\right)^2 + \ldots,$$

(5.136)

which may be substituted accordingly into Eqs. (5.131) through (5.135). Related to an observer at rest,

$$\frac{\Delta t' - \Delta t}{\Delta t} = \frac{\Delta x' - \Delta x}{\Delta x} = -\frac{f' - f}{f} = \frac{m' - m}{m} = -\frac{1}{2}\left(\frac{v}{c}\right)^2 \qquad (5.137)$$

accounts for the mentioned effects of the special relativity in one formula.

5.4.2 General relativity

The theory of general relativity includes accelerated reference systems too, where the gravitational field plays the key role. Formulas analogous to (5.137) may be derived if the kinetic energy $v^2/2$ in special relativity is replaced by the potential energy ΔU. Thus,

$$\frac{\Delta t' - \Delta t}{\Delta t} = \frac{\Delta x' - \Delta x}{\Delta x} = -\frac{f' - f}{f} = \frac{m' - m}{m} = -\frac{\Delta U}{c^2} \qquad (5.138)$$

represents the relations in general relativity, where ΔU is the difference of the gravitational potential in the two reference frames under consideration.

5.4.3 Relevant relativistic effects for GNSS

The reference frame (relatively) at rest is located in the center of the earth and an accelerated reference frame is attached to each GNSS satellite. Therefore, the theory of special and general relativity must be taken into account. Relativistic effects are relevant for the satellite orbit, the satellite signal propagation, and both the satellite and receiver clock. An overview of all these effects is given for example in Zhu and Groten (1988). The relativistic effects on rotating and gravitating clocks is also treated in Grafarend and Schwarze (1991). With respect to general relativity, Ashby (1987) shows that only the gravitational field of the earth must be considered. The relativistic influence of sun and moon and consequently all other masses in the solar system is negligible. Deines (1992) investigates the uncompensated effects if the noninertial GNSS observations are not transformed to an inertial frame.

5.4 Relativistic effects

Ashby (2003) extensively discusses the relativistic principles and effects which must be considered for space-based navigation. Among others, frequency jumps arising from orbit adjustments have been identified as relativistic effects. Several secondary relativistic effects at the level of a few centimeters (which corresponds to 100 picoseconds of delay) are mentioned: the Shapiro signal propagation delay, the space-time curvature effect on a geodetic distance, the effect of other solar system bodies. Furthermore, the phase wind-up (sometimes also denoted as wrap-up) is another secondary effect. The electric field vector of the satellite-transmitted signal rotates with an angular frequency. Supposing a rapidly spinning receiver (with another angular frequency), the received frequency is composed of the two angular frequencies. This also translates to an accumulation of phase and is, therefore, called phase wind-up.

Relativity affecting the satellite orbit

The gravitational field of the earth causes relativistic perturbations in the satellite orbits. An approximate formula for the disturbing acceleration is given by Eq. (3.32). For more details see Zhu and Groten (1988).

Relativity affecting the satellite signal

The gravitational field gives rise to a space-time curvature of the satellite signal. Therefore, a propagation correction must be applied to get the Euclidean range for instance. The range correction (expressed in meters) may be represented in the form

$$\delta^{rel} = \frac{2\mu}{c^2} \ln \frac{\varrho^s + \varrho_r + \varrho_r^s}{\varrho^s + \varrho_r - \varrho_r^s}, \tag{5.139}$$

where $\mu = 3\,986\,004.418 \cdot 10^8\, m^3\, s^{-2}$ is the earth's gravitational constant (see Sect. 3.2.1). The geocentric distances of satellite s and observing receiver site r are denoted ϱ^s and ϱ_r, and ϱ_r^s is the distance between the satellite and the observing receiver site. In order to estimate the maximum effect for a point on the surface of the earth take the mean radius of the earth $R_e = 6\,370$ km and an altitude of $h = 20\,000$ km for the satellites. The maximum distance $\varrho_r^s = \sqrt{(R_e + h)^2 - R_e^2}$ results from the Pythagorean theorem and is about 25 600 km. Substituting these values, the maximum range error $\delta^{rel} = 18.6$ mm results from (5.139). Note that this maximum value only applies to point positioning. In relative positioning, the effect is much smaller and amounts to 0.001 ppm (Zhu and Groten 1988).

Relativity affecting the satellite clock

Assume a nominal frequency $f_0 = 10.23$ MHz of the satellite clock. This frequency is influenced by the motion of the satellite and by the difference of the gravitational field at the satellite and the observing site. The corresponding effects of special and general relativity are small and may be linearly superposed. Thus,

$$\delta^{rel} \equiv \frac{f'_0 - f_0}{f_0} = \underbrace{\frac{1}{2}\left(\frac{v}{c}\right)^2 + \frac{\Delta U}{c^2}}_{\text{special and general relativity}} \quad (5.140)$$

is the effect on the frequency of the satellite clock, where Eqs. (5.137) and (5.138) have been used. To get a numerical value, circular orbits and a spherical earth with the observing site on its surface are assumed. Introducing these simplifications, (5.140) takes the form

$$\delta^{rel} \equiv \frac{f'_0 - f_0}{f_0} = \frac{1}{2}\left(\frac{v}{c}\right)^2 + \frac{\mu}{c^2}\left[\frac{1}{R_e + h} - \frac{1}{R_e}\right], \quad (5.141)$$

with v being the mean velocity of the satellite. Substituting the numerical value $h = 20\,000$ km, which according to (3.9) corresponds to $v \approx 3.9$ km s^{-1}, yields

$$\frac{f'_0 - f_0}{f_0} = -4.464 \cdot 10^{-10},$$

which, despite the simplifications, is sufficiently accurate. The influence of the earth's oblateness, investigated by Ashby (2001), causes a periodic fractional frequency shift with a period of almost six hours and an amplitude of $0.695 \cdot 10^{-14}$. Recall that f'_0 is the emitted frequency and f_0 is the frequency received at the observation site. Thus, it can be seen that the satellite-transmitted nominal frequency would be increased by $df = 4.464 \cdot 10^{-10} f_0 = 4.57 \cdot 10^{-3}$ Hz. However, it is desired to receive the nominal frequency. This is achieved by an offset df in the satellite clock frequency, so that 10.22999999543 MHz are emitted.

Another periodic effect arises due to the assumption of a circular orbit and may be denoted as eccentricity correction. An adequate formula for the correction (expressed in seconds) is given by

$$\delta^{rel} = -\frac{2}{c^2}\sqrt{\mu a}\,(e \sin E), \quad (5.142)$$

where e denotes the eccentricity, a the semimajor axis, and E the eccentric anomaly of the satellite orbit.

Substituting (3.36) into (5.142) yields an alternative but equivalent form

$$\delta^{rel} = -\frac{2}{c^2}\,\varrho^s \cdot \dot{\varrho}^s, \quad (5.143)$$

where ϱ^s and $\dot{\varrho}^s$ are the instantaneous satellite position and velocity vector, respectively.

5.4 Relativistic effects

A further representation of (5.142) is

$$\delta^{rel} = F\, e\, \sqrt{a}\, \sin E, \tag{5.144}$$

where $F = -2\sqrt{\mu}/c^2$ has been introduced.

This relativistic effect may be included in the clock polynomial (see Eq. (6.4)) broadcast via the navigation message, where the time-dependent eccentric anomaly E is expanded into a Taylor series. However, this means that the correction must be implemented in the receiver software. Ashby (2003) would prefer an incorporation of this correction into the time broadcast by the satellites.

In the case of relative positioning, the effect cancels (Zhu and Groten 1988).

Relativity affecting the receiver clock

A receiver clock located on the surface of the earth is rotating with respect to the resting reference frame at the geocenter. The associated linear velocity at the equator is approximately $0.5\,\mathrm{km\,s^{-1}}$ and, thus, roughly one tenth of the velocity of the satellite. Substituting this value into the special relativistic part of Eq. (5.140) yields a relative frequency shift in the order of 10^{-12} which after 3 hours corresponds to a clock error of 10 nanoseconds (1 ns = 10^{-9} s $\doteq$ 30 cm).

Due to the rotation of the earth (causing a rotation of the receiver clock) while the signal is propagating from the satellite to the receiver on the earth, a relativistic effect is introduced known as Sagnac effect (Conley et al. 2006: p. 307). Following Su (2001), the Sagnac effect can be modeled by

$$\delta^{rel} = \frac{1}{c}(\varrho_r - \varrho^s) \cdot (\omega_e \times \varrho_r), \tag{5.145}$$

where c is the speed of light, ϱ_r and ϱ^s are the geocentric position vectors of the receiver and the satellite, respectively (Fig. 1.1), and ω_e is the earth's rotation vector. Introducing

$$\mathbf{S} = \tfrac{1}{2}(\varrho^s \times \varrho_r), \tag{5.146}$$

the area of the triangle with vertices at the satellite, the receiver, and the center of the earth, the Sagnac effect may also be written as

$$\delta^{rel} = \frac{2}{c}\mathbf{S} \cdot \omega_e. \tag{5.147}$$

Corrections of the Sagnac effect are also referred to as earth rotation corrections.

5.5 Antenna phase center offset and variation

5.5.1 General remarks

Assuming an idealized situation, then the electrical phase center of the antenna is the point to which all measurements derived from received GNSS signals refer. Usually, this will not be a point which may be accessed (e.g., by a tape measurement). Therefore, a geometrical point on the antenna denoted as antenna reference point (ARP) is introduced (Fig. 5.6). The IGS has defined the ARP as the intersection of the vertical antenna axis of symmetry with the bottom of the antenna.

However, this idealized situation does not reflect the reality because the electrical antenna phase center varies with elevation, azimuth, intensity of the satellite signal, and is also frequency-dependent. In other terms, each incoming signal has its own electrical antenna phase center. Therefore, a mean position of the electrical antenna phase center is determined for the purpose of the offset calibration.

The antenna phase center offset (PCO) defines the difference between the ARP and the mean electrical antenna phase center.

Usually, the antenna PCO is given by three-dimensional coordinates of the electrical antenna phase center referring to the ARP and should be provided by the manufacturer; if not, the determination of these coordinates is carried out by a calibration procedure (Görres et al. 2006). Note that due to the dependence on frequency the antenna PCO must be given for each carrier frequency.

Comparing now the electrical antenna phase center of an individual measurement with the mean electrical antenna phase center, a deviation will arise. These

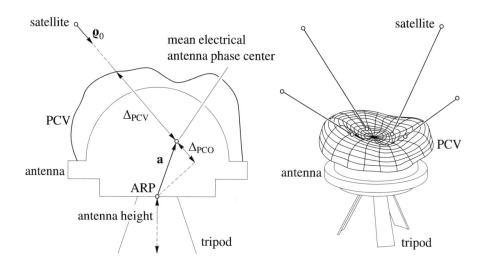

Fig. 5.6. Electrical phase center and antenna reference point

5.5 Antenna phase center offset and variation

deviations are denoted as antenna phase center variations (PCV). The azimuth- and elevation-dependent PCV define the phase pattern (individually for each carrier frequency).

The total antenna phase center correction for an individual phase measurement is composed of the influence by the PCO plus the azimuth- and elevation-dependent PCV.

Introducing for the PCO the vector **a** as indicated in Fig. 5.6 and the unit vector $\boldsymbol{\varrho}_0$ between the satellite and the receiver, Δ_{PCO}, the influence of the PCO on the phase measurement, may be obtained as the projection of **a** onto the unit vector $\boldsymbol{\varrho}_0$ between the satellite and the receiver, thus

$$\Delta_{PCO} = \mathbf{a} \cdot \boldsymbol{\varrho}_0 . \tag{5.148}$$

The influence of PCV on the phase pseudorange, denoted as Δ_{PCV}, is described by a function depending on the azimuth α and the zenith angle z of the satellite, and of the carrier frequency f:

$$\Delta_{PCV} = \Delta_{PCV}(\alpha, z, f) . \tag{5.149}$$

The total correction of the phase pseudorange due to PCO and PCV is the combined effect $\Delta_{PCO} + \Delta_{PCV}$. Applying this total correction, the phase pseudoranges refer to the ARP. In other terms, the coordinates of the ARP will result after processing the measured data properly. As seen from Fig. 5.6, the resulting height component must still be reduced by the antenna height.

The PCV is systematic and can be determined by test series. Variations can amount to 1–2 cm horizontally and up to 10 cm vertically (Mader 1999). However, it is fairly difficult to model the PCV because it is different for each antenna and also for various types. Geiger (1988) shows the different characteristics of conical spiral antennas, microstrip antennas, dipole antennas, and helices. As a consequence, the direct computation of the antenna effects on the distance measurements with respect to azimuth and elevation was proposed. Simple functions for an appropriate modeling may also be found by laboratory tests (Schupler and Clark 1991), e.g., in an anechoic chamber.

The geometric effect of the antenna orientation on the carrier phase is investigated by Wu et al. (1993). The observed carrier phase depends on the orientation of the antennas of the transmitter and the receiver as well as the direction of the line of sight. Wu et al. (1993) demonstrate that the effect does not cancel for double-differences and may amount to 1 part in 10^9.

Campbell et al. (2004) investigate the accuracy of antenna calibrations using laboratory measurements. Rothacher (2001b) compares relative and absolute antenna PCV, which is treated in some more detail in the next two sections.

5.5.2 Relative antenna calibration

Now the question arises how to establish proper calibration models. One option is a relative antenna calibration which means that field measurements are used to determine the relative antenna phase center position and PCV with respect to a reference antenna as carried out at the US National Geodetic Survey (Mader 1999).

The principle of the relative antenna calibration is simple. The test range is a 5 m baseline equipped with two stable concrete piers. Antenna-mounting plates are permanently attached to these piers. The reference antenna (always the same!) is placed on one of the piers, the test antenna on the other.

As will be shown in Sect. 5.5.4 in a numerical investigation, the average phase center location is a function of the elevation cutoff angle. For each frequency individually, the average phase center location (relatively to an a priori known position) is determined where no PCV or tropospheric scale factor is taken into account.

Where does the a priori information come from? Note that the reference antenna has been placed on the test pier in order to determine the location of this antenna's phase centers (separately for frequency 1 and 2). These positions are then used as the a priori positions for the frequency 1 and frequency 2 phase centers of the test antenna (Mader 1999).

Referring to the a priori values, the average phase center location may be regarded as relative PCO.

As mentioned earlier, Mader (1999) does not determine an azimuthal component of the PCV, only elevation dependence is considered. Without describing details, basically single-difference phase residuals are formed by constraining the test antenna to its previously determined mean (average) PCO. Then a least-squares solution for a fourth-order polynomial is used for each measurement epoch to account for a clock offset and the elevation dependence. Again, this procedure is separated for each frequency.

The polynomial coefficients for each measurement are now the tool to correct the observed phase data. Using the elevation of the satellite and the corresponding measurement epoch, the correction quantity may be calculated from the fourth-order polynomial and applied to the measured phase.

5.5.3 Absolute antenna calibration

The relation to a reference antenna was the key feature of the relative antenna calibration. Considering absolute antenna calibration in the sense of Wübbena et al. (1997, 2000) and Menge et al. (1998), the term "absolute" indicates that the PCV are determined independently from a reference antenna. However, the size of the absolute phase pattern cannot be determined, "only the topology" (Wübbena et al. 1997). The reason is that relative observables are used. To model the PCV,

5.5 Antenna phase center offset and variation

a continuous and periodic function in the horizontal and vertical directions (referring to a local-level coordinate system) is required to describe the PCV depending on the azimuth and the elevation (or the zenith angle, respectively) of the satellite. Rothacher et al. (1995) propose a spherical harmonic function in the form

$$\Delta_{\text{PCV}}(\alpha, z) = \sum_{n=0}^{\infty} \sum_{m=0}^{n} (A_{nm} \cos m\alpha + B_{nm} \sin m\alpha) P_{nm}(\cos z), \quad (5.150)$$

where $\Delta_{\text{PCV}}(\alpha, z)$ is the PCV depending on the azimuth α and the zenith angle z. On the right-hand side, A_{nm} and B_{nm} are the coefficients to be determined and $P_{nm}(\cos z)$ are Legendre's functions. For more details on harmonic functions and Legendre's functions see Hofmann-Wellenhof and Moritz (2006: Chap. 1).

Supposing for a moment known coefficients A_{nm} and B_{nm}, then $\Delta_{\text{PCV}}(\alpha, z)$ may be calculated for arbitrary α and z. If there are sufficient measurement quantities $\Delta_{\text{PCV}}(\alpha, z)$ available, the coefficients can be estimated by the least-squares adjustment method.

However, multipath must be taken into account. This effect can be tackled by the basic idea to use repeated satellite constellation (e.g., after one sidereal day in the case of GPS). As will be shown in Sect. 5.6, if the site conditions remain unchanged, the multipath effect repeats with the same satellite ground track repeat periods. Therefore, forming differences of the observations with repeated satellite constellation, then the effect of the multipath will be removed. However, the PCV would also be eliminated if the satellite signal is received with the same antenna orientation. This is avoided by tilting and rotating the antenna on one of the two days. Now the difference of the PCV values of the two days is the measurement value (which will be in general not zero). The measurement of the first day is regarded as "zero position" of the reference day. Thus, the resulting measurement quantity, the input for the left-hand side of (5.150), is a difference of two antenna orientations PCV (which indicates the problematic interpretation of the term "absolute").

As shown in Wübbena et al. (2000), the rotations and tilts of the antenna have been automated by a precisely controlled motion of a calibrated robot. This automation process enables several thousand different antenna orientations to eliminate the multipath and determine the PCV adequately. Additionally, the PCV values are independent of the "polar holes" (Seeber 2003: p. 322) which otherwise occur as may be seen from Fig. 5.7. The high amount of orientation measurements is required to determine a high-resolution PCV model according to (5.150). Wübbena et al. (2000) use 6 000 to 8 000 different orientations for one calibration process.

For some antenna types, large azimuthal PCV have been demonstrated. The method depending on elevation only as described in the previous section does not account for the azimuth-dependent information, which might be disadvantageous in case of very high accuracy requirements.

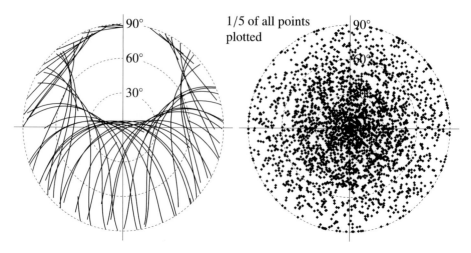

Fig. 5.7. Coverage of the antenna hemisphere with 24-hour observation data with static antenna (left) and rotating and tilting antenna (right), courtesy Wübbena et al. (2000)

The result of the antenna calibration is stored in a file containing the horizontal and the vertical offsets and the PCV given by elevation- and azimuth-dependent corrections. The PCV must directly be applied to the phase pseudoranges. The antenna PCO must also be taken into account.

5.5.4 Numerical investigation

The objective of this section is to illustrate numerically the effects of the electrical antenna PCO and PCV.

As mentioned before, the location of the electrical antenna phase center depends apart from the carrier frequency on the direction of the received signal, which may be decomposed into azimuth and elevation. It also depends on the signal intensity. Since each measurement refers to an individual antenna phase center, the mean phase center location may be obtained by a weighted average of these individual phase centers. Assuming a known average defining the phase center, then there is still the need to relate this phase center to the antenna reference point (ARP) which is geometrically defined and may be accessed by, e.g., a tape measurement. The relation may be established by a three-dimensional vector in a local north, east, up system where the origin lies in the ARP.

Mader (1999) illustrates the problem by measuring a very short baseline with baseline components n, e, u in a local system using three different solutions: single-frequency f_1, f_2 data and the ionosphere-free combination with frequency f_3. In Table 5.6, each of the three baseline solutions used the same 24-hour data set

5.5 Antenna phase center offset and variation

Table 5.6. Baseline components and their deviations from the respective means without applying PCO and PCV corrections

Frequency	n [m]	e [m]	u [m]	$n - \mu_n$	$e - \mu_e$	$u - \mu_u$
f_1	4.9712	0.0736	0.0371	0.0002	−0.0008	0.0035
f_2	4.9724	0.0694	0.0562	0.0014	−0.0050	0.0226
f_3	4.9693	0.0802	0.0074	−0.0017	0.0058	−0.0262
μ_n, μ_e, μ_u	4.9710	0.0744	0.0336			

and no tropospheric unknowns were estimated. Considering the arithmetic means μ_n, μ_e, μ_u of the three solutions for the individual baseline components, the differences of the north components amount to 1 mm and are negligible, two of the east components deviate from the respective mean by some 5 mm, but two up components show a difference of more than 2 cm.

In comparison with these results, PCO were determined and, after applying them to the same measurements as used previously, the quantities in Table 5.7 result. This illustrates the success of applying calibration values to account for the PCO.

So far the PCO only has been considered but not yet the PCV depending on the direction of the received signal. Mader (1999) does not separate into azimuth- and elevation-dependent influences; assuming azimuthally symmetric antennas, the dominant phase variation arises from elevation, which may be demonstrated by varying the elevation cutoff angle, i.e., data received below the cutoff angle are omitted. Table 5.8 shows the impact of various cutoff angles. Note that PCO has been applied. The changes of the north and east components are almost negligible, but the up component of the short baseline varies by about 1 cm as the elevation cutoff angle changes from 10 to 25 degrees. This enables to account for the information describing the PCV as a function of elevation yielding Table 5.9. Mader (1999) in-

Table 5.7. Baseline components and their deviations from the respective means with PCO applied

Frequency	n [m]	e [m]	u [m]	$n - \mu_n$	$e - \mu_e$	$u - \mu_u$
f_1	4.9727	0.0724	0.0022	−0.0003	−0.0002	−0.0026
f_2	4.9714	0.0710	0.0026	−0.0016	−0.0016	−0.0022
f_3	4.9748	0.0745	0.0095	0.0018	0.0019	0.0047
μ_n, μ_e, μ_u	4.9730	0.0726	0.0048			

Table 5.8. Baseline components of ionosphere-free solutions as functions of the elevation cutoff angle with PCO applied

Cutoff [°]	n [m]	e [m]	u [m]
10	4.9741	0.0741	0.0122
15	4.9748	0.0745	0.0095
20	4.9753	0.0735	0.0064
25	4.9763	0.0731	0.0025

Table 5.9. Baseline components of ionosphere-free solutions as functions of the elevation cutoff angle with PCO and PCV applied

Cutoff [°]	n [m]	e [m]	u [m]
10	4.9736	0.0754	−0.0001
15	4.9743	0.0759	−0.0014
20	4.9745	0.0748	0.0003
25	4.9754	0.0745	0.0015

terprets the results as a systematic shift by about 1 mm and the "wandering" of the up component with changing cutoff angle is now reduced to about 3 mm.

5.6 Multipath

5.6.1 General remarks

The effect is well described by its name: a satellite-emitted signal arrives at the receiver by more than one path. Multipath is mainly caused by reflecting surfaces near the receiver (Fig. 5.8). Secondary effects are reflections at the satellite during signal transmission.

Referring to Fig. 5.8, the satellite signal arrives at the receiver on three different paths, one direct and two indirect ones. As a consequence, the received signals have relative phase offsets and the phase differences are proportional to the differences of the path lengths. There is no general model of the multipath effect because of the time- and location-dependent geometric situation. The influence of the multipath, however, can be estimated by using a combination of f_1 and f_2 code and carrier phase measurements. The principle is based on the fact that the troposphere, clock errors, and relativistic effects influence code and carrier phases by the same amount.

5.6 Multipath

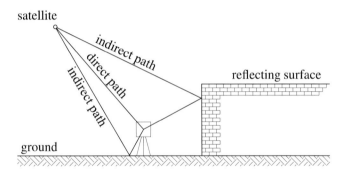

Fig. 5.8. Multipath effect

This is not true for ionospheric refraction and multipath, which are frequency dependent. Taking ionosphere-free code ranges and carrier phases, and forming corresponding differences, all mentioned effects except for multipath are canceled. The residuals, apart from the noise level, reflect the multipath effect. Tranquilla and Carr (1990/91) group the multipath errors of pseudoranges into three classes: (1) diffuse forward scattering from a widely distributed area (e.g., the signal passes through a cluttered metallic environment), (2) specular reflection from well-defined objects or reflective surfaces in the vicinity of the antenna, and (3) fluctuations of very low frequency, usually associated with reflection from the surface of water.

Purely from geometry it is clear that signals received from low satellite elevations are more susceptible to multipath than signals from high elevations. Note also that code ranges are more affected by multipath than carrier phases. Comparing single epochs, the multipath effect may amount to 10–20 m for code pseudoranges (Wells et al. 1987). Under certain extreme circumstances, the error resulting from multipath may grow to about 100 m in the vicinity of buildings (Nee 1992). In severe cases of multipath, loss of lock may even occur.

The multipath effects on carrier phases for relative positioning with short baselines, should, generally, not be greater than about 1 cm (good satellite geometry and a reasonably long observation interval). But even in those cases, a simple change of the height of the receiver may increase the multipath and, thus, deteriorate the results. When performing static surveys where the observation times are relatively long, intermittent periods of multipath contamination are not a problem. Such situations occur when the receiver is set up in the center of a highway and large metal trucks continually pass by the antenna. Rapid static surveys (i.e., surveys with very short observation times) may be more contaminated in such cases, and longer observation times would be appropriate.

5.6.2 Mathematical model

The effect of multipath on carrier phases may be estimated by the following considerations (Fig. 5.8). The direct and indirect signals interfere at the antenna phase center and may be represented by

$$a \cos \varphi \quad \ldots \quad \text{direct signal}, \\ \beta a \cos (\varphi + \Delta \varphi) \quad \ldots \quad \text{indirect signal}, \tag{5.151}$$

where a and φ denote the amplitude and the phase of the direct signal. The amplitude of the indirect signal is affected by the damping factor β because of the reflection at a surface (Seeber 2003: p. 317). This damping factor is in general in the range $0 \leq \beta \leq 1$ and covers the full range from no reflection ($\beta = 0$) to full reflection ($\beta = 1$) with the reflected signal as strong as the direct signal. The phase of the indirect signal is delayed by the phase shift $\Delta \varphi$, which is a function of the geometric configuration. The superposition of the signals in (5.151) is represented by

$$a \cos \varphi + \beta a \cos (\varphi + \Delta \varphi). \tag{5.152}$$

Applying the cosine-theorem yields

$$a \cos \varphi + \beta a \cos \varphi \cos \Delta \varphi - \beta a \sin \varphi \sin \Delta \varphi, \tag{5.153}$$

which is slightly rearranged to

$$(1 + \beta \cos \Delta \varphi) a \cos \varphi - (\beta \sin \Delta \varphi) a \sin \varphi. \tag{5.154}$$

This resultant signal may be represented (Joos 1956: p. 44) in the form

$$\beta_M a \cos (\varphi + \Delta \varphi_M), \tag{5.155}$$

where the subscript M indicates multipath. The cosine-theorem gives

$$(\beta_M \cos \Delta \varphi_M) a \cos \varphi - (\beta_M \sin \Delta \varphi_M) a \sin \varphi. \tag{5.156}$$

Comparing the coefficients for $a \sin \varphi$ and $a \cos \varphi$ of Eqs. (5.154) and (5.156) leads to the relations

$$\beta_M \sin \Delta \varphi_M = \beta \sin \Delta \varphi, \\ \beta_M \cos \Delta \varphi_M = 1 + \beta \cos \Delta \varphi, \tag{5.157}$$

which represent two equations for the desired quantities β_M and $\Delta \varphi_M$. An explicit expression for β_M follows by squaring and adding the two equations. Thus,

$$\beta_M = \sqrt{1 + \beta^2 + 2\beta \cos \Delta \varphi} \tag{5.158}$$

5.6 Multipath

is obtained. An explicit expression for $\Delta\varphi_M$ follows by dividing the two equations in (5.157). Thus,

$$\tan \Delta\varphi_M = \frac{\beta \sin \Delta\varphi}{1 + \beta \cos \Delta\varphi} \qquad (5.159)$$

is the solution.

As indicated above, the damping factor β may vary between 0 and 1. The substitution of $\beta = 0$ (i.e., there is no reflected signal and no multipath) into (5.158) and (5.159) gives $\beta_M = 1$ and $\Delta\varphi_M = 0$. This means that the "resultant" signal is identical to the direct signal. The substitution of $\beta = 1$ into (5.158) and (5.159) leads to

$$\beta_M = \sqrt{2(1 + \cos \Delta\varphi)} = 2 \cos \frac{\Delta\varphi}{2} \qquad (5.160)$$

and

$$\tan \Delta\varphi_M = \frac{\sin \Delta\varphi}{1 + \cos \Delta\varphi} = \tan \frac{\Delta\varphi}{2} \qquad (5.161)$$

yielding

$$\Delta\varphi_M = \tfrac{1}{2} \Delta\varphi. \qquad (5.162)$$

Examples for numerical values for β_M and $\Delta\varphi_M$ as a function of $\Delta\varphi$ are

$\Delta\varphi$	β_M	$\Delta\varphi_M$
0°	2	0°
90°	$\sqrt{2}$	45°
180°	0	90°

which shows that the maximum effect of multipath on phase measurements occurs for $\Delta\varphi_M = 90° = 1/4$ cycle. Converting this phase shift to range gives $\lambda/4$ or, with $\lambda = 20$ cm, the maximum change in range of about 5 cm. However, it should be noted that this value may increase if linear phase combinations are used.

The phase shift $\Delta\varphi$ can be expressed as a function of the extra path length Δs. In the case of a horizontal reflector (ground),

$$\Delta\varphi = \frac{1}{\lambda} \Delta s = \frac{2h}{\lambda} \sin E \qquad (5.163)$$

is obtained, where the phase shift is expressed in cycles. The parameter h denotes the vertical distance between the antenna and the ground and E is the elevation of the satellite (Fig. 5.9). Multipath is periodic because E varies with time. The

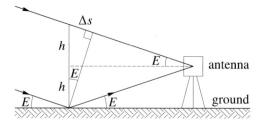

Fig. 5.9. Geometry of multipath

frequency of multipath is

$$f = \frac{d(\Delta\varphi)}{dt} = \frac{2h}{\lambda} \cos E \frac{dE}{dt}. \tag{5.164}$$

Substituting typical values like $E = 45°$ and $dE/dt = 0.07$ mrad per second leads for a carrier with a frequency of 1.5 GHz to the approximation

$$f = 0.521 \cdot 10^{-3} h, \tag{5.165}$$

where f is obtained in hertz if h is given in meters (Wei and Schwarz 1995a). Thus, an antenna height of 2 m leads to an approximate period of 16 minutes for the multipath error.

5.6.3 Multipath reduction

To reduce or estimate the multipath effects, various methods were developed that are classified by Ray et al. (1999) as (1) antenna-based mitigation, (2) improved receiver technology, and (3) signal and data processing.

Among the antenna-based mitigation methods, improving the antenna gain pattern by choke rings, creating special antenna designs and arrays are very effective (Moelker 1997, Bartone and Graas 1998). The elimination of multipath signals is possible by selecting an antenna that takes advantage of the signal polarization. If the transmitted GNSS signals are right-handed circularly polarized, then the reflected signals are left-handed polarized. A reduction of the multipath effect may also be achieved by digital filtering, wideband antennas, antenna ground planes absorbing radio frequencies, choke ring antennas including the advanced dual-frequency choke ring design (Philippov et al. 1999). A refined version of the choke ring idea uses spiral arms. The advantage of the more recent developments over the choke rings is the sharper radiation pattern roll-off (which reduces the multipath susceptibility), no phase center offset between two carriers, no necessity for any alignment, e.g., to the north direction, due to its symmetry, and its planar structure (Kunysz 2000). The absorbent antenna ground plane reduces the interference

5.6 Multipath

of satellite signals with low or even negative elevation angles which occur in case of multipath.

Improving the receiver technology for multipath reduction includes narrow correlation spacing, extending the multipath estimation delay lock loop, enhancing the strobe correlator multipath rejection; more details are given in Dierendonck and Braasch (1997), Garin and Rousseau (1997). Examples for the ongoing research for multipath reduction are approaches as the multipath estimating delay lock loop (MEDLL) (Townsend et al. 1995, 2000). This technique separates the incoming signal into the direct line-of-sight component and the indirect signal by using an array of correlators and measuring the received correlation function. Tests with MEDLL have shown an error reduction up to 90% (Fenton and Townsend 1994).

Numerous methods investigate multipath mitigation by signal and data processing: exploring the signal-to-noise ratio, smoothing carrier phases, or using data combinations.

The detection and reduction of multipath in the spectral domain is proposed by Li et al. (1993). The measured data are transformed by the Fourier transformation into the spectral domain. The detection and reduction of multipath is carried out by amplitude filtering. The inverse Fourier transformation outputs the filtered data.

As another example, Phelts and Enge (2000) locate the multipath invariant points using one or more additional correlator pairs and by the tracking error compensator which corrects for the code tracking error caused by code multipath and thermal noise.

The relation (5.159) expressing the carrier phase error due to multipath is investigated by Ray et al. (1999) to mitigate multipath effects using multiple closely spaced antennas for static applications. The nearby placed antennas cause a strong correlation of the reflected signals. The phase of the reflected signal at each antenna phase center depends on the signal direction which may be expressed by azimuth and elevation. Also the geometry of the antennas must be taken into account. Ray et al. (1999) introduce a reference antenna and five antennas assembled around it. For each satellite, a Kalman filter is implemented. The four-element state vector of the estimator comprises the damping factor β, the reflected signal phase at the antenna, and azimuth and elevation of the reflected signal. By individually combining the data of the reference antenna with the others, profiting from the known geometry of the antennas and using a single external stable clock to get a negligible receiver clock bias, the model for the measurements may strongly be simplified so that it mainly reflects the oscillatory multipath error and the random carrier phase noise. The results of the Kalman filter estimation may be adapted to finally apply to (5.159) allowing for the multipath error determination in the carrier phase at each antenna. Neglecting the filter convergence period, test measurements show an improvement of about 70%.

The most effective countermeasure to multipath is to avoid sites where it could be a problem (e.g., near chain-link fence). Considering Fig. 5.8, placing the antenna directly on the reflecting ground without a tripod would eliminate one of the two indirect paths; however, a vertical reflecting surface would still contaminate the results. The general recommendation is, therefore, to avoid, as far as possible, reflecting surfaces in the neighborhood of the receivers.

Nowadays, multipath analysis and mitigation is no longer restricted to high-precision (static) applications. Car navigation is one example using multiple antennas to isolate and detect multipath on code measurements (Nayak et al. 2000). Multipath on code measurements remains the most significant error source for differential GNSS vehicle navigation. Compared to static applications, the positions of various reflectors are changing rapidly, increasing the difficulty of a proper model. For the multipath affecting code measurements, the residuals of code and phase may be analyzed since the carrier receiver noise and the multipath affecting phases are very small compared to the corresponding code values. Successful identification and elimination of the multipath-corrupted measurement is the final objective being demonstrated in some experiments by Nayak et al. (2000). The gain in position accuracy improvement depends on the size of the multipath errors. Even with high-performance correlator receivers, multipath errors of several meters frequently occur.

6 Mathematical models for positioning

6.1 Point positioning

6.1.1 Point positioning with code ranges

Code range model

The code pseudorange at an epoch t can be modeled, cf. Eq. (5.2), by

$$R_r^s(t) = \varrho_r^s(t) + c\,\Delta\delta_r^s(t). \tag{6.1}$$

Here, $R_r^s(t)$ is the measured code pseudorange between the observing receiver site r and the satellite s, the term $\varrho_r^s(t)$ is the geometric distance between the observing point and the satellite, and c is the speed of light. The last item to be explained is $\Delta\delta_r^s(t)$. This clock bias represents the combined clock offsets of the receiver and the satellite clock with respect to system time, cf. Eq. (5.1).

Examining Eq. (6.1), the desired coordinates of the receiver site to be determined are implicit in the distance $\varrho_r^s(t)$, which can explicitly be written as

$$\varrho_r^s(t) = \sqrt{(X^s(t) - X_r)^2 + (Y^s(t) - Y_r)^2 + (Z^s(t) - Z_r)^2}, \tag{6.2}$$

where $X^s(t)$, $Y^s(t)$, $Z^s(t)$ are the components of the geocentric position vector of the satellite at epoch t, and X_r, Y_r, Z_r are the three ECEF coordinates of the (stationary) observing receiver site. Now, the clock bias $\Delta\delta_r^s(t)$ must be investigated in more detail. For the moment consider a single epoch; a single position r is automatically implied. Each satellite contributes one unknown clock bias which can be recognized from the superscript s at the clock term. Neglecting, for the present, the site r clock bias, the pseudorange equation for the first satellite would have four unknowns. These are the three site coordinates and one clock bias of this satellite. Each additional satellite adds one equation with the same site coordinates but with a new satellite clock bias. Thus, there would always be more unknowns than measurements. Even when an additional epoch is considered, new satellite clock biases must be modeled due to clock drift. Fortunately, the satellite clock information is known with sufficient accuracy and transmitted via the broadcast navigation message, e.g., in the form of three polynomial coefficients a_0, a_1, a_2 with a reference time t_c, cf. Eq. (3.56). Therefore, the equation

$$\delta^s(t) = a_0 + a_1(t - t_c) + a_2(t - t_c)^2 \tag{6.3}$$

enables the calculation of the satellite clock bias at epoch t. It should be noted that the polynomial (6.3) removes a great deal of the satellite clock uncertainty, but a small amount of (random) error remains. It should also be noted that the relativistic effects are not included in the polynomial. Therefore, for a more complete user algorithm for satellite clock correction, the term (5.144) must also be taken into account by

$$\delta^s(t) = a_0 + a_1(t - t_c) + a_2(t - t_c)^2 + \delta^{\text{rel}}, \tag{6.4}$$

cf. ARINC Engineering Services (2006a).

The combined bias term $\Delta\delta_r^s(t)$ is split into two parts by

$$\Delta\delta_r^s(t) = \delta_r(t) - \delta^s(t), \tag{6.5}$$

where the satellite-related part $\delta^s(t)$ is known and the receiver-related term $\delta_r(t)$ remains unknown. Substituting (6.5) into (6.1) and shifting the satellite clock bias to the left side of the equation yields

$$R_r^s(t) + c\,\delta^s(t) = \varrho_r^s(t) + c\,\delta_r(t). \tag{6.6}$$

Note that the left side of the equation contains observed or known quantities, while the terms on the right side are unknown.

Basic configurations

Basic configurations are defined by the condition that the number of observations must be equal to or greater than the number of unknowns. This condition is sufficient but does not necessarily give a solution. The reason for this is that inherent rank deficiencies may prevent a numerical solution because of a singularity. More explanations are given later when the rank deficiency becomes an issue.

The number of observations is $n_s\,n_t$, where n_s denotes the number of satellites and n_t the number of epochs.

For static point positioning, the three coordinates of the observing site and the receiver clock bias for each observation epoch are unknown. Thus, the number of unknowns is $3 + n_t$. The basic configuration is defined by

$$n_s\,n_t \geq 3 + n_t, \tag{6.7}$$

which yields the explicit relation

$$n_t \geq \frac{3}{n_s - 1}. \tag{6.8}$$

The minimum number of satellites to get a solution is $n_s = 2$, leading to $n_t \geq 3$ observation epochs. For $n_s = 4$, the solution $n_t \geq 1$ is obtained. This solution

6.1 Point positioning

reflects the instantaneous positioning capability of GNSS, where the four unknowns at any epoch are solved if at least four satellites are tracked.

For kinematic point positioning, the basic configuration can be directly derived from the following consideration. Due to the motion of the receiver, the number of the unknown station coordinates is $3n_t$. Adding the n_t unknown receiver clock biases, the total number of unknowns is $4n_t$. Hence, the basic configuration is defined by Eq. (6.7),

$$n_s\, n_t \geq 4 n_t \tag{6.9}$$

yielding $n_s \geq 4$. In other words, the position (and system time) of a moving receiver can be determined at any instant as long as at least four satellites are tracked. Geometrically, the solution is represented by the intersection of four pseudoranges. For the rigorous analytical solution see Kleusberg (1994) or Lichtenegger (1995).

The basic configurations must be considered from a theoretical point of view. The solution $n_s = 2$, $n_t \geq 3$ for static point positioning, for example, means that simultaneous observations of two satellites over three epochs would theoretically suffice. In practice, however, this situation would yield unacceptable results or the computation would fail because of an ill-conditioned system of observation equations unless the epochs were widely spaced (e.g., hours). A solution is also possible if observations of three epochs for two satellites are made, followed by three additional epochs (e.g., seconds apart) for two other satellites. Such an application will be rare but is imaginable under special circumstances (e.g., in urban areas).

6.1.2 Point positioning with carrier phases

Phase range models

Pseudoranges can also be obtained from carrier phase measurements. The mathematical model for these measurements, cf. Eq. (5.9), is given by

$$\Phi_r^s(t) = \frac{1}{\lambda^s} \varrho_r^s(t) + N_r^s + \frac{c}{\lambda^s} \Delta\delta_r^s(t), \tag{6.10}$$

where $\Phi_r^s(t)$ is the measured carrier phase expressed in cycles, λ^s is the wavelength, and $\varrho_r^s(t)$ is the same as for the code range model. The time-independent phase ambiguity N_r^s is an integer number and, therefore, often called integer ambiguity or integer unknown or simply ambiguity. The term c denotes the speed of light and $\Delta\delta_r^s(t)$ is the combined receiver and satellite clock bias.

Inserting Eq. (6.5) into Eq. (6.10) and shifting the (known) satellite clock bias to the left side of the equation yields

$$\Phi_r^s(t) + f^s\, \delta^s(t) = \frac{1}{\lambda^s} \varrho_r^s(t) + N_r^s + f^s\, \delta_r(t), \tag{6.11}$$

where the frequency of the satellite carrier $f^s = c/\lambda^s$ has been substituted.

Basic configurations

Using the same notations as before, the number of observations is again $n_s\, n_t$. The number of unknowns, however, is increased by the number n_s because of the ambiguities.

For static point positioning, the number of unknowns is composed of 3 coordinates of the observing station, n_s unknown ambiguities, and n_t unknown receiver clock biases. Referring to (6.11), the problem of rank deficiency is encountered. Mathematically less interested readers may skip the next paragraph.

A few basics on rank and rank deficiency are given here. Deeper insight may be obtained from Koch (1987: Sects. 132, 333). Assume a large number of equations of type (6.11) being prepared to be solved for the unknowns. This implies a matrix-vector representation where the right side is composed of a product of a design matrix $\mathbf{A}$ and a vector comprising the unknowns in linear form. The rank of the design matrix is equal to the order of the largest nonsingular matrix that can be formed inside $\mathbf{A}$. Formulated differently: the maximum number of the linearly independent rows of matrix $\mathbf{A}$ is called the rank of the matrix and is denoted by rank $\mathbf{A}$. Linear dependence of two rows means that their linear combination yields zero. The word "rows" in this definition may also be replaced by the word "columns". For a simpler discussion, assume a quadratic matrix with $m \times m$ rows and columns. Thus, if the largest nonsingular matrix is the matrix $\mathbf{A}$ itself, the rank equals rank $\mathbf{A} = m$ and the matrix is regular, i.e., it may be inverted without troubles. On the other hand, if the largest nonsingular matrix inside $\mathbf{A}$ is a matrix with, e.g., $(m-2)\times(m-2)$ rows and columns, the rank would be $m-2$ and implies a rank deficiency of $m-\text{rank}\,\mathbf{A}$ which turns out to be $m-(m-2)$ which amounts to 2. As a consequence, the singular system becomes regularly solvable if two unknowns (also denoted as parameters) are arbitrarily chosen. This equals the "fixing" of two parameters. Figuratively speaking, two of the parameters may be transferred to the left side of the matrix-vector system comprising the measurements. This transfer reduces on the other hand the columns of the matrix on the right side by the amount of the rank deficiency, i.e., by two in the example discussed. This concludes the short discussion on rank and rank deficiency.

The model in the form (6.11) comprises a rank deficiency of 1, this means that one of the unknown parameters may (and must) be arbitrarily chosen. Suppose that a receiver clock bias at one epoch is chosen, then, instead of n_t unknown receiver clock biases, only $n_t - 1$ clock biases remain. Therefore, the basic configuration for static point positioning without rank deficiency is defined by the relation

$$n_s\, n_t \geq 3 + n_s + (n_t - 1), \tag{6.12}$$

which yields explicitly the required number of epochs as

$$n_t \geq \frac{n_s + 2}{n_s - 1}. \tag{6.13}$$

6.1 Point positioning

The minimum number of satellites to get a solution is $n_s = 2$ leading to $n_t \geq 4$ observation epochs. Another integer solution pair is $n_s = 4$, $n_t \geq 2$.

For kinematic point positioning with phases, $3n_t$ unknown station coordinates must be considered because of the roving receiver compared to the 3 unknowns in (6.12). The other considerations including the discussion on the rank deficiency remain unchanged. Therefore, the basic configuration is defined by

$$n_s\, n_t \geq 3n_t + n_s + (n_t - 1) \tag{6.14}$$

yielding the explicit relation

$$n_t \geq \frac{n_s - 1}{n_s - 4}. \tag{6.15}$$

The minimum number of satellites to get a solution is $n_s = 5$ which have to be tracked for $n_t \geq 4$ epochs. Another integer solution pair is $n_s = 7$, $n_t \geq 2$.

Note that solutions for a single epoch (i.e., $n_t = 1$) do not exist for point positioning with carrier phases. As a consequence, kinematic point positioning with phases is only possible if the n_s phase ambiguities are known from some initialization. In this case, the phase range model converts to the code range model.

6.1.3 Point positioning with Doppler data

The mathematical model for Doppler data, cf. Eq. (5.11), is

$$D_r^s(t) = \dot{\varrho}_r^s(t) + c\,\Delta\dot{\delta}_r^s(t) \tag{6.16}$$

and may be considered as time derivative of a code or phase pseudorange. In this equation, $D_r^s(t)$ denotes the observed Doppler shift scaled to range rate, $\dot{\varrho}_r^s(t)$ is the instantaneous radial velocity between the satellite and the receiver, and $\Delta\dot{\delta}_r^s(t)$ is the time derivative of the combined clock bias term.

The radial velocity for a stationary receiver, cf. Eq. (3.34),

$$\dot{\varrho}_r^s(t) = \frac{\boldsymbol{\varrho}^s(t) - \boldsymbol{\varrho}_r}{\|\boldsymbol{\varrho}^s(t) - \boldsymbol{\varrho}_r\|} \cdot \dot{\boldsymbol{\varrho}}^s(t) \tag{6.17}$$

relates the unknown position vector $\boldsymbol{\varrho}_r$ of the receiver to the instantaneous position vector $\boldsymbol{\varrho}^s(t)$ and velocity vector $\dot{\boldsymbol{\varrho}}^s(t)$ of the satellite. These vectors can be calculated from the satellite ephemerides. Introducing on the one hand $\varrho = \|\boldsymbol{\varrho}^s(t) - \boldsymbol{\varrho}_r\|$ according to (3.33) and on the other hand the components $X^s(t), Y^s(t), Z^s(t)$ of the vector $\boldsymbol{\varrho}^s(t)$, likewise X_r, Y_r, Z_r as the components of the vector $\boldsymbol{\varrho}_r$, and $\dot{X}^s(t), \dot{Y}^s(t), \dot{Z}^s(t)$ for the vector $\dot{\boldsymbol{\varrho}}^s(t)$ accordingly, the radial velocity may also be written as

$$\dot{\varrho}_r^s(t) = \frac{X^s(t) - X_r}{\varrho}\dot{X}^s(t) + \frac{Y^s(t) - Y_r}{\varrho}\dot{Y}^s(t) + \frac{Z^s(t) - Z_r}{\varrho}\dot{Z}^s(t) \tag{6.18}$$

after evaluating the inner product.

The contribution of the satellite clock to $\Delta \dot{\delta}_r^s(t)$ is given by, cf. Eq. (6.3),

$$\dot{\delta}^s(t) = a_1 + 2a_2(t - t_c) \tag{6.19}$$

and is known. Summarizing, the observation equation (6.16) contains four unknowns. These unknowns are the three coordinates of ϱ_r and the receiver clock drift $\dot{\delta}_r(t)$. Hence, compared to the code range model, the Doppler equation contains the receiver clock drift instead of the receiver clock offset.

The concept of combined code pseudorange and Doppler data processing leads to a total of five unknowns. These unknowns are the three point coordinates, the receiver clock offset, and the receiver clock drift. Each satellite contributes two equations, one code pseudorange and one Doppler equation. Therefore, three satellites are sufficient to solve for the five unknowns.

The similarity of the pseudorange and the Doppler equation gives rise to the question of a linear dependence of the equations. However, it can be shown that the surfaces of constant pseudoranges and the surfaces of constant Doppler are orthogonal and hence independent (Levanon 1999).

6.1.4 Precise point positioning

Basic model

Considering the methods of point positioning described in the previous sections, the main limiting factors with respect to the achievable accuracy are the orbit errors, the clock errors, and the atmospheric influences (ionospheric and tropospheric refraction). Therefore, following Witchayangkoon (2000: p. 2), precise point positioning (PPP) uses accurate orbital data and accurate satellite clock data (as provided, e.g., by the IGS), and dual-frequency code pseudoranges and/or carrier phase observations by definition. The preferred model is based on an ionosphere-free combination of code pseudoranges and carrier phases as well.

The respective equation for the code pseudoranges is obtained from (5.83) and reads

$$\left[R_1 - \frac{f_2^2}{f_1^2} R_2 \right] \frac{f_1^2}{f_1^2 - f_2^2} = \varrho + c\,\Delta\delta + \Delta^{\mathrm{Trop}}, \tag{6.20}$$

where a term to model the tropospheric delay has been added.

The ionosphere-free carrier phase relation as given in (5.80) reads

$$\left[\Phi_1 - \frac{f_2}{f_1} \Phi_2 \right] \frac{f_1^2}{f_1^2 - f_2^2} = \frac{f_1}{c} \varrho + f_1 \Delta\delta + \left[N_1 - \frac{f_2}{f_1} N_2 \right] \frac{f_1^2}{f_1^2 - f_2^2}. \tag{6.21}$$

6.1 Point positioning

This equation is now multiplied by the factor c/f_1 yielding

$$\left[\Phi_1 - \frac{f_2}{f_1}\Phi_2\right]\frac{c\,f_1}{f_1^2 - f_2^2} = \varrho + c\,\Delta\delta + \left[N_1 - \frac{f_2}{f_1}N_2\right]\frac{c\,f_1}{f_1^2 - f_2^2} \tag{6.22}$$

or, by substituting $c = \lambda_1 f_1$ and adding the tropospheric delay,

$$\left[\Phi_1 - \frac{f_2}{f_1}\Phi_2\right]\frac{\lambda_1 f_1^2}{f_1^2 - f_2^2} = \varrho + c\,\Delta\delta + \Delta^{\text{Trop}} + \left[N_1 - \frac{f_2}{f_1}N_2\right]\frac{\lambda_1 f_1^2}{f_1^2 - f_2^2} \tag{6.23}$$

is obtained. This formula yields after a slight rearrangement and by using $c = \lambda_2 f_2$

$$\frac{\lambda_1 \Phi_1 f_1^2}{f_1^2 - f_2^2} - \frac{\lambda_2 \Phi_2 f_2^2}{f_1^2 - f_2^2} = \varrho + c\,\Delta\delta + \Delta^{\text{Trop}} + \frac{\lambda_1 N_1 f_1^2}{f_1^2 - f_2^2} - \frac{\lambda_2 N_2 f_2^2}{f_1^2 - f_2^2}. \tag{6.24}$$

In summary, (6.20) and (6.24) are the desired ionosphere-free combinations of code pseudoranges and carrier phases for PPP:

$$\begin{aligned}\frac{R_1 f_1^2}{f_1^2 - f_2^2} - \frac{R_2 f_2^2}{f_1^2 - f_2^2} &= \varrho + c\,\Delta\delta + \Delta^{\text{Trop}}, \\ \frac{\lambda_1 \Phi_1 f_1^2}{f_1^2 - f_2^2} - \frac{\lambda_2 \Phi_2 f_2^2}{f_1^2 - f_2^2} &= \varrho + c\,\Delta\delta + \Delta^{\text{Trop}} + \frac{\lambda_1 N_1 f_1^2}{f_1^2 - f_2^2} - \frac{\lambda_2 N_2 f_2^2}{f_1^2 - f_2^2},\end{aligned} \tag{6.25}$$

where (6.20) has been slightly rearranged. The unknown parameters to be determined are the point position contained in ϱ, the receiver clock error contained in $\Delta\delta$ (see Eq. (5.1)), the tropospheric delay Δ^{Trop}, and the ambiguities. Based on this model, PPP may be applied either in static or in kinematic mode.

To solve for the mentioned unknowns, several methods are possible. Deo et al. (2003) apply a sequential least-squares adjustment, (extended) Kalman filtering is another frequently used method.

Apart from the PPP model given in (6.25), different strategies may be found in the literature, e.g., with respect to the tropospheric term. Witchayangkoon (2000) and Kouba and Héroux (2001) estimate the total tropospheric zenith path delay as above, whereas Gao and Shen (2001) model the dry tropospheric zenith path delay and estimate the wet component as a parameter.

Model refinements

To exploit the full potential of PPP, a refinement of the model must be performed. Additional terms are necessary to account for the Sagnac effect, the solid earth tides, the ocean loading, the atmospheric loading (caused by the atmospheric pressure variation), polar motion, earth orientation effects, crustal motion and other earth deformation effects (Kouba and Héroux 2001).

Also the antenna phase center offset (at the satellite and at the receiver) and antenna phase wind-up error (Witchayangkoon 2000: pp. 24–26) should be taken into account.

The proper weighting of the observations is also a key to improve the accuracy. Numerous investigations on different weighting schemes exist. Among them, Witchayangkoon (2000: Sect. 7.3.3.4) mentions exponential weighting schemes taking into account that observations from satellites near the horizon get a lower weight (Euler and Goad 1991); using weights reflecting the signal-to-noise (S/N) ratio values (Collins and Langley 1999, Hartinger and Brunner 1999); Langley (1997) derives carrier-to-noise power density (C/N_0) ratios varying with the elevations of the arriving signal; weighting as a cosecant or square of a cosecant function of the satellite elevation angle E (Vermeer 1997, Collins and Langley 1999, Hartinger and Brunner 1999) which is justified by the cosecant shape of the various models of the tropospheric mapping function. Wieser (2007a, b) compares identical variances σ^2 for all observations, elevation-dependent variances $\sigma_0^2/\sin^2 E$, and SIGMA-ε variances defined by $k \cdot 10^{-(C/N_0)/10}$, where the model parameters σ, σ_0, and the factor k depend on the receiver and antenna types and can be determined in advance. As outlined in Wieser et al. (2005), the measured C/N_0 is a quality indicator because there is a functional relation between this quantity and the tracking loop noise.

Numerical results

Gao and Chen (2004) use real-time precise orbit and clock corrections (accurate to 20 cm and 0.5 ns, respectively) provided by JPL and present results for different positioning modes. For a static observation, they demonstrate that all position components (latitude, longitude, height) "converge to centimeter level" after 20 minutes. The problem of the convergence arises from the ambiguities. After this convergence has been achieved, the results remain even below the subcentimeter level. Therefore, they conclude that PPP is capable to provide real-time centimeter-level accuracy for static surveys.

An additional remark on the ambiguities is appropriate here: because of unknown receiver and also transmitter-specific phase delays which in addition vary with time, the ambiguities are not integers (Zumberge et al. 1997); only the double-difference ambiguities are integers. Note, however, that for "applications that do not require accuracies better than a few millimeters in the horizontal dimension and approximately 1 cm in the vertical dimension, ambiguity resolution is not necessary, provided that the observation time is of the order of 1 day" (Zumberge et al. 1997).

For kinematic applications, also real-time centimeter-level accuracy is demonstrated for a car and an airplane by Gao and Chen (2004), again based on the JPL orbit and clock corrections.

Note, however, these excellent results mainly profit from the JPL input data for the satellite orbit and, even more important, the satellite clock correction. These JPL data are available for commercial applications (Gao and Chen 2004). When relying on freely available products like the predicted ultrarapid orbit product as provided by the IGS, with an even better (compared to the JPL product) orbit accuracy of 10 cm but a significantly worse accuracy of 5 ns for the clock (Gao and Chen 2004), then this accuracy is insufficient for real-time decimeter-level PPP (Deo et al. 2003). Kinematic results based on the use of IGS final orbit and clock corrections are given in Abdel-salam (2005). The position accuracy is better than 3 decimeters for many cases including land vehicle, marine, and airborne applications.

Witchayangkoon (2000) gives a detailed model and incorporates corrections for solid earth tides, relativity, and satellite antenna phase center offsets and reports some results from numerical examples. In cases of low impact by multipath, "single-frequency ionosphere-free PPP solutions are equivalent to the dual-frequency solutions".

Thus, a future trend is PPP using single-frequency data only. Even with a simple model as proposed by Satirapod and Kriengkraiwasin (2006), which uses single-frequency ionospherically corrected code and phase observations corrected, introduces the Saastamoinen troposphere model to calculate the total tropospheric zenith delay which is mapped to the line-of-sight delay, a horizontal accuracy of 1–4 m can be achieved with data sessions ranging from 5–30 minutes (but being based on the precise orbit files of the IGS).

6.2 Differential positioning

6.2.1 Basic concept

Differential positioning with GNSS, abbreviated by DGNSS, is a real-time positioning technique where two or more receivers are used. One receiver, usually at rest, is located at the reference or base station with (assumed) known coordinates and the remote receivers are fixed or roving and their coordinates are to be determined (Fig. 6.1). The reference station commonly calculates pseudorange corrections (PRC) and range rate corrections (RRC) which are transmitted to the remote receiver in real time. The remote receiver applies the corrections to the measured pseudoranges and performs point positioning with the corrected pseudoranges. The use of the corrected pseudoranges improves the position accuracy with respect to the base station.

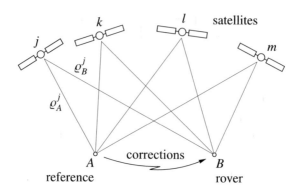

Fig. 6.1. Basic concept of differential positioning

6.2.2 DGNSS with code ranges

Generalizing (6.6) and following Lichtenegger (1998), the code range at base station A to satellite s measured at epoch t_0 may be modeled by

$$R_A^s(t_0) = \varrho_A^s(t_0) + \Delta\varrho_A^s(t_0) + \Delta\varrho^s(t_0) + \Delta\varrho_A(t_0), \tag{6.26}$$

where $\varrho_A^s(t_0)$ is the geometric range, the term $\Delta\varrho_A^s(t_0)$ denotes range biases depending on the terrestrial base position and satellite position as well (e.g., radial orbital error, refraction effects), the range bias $\Delta\varrho^s(t_0)$ is purely satellite-dependent (e.g., effect of satellite clock error), and the range bias $\Delta\varrho_A(t_0)$ is purely receiver-dependent (e.g., effect of receiver clock error, multipath). Note that noise has been neglected in (6.26).

The pseudorange correction for satellite s at reference epoch t_0 is defined by the relation

$$\begin{aligned} \mathrm{PRC}^s(t_0) &= \varrho_A^s(t_0) - R_A^s(t_0) \\ &= -\Delta\varrho_A^s(t_0) - \Delta\varrho^s(t_0) - \Delta\varrho_A(t_0) \end{aligned} \tag{6.27}$$

and can be calculated since the geometric range $\varrho_A^s(t_0)$ is obtained from the known position of the reference station and the broadcast ephemerides and $R_A^s(t_0)$ is the measured quantity. In addition to the pseudorange correction $\mathrm{PRC}^s(t_0)$, the time derivative or range rate correction $\mathrm{RRC}^s(t_0)$ is determined at the base station.

Range and range rate corrections referring to the reference epoch t_0 are transmitted to the rover site B in real time. At B the pseudorange corrections are predicted for the observation epoch t using the relation

$$\mathrm{PRC}^s(t) = \mathrm{PRC}^s(t_0) + \mathrm{RRC}^s(t_0)(t - t_0), \tag{6.28}$$

where $t - t_0$ is defined as latency. The achievable accuracy increases for smaller variations of the pseudorange corrections and for smaller latencies.

6.2 Differential positioning

Adapting (6.26) to the rover site B and epoch t, the code pseudorange measured at the rover can be modeled by

$$R_B^s(t) = \varrho_B^s(t) + \Delta\varrho_B^s(t) + \Delta\varrho^s(t) + \Delta\varrho_B(t). \tag{6.29}$$

Applying the predicted pseudorange correction $\text{PRC}^s(t)$, cf. Eq. (6.28), to the measured pseudorange $R_B^s(t)$ yields

$$R_B^s(t)_{\text{corr}} = R_B^s(t) + \text{PRC}^s(t) \tag{6.30}$$

or, after substitution of (6.29) and the pseudorange correction according to (6.27) and (6.28), respectively,

$$R_B^s(t)_{\text{corr}} = \varrho_B^s(t) + [\Delta\varrho_B^s(t) - \Delta\varrho_A^s(t)] + [\Delta\varrho_B(t) - \Delta\varrho_A(t)], \tag{6.31}$$

where the satellite-dependent bias has canceled out. For moderate distances between the base and the rover site, the satellite–receiver-specific biases are highly correlated. Therefore, the influence of radial orbital errors and of refraction is significantly reduced. Neglecting these biases, Eq. (6.31) simplifies to

$$R_B^s(t)_{\text{corr}} = \varrho_B^s(t) + \Delta\varrho_{AB}(t), \tag{6.32}$$

where $\Delta\varrho_{AB}(t) = \Delta\varrho_B(t) - \Delta\varrho_A(t)$. If multipath is neglected, this term converts to the combined receiver clock bias scaled to range, i.e., $\Delta\varrho_{AB}(t) = c\,\delta_{AB}(t) = c\,\delta_B(t) - c\,\delta_A(t)$. If no latency exists, the equation is identical with the between-receiver single-difference of code ranges measured at A and B, and differential positioning converts to relative positioning (Sect. 6.3).

Positioning at the rover site B is performed with the corrected code pseudoranges $R_B^s(t)_{\text{corr}}$ leading to improved position accuracies. The basic configuration for DGNSS with code ranges is identical with that for kinematic point positioning with code ranges, cf. Eq. (6.9).

6.2.3 DGNSS with phase ranges

Generalizing (6.10) and following Lichtenegger (1998), the phase pseudorange measured at the base station A at epoch t_0 can be modeled by

$$\lambda^s\,\Phi_A^s(t_0) = \varrho_A^s(t_0) + \Delta\varrho_A^s(t_0) + \Delta\varrho^s(t_0) + \Delta\varrho_A(t_0) + \lambda^s N_A^s, \tag{6.33}$$

where, in analogy to the code range model, $\varrho_A^s(t_0)$ is the geometric range, $\Delta\varrho_A^s(t_0)$ is the satellite–receiver-dependent bias, $\Delta\varrho^s(t_0)$ is purely satellite-dependent, $\Delta\varrho_A(t_0)$ is purely receiver-dependent. Finally, N_A^s is the phase ambiguity. Consequently, the phase range correction at reference epoch t_0 is given by

$$\begin{aligned}\text{PRC}^s(t_0) &= \varrho_A^s(t_0) - \lambda^s\,\Phi_A^s(t_0), \\ &= -\Delta\varrho_A^s(t_0) - \Delta\varrho^s(t_0) - \Delta\varrho_A(t_0) - \lambda^s N_A^s.\end{aligned} \tag{6.34}$$

The formulation of range rate corrections at the base station A as well as the application of predicted range corrections to the observed phase ranges at the rover site B is carried out in full analogy to the previously described code range procedure. Therefore,

$$\lambda^s \, \Phi_B^s(t)_{\text{corr}} = \varrho_B^s(t) + \Delta\varrho_{AB}(t) + \lambda^s N_{AB}^s \tag{6.35}$$

results for the corrected phase ranges, where $\Delta\varrho_{AB}(t) = \Delta\varrho_B(t) - \Delta\varrho_A(t)$ and $N_{AB}^s = N_B^s - N_A^s$ is the (single-) difference of the phase ambiguities. As in the code range model, if multipath is neglected, the term $\Delta\varrho_{AB}(t)$ converts to the combined receiver clock bias scaled to range, i.e., $\Delta\varrho_{AB}(t) = c\,\delta_{AB}(t) = c\,\delta_B(t) - c\,\delta_A(t)$.

Point positioning at the rover site B is performed with the corrected phase pseudoranges $\lambda^s \, \Phi_B^s(t)_{\text{corr}}$. The basic configuration for DGNSS with phase ranges is identical with that for kinematic point positioning with phase ranges, cf. Eq. (6.15).

DGNSS with phase ranges, sometimes denoted as carrier phase differential technique, is used for most precise kinematic applications. For this mode of operation, on-the-fly (OTF) techniques are required to resolve the ambiguities. More details on OTF are given in Sect. 7.2.3.

Note that DGNSS with phases converts to relative positioning with phases if the latency becomes zero. This method is usually denoted real-time kinematic (RTK) technique.

6.2.4 Local-area DGNSS

An extension of DGNSS is the local-area DGNSS (LADGNSS) which uses a network of GNSS reference stations. As the name implies, LADGNSS covers a larger territory than can be reasonably accommodated by a single reference station. One of the main advantages of LADGNSS is that a more consistent accuracy can be achieved throughout the region supported by the network. In the case of DGNSS with a single reference station, the accuracy decreases as a function of distance from the reference station at a rate of approximately 1 cm per 1 km. Other advantages of LADGNSS are that inaccessible regions can be covered, e.g., large bodies of water, and that in case of a failure in one of the reference stations, the network will still maintain a relatively high level of integrity and reliability compared to a collection of individual DGNSS reference stations.

Apart from the monitor stations, the LADGNSS network includes (at least) one master station. This station collects the range corrections from the monitor stations, processes these data to form LADGNSS corrections which are transmitted to the user community as well as to the monitor stations (Mueller 1994). The networks may cause slight additional delay beyond regular DGNSS due to the additional communication required between the monitor stations and the master station.

6.3 Relative positioning

Since the reference stations of the LADGNSS network may be very distant from the user location, the virtual reference station (VRS) concept (Sect. 6.3.7) has been developed (Wanninger 1999). Here, the user gets range corrections or even the observables of a nonexistent (i.e., virtual) reference station at a user-specified position. This concept is a prerequisite mainly for RTK applications which require short distances to reference stations to facilitate ambiguity resolution.

6.3 Relative positioning

6.3.1 Basic concept

The objective of relative positioning is to determine the coordinates of an unknown point with respect to a known point which, for most applications, is stationary. In other words, relative positioning aims at the determination of the vector between the two points, which is often called the baseline vector or simply baseline (Fig. 6.2). Let A denote the (known) reference point, B the unknown point, and $\mathbf{b}_{AB}$ the baseline vector. Introducing the corresponding position vectors $\mathbf{X}_A$, $\mathbf{X}_B$, the relation

$$\mathbf{X}_B = \mathbf{X}_A + \mathbf{b}_{AB} \tag{6.36}$$

may be formulated, and the components of the baseline vector $\mathbf{b}_{AB}$ are

$$\mathbf{b}_{AB} = \begin{bmatrix} X_B - X_A \\ Y_B - Y_A \\ Z_B - Z_A \end{bmatrix} = \begin{bmatrix} \Delta X_{AB} \\ \Delta Y_{AB} \\ \Delta Z_{AB} \end{bmatrix}. \tag{6.37}$$

The coordinates of the reference point must be given and can be approximated by a code range solution. More often the coordinates are precisely known based upon GNSS or other methods.

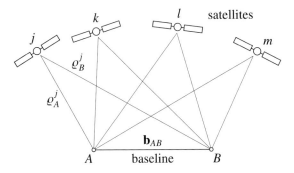

Fig. 6.2. Basic concept of relative positioning

Relative positioning can be performed with code ranges, cf. Eq. (6.6), or with phase ranges, cf. Eq. (6.11). Subsequently, only phase ranges are explicitly considered because solutions based on phase ranges are far more accurate. Relative positioning requires simultaneous observations at both the reference and the unknown point. This means that the observation time tags for the two points must be the same. Assuming such simultaneous observations at the two points A and B to satellites j and k, linear combinations can be formed leading to single-differences, double-differences, and triple-differences. Differencing can basically be accomplished in three different ways: across receivers, across satellites, across time (Logsdon 1992: p. 96). Instead of "across" frequently "between" is used. In order to avoid overburdened expressions, shorthand notations will be used throughout the textbook with the following meanings: single-difference corresponds to across-receiver difference (or between-receiver difference), double-difference corresponds to across-receiver and across-satellite difference, and triple-difference corresponds to across-receiver and across-satellite and across-time difference. Most postprocessing software uses these three difference techniques, so their basic mathematical modeling is shown in the following sections.

6.3.2 Phase differences

Single-differences

Two receivers and one satellite are involved. Denoting the receiver sites by A and B and the satellite by j and using Eq. (6.11), the phase equations for the two points are

$$\Phi_A^j(t) + f^j \, \delta^j(t) = \frac{1}{\lambda^j} \varrho_A^j(t) + N_A^j + f^j \, \delta_A(t),$$
$$\Phi_B^j(t) + f^j \, \delta^j(t) = \frac{1}{\lambda^j} \varrho_B^j(t) + N_B^j + f^j \, \delta_B(t) \qquad (6.38)$$

and the difference of the two equations is

$$\Phi_B^j(t) - \Phi_A^j(t) = \frac{1}{\lambda^j} \left[\varrho_B^j(t) - \varrho_A^j(t)\right] + N_B^j - N_A^j + f^j \left[\delta_B(t) - \delta_A(t)\right]. \quad (6.39)$$

Equation (6.39) is referred to as single-difference equation. This equation stresses one aspect of the solution for the unknowns on the right side. A system of such equations would lead to a rank deficiency even in the case of an arbitrarily large redundancy. This means that the design matrix of the adjustment has linearly dependent columns and a rank deficiency exists. Therefore, the relative quantities

$$N_{AB}^j = N_B^j - N_A^j,$$
$$\delta_{AB}(t) = \delta_B(t) - \delta_A(t) \qquad (6.40)$$

6.3 Relative positioning

are introduced. Using additionally the shorthand notations

$$\Phi^j_{AB}(t) = \Phi^j_B(t) - \Phi^j_A(t),$$
$$\varrho^j_{AB}(t) = \varrho^j_B(t) - \varrho^j_A(t) \tag{6.41}$$

and substituting (6.40) and (6.41) into (6.39) gives

$$\Phi^j_{AB}(t) = \frac{1}{\lambda^j} \varrho^j_{AB}(t) + N^j_{AB} + f^j \delta_{AB}(t), \tag{6.42}$$

which is the final form of the single-difference equation. Note that the satellite clock bias has canceled, compared to the phase equation (6.11).

Double-differences

Assuming the two points A, B, and the two satellites j, k, two single-differences according to Eq. (6.42) may be formed:

$$\Phi^j_{AB}(t) = \frac{1}{\lambda^j} \varrho^j_{AB}(t) + N^j_{AB} + f^j \delta_{AB}(t),$$
$$\Phi^k_{AB}(t) = \frac{1}{\lambda^k} \varrho^k_{AB}(t) + N^k_{AB} + f^k \delta_{AB}(t). \tag{6.43}$$

To obtain a double-difference, these single-differences are subtracted. Two cases must be considered.

Case 1
Assuming equal frequencies $f = f^j = f^k$ for the satellite signals, the result is

$$\Phi^k_{AB}(t) - \Phi^j_{AB}(t) = \frac{1}{\lambda} \left[\varrho^k_{AB}(t) - \varrho^j_{AB}(t) \right] + N^k_{AB} - N^j_{AB}. \tag{6.44}$$

Using shorthand notations for the satellites j and k analogously to (6.41), the final form of the double-difference equation is

$$\Phi^{jk}_{AB}(t) = \frac{1}{\lambda} \varrho^{jk}_{AB}(t) + N^{jk}_{AB}, \tag{6.45}$$

where $\lambda = \lambda^j = \lambda^k$. The elimination of the receiver clock biases is the main reason why double-differences are preferably used. This cancellation resulted from the assumptions of simultaneous observations and equal frequencies of the satellite signals.

Symbolically, the convention

$$*^{jk}_{AB} = *^k_{AB} - *^j_{AB} \tag{6.46}$$

has been introduced, where the asterisk may be replaced by Φ, ϱ, or N. Note that these terms comprising two subscripts and two superscripts are actually composed of four terms. The symbolic notation

$$*_{AB}^{jk} = *_B^k - *_B^j - *_A^k + *_A^j \tag{6.47}$$

characterizes, in detail, the terms in the double-difference equation:

$$\begin{aligned}\Phi_{AB}^{jk}(t) &= \Phi_B^k(t) - \Phi_B^j(t) - \Phi_A^k(t) + \Phi_A^j(t), \\ \varrho_{AB}^{jk}(t) &= \varrho_B^k(t) - \varrho_B^j(t) - \varrho_A^k(t) + \varrho_A^j(t), \\ N_{AB}^{jk} &= N_B^k - N_B^j - N_A^k + N_A^j.\end{aligned} \tag{6.48}$$

Case 2

Now different frequencies $f^j \neq f^k$ are considered. Referring to Eq. (6.38), the model equations for carrier phases measured at the two sites A and B to the satellite j are now given by

$$\begin{aligned}\Phi_A^j(t) + f^j\,\delta^j(t) &= \frac{1}{\lambda^j}\varrho_A^j(t) + N_A^j + f^j\,\delta_A(t), \\ \Phi_B^j(t) + f^j\,\delta^j(t) &= \frac{1}{\lambda^j}\varrho_B^j(t) + N_B^j + f^j\,\delta_B(t).\end{aligned} \tag{6.49}$$

The measured phases are scaled to ranges by

$$\tilde{\Phi}^j(t) = \lambda^j\,\Phi^j(t), \tag{6.50}$$

where $\tilde{\Phi}^j(t)$ is output by the receiver. The single-difference of the two equations (6.49) leads to

$$\tilde{\Phi}_B^j(t) - \tilde{\Phi}_A^j(t) = \varrho_B^j(t) - \varrho_A^j(t) + \lambda^j\,[N_B^j - N_A^j] + c\,[\delta_B(t) - \delta_A(t)] \tag{6.51}$$

with $c = \lambda^j f^j$ being the speed of light. Introducing the shorthand notations of (6.40) and (6.41), i.e., symbolically $*_{AB}^j = *_B^j - *_A^j$, a more compact form is achieved by

$$\tilde{\Phi}_{AB}^j(t) = \varrho_{AB}^j(t) + \lambda^j\,N_{AB}^j + c\,\delta_{AB}(t). \tag{6.52}$$

Assuming two satellites j, k gives rise to two single-differences (6.52). From these the double-difference

$$\tilde{\Phi}_{AB}^k(t) - \tilde{\Phi}_{AB}^j(t) = \varrho_{AB}^k(t) - \varrho_{AB}^j(t) + \lambda^k\,N_{AB}^k - \lambda^j\,N_{AB}^j \tag{6.53}$$

6.3 Relative positioning

is obtained. Introducing again shorthand notations, symbolically $*_{AB}^{jk} = *_{AB}^{k} - *_{AB}^{j}$, there results

$$\tilde{\Phi}_{AB}^{jk}(t) = \varrho_{AB}^{jk}(t) + \lambda^k N_{AB}^k - \lambda^j N_{AB}^j, \qquad (6.54)$$

which may be rearranged by "adding zero" in the form of $-\lambda^k N_{AB}^j + \lambda^k N_{AB}^j$ so that

$$\tilde{\Phi}_{AB}^{jk}(t) = \varrho_{AB}^{jk}(t) + \lambda^k N_{AB}^{jk} + N_{AB}^j(\lambda^k - \lambda^j) \qquad (6.55)$$

is finally obtained. This equation differs from the double-difference equation (6.45) by the "single-difference bias" $b_{SD} = N_{AB}^j(\lambda^k - \lambda^j)$. The unknown single-difference N_{AB}^j can be estimated from single-point positioning with an accuracy of about 10 m corresponding to 50 cycles (for a typical wavelength of some 20 cm). If the wavelength difference for two carriers corresponds to 0.000 351 cycles (which is typical for GLONASS carriers being separated by one carrier number, cf. Eq. (10.2)), then the result $b_{SD} = 0.02$ cycles is obtained. This shows that for small frequency differences b_{SD} acts as a nuisance parameter. For larger differences, iterative processing has been proposed. In the first step, only satellites with small wavelength differences are considered. Hence, the double-differenced ambiguities of these satellites can be resolved and an improved position is obtained leading to a more accurate estimation for N_{AB}^j. The procedure is then continued and stepwise extended to all satellites until all ambiguities have been resolved. More details on this subject can be found in Habrich et al. (1999), Han et al. (1999).

Triple-differences

So far only one epoch t has been considered. To eliminate the time-independent ambiguities, Remondi (1984) has suggested differencing double-differences between two epochs. Subsequently, only the case $f^j = f^k$ is considered explicitly. Denoting the two epochs in (6.45) by t_1 and t_2, then

$$\Phi_{AB}^{jk}(t_1) = \frac{1}{\lambda} \varrho_{AB}^{jk}(t_1) + N_{AB}^{jk},$$
$$\Phi_{AB}^{jk}(t_2) = \frac{1}{\lambda} \varrho_{AB}^{jk}(t_2) + N_{AB}^{jk} \qquad (6.56)$$

are the two double-differences, and

$$\Phi_{AB}^{jk}(t_2) - \Phi_{AB}^{jk}(t_1) = \frac{1}{\lambda}\left[\varrho_{AB}^{jk}(t_2) - \varrho_{AB}^{jk}(t_1)\right] \qquad (6.57)$$

is the triple-difference, which may be written in the simplified form

$$\Phi_{AB}^{jk}(t_{12}) = \frac{1}{\lambda} \varrho_{AB}^{jk}(t_{12}) \qquad (6.58)$$

if the symbolic formula

$$*(t_{12}) = *(t_2) - *(t_1) \tag{6.59}$$

is applied to the terms Φ and ϱ. It should be noted that both $\Phi_{AB}^{jk}(t_{12})$ and $\varrho_{AB}^{jk}(t_{12})$ are actually composed of eight terms each. Resubstituting (6.57) and either (6.47) or (6.48) yields

$$\begin{aligned}\Phi_{AB}^{jk}(t_{12}) = &+ \Phi_B^k(t_2) - \Phi_B^j(t_2) - \Phi_A^k(t_2) + \Phi_A^j(t_2) \\ &- \Phi_B^k(t_1) + \Phi_B^j(t_1) + \Phi_A^k(t_1) - \Phi_A^j(t_1)\end{aligned} \tag{6.60}$$

and

$$\begin{aligned}\varrho_{AB}^{jk}(t_{12}) = &+ \varrho_B^k(t_2) - \varrho_B^j(t_2) - \varrho_A^k(t_2) + \varrho_A^j(t_2) \\ &- \varrho_B^k(t_1) + \varrho_B^j(t_1) + \varrho_A^k(t_1) - \varrho_A^j(t_1).\end{aligned} \tag{6.61}$$

It may be proved by the reader that for the case $f^j \neq f^k$ the equation

$$\tilde{\Phi}_{AB}^{jk}(t_{12}) = \varrho_{AB}^{jk}(t_{12}) \tag{6.62}$$

is obtained instead of (6.58).

The advantage of triple-differences is the canceling effect for the ambiguities, which eliminates the need to determine them.

6.3.3 Correlations of the phase combinations

In general, there are two groups of correlations, (1) the physical and (2) the mathematical correlations. The phases from one satellite received at two points, for example, $\Phi_A^j(t)$ and $\Phi_B^j(t)$, are physically correlated since they refer to the same satellite. Usually, the physical correlation is not taken into account. The main interest is directed to the mathematical correlations introduced by differencing.

The assumption may be made that the phase errors show a random behavior resulting in a normal distribution with expectation value zero and variance σ^2, where the variance is estimated by the UERE. Measured (or raw) phases are, therefore, linearly independent or uncorrelated. Introducing a vector Φ containing the phases and assuming equal accuracy, then

$$\Sigma_\Phi = \sigma^2 \mathbf{I} \tag{6.63}$$

is the covariance matrix for the phases, where $\mathbf{I}$ is the unit matrix.

6.3 Relative positioning

Correlation of single-differences

Considering the two points A, B and the satellite j at epoch t gives

$$\Phi_{AB}^{j}(t) = \Phi_{B}^{j}(t) - \Phi_{A}^{j}(t) \tag{6.64}$$

as the corresponding single-difference. Forming a second single-difference for the same two points but with another satellite k at the same epoch yields

$$\Phi_{AB}^{k}(t) = \Phi_{B}^{k}(t) - \Phi_{A}^{k}(t). \tag{6.65}$$

The two single-differences may be computed from the matrix-vector relation

$$\mathbf{S} = \mathbf{C}\,\mathbf{\Phi}, \tag{6.66}$$

where

$$\mathbf{S} = \begin{bmatrix} \Phi_{AB}^{j}(t) \\ \Phi_{AB}^{k}(t) \end{bmatrix}, \quad \mathbf{C} = \begin{bmatrix} -1 & 1 & 0 & 0 \\ 0 & 0 & -1 & 1 \end{bmatrix}, \quad \mathbf{\Phi} = \begin{bmatrix} \Phi_{A}^{j}(t) \\ \Phi_{B}^{j}(t) \\ \Phi_{A}^{k}(t) \\ \Phi_{B}^{k}(t) \end{bmatrix}. \tag{6.67}$$

The covariance law applied to Eq. (6.66) gives

$$\Sigma_{\mathbf{S}} = \mathbf{C}\,\Sigma_{\mathbf{\Phi}}\,\mathbf{C}^{\mathrm{T}} \tag{6.68}$$

and, by substituting Eq. (6.63),

$$\Sigma_{\mathbf{S}} = \mathbf{C}\,\sigma^{2}\,\mathbf{I}\,\mathbf{C}^{\mathrm{T}} = \sigma^{2}\,\mathbf{C}\,\mathbf{C}^{\mathrm{T}} \tag{6.69}$$

is obtained. Taking $\mathbf{C}$ from (6.67), the matrix product

$$\mathbf{C}\,\mathbf{C}^{\mathrm{T}} = 2 \begin{bmatrix} 1 & 0 \\ 0 & 1 \end{bmatrix} = 2\,\mathbf{I} \tag{6.70}$$

substituted into (6.69) leads to the covariance of the single-differences

$$\Sigma_{\mathbf{S}} = 2\sigma^{2}\,\mathbf{I}. \tag{6.71}$$

This shows that single-differences are uncorrelated. Note that the dimension of the unit matrix in (6.71) corresponds to the number of single-differences at epoch t, whereas the factor 2 does not depend on the number of single-differences. Considering more than one epoch, the covariance matrix is again a unit matrix with the dimension equivalent to the total number of single-differences.

Correlation of double-differences

Now, three satellites j, k, ℓ with j as reference satellite are considered. For the two points A, B and epoch t, the double-differences

$$\Phi_{AB}^{jk}(t) = \Phi_{AB}^{k}(t) - \Phi_{AB}^{j}(t),$$
$$\Phi_{AB}^{j\ell}(t) = \Phi_{AB}^{\ell}(t) - \Phi_{AB}^{j}(t) \tag{6.72}$$

can be derived from the single-differences. These two equations can be written in the matrix-vector form

$$\mathbf{D} = \mathbf{C}\,\mathbf{S}, \tag{6.73}$$

where

$$\mathbf{D} = \begin{bmatrix} \Phi_{AB}^{jk}(t) \\ \Phi_{AB}^{j\ell}(t) \end{bmatrix},$$

$$\mathbf{C} = \begin{bmatrix} -1 & 1 & 0 \\ -1 & 0 & 1 \end{bmatrix}, \quad \mathbf{S} = \begin{bmatrix} \Phi_{AB}^{j}(t) \\ \Phi_{AB}^{k}(t) \\ \Phi_{AB}^{\ell}(t) \end{bmatrix} \tag{6.74}$$

have been introduced. The covariance matrix for the double-differences is given by

$$\Sigma_\mathbf{D} = \mathbf{C}\,\Sigma_\mathbf{S}\,\mathbf{C}^\mathrm{T} \tag{6.75}$$

and substituting (6.71) leads to

$$\Sigma_\mathbf{D} = 2\sigma^2\,\mathbf{C}\,\mathbf{C}^\mathrm{T} \tag{6.76}$$

or, explicitly, using $\mathbf{C}$ from (6.74),

$$\Sigma_\mathbf{D} = 2\sigma^2 \begin{bmatrix} 2 & 1 \\ 1 & 2 \end{bmatrix}. \tag{6.77}$$

This shows that double-differences are correlated. The weight or correlation matrix $\mathbf{P}(t)$ is obtained from the inverse of the covariance matrix

$$\mathbf{P}(t) = \Sigma_\mathbf{D}^{-1} = \frac{1}{2\sigma^2}\,\frac{1}{3}\begin{bmatrix} 2 & -1 \\ -1 & 2 \end{bmatrix}, \tag{6.78}$$

where two double-differences at one epoch were used. Generally, with n_D being the number of double-differences at epoch t, the correlation matrix is given by

$$\mathbf{P}(t) = \frac{1}{2\sigma^2}\,\frac{1}{n_D + 1}\begin{bmatrix} n_D & -1 & -1 & \cdots \\ -1 & n_D & -1 & \cdots \\ -1 & & & \\ \vdots & \cdots & & n_D \end{bmatrix}, \tag{6.79}$$

6.3 Relative positioning

where the dimension of the matrix is $n_D \times n_D$. For a better illustration, assume four double-differences. In this case, the 4×4 matrix

$$\mathbf{P}(t) = \frac{1}{2\sigma^2} \frac{1}{5} \begin{bmatrix} 4 & -1 & -1 & -1 \\ -1 & 4 & -1 & -1 \\ -1 & -1 & 4 & -1 \\ -1 & -1 & -1 & 4 \end{bmatrix} \qquad (6.80)$$

is the correlation matrix. So far only one epoch has been considered. For epochs $t_1, t_2, t_3, \ldots$, the correlation matrix becomes a block-diagonal matrix

$$\mathbf{P}(t) = \begin{bmatrix} \mathbf{P}(t_1) & & & \\ & \mathbf{P}(t_2) & & \\ & & \mathbf{P}(t_3) & \\ & & & \ddots \end{bmatrix}, \qquad (6.81)$$

where each "element" of the matrix is itself a matrix. The matrices $\mathbf{P}(t_1)$, $\mathbf{P}(t_2)$, $\mathbf{P}(t_3)$, ... do not necessarily have to be of the same dimension because there may be different numbers of double-differences at different epochs.

Correlation of triple-differences

The triple-difference equations are slightly more complicated because several different cases must be considered. The covariance of a single triple-difference is computed by applying the covariance propagation law to the relation, cf. Eqs. (6.60) and (6.64),

$$\Phi_{AB}^{jk}(t_{12}) = \Phi_{AB}^{k}(t_2) - \Phi_{AB}^{j}(t_2) - \Phi_{AB}^{k}(t_1) + \Phi_{AB}^{j}(t_1). \qquad (6.82)$$

Now, two triple-differences with the same epochs and sharing one satellite are considered. The first triple-difference using the satellites j, k is given by Eq. (6.82). The second triple-difference corresponds to the satellites j, ℓ:

$$\begin{aligned} \Phi_{AB}^{jk}(t_{12}) &= \Phi_{AB}^{k}(t_2) - \Phi_{AB}^{j}(t_2) - \Phi_{AB}^{k}(t_1) + \Phi_{AB}^{j}(t_1), \\ \Phi_{AB}^{j\ell}(t_{12}) &= \Phi_{AB}^{\ell}(t_2) - \Phi_{AB}^{j}(t_2) - \Phi_{AB}^{\ell}(t_1) + \Phi_{AB}^{j}(t_1). \end{aligned} \qquad (6.83)$$

By introducing

$$\mathbf{T} = \begin{bmatrix} \Phi_{AB}^{jk}(t_{12}) \\ \Phi_{AB}^{j\ell}(t_{12}) \end{bmatrix}, \qquad \mathbf{S} = \begin{bmatrix} \Phi_{AB}^{j}(t_1) \\ \Phi_{AB}^{k}(t_1) \\ \Phi_{AB}^{\ell}(t_1) \\ \Phi_{AB}^{j}(t_2) \\ \Phi_{AB}^{k}(t_2) \\ \Phi_{AB}^{\ell}(t_2) \end{bmatrix} \qquad (6.84)$$

$$\mathbf{C} = \begin{bmatrix} 1 & -1 & 0 & -1 & 1 & 0 \\ 1 & 0 & -1 & -1 & 0 & 1 \end{bmatrix},$$

Table 6.1. Symbolic composition of triple-differences

Epoch	t_1			t_2		
Satellite	j	k	ℓ	j	k	ℓ
$\Phi_{AB}^{jk}(t_{12})$	1	−1	0	−1	1	0
$\Phi_{AB}^{j\ell}(t_{12})$	1	0	−1	−1	0	1

the matrix-vector relation

$$\mathbf{T} = \mathbf{C}\,\mathbf{S} \tag{6.85}$$

can be formed, and the covariance for the triple-difference follows from

$$\Sigma_T = \mathbf{C}\,\Sigma_S\,\mathbf{C}^T \tag{6.86}$$

or, by substituting (6.71),

$$\Sigma_T = 2\sigma^2\,\mathbf{C}\,\mathbf{C}^T \tag{6.87}$$

is obtained, which, using (6.84), yields

$$\Sigma_T = 2\sigma^2 \begin{bmatrix} 4 & 2 \\ 2 & 4 \end{bmatrix} \tag{6.88}$$

for the two triple-differences (6.83). The tedious derivation may be abbreviated by setting up Table 6.1.

It can be seen that the triple-difference $\Phi_{AB}^{jk}(t_{12})$, for example, is composed of the two single-differences (with the signs according to Table 6.1) for the satellites j and k at epoch t_1 and of the two single-differences for the same satellites but epoch t_2. Accordingly, the same applies for the other triple-difference $\Phi_{AB}^{j\ell}(t_{12})$. Thus, the coefficients of Table 6.1 are the same as those of matrix $\mathbf{C}$ in Eq. (6.84). Finally, the product $\mathbf{C}\,\mathbf{C}^T$, appearing in Eq. (6.87), is also aided by referring to Table 6.1. All combinations of inner products of the two rows (one row represents one triple-difference) must be taken. The inner product (row 1 · row 1) yields the first-row, first-column element of $\mathbf{C}\,\mathbf{C}^T$, the inner product (row 1 · row 2) yields the first-row, second-column element of $\mathbf{C}\,\mathbf{C}^T$, etc. Based on the general formula (6.82) and Table 6.1, arbitrary cases may be derived systematically. Table 6.2 shows the second group of triple-difference correlations if adjacent epochs t_1, t_2, t_3 are taken. Two cases are considered.

It can be seen from Table 6.2 that an exchange of the satellites for one triple-difference causes a change of the sign in the off-diagonal elements of the matrix

6.3 Relative positioning

Table 6.2. Triple-difference correlations

Epoch	t_1			t_2			t_3			$\mathbf{C\,C}^T$	
Satellite	j	k	ℓ	j	k	ℓ	j	k	ℓ		
$\Phi_{AB}^{jk}(t_{12})$	1	−1	0	−1	1	0	0	0	0	4	−2
$\Phi_{AB}^{jk}(t_{23})$	0	0	0	1	−1	0	−1	1	0	−2	4
$\Phi_{AB}^{jk}(t_{12})$	1	−1	0	−1	1	0	0	0	0	4	−1
$\Phi_{AB}^{j\ell}(t_{23})$	0	0	0	1	0	−1	−1	0	1	−1	4

$\mathbf{C\,C}^T$. Therefore, the correlation of $\Phi_{AB}^{kj}(t_{12})$ and $\Phi_{AB}^{j\ell}(t_{23})$ produces +1 as off-diagonal element. Based on a table such as Table 6.2, each case may be handled with ease. According to Remondi (1984: p. 142), computer program adaptations require only a few simple rules. These are the basic mathematical correlations for single-, double-, and triple-differences.

More sophisticated models are investigated in Euler and Goad (1991), Gerdan (1995), Jin and Jong (1996) by taking into account the elevation dependence of the observation variances. Gianniou (1996) introduces variable weights by forming differences, applying polynomial fitting, and by using the signal-to-noise ratio for code ranges as well as for phases. Jonkman (1998) and Tiberius (1998) consider time correlation and crosscorrelation of the code ranges and the phases.

6.3.4 Static relative positioning

In a static survey of a single baseline vector between points A and B, the two receivers must stay stationary during the entire observation session. In the following, the single-, double-, and triple-differencing are investigated with respect to the number of observation equations and unknowns. It is assumed that the two sites A and B are able to observe the same satellites at the same epochs. The practical problem of satellite blockage is not considered here. The number of epochs is again denoted by n_t, and n_s denotes the number of satellites.

The undifferenced phase as shown in Eq. (6.11) (where the satellite clock is assumed to be known) is not included here, because there would be no connection (no common unknown) between point A and point B. The two data sets could be solved separately, which would be equivalent to point positioning.

A single-difference may be expressed for each satellite and for each epoch. The number of measurements is, therefore, $n_s n_t$. The number of unknowns is written below the corresponding terms of the single-difference equation, cf. Eq. (6.42):

$$\Phi_{AB}^j(t) = \frac{1}{\lambda^j} \varrho_{AB}^j(t) + N_{AB}^j + f^j \, \delta_{AB}(t) \,, \tag{6.89}$$

$$n_s \, n_t \geq \quad 3 \quad + n_s + (n_t - 1) \,.$$

The $n_t - 1$ unknown clock biases indicate a rank deficiency of 1. The explanation is the same as for static point positioning, cf. Eq. (6.12). From above, the relation

$$n_t \geq \frac{n_s + 2}{n_s - 1} \tag{6.90}$$

may be derived. Although this equation is equivalent to Eq. (6.13), it is useful to repeat the (theoretically) minimum requirements for a solution. A single satellite does not provide a solution because the denominator of (6.90) becomes zero. With two satellites, there results $n_t \geq 4$, and for the normal case of four satellites, $n_t \geq 2$ is obtained.

For double-differences, the relationship of measurements and unknowns is obtained using the same logic. Note that for one double-difference two satellites are necessary. For n_s satellites, therefore, $n_s - 1$ double-differences are obtained at each epoch so that the total number of double-differences is $(n_s - 1) n_t$. The number of unknowns is written below the corresponding terms of the double-difference equation, cf. Eq. (6.45):

$$\Phi_{AB}^{jk}(t) = \frac{1}{\lambda} \varrho_{AB}^{jk}(t) + N_{AB}^{jk} \,, \tag{6.91}$$

$$(n_s - 1) n_t \geq \quad 3 \quad + (n_s - 1) \,.$$

From above, the relation

$$n_t \geq \frac{n_s + 2}{n_s - 1} \tag{6.92}$$

is obtained, which is identical with Eq. (6.90) and, therefore, the basic configurations are again given by the pairs $n_s = 2, n_t \geq 4$ and $n_s = 4, n_t \geq 2$. To avoid linearly dependent equations when forming double-differences, a reference satellite is used, against which the measurements of the other satellites are differenced. For example, take the case where measurements are made to the satellites 6, 9, 11, and 12 and 6 is used as reference satellite. Then, at each epoch the following double-differences can be formed: (9-6), (11-6), and (12-6). Other double-differences are linear combinations and, thus, linearly dependent. For instance, the double-difference (11-9) can be formed by subtracting (11-6) and (9-6).

Note that relation (6.92) also applies if the frequencies of the satellite signals are not equal. Referring to (6.54), the number of single-difference ambiguities corresponds to that of (6.89) and amounts to n_s, which may be combined to $n_s - 1$

6.3 Relative positioning

double-difference ambiguities if the single-difference ambiguities of one satellite are taken as reference.

The triple-difference model includes only the three unknown point coordinates. For a single triple-difference, two epochs are necessary. Consequently, in the case of n_t epochs, $n_t - 1$ linearly independent epoch combinations are possible. Thus,

$$\Phi_{AB}^{jk}(t_{12}) = \frac{1}{\lambda} \varrho_{AB}^{jk}(t_{12}), \qquad (6.93)$$

$$(n_s - 1)(n_t - 1) \geq 3$$

are the resulting equations. From above, the relation

$$n_t \geq \frac{n_s + 2}{n_s - 1} \qquad (6.94)$$

is obtained. This equation is identical with Eq. (6.90) and, hence, the basic configurations are again given by the pairs $n_s = 2, n_t \geq 4$ and $n_s = 4, n_t \geq 2$.

This completes the discussion on static relative positioning. As shown, each of the mathematical models – single-difference, double-difference, triple-difference – may be used. The relationships between the number of observation equations and the number of unknowns will be referred to again in the discussion of the kinematic case.

6.3.5 Kinematic relative positioning

In kinematic relative positioning, the receiver on the known point A of the baseline vector remains fixed. The second receiver moves, and its position is to be determined for arbitrary epochs. The models for single-, double-, and triple-difference implicitly contain the motion in the geometric distance. Considering point B and satellite j, the geometric distance in the static case is given by, cf. Eq. (6.2),

$$\varrho_B^j(t) = \sqrt{(X^j(t) - X_B)^2 + (Y^j(t) - Y_B)^2 + (Z^j(t) - Z_B)^2} \qquad (6.95)$$

and in the kinematic case by

$$\varrho_B^j(t) = \sqrt{(X^j(t) - X_B(t))^2 + (Y^j(t) - Y_B(t))^2 + (Z^j(t) - Z_B(t))^2}, \qquad (6.96)$$

where the time dependence for point B appears. In this mathematical model, three coordinates are unknown at each epoch. Thus, the total number of unknown site coordinates is $3 n_t$ for n_t epochs. The relations between the number of observations and the number of unknowns for the kinematic case follow from the static single- and double-difference models, cf. Eqs. (6.89), (6.91):

$$\begin{aligned} \text{single-difference:} \quad & n_s n_t \geq 3 n_t + n_s + (n_t - 1), \\ \text{double-difference:} \quad & (n_s - 1) n_t \geq 3 n_t + (n_s - 1). \end{aligned} \qquad (6.97)$$

For example, the relation

$$n_t \geq \frac{n_s - 1}{n_s - 4} \tag{6.98}$$

is the basic configuration for single-differences, which is equivalent to Eq. (6.15).

The continuous motion of the roving receiver restricts the available data for the determination of its position to one epoch. But none of the above two models provides a useful solution for $n_t = 1$. Thus, these models are modified: the number of unknowns is reduced by omitting the ambiguity unknowns, i.e., the ambiguities are assumed to be known. For the single-difference case, this has a twofold effect: first, the n_s ambiguities may be omitted and, second, the rank deficiency vanishes because of the known ambiguities so that n_t unknown clock biases have to be determined. The modified observation requirement for the single-difference is therefore $n_s n_t \geq 4 n_t$ and reduces to $n_s \geq 4$ for a single epoch. Similarly, for the double-difference $n_s - 1$ ambiguities are omitted in (6.97) so that $(n_s - 1) n_t \geq 3 n_t$ results, which reduces to $n_s \geq 4$ for a single epoch. Hence, the single-difference and the double-difference models end up again with the fundamental requirement of four simultaneously observable satellites.

The use of triple-differences for kinematic cases is strongly restricted. In principle, the definition of triple-differences with two satellites at two epochs and two stations at – with respect to the two epochs – fixed positions exclude any application since the rover position changes epoch by epoch. However, triple-differences could be used if, e.g., the coordinates of the roving receiver were known at the reference epoch. In this case, adapting (6.93) to the kinematic case with $3 n_t$ unknowns and reducing the number of unknown rover positions by 3 because of the known rover position at the reference epoch, the relationship obtained would be $(n_s - 1)(n_t - 1) \geq 3 (n_t - 1)$. This leads to $n_s \geq 4$, which is the same requirement as for the ambiguity-reduced single- and double-differences.

Omitting the ambiguities for single- and double-difference means that they must be known. The corresponding equations are simply obtained by rewriting (6.89) and (6.91) with the ambiguities shifted to the left side of the equations. The single-differences become

$$\Phi^j_{AB}(t) - N^j_{AB} = \frac{1}{\lambda^j} \varrho^j_{AB}(t) + f^j \delta_{AB}(t) \tag{6.99}$$

and the double-differences

$$\Phi^{jk}_{AB}(t) - N^{jk}_{AB} = \frac{1}{\lambda} \varrho^{jk}_{AB}(t), \tag{6.100}$$

where the unknowns now appear only on the right sides.

If the frequencies of the satellite signals are not equal, an analogous relation for the double-difference is obtained. Referring to (6.55), the two terms containing

6.3 Relative positioning

ambiguities may be shifted to the left side to indicate that they are known. Then the only remaining term on the right side is $\varrho_{AB}^{jk}(t)$.

Thus, all of the equations can be solved if one position of the moving receiver is known. Preferably (but not necessarily), this will be the starting point of the moving receiver. The baseline related to this starting point is denoted as the starting vector. With a known starting vector, the ambiguities are determined and are known for all subsequent positions of the roving receiver as long as no loss of signal lock occurs and a minimum of four satellites is in view.

Static initialization

Three methods are available for the static determination of the starting vector. In the first method, the moving receiver is initially placed at a known point, creating a known starting vector. The ambiguities can then be calculated from the double-difference model (6.91) as real values and are then fixed to integers. A second method is to perform a static determination of the starting vector. The third initialization technique is the antenna swap method according to B.W. Remondi. The antenna swap is performed as follows: denoting the reference mark as A and the starting position of the moving receiver as B, a few measurements are taken in this configuration, and with continuous tracking, the receiver at A is moved to B, while the receiver at B is moved to A, where again a few measurements are taken. This is sufficient to precisely determine the starting vector in a very short time (e.g., 30 seconds). Often, a second antenna swap is performed by moving the receivers to their starting positions.

Kinematic initialization

Special applications require kinematic GNSS without static initialization since the moving object whose position is to be calculated is in a permanent motion (e.g., a buoy or an airplane while flying). Translated to model equations, this means that the most challenging case is the determination of the ambiguities on-the-fly (OTF). The solution requires an instantaneous ambiguity resolution or an instantaneous positioning (i.e., for a single epoch). This strategy sounds very simple but it can require advanced methods. A vast literature has been written on this important topic. The main problem is to find the position as fast and as accurately as possible. This is achieved by starting with approximations for the position and improving them by least-squares adjustments or search techniques.

6.3.6 Pseudokinematic relative positioning

The pseudokinematic method can be identified as static surveying with large data gaps (Kleusberg 1990). The mathematical model, e.g., for double-differences, corresponds to Eq. (6.91) where generally two sets of phase ambiguities must be resolved since the point is occupied at different times. B.W. Remondi also has applied the triple-difference method followed by the ambiguity function method (Sect. 7.2.3) to avoid ambiguities altogether. Processing of the data could start with a triple-difference solution for the few minutes of data collected during the two occupations of a site. Based on this solution, the connection between the two ambiguity sets is computed (Remondi 1990b). After the successful ambiguity connection, the normal double-difference solutions are performed.

The time span between the two occupations is an important factor affecting accuracy. Willis and Boucher (1990) investigate the accuracy improvements by an increasing time span between the two occupations. As a rule of thumb, the minimum time span should be one hour.

Note that the pseudokinematic relative positioning method is rarely used today.

6.3.7 Virtual reference stations

When processing a baseline, the effects of orbit errors, ionospheric and tropospheric refraction are reduced by forming differences of the observables, e.g., double-differences. These effects grow with increasing baseline length. Therefore, it is good practice to use short baselines requiring a reference station close to the rover. These basic considerations have led to reference station networks like the Austrian positioning service (APOS), the German satellite positioning service (SAPOS), and several others. After the establishment of such networks, some new ideas have evolved to exploit the available data accordingly. Among many others, real-time differential error modeling in reference station networks (Wanninger 1997), multi-base real-time kinematic (RTK) positioning using virtual reference stations (Vollath et al. 2000), and network-based techniques for RTK applications (Wübbena et al. 2001) are respective examples.

Even in case of an existing reference station network it is desirable to further reduce the baseline length. The idea is to generate "observation data" for a nonexisting station, i.e., a virtual station, from real observations of a multiple reference station network and to transmit these data to the rover station. This is the basic principle of the virtual reference station (VRS) concept. Usually, the data of three or more reference stations surrounding the VRS are taken to calculate the observation data for the VRS. The result of this concept yields a horizontal accuracy at the level of 5 cm for baselines up to 35 km (Retscher 2002).

Understanding the VRS principle by an elementary approach

The objective is to transform measurements made at real reference stations to the location of the VRS, i.e., to a different location. This implies that all terms of the observation equation model depending on the reference receiver location have to

6.3 Relative positioning

be corrected to account for the new location. To keep it as simple as possible, the phase pseudorange

$$\Phi_r^s(t) = \frac{1}{\lambda^s} \varrho_r^s(t) + N_r^s + f^s \, \Delta\delta_r^s(t), \tag{6.101}$$

as given in (6.10) after substitution of $f^s = c/\lambda^s$, is considered. On the left side, $\Phi_r^s(t)$ is the measured carrier phase which is modeled by $\varrho_r^s(t)$, the geometric distance between receiver and satellite, the time-independent integer ambiguity N_r^s, and $\Delta\delta_r^s(t)$, the combined receiver and satellite clock bias. Now the key question is which of the terms is location-dependent? In other words, which of the terms changes if the same receiver is assumed at another location but considered at the same epoch t? The answer is $\Phi_r^s(t)$ and $\varrho_r^s(t)$, because the other two terms do not change with varying location.

Now it is assumed that receiver r is once located at the real reference station A represented by the coordinate vector $\mathbf{X}_A$ and once at the virtual reference station (VRS) represented by $\mathbf{X}_V$. Then there result from these two locations from (6.101) the two equations

$$\begin{aligned}\Phi_r^s(\mathbf{X}_A, t) &= \frac{1}{\lambda^s} \varrho_r^s(\mathbf{X}_A, t) + N_r^s + f^s \, \Delta\delta_r^s(t), \\ \Phi_r^s(\mathbf{X}_V, t) &= \frac{1}{\lambda^s} \varrho_r^s(\mathbf{X}_V, t) + N_r^s + f^s \, \Delta\delta_r^s(t),\end{aligned} \tag{6.102}$$

where the location dependence is indicated accordingly. Forming the difference of the two equations yields

$$\Phi_r^s(\mathbf{X}_V, t) - \Phi_r^s(\mathbf{X}_A, t) = \frac{1}{\lambda^s} \varrho_r^s(\mathbf{X}_V, t) - \frac{1}{\lambda^s} \varrho_r^s(\mathbf{X}_A, t), \tag{6.103}$$

where the ambiguity and the clock error have vanished. After a slight rearrangement,

$$\Phi_r^s(\mathbf{X}_V, t) = \Phi_r^s(\mathbf{X}_A, t) + \frac{1}{\lambda^s} [\varrho_r^s(\mathbf{X}_V, t) - \varrho_r^s(\mathbf{X}_A, t)] \tag{6.104}$$

results. The left-hand side is the desired "measurement quantity" at the virtual reference station. Therefore, if all terms on the right-hand side may be obtained, then there is no need to actually measure it. The term $\Phi_r^s(\mathbf{X}_A, t)$ refers to the real reference station A and is measured. Accordingly, $\varrho_r^s(\mathbf{X}_A, t)$ is known since the station coordinates of A and the satellite coordinates of s are known and calculable, respectively. The only remaining term to be discussed is $\varrho_r^s(\mathbf{X}_V, t)$, which comprises the coordinates of the virtual reference station. Figure 6.3 shows the network of reference stations $A, B, C, \ldots$, the virtual reference station, and the user receiver position indicated by r. In principle, the location of the virtual reference station is arbitrary;

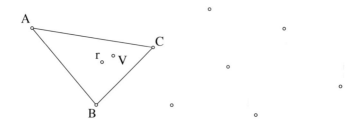

Fig. 6.3. Network of reference stations with the VRS denoted as V and the roving user receiver indicated by r

however, the baseline between the receiver r and the virtual reference station V should be smaller than the baselines between r and any other of the real reference stations. A very useful and convenient location is therefore an approximate position of the user receiver which may readily be obtained by point positioning with code measurements, yielding $\mathbf{X}_V$, the coordinates of the VRS. Once the coordinates of the virtual station are known, they may remain fixed for the subsequent epochs (unless the roving user receiver moves too far from the VRS).

In summary, the right-hand side of (6.104) is now fully determined. Thus, the observable $\Phi_r^s(\mathbf{X}_V, t)$ of the VRS may be obtained without actually measuring it. Since $\mathbf{X}_A$ and $\mathbf{X}_V$ are known as well, the computation of $\varrho_r^s(\mathbf{X}_A, t)$ and $\varrho_r^s(\mathbf{X}_V, t)$ is straightforward for the subsequent epochs; and $\Phi_r^s(\mathbf{X}_A, t)$ is measured at the real reference station A.

The more complex reality

Is it really that simple? In theory "yes", but in practice "no" since (6.101) and (6.102) are idealized models without taking into account errors like satellite orbit errors, ionospheric and tropospheric refraction. These errors may be considered for the real reference station A by the collective term

$$\Delta_r^s(\mathbf{X}_A, t) = \Delta^{\mathrm{Orbit}}(\mathbf{X}_A, t) + \Delta^{\mathrm{Iono}}(\mathbf{X}_A, t) + \Delta^{\mathrm{Trop}}(\mathbf{X}_A, t). \tag{6.105}$$

Consequently, model (6.102) for the measurement equation of the reference station A is expanded to

$$\Phi_r^s(\mathbf{X}_A, t) = \frac{1}{\lambda^s} \varrho_r^s(\mathbf{X}_A, t) + N_r^s + f^s \Delta\delta_r^s(t) + \Delta_r^s(\mathbf{X}_A, t) \tag{6.106}$$

and in analogous form for all other real reference stations. To estimate the error term properly, all baselines of the reference network are solved. Note that this also requires the correct ambiguity determination (which should in general not be a major problem because the station coordinates are known). The desired results of

6.3 Relative positioning

the network solution are the error residuals $\Delta_r^s(\mathbf{X}_A, t)$, $\Delta_r^s(\mathbf{X}_B, t)$, ... for all reference stations and at every epoch individually.

Similarly, the relation (6.104) for the VRS is improved by

$$\Phi_r^s(\mathbf{X}_V, t) = \Phi_r^s(\mathbf{X}_A, t) + \frac{1}{\lambda^s} \left[\varrho_r^s(\mathbf{X}_V, t) - \varrho_r^s(\mathbf{X}_A, t) \right] + \Delta_r^s(\mathbf{X}_V, t), \quad (6.107)$$

but now the problem arises to determine $\Delta_r^s(\mathbf{X}_V, t)$ for the VRS.

An intuitively simple approach is to take the error residuals $\Delta_r^s(\mathbf{X}_A, t)$, $\Delta_r^s(\mathbf{X}_B, t)$, $\Delta_r^s(\mathbf{X}_C, t)$ of three reference stations A, B, C surrounding the VRS and to compute $\Delta_r^s(\mathbf{X}_V, t)$ by a weighted mean for which the weights depend inversely on the distance between the virtual reference station and the respective real reference station.

Another approach is to model the error residuals at the reference stations i by

$$\Delta_r^s(\mathbf{X}_i, t) = a X_i + b Y_i + c Z_i, \quad (6.108)$$

where X_i, Y_i, Z_i are the coordinates of the reference stations (e.g., ECEF coordinates or plane coordinates supplemented by a height component). Assuming three real reference stations A, B, C, the coefficients a, b, c can be calculated.

To determine the error residual $\Delta_r^s(\mathbf{X}_V, t)$ for the virtual reference station, use (6.108) by substituting the coefficients a, b, c and the coordinates of the VRS.

If more than three real reference stations are available, the model (6.108) can either be extended or a least-squares adjustment applied.

A few more words should be spent on the modeling of the error term (6.105). First, it may further be expanded by taking into account additional error influences like antenna phase center offset and variation or multipath. Note, however, these purely station-dependent terms are uncorrelated between stations and it does not make sense to transfer their influence to the VRS by the approach as shown above. Therefore, these error influences must be reduced or corrected by proper modeling or simply neglected.

Second, the influence of the ionosphere and the troposphere in (6.105) may also be reduced by data combinations (Sect. 5.2) or modeling (Sects. 5.3.2, 5.3.3). Nevertheless, remaining residuals will contribute to the error term.

Several other approaches for modeling the error term are proposed. Wübbena et al. (2001) and Wanninger (2002) propose area correction parameters, Landau et al. (2002) mention a weighted linear approximation approach and a least-squares collocation. Dai et al. (2001) compare several interpolation algorithms like linear combination model, distance-based linear interpolation method, linear interpolation method, low-order surface model, and least-squares collocation and find out comparable performances.

7 Data processing

7.1 Data preprocessing

7.1.1 Data handling

Downloading

Both the observables and the navigation message and additional information are generally stored in a binary (and receiver-dependent) format. The downloading of the data from the receiver is necessary before postprocessing can begin.

Most GNSS manufacturers have designed a data management system which they recommend for data processing. Individual software is fully documented in the manuals of the manufacturers and will not be covered here.

Data exchange

Although the binary receiver data may have been converted into computer-independent ASCII format during downloading, the data are still receiver dependent. Also, each GNSS processing software has its own format which necessitates the conversion of specific data into a software-independent format when they are processed with a different type of program.

From the preceding, one may conclude that a receiver-independent format of GNSS data promotes data exchange. This has been realized by the receiver-independent exchange (RINEX) format. This format was first defined for GPS data in 1989 and has been published in a second version by Gurtner and Mader (1990). Later, several minor changes were adopted and in 1997 the format was extended to also account for GLONASS. Further updates are covered in the versions 2.10 and 2.11. As of 2006, RINEX 3.0 is the latest version. Gurtner and Estey (2006) describe this version in full detail.

The data of the measurement campaign is commonly stored in three ASCII file types: (1) the observation data file, (2) the navigation message file, and (3) the meteorological data file. Each file comprises a header section and a data section. The header section contains generic file information, and the data section contains the actual data.

Basically, the observation and meteorological data files must be created for each site of the session. The RINEX (version 2 and higher) also permits the inclusion of observation data from more than one site subsequently occupied by a roving receiver. However, according to Gurtner and Estey (2006) it is not recommended to assemble data of more than one receiver (or antenna) into the same file.

The navigation message file is more or less site independent. In order to avoid the collection of identical satellite navigation messages from different receivers, one navigation message file only is created containing nonredundant information and being possibly composite from several receivers.

To demonstrate the potential of RINEX, a few details of the recommended specification are described. RINEX uses the filename convention "ssssdddf.yyt". The first four characters of the sequence (ssss) are the site identifier, the next three (ddd) indicate the day of year, and the eighth character (f) is the session indicator. The first two file extension characters (yy) denote the last two digits of the current year, and the file type (t) is given by the last character. This file type may indicate an observation file, a GPS, GLONASS, or Galileo navigation message file, a meteorological data file, a mixed GNSS navigation message file, a space-based augmentation system (SBAS) payload navigation message file, an SBAS broadcast data file, etc.

The satellite designation is defined in the form "snn". The first character (s) is an identifier of the satellite system, and the remaining two digits denote the satellite number. Examples for the identifier s of the satellite system are GPS, GLONASS, SBAS payload, Galileo. Thus, the RINEX format enables the combination of observations of different satellite types.

At present, RINEX is the most favored format. As a consequence, all receiver manufacturers implement software for the conversion of their receiver-dependent format into RINEX.

The software-independent exchange (SINEX) format is mentioned for the sake of completeness. This format enables the exchange of processing results and is used, for example, by the IGS (Mervart 1999). Further information is available at http://tau.fesg.tu-muenchen.de/~iers/web/sinex/format.php (SINEX versions 2.00, 2.01). Recently, the SINEX version 2.10 was proposed in the International Earth Rotation Service (IERS) Message no. 96 to include Galileo regarding the station and satellite information for the specific frequencies. Some more new parameters were added. This proposal can be found under http://www.iers.org/documents/ac/sinex/sinex_v210_proposal.pdf.

7.1.2 Cycle slip detection and repair

Definition of cycle slips

When a receiver is turned on, the fractional part of the beat phase (i.e., the difference between the satellite-transmitted carrier and a receiver-generated replica) is observed and an integer counter is initialized. During tracking, the counter is incremented by one cycle whenever the fractional phase changes from 2π to 0. Thus, at a given epoch, the observed accumulated phase $\Delta\varphi$ is the sum of the fractional phase φ and the integer count n. The initial integer number N of cycles between the

7.1 Data preprocessing

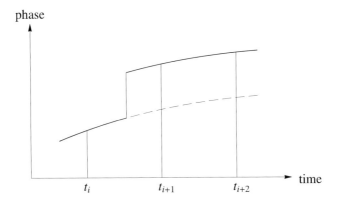

Fig. 7.1. Schematic representation of a cycle slip

satellite and the receiver is unknown. This phase ambiguity N remains constant as long as no loss of the signal lock occurs. In this event, the integer counter is reinitialized, which causes a jump in the instantaneous accumulated phase by an integer number of cycles. This jump is called a cycle slip which, of course, is restricted to phase measurements.

A schematic representation of a cycle slip is given in Fig. 7.1. When the measured phases are plotted versus time, a fairly smooth curve should be obtained. In the case of a cycle slip, a sudden jump appears in the plotted curve. Three sources for cycle slips can be distinguished. First, cycle slips are caused by obstructions of the satellite signal due to trees, buildings, bridges, mountains, etc. This source is the most frequent one (particularly for kinematic activities based upon the carrier phase). The second source for cycle slips is a low S/N due to bad ionospheric conditions, multipath, high receiver dynamics, or low satellite elevation. A third source is a failure in the receiver software (Hein 1990b), which leads to incorrect signal processing. Cycle slips could also be caused by malfunctioning satellite oscillators, but these cases are rare.

As seen from Fig. 7.1, cycle slip detection and repair requires the location of the jump (i.e., cycle slip) and the determination of its size. Detection is accomplished by a testing quantity. Repairs of cycle slips are made by correcting all subsequent phase observations for this satellite and this carrier by a fixed integer number of cycles. The determination of the cycle slip size and the correction of the phase data is often denoted as cycle slip repair or fixing.

Testing quantities

For a single site, the testing quantities are raw phases, phase combinations, combinations of phases and code ranges, or combinations of phases and integrated

Doppler frequencies. Single receiver tests are important because they enable in situ cycle slip detection and repair by the internal software of the receiver.

When the data of two sites are combined, single-, double-, or triple-differences can be used for cycle slip detection. This means that, in a first step, unrepaired phase combinations are used to process an approximate baseline vector. The corresponding residuals are then tested. Quite often several iterations are necessary to improve the baseline solution. Note that triple-differences can achieve convergence and rather high accuracy without fixing cycle slips. Note also that triple-differences make the amount of double-difference cycle slips very clear – in the static case.

Subsequently, the testing quantities for a single site are treated in more detail.

Raw phases
The measured raw phase $\Phi_r^s(t)$ can be modeled by

$$\lambda \Phi_r^s(t) = \varrho_r^s(t) + \lambda N_r^s + c \Delta\delta_r^s(t) - \Delta^{\text{Iono}}(t) + \ldots, \tag{7.1}$$

where r and s denote the receiver site and the satellite, respectively. Note that the phase model contains a number of time-dependent terms on the right side of (7.1) which may prevent cycle slip detection.

Phase combinations
The model for the dual-frequency phase combination is developed considering a single site, a single satellite, and a single epoch. Thus, the sub- and superscripts and even the time dependency in Eq. (7.1) may be omitted. According to Eq. (5.76), the phases are modeled by

$$\begin{aligned}\Phi_1 &= a f_1 + N_1 - \frac{b}{f_1}, \\ \Phi_2 &= a f_2 + N_2 - \frac{b}{f_2},\end{aligned} \tag{7.2}$$

where the frequency dependency is shown explicitly by the subscripts 1 and 2.

In order to eliminate the geometry term a, the first equation of (7.2) is multiplied by f_2 and the second by f_1. Subtracting the resulting equations yields

$$f_2 \Phi_1 - f_1 \Phi_2 = f_2 N_1 - f_1 N_2 - b \left(\frac{f_2}{f_1} - \frac{f_1}{f_2} \right) \tag{7.3}$$

and dividing the difference by f_2 gives

$$\Phi_1 - \frac{f_1}{f_2} \Phi_2 = N_1 - \frac{f_1}{f_2} N_2 - \frac{b}{f_2}\left(\frac{f_2}{f_1} - \frac{f_1}{f_2} \right) \tag{7.4}$$

7.1 Data preprocessing

or, by extracting f_2/f_1 from the term in parentheses on the right side of the equation, the final form of the geometry-free phase combination

$$\Phi_1 - \frac{f_1}{f_2}\Phi_2 = N_1 - \frac{f_1}{f_2}N_2 - \frac{b}{f_1}\left(1 - \frac{f_1^2}{f_2^2}\right) \tag{7.5}$$

is obtained. The left side of Eq. (7.5) is identical with the ionospheric residual, cf. Eq. (5.21). The right side shows that the only time-varying quantity is the ionosphere term b. In comparison to the influence on the raw phases in Eq. (7.1), the influence of the ionosphere on the dual-frequency combination is affected by the factor $(1 - f_1^2/f_2^2)$. Substituting typical GNSS values, $f_1 = 1.6\,\text{GHz}$ and $f_2 = 1.2\,\text{GHz}$, this factor is -0.78.

If there are no cycle slips, the temporal variations of the ionospheric residual would be small for normal ionospheric conditions and for short baselines. Indicators of cycle slips are sudden jumps in successive values of the ionospheric residual. The remaining problem is to determine whether the cycle slip was on phase data referring to f_1 or f_2 or both. This will be investigated in the next paragraphs.

Phase and code range combinations
Another testing quantity follows from a phase and code range combination. Modeling the carrier phase and the code pseudoranges by

$$\begin{aligned}\lambda\,\Phi_r^s(t) &= \varrho_r^s(t) + \lambda\,N_r^s + c\,\Delta\delta_r^s(t) - \Delta^{\text{Iono}}(t) + \Delta^{\text{Trop}}(t), \\ R_r^s(t) &= \varrho_r^s(t) \hspace{2.2cm} + c\,\Delta\delta_r^s(t) + \Delta^{\text{Iono}}(t) + \Delta^{\text{Trop}}(t)\end{aligned} \tag{7.6}$$

and forming the difference

$$\lambda\,\Phi_r^s(t) - R_r^s(t) = \lambda\,N_r^s - 2\,\Delta^{\text{Iono}}(t) \tag{7.7}$$

provides a formula where the time-dependent terms (except the ionospheric refraction) vanish from the right side of the equation. Thus, the phase and code range combination could also be used as testing quantity. The ionospheric influence may either be modeled or neglected. The change of $\Delta^{\text{Iono}}(t)$ will be fairly small between closely spaced epochs; this might justify neglecting the ionospheric term. It may also be neglected when using double-differences.

The simple testing quantity (7.7) has a shortcoming which is related to the noise level. The phase and code range combinations have a noise level in the range of 5 cycles. This noise is mainly caused by the noise level of the code measurements and to a minor extent by the ionosphere. The noise of code measurements is larger than the noise for phase measurements because resolution and multipath are proportional to the wavelength. Traditionally, the measurement resolution was

$\lambda/100$; today, receiver hardware is achieving improved measurement resolutions approaching $\lambda/1000$. In other words, this leads to code range noise levels of a few centimeters. Hence, the phase and code range combination could be an ideal testing quantity for cycle slip detection.

Combination of phases and integrated Doppler
Comparing differences of measured phases with phase differences derived from integrated Doppler which has the advantage of being immune from cycle slips is a further possibility for a testing quantity.

Detection and repair
Each of the described testing quantities allows the location of cycle slips by checking the difference of two consecutive epoch values. This also yields an approximate size of the cycle slip. To find the correct size, the time series of the testing quantity must be investigated in more detail. Note that for all previously mentioned testing quantities, except the ionospheric residual, the detected cycle slip must be an integer. In certain processing scenarios, cycle slips are easy to detect and repair without knowing which satellite or receiver had the problem.

One of the methods for cycle slip detection is the scheme of differences. The principle can be seen from the example in Table 7.1. Assume a time series $y(t_i)$, $i = 1, 2, \ldots, 7$, for a signal which contains a jump of ε at epoch t_4. Note that any of the described testing quantities may be used as signal for the scheme of differences.

Table 7.1. Scheme of differences

t_i	$y(t_i)$	y^1	y^2	y^3	y^4
t_1	0				
		0			
t_2	0		0		
		0		ε	
t_3	0		ε		-3ε
		ε		-2ε	
t_4	ε		$-\varepsilon$		3ε
		0		ε	
t_5	ε		0		$-\varepsilon$
		0		0	
t_6	ε		0		
		0			
t_7	ε				

7.1 Data preprocessing

The terms y^1, y^2, y^3, y^4 denote the first-order, second-order, third-order, and fourth-order differences. The important property in the context of data irregularities is the amplification of a jump in higher-order differences and, thus, the improved possibility of detecting the jump. The theoretical reason implied is the fact that differences are generated by subtractive filters. These are high-pass filters damping low frequencies and eliminating constant parts. High-frequency constituents such as a jump are amplified. Replacing the signal $y(t_i)$, for example, by the phase and assuming ε to be a cycle slip, the effect of the scheme of differences becomes evident.

A method to determine the size of a cycle slip is to fit a curve through the testing quantities before and after the cycle slip. The size of the cycle slip is found from the shift between the two curves. The fits may be obtained from a simple linear regression or from more realistic least-squares models. These methods are generally called interpolation techniques. Other possibilities are prediction methods such as Kalman filtering (Sect. 7.3). At a certain epoch, the function value (i.e., one of the testing quantities) for the next epoch is predicted based on the information obtained from preceding function values. The predicted value is then compared with the observed value to detect a cycle slip. The application of Kalman filtering for cycle slip detection is demonstrated by, e.g., Landau (1988). In static processing, the best method of detecting the amount of the double-difference integer jump is the triple-difference method. While this method, by itself, may not indicate which satellite or which epoch or which receiver caused the missing integer, it does indicate how to fix the double-difference integer exactly. Once fixed, double-difference processing can proceed.

When a cycle slip has been detected, the testing quantities can be corrected by adding the size of the cycle slip to each of the subsequent quantities. The assignment of the detected cycle slip to a single-phase observation is ambiguous if the testing quantities were phase combinations. An exception is the ionospheric residual. Under special circumstances, this testing quantity permits a unique separation. Consider Eq. (7.5) and assume ambiguity changes ΔN_1 and ΔN_2 caused by cycle slips. Consequently, a jump ΔN in the ionospheric residual would be detected. This jump is equivalent to

$$\Delta N = \Delta N_1 - \frac{f_1}{f_2} \Delta N_2, \qquad (7.8)$$

where ΔN is no longer an integer. Equation (7.8) represents a diophantine equation for the two integer unknowns ΔN_1 and ΔN_2. This means, there is one equation with two unknowns; hence, there is no unique solution. This can be seen by solving for integer values ΔN_1 and ΔN_2 such that ΔN becomes zero. To get $\Delta N = 0$ and considering as an example $f_1/f_2 = 77/60$, the condition

$$\Delta N_1 = \frac{f_1}{f_2} \Delta N_2 = \frac{77}{60} \Delta N_2 \tag{7.9}$$

must be fulfilled. This means that $\Delta N_1 = 77$ and $\Delta N_2 = 60$ cannot be distinguished from $\Delta N_1 = 154$ and $\Delta N_2 = 120$ since both solutions satisfy Eq. (7.9). However, the solution would be unambiguous if ΔN_1 is less than 77 cycles. So far the consideration assumed error-free measurements. To be more realistic, the effect of measurement noise must be taken into account. A simple model for the phase measurement noise is

$$\sigma_\Phi = 0.01 \text{ cycles}, \tag{7.10}$$

which corresponds to a resolution of $\lambda/100$. The same model is applied to both carriers and, thus, frequency-dependent noise such as multipath is neglected. The assumption is not correct for codeless or quasi-codeless receivers since additional noise is introduced during signal processing.

The value ΔN, in principle, is derived from two consecutive ionospheric residuals. Hence,

$$\Delta N = \Phi_1(t + \Delta t) - \frac{f_1}{f_2} \Phi_2(t + \Delta t) - \left[\Phi_1(t) - \frac{f_1}{f_2} \Phi_2(t) \right] \tag{7.11}$$

and applying to this equation the error propagation law gives

$$\sigma_{\Delta N} = 2.3 \, \sigma_\Phi = 0.023 \text{ cycles}. \tag{7.12}$$

The 3σ error yields approximately 0.07 cycles. This may be interpreted as the resolution of ΔN. The conclusion is that two ΔN calculated by (7.8) and using arbitrary integers, ΔN_1 and ΔN_2 must differ by at least 0.07 cycles in order to be uniquely separable. A systematic investigation of the lowest values for ΔN_1, ΔN_2 is given in Table 7.2. For ΔN_1 and ΔN_2 the values 0, ±1, ±2, ..., ±5 have been permutated and ΔN calculated by (7.8). Table 7.2 is sorted with increasing ΔN in the first column. In the second column, the first-order differences of the function values ΔN are given. To shorten the length of the table, only the negative function values ΔN and zero are displayed. For supplementing with positive function values, the signs in the first, third, and fourth column must be reversed.

Those lines in Table 7.2 being marked with an asterisk do not fulfill the criterion of an at least 0.07 cycle difference. For these values, an unambiguous separation is not possible because the measurement noise is larger than the separation value. Consider the next to the last line in Table 7.2. A jump in the ionospheric residual of about 0.14 cycle could result from the pair of cycle slips $\Delta N_1 = -4$, $\Delta N_2 = -3$ or $\Delta N_1 = 5$, $\Delta N_2 = 4$; however, notice that for the marked lines either ΔN_1 or ΔN_2

7.1 Data preprocessing

Table 7.2. Resulting ΔN by permutating ambiguity changes of ΔN_1 and ΔN_2

ΔN	Diff.	ΔN_1	ΔN_2	ΔN	Diff.	ΔN_1	ΔN_2
−11.42	1.00	−5	5	−3.72	0.16	−5	−1
−10.42	0.29	−4	5	−3.56	0.14	−1	2
−10.13	0.71	−5	4	−3.42	0.14	3	5
−9.42	0.29	−3	5	−3.28	0.15	−2	1
−9.13	0.28	−4	4	−3.13	0.13	2	4
−8.85	0.43	−5	3	−3.00	0.15	−3	0
−8.42	0.29	−2	5	−2.85	0.13	1	3
−8.13	0.28	−3	4	−2.72	0.16	−4	−1
−7.85	0.29	−4	3	−2.56	0.12	0	2
−7.56	0.14	−5	2	−2.44	0.02 *	−5	−2
−7.42	0.29	−1	5	−2.42	0.14	4	5
−7.13	0.28	−2	4	−2.28	0.15	−1	1
−6.85	0.29	−3	3	−2.13	0.13	3	4
−6.56	0.14	−4	2	−2.00	0.15	−2	0
−6.42	0.14	0	5	−1.85	0.13	2	3
−6.28	0.15	−5	1	−1.72	0.16	−3	−1
−6.13	0.28	−1	4	−1.56	0.12	1	2
−5.85	0.29	−2	3	−1.44	0.02 *	−4	−2
−5.56	0.14	−3	2	−1.42	0.14	5	5
−5.42	0.14	1	5	−1.28	0.13	0	1
−5.28	0.15	−4	1	−1.15	0.02 *	−5	−3
−5.13	0.13	0	4	−1.13	0.13	4	4
−5.00	0.15	−5	0	−1.00	0.15	−1	0
−4.85	0.29	−1	3	−0.85	0.13	3	3
−4.56	0.14	−2	2	−0.72	0.16	−2	−1
−4.42	0.14	2	5	−0.56	0.12	2	2
−4.28	0.15	−3	1	−0.44	0.16	−3	−2
−4.13	0.13	1	4	−0.28	0.13	1	1
−4.00	0.15	−4	0	−0.15	0.02 *	−4	−3
−3.85	0.13	0	3	−0.13	0.13	5	4
−3.72		−5	−1	0.00		0	0

equals 5 (plus or minus). Therefore, omitting the values for $\Delta N_1 = \pm 5$ and $\Delta N_2 = \pm 5$ creates uniqueness in the sense of separability. Up to ± 4 cycles the function values ΔN are discernible by 0.12 cycles.

The conclusions for cycle slip repair using the ionospheric residual are as follows. Based on the measurement noise assumption in (7.10), the separation of the cycle slips is unambiguously possible for up to ± 4 cycles. A smaller measurement noise increases the separability. For larger cycle slips, another method should be used in order to avoid wrong choices in ambiguous situations.

Most often, there will be more than one cycle slip. In these cases, each cycle slip must be detected and corrected individually. The corrected phases, single-, double-, or triple-differences are then used to process the baseline.

In recent years, fixing cycle slips by combining GNSS data with data of other sensors, mainly inertial navigation systems (INS), succeeded to some extent. As Colombo et al. (1999) demonstrate, even a moderately accurate (and low-cost) INS, which is also small, lightweight, and portable, can substantially enhance the ability to detect and correct cycle slips. If the INS data must bridge GNSS data gaps in a stand-alone mode, this bridging time is the critical factor for keeping the desired high accuracy. It depends on several factors as, e.g., the type of application, the baseline length, the accuracy of the INS. Accordingly, the bridging time may be limited from a few seconds only to a few minutes. Details on the GNSS/INS data modeling and on tests are given in Schwarz et al. (1994), Colombo et al. (1999), Altmayer (2000), El-Sheimy (2000), Alban (2004), Kim and Sukkarieh (2005).

7.2 Ambiguity resolution

7.2.1 General aspects

The ambiguity inherent with phase measurements depends upon both the receiver and the satellite. There is no time dependency as long as tracking is maintained without interruption. In the model for the phase,

$$\Phi = \frac{1}{\lambda}\varrho + f\,\Delta\delta + N - \frac{1}{\lambda}\Delta^{\text{Iono}}, \tag{7.13}$$

the ambiguity is denoted by N. As soon as the ambiguity is determined as an integer value, the ambiguity is said to be resolved or fixed. In general, ambiguity fixing strengthens the baseline solution. Joosten and Tiberius (2000) give an illustrative example. First, a short baseline is computed conventionally and the ambiguities are resolved. Then, introducing the float ambiguities (i.e., real values) and the fixed ambiguities (i.e., integer values), respectively, as given quantities, single-point positions on an epoch-by-epoch basis show a strong difference in the precision: with real-valued ambiguities, the solutions are scattering in the meter range in the coordinate components north, east, and up. For the integer-fixed solution, the precision of the coordinates is below the 1 cm level. But sometimes solutions with fixed ambiguities and float ambiguities may agree within a few millimeters.

The use of double-differences instead of single-differences for carrier phase processing is important. The reason is that in the case of single-differences an additional unknown parameter for the receiver clock offset must be considered, which prevents an effective separation of the integer ambiguities from the clock offset. In the case of double-differences, the clock terms have been eliminated and the isolation of the ambiguities is possible.

In order to fully exploit the high accuracy of the carrier phase observable, the ambiguities must be resolved to their correct integer value since one cycle

7.2 Ambiguity resolution

may translate to a range error of some decimeters for GNSS carriers. It should be stressed here that integer ambiguity resolution may not always be possible. One of the reasons is the baseline length. When considering short baselines (e.g., sometimes <20 km), the model for double-difference phases may be simplified to

$$\lambda \Phi_{AB}^{jk}(t) = \varrho_{AB}^{jk}(t) + \lambda N_{AB}^{jk} + \text{noise} \tag{7.14}$$

since the effects of the ionosphere, the troposphere, and other minor effects may in general be neglected. Any significant residual error from these neglected terms will spill over into the unknown parameters, namely station coordinates and ambiguities, and has the effect of degrading both the position accuracy and the integer nature of the ambiguities. Thus, if applications require a long range from the reference station, there may be a need to install several reference stations for integer ambiguity resolution or to apply the concept of virtual reference stations (Sect. 6.3.7).

Another important aspect of ambiguity resolution is the satellite geometry, which can be viewed from two points. First, an increasing number of satellites tracked at any instant translates in general into a better dilution of precision (DOP) value. Thus, all-in-view receivers with the ability to track all visible satellites are preferable since redundant satellites aid in the efficiency and reliability of ambiguity resolution. The second point with respect to geometry is the length of time required to resolve ambiguities. The information content of the carrier phase is a function of time which is directly correlated to the movement of the satellite. This last point can be illustrated through an example. Suppose two datasets. The first one consists of observations collected every 15 seconds for one hour, for a total of 240 measurements per satellite. Measurements for the second dataset are collected every second for four minutes, for a total of 240 measurements per satellite. Although the number of measurements is the same, the information content clearly is not. The first dataset has a higher probability of correct ambiguity resolution since the elapsed time is longer. The time is a critical component of ambiguity resolution even under good geometric conditions.

Multipath is also a critical factor for ambiguity resolution. Since multipath is station dependent, it may be significant even for short baselines. As in the case of atmospheric and orbital errors for long baselines, multipath has the effect of both contaminating the station coordinates and ambiguities.

Ambiguity resolution involves three major steps. The first step is the generation of potential integer ambiguity combinations that should be considered by the algorithm. A combination is composed of an integer ambiguity for, e.g., each of the double-difference satellite pairs. In order to determine these combinations, a search space must be constructed. The search space is the volume of uncertainty which surrounds the approximate coordinates of the unknown antenna location. Since the search space dictates which integer ambiguities will be considered, it should be conservatively selected since it must contain the true antenna location. In the case of static positioning, this search space can be realized from the so-called float ambiguity solution, while for kinematic positioning it is realized from a code range solution. An important aspect of this first step in ambiguity resolution is that the size of the search space will affect the efficiency, i.e., computational speed, of the process. A larger search space gives a higher number of potential integer ambiguity combinations to assess, which in turn increases the computational burden. This is important for kinematic applications where a real-time implementation may be sought. It is, therefore, necessary to balance computational load with a conservative search space size.

The second major step in the ambiguity resolution process is the identification of the correct integer ambiguity combination. The criterion used by many ambiguity resolution techniques is the selection of the integer combination which minimizes the sum of squared residuals in the sense of least-squares adjustment. The reasoning here comes from the argument that the combination which best fits the data should be the correct result. However, this can be problematic if there are not enough redundant satellites.

The third step in the ambiguity resolution process should be a validation (or verification) of the ambiguities. The assessment of the correctness of the integer numbers obtained should gain more attention (Verhagen 2004). The ambiguity success rate as defined in Joosten and Tiberius (2000) may be used as a tool for determining the probability of correct integer estimation. The ambiguity success rate depends on three factors: the observation equations (i.e., the functional model), the precision of the observables (i.e., the stochastic model), and the method of integer ambiguity estimation.

Although the last two steps based on residual analysis are rather straightforward, a few remarks should be made with respect to some of the potential difficulties of this approach. The first issue is the basic assumption in least-squares theory that the residuals should be normally distributed. In many cases, this assumption is not fulfilled due to systematic effects from multipath, orbital errors, and atmospheric errors. This is the reason why ambiguity resolution generally fails for long baselines; however, if strong multipath exists, it may even fail for short baselines. A second related issue is the need for statistical significance when the integer ambiguity decision is made. This means that the integer ambiguity combination which best fits the measurements should do so significantly better than all the other combinations. Statistical criteria can be used for this decision as will be discussed in some of the following subsections. Remaining systematic effects mentioned above play a role here as well as the aspect of time, i.e., ambiguity resolution is more difficult for shorter time intervals.

This three-step approach, (1) generation of potential integer ambiguity combinations, (2) identification of the optimum integer ambiguity combination, and (3) validation of the ambiguities, may also be refined and expanded. Han and Ri-

7.2 Ambiguity resolution

zos (1997) propose six general classes and include the ambiguity recovery techniques (to reestimate ambiguities when cycle slips occur) as well as integrated models using GNSS measurements and data from other sensors.

Hatch and Euler (1994) propose a respective partitioning into three classes which is similarly adopted by Kim and Langley (2000):

1. Ambiguity resolution in the measurement domain.
2. Search technique in the coordinate domain.
3. Search technique in the ambiguity domain.

Following this classification, a few key principles from the numerous kinds of ambiguity resolution techniques will be demonstrated subsequently. Many variations may be derived (e.g., Mervart 1995, Kim and Langley 2000).

The basic approaches as given in Sect. 7.2.2 belong to the ambiguity resolution methods in the measurement domain (but are usually combined with a search technique in the ambiguity domain).

Despite its relatively poor computational efficiency, the ambiguity function method in Sect. 7.2.3 is one representative example of a search technique in the coordinate domain.

The overwhelming part of current research is dedicated to the third class, the search technique in the ambiguity domain. Some examples are given in Sect. 7.2.3. This class of ambiguity resolution mainly refers to the integer least-squares method which is theoretically established by the fact that it will yield the optimal solution in the sense that the probability of correct integer estimation is maximized (Teunissen 1999a, b). Techniques using the integer least-squares method are usually based on three steps: (1) the float solution, (2) the integer ambiguity estimation, and (3) the fixed solution. The variance-covariance matrix resulting from the float solution in the first step is employed for different ambiguity search processes (Kim and Langley 2000). Representative methods, some of them described in Sect. 7.2.3, are given in Table 7.3. Sometimes these methods are very similar to each other, e.g., OMEGA may be regarded as a refined version of LSAST. Note that Table 7.3 does not contain methods based on simulations for multiple (more than two) frequency methods like the three-carrier ambiguity resolution (TCAR) (Forssell et al. 1997, Vollath et al. 1999).

7.2.2 Basic approaches

Single-frequency phase data

When phase measurements for only one frequency are available, the most direct approach is as follows. The measurements are modeled by Eq. (7.13), and the linearized equations are processed. Depending on the model chosen, a number of unknowns (e.g., point coordinates, clock parameters) is estimated along with N in

Table 7.3. Some representative ambiguity determination methods

Acronym	Method	Principal reference(s)
LSAST	Least-squares ambiguity search technique	Hatch (1990)
FARA	Fast ambiguity resolution approach	Frei and Beutler (1990), Frei (1991)
—	Modified Cholesky decomposition method	Euler and Landau (1992)
LAMBDA	Least-squares ambiguity decorrelation adjustment	Teunissen (1993, 1995a)
—	Null space method	Martín-Neira et al. (1995)
FASF	Fast ambiguity search filter	Chen and Lachapelle (1994)
OMEGA	Optimal method for estimating GPS ambiguities	Kim and Langley (1999)

a common adjustment. In this approach, the unmodeled errors affect all estimated parameters. Therefore, the integer nature of the ambiguities is not exploited, and they are estimated as real values. To fix ambiguities as integer values, a sequential adjustment could be performed. After an initial adjustment, the ambiguity with a computed value closest to an integer and with minimum standard error is considered to be determined most reliably. This bias is then fixed, and the adjustment is repeated (with one less unknown) to fix another ambiguity and so on. When using double-differences over short baselines, this approach is usually successful. The critical factor is the ionospheric refraction, which must be modeled and which may prevent a correct resolution of all ambiguities.

For kinematic applications, the initialization, i.e., the ambiguity determination, is a necessary initial step. Three static methods have been described in the subsection "Static initialization" of Sect. 6.3.5: (1) using a known (and usually short) baseline (the coordinates of both sites are known), which allows ambiguity resolution after a few observation epochs; (2) static determination of the first baseline; (3) the antenna swap method.

The kinematic initialization (Sect. 6.3.5) is the on-the-fly (OTF) method. This is the most advanced technique to resolve phase ambiguities and is described in more detail in Sect. 7.2.3.

Dual-frequency phase data

The situation for the ambiguity resolution improves significantly when using dual-frequency phase data. There are many advantages implied in dual-frequency data

7.2 Ambiguity resolution

because of the various possible linear combinations that can be formed like the wide-lane and narrow-lane techniques. Denoting the phase data referring to the frequencies f_1 and f_2 by Φ_1 and Φ_2, then, according to Eq. (5.17),

$$\Phi_{21} = \Phi_1 - \Phi_2 \tag{7.15}$$

is the wide-lane signal. The frequency of this signal is $f_{21} = f_1 - f_2$ and the corresponding wavelength is increased compared to the original wavelengths. The increased wide-lane wavelength λ_{21} provides an increased ambiguity spacing. This is the key to an easier resolution of the integer ambiguities. To show the principle, consider the phase models in the modified form, cf. Eq. (7.2):

$$\Phi_1 = a\, f_1 + N_1 - \frac{b}{f_1},$$
$$\Phi_2 = a\, f_2 + N_2 - \frac{b}{f_2}, \tag{7.16}$$

with the geometry term a and the ionosphere term b as known from (5.77). The difference of the two equations gives

$$\Phi_{21} = a\, f_{21} + N_{21} - b\left(\frac{1}{f_1} - \frac{1}{f_2}\right), \tag{7.17}$$

with the wide-lane quantities

$$\Phi_{21} = \Phi_1 - \Phi_2,$$
$$f_{21} = f_1 - f_2, \tag{7.18}$$
$$N_{21} = N_1 - N_2.$$

The adjustment based on the wide-lane model gives wide-lane ambiguities N_{21}, which are more easily resolved than the base carrier ambiguities.

To compute the ambiguities for the measured phases (i.e., N_1 for Φ_1 and N_2 for Φ_2), divide the first equation of (7.16) by f_1 and (7.17) by f_{21}:

$$\frac{\Phi_1}{f_1} = a + \frac{N_1}{f_1} - \frac{b}{f_1^2},$$
$$\frac{\Phi_{21}}{f_{21}} = a + \frac{N_{21}}{f_{21}} - \frac{b}{f_{21}}\left(\frac{1}{f_1} - \frac{1}{f_2}\right), \tag{7.19}$$

and the difference of the two equations gives

$$\frac{\Phi_1}{f_1} - \frac{\Phi_{21}}{f_{21}} = \frac{N_1}{f_1} - \frac{N_{21}}{f_{21}} - \frac{b}{f_1^2} + \frac{b}{f_{21}}\left(\frac{1}{f_1} - \frac{1}{f_2}\right). \tag{7.20}$$

The desired ambiguity N_1 follows explicitly after rearranging and multiplying the equation above by f_1:

$$N_1 = \Phi_1 - \frac{f_1}{f_{21}}(\Phi_{21} - N_{21}) + \frac{b}{f_1} - \frac{b}{f_{21}}\left(1 - \frac{f_1}{f_2}\right). \tag{7.21}$$

The terms reflecting the ionospheric influence may be treated as follows:

$$\begin{aligned}\frac{b}{f_1} - \frac{b}{f_{21}}\left(1 - \frac{f_1}{f_2}\right) &= b\,\frac{f_{21} f_2 - f_1 f_2 + f_1^2}{f_1 f_{21} f_2} \\ &= b\,\frac{f_{21} f_2 + f_1 (f_1 - f_2)}{f_1 f_{21} f_2} \\ &= b\,\frac{f_2 + f_1}{f_1 f_2}, \end{aligned} \tag{7.22}$$

where on the right side the term in parentheses was replaced by the wide-lane frequency f_{21}, which then canceled. Therefore, the phase ambiguity N_1 in (7.21) can be calculated from the wide-lane ambiguity by

$$N_1 = \Phi_1 - \frac{f_1}{f_{21}}(\Phi_{21} - N_{21}) + b\,\frac{f_1 + f_2}{f_1 f_2} \tag{7.23}$$

and, in an analogous way, for N_2 by exchanging the roles of f_1 and f_2 in the equation above accordingly. Equation (7.23) represents the so-called geometry-free linear phase combination since the geometric distance ϱ and the clock bias term $\Delta\delta$ do not appear explicitly. Note, however, that these terms are implicitly contained in N_{21}, cf. Eq. (7.17). The ionospheric term is most annoying. The influence of this term will be negligible for short baselines with similar ionospheric refraction at both sites (using differenced phases). For long baselines or irregular ionospheric conditions, however, the ionospheric term may cause problems.

To eliminate the ionosphere-dependent term b in the computation of the ambiguities for the measured phases (e.g., referring to f_1), one could proceed as follows. Start again with the phase equations (7.16) and multiply the first equation by f_1 and the second by f_2. Form the differences of the resulting equations and, thus,

$$f_2 \Phi_2 - f_1 \Phi_1 = a\,(f_2^2 - f_1^2) + f_2 N_2 - f_1 N_1 \tag{7.24}$$

is obtained. Eliminating N_2 via the relation $N_2 = N_1 - N_{21}$ leads to

$$f_2 \Phi_2 - f_1 \Phi_1 = a\,(f_2^2 - f_1^2) - f_2 N_{21} + N_1 (f_2 - f_1) \tag{7.25}$$

or, introducing $f_{21} = f_1 - f_2$ and dividing the equation by this relation,

$$N_1 = \frac{f_1}{f_{21}}\Phi_1 - \frac{f_2}{f_{21}}(\Phi_2 + N_{21}) - a\,(f_1 + f_2) \tag{7.26}$$

7.2 Ambiguity resolution

results. By simple linear algebra it may be verified that Eq. (7.26) is another representation of the ionosphere-free phase combination, cf. Eq. (5.79).

A final remark concerning the ambiguities is appropriate. Combining the terms containing N_1 and N_2 into a single term in the geometry-free or ionosphere-free combination destroys the integer nature of the term. This is a kind of vicious circle: either the ambiguities may be resolved where the ionosphere is a problem or the ionospheric influence is eliminated which destroys the integer nature of the ambiguities. The integer nature can be preserved by separately calculating the ambiguities, first N_{21} and then N_1 by (7.23) or (7.26).

Combining dual-frequency carrier phase and code data

The most unreliable factor of the wide-lane technique described in the previous paragraph is the influence of the ionosphere, which increases with baseline length. This drawback can be partially overcome by a combination of phase and code data. The models for dual-frequency carrier phases and code ranges, both expressed in cycles of the corresponding carrier, can be written in the form

$$\begin{aligned}
\Phi_1 &= a\, f_1 - \frac{b}{f_1} + N_1, \\
\Phi_2 &= a\, f_2 - \frac{b}{f_2} + N_2, \\
R_1 &= a\, f_1 + \frac{b}{f_1}, \\
R_2 &= a\, f_2 + \frac{b}{f_2},
\end{aligned} \qquad (7.27)$$

with the geometry term a and the ionosphere term b as known from (5.77). Note that four equations are available with four unknowns for each epoch. The unknowns are a, b, and the ambiguities N_1, N_2 and may be expressed explicitly as a function of the measured quantities by inverting the system represented by (7.27).

Multiplying the third equation of (7.27) by f_1 and the fourth by f_2 and differencing the resulting equations yields the geometry term

$$a = \frac{1}{f_2^2 - f_1^2} (R_2 f_2 - R_1 f_1). \qquad (7.28)$$

Multiplying now the third equation of (7.27) by f_2 and the fourth by f_1 and differencing the resulting equations yields the ionosphere term

$$b = \frac{f_1 f_2}{f_2^2 - f_1^2} (R_1 f_2 - R_2 f_1). \qquad (7.29)$$

Substituting (7.28) and (7.29) into the first two equations of (7.27) leads to explicit expressions for the phase ambiguities

$$N_1 = \Phi_1 + \frac{f_2^2 + f_1^2}{f_2^2 - f_1^2} R_1 - \frac{2 f_1 f_2}{f_2^2 - f_1^2} R_2 ,$$

$$N_2 = \Phi_2 + \frac{2 f_1 f_2}{f_2^2 - f_1^2} R_1 - \frac{f_2^2 + f_1^2}{f_2^2 - f_1^2} R_2 .$$

(7.30)

By forming the difference $N_{21} = N_1 - N_2$, finally

$$N_{21} = \Phi_{21} - \frac{f_1 - f_2}{f_1 + f_2} (R_1 + R_2)$$

(7.31)

is obtained. This rather elegant equation allows for the determination of the wide-lane ambiguity N_{21} for each epoch and each site. It is independent of the baseline length and of the ionospheric effects. Even if all modeled systematic effects cancel out in (7.31), the multipath effect remains and affects phase and code differently. Multipath is almost exclusively responsible for a variation of N_{21} by several cycles from epoch to epoch. These variations may be overcome by averaging over a longer period.

According to Euler and Goad (1991) and Euler and Landau (1992), the ambiguity resolution for the combination of dual-frequency code data with a reasonably low noise level and phase data will be possible "under all circumstances" with a few epochs of data. The approach described is even appropriate for instantaneous ambiguity resolution in kinematic applications. Hatch (1990) mentions that a single-epoch solution is usually possible for short baselines if seven or more satellites can be tracked. Note that several variations of the technique are known.

Combining triple-frequency carrier phase and code data

The technique based on three carriers is denoted as three-carrier ambiguity resolution (TCAR). Before pointing out the model equations, a few remarks are appropriate when comparing TCAR with the previously described dual-frequency carrier phase and code data ambiguity resolution. Theoretically, the four unknowns a, b, N_1, N_2 of (7.27) can be determined instantaneously by solving the four equations. Thus, in principle the unknowns can be determined epoch by epoch. In reality, fixing the ambiguities N_1, N_2 to their correct values will be very unlikely even for short baselines because of the magnification of noise associated with the algebraic solution. Therefore, a detour via the wide-lane ambiguities is taken.

Similarly, it may be expected that an instantaneous TCAR solution is also possible. This expectation becomes true as it may be seen immediately from the triple-

7.2 Ambiguity resolution

frequency carrier phase and code data model

$$\begin{aligned}
\Phi_1 &= a\,f_1 - \frac{b}{f_1} + N_1\,, \\
\Phi_2 &= a\,f_2 - \frac{b}{f_2} + N_2\,, \\
\Phi_3 &= a\,f_3 - \frac{b}{f_3} + N_3\,, \\
R_1 &= a\,f_1 + \frac{b}{f_1}\,, \\
R_2 &= a\,f_2 + \frac{b}{f_2}\,, \\
R_3 &= a\,f_3 + \frac{b}{f_3}\,,
\end{aligned} \tag{7.32}$$

where, apart from the two carrier phase data on f_1, f_2, the third carrier phase on f_3 is introduced. This system of six equations contains five unknowns: the geometry term a, the ionosphere term b, and the ambiguities N_1, N_2, N_3. Therefore, the system has the redundancy 1 and could be solved by least-squares adjustment. Note, however, referring to the estimated ambiguities, Sjöberg (1997, 1998) indicates "that these estimates are too poor to be useful". By contrast, it is possible to determine a wide-lane ambiguity accurately. In Table 7.4, specific values (typical for GNSS) for the three frequencies and wide-lane combinations are given.

From the dual-frequency approach, cf. Eq. (7.31), the result

$$N_{21} = \Phi_{21} - \frac{f_1 - f_2}{f_1 + f_2}(R_1 + R_2) \tag{7.33}$$

Table 7.4. GNSS frequencies and wide-lane combinations

Frequency	MHz	Wavelength [m]
f_1	1580	0.19
f_2	1230	0.24
f_3	1180	0.25
$f_1 - f_3$	400	0.75
$f_1 - f_2$	350	0.86
$f_2 - f_3$	50	6.00

is obtained for the $f_1 - f_2$ combination and

$$N_{31} = \Phi_{31} - \frac{f_1 - f_3}{f_1 + f_3}(R_1 + R_3). \tag{7.34}$$

for the $f_1 - f_3$ combination. Following from the wide-lane definitions $N_{21} = N_1 - N_2$ and $N_{31} = N_1 - N_3$, the individual ambiguities are

$$\begin{aligned} N_2 &= N_1 - N_{21}, \\ N_3 &= N_1 - N_{31}, \end{aligned} \tag{7.35}$$

where N_1 is still unknown and to be determined. These equations are resubstituted into (7.32), the initial set of model equations:

$$\begin{aligned}
\Phi_1 &= a\,f_1 - \frac{b}{f_1} + N_1, \\
\Phi_2 + N_{21} &= a\,f_2 - \frac{b}{f_2} + N_1, \\
\Phi_3 + N_{31} &= a\,f_3 - \frac{b}{f_3} + N_1, \\
R_1 &= a\,f_1 + \frac{b}{f_1}, \\
R_2 &= a\,f_2 + \frac{b}{f_2}, \\
R_3 &= a\,f_3 + \frac{b}{f_3},
\end{aligned} \tag{7.36}$$

where the known wide-lane ambiguities N_{21} and N_{31} have been shifted to the left side of the equations. This system of six equations comprises only three unknowns: a, b and N_1, thus the redundancy amounts to 3. Inherently, this combined data set of code and phase measurements reflects two accuracy classes because the last three code range equations are much less accurate compared to the first three mainly phase-derived equations. Sjöberg (1999) neglects the three code range equations by arguing that they contribute little to the least-squares solution. With the remaining phase equations, the calculation of the three unknowns is still possible for a single epoch.

After the successful computation of the N_1 ambiguity, the same procedure may be applied accordingly to get the other two carrier ambiguities N_2 and N_3.

Vollath et al. (1999) use the same set of equations as given in (7.36) but with an extended modeling of the ionospheric influence and apply a recursive least-squares adjustment.

7.2 Ambiguity resolution

Hatch et al. (2000) conclude that over short baselines the ambiguities may be resolved much more quickly (often in a single epoch), whereas for longer baselines there is limited gain from the third frequency. Vollath et al. (1999) conclude similarly that the TCAR procedure will generally not suffice to resolve the ambiguities instantaneously, i.e., using data of a single epoch unless very short baselines are considered. Accumulating several epochs will, on the one hand, reduce the noise but, on the other hand, the main error components caused by the ionosphere and multipath remain because of their long correlation times. Therefore, a search for the optimal solution along with a validation is still required. However, the number of possible candidates for this optimal result is substantially reduced.

Several other procedures exist like the integrated three-carrier ambiguity resolution (ITCAR), the cascade integer resolution (CIR) (Jung et al. 2000), which is essentially the same as ITCAR, or the extension of the null space method from the dual-frequency method to the triple-frequency approach (Fernández-Plazaola et al. 2004). Verhagen and Joosten (2004) analyse the concepts and performances of TCAR, ITCAR, CIR, LAMBDA, and the null space method.

Martín-Neira et al. (2003) mention the investigation of the multiple carrier ambiguity resolution (MCAR), which is manifested by Werner and Winkel (2003) and briefly described in the next paragraph.

Multiple carrier ambiguity resolution

In the near future of GNSS, the data combination will no longer be restricted to dual and triple frequencies because more than a single global positioning system will be available, i.e, the modernized GPS, GLONASS, and Galileo. MCAR will be the future. The somehow misleading term (dual and triple frequency being also "multiple" are not included) must be understood in the continuous development of an increased number of available frequencies. Numerous simulations focus therefore on possible benefits arising from the potential of a combined use of modernized GPS and Galileo, e.g., Werner and Winkel (2003), Zhang et al. (2003), Julien et al. (2004a), Sauer et al. (2004). The investigated benefits address the ambiguity resolution itself, the initialization performance, reliability, accuracy, and other aspects. Not only single baselines are considered but also the influence on a network of baselines is discussed (Landau et al. 2004).

Feng and Rizos (2005) summarize these benefits in a generic way as follows:

- allowing ambiguity resolution over long distances,
- allowing the fixing of correct integer solutions within much shorter periods,
- achieving highly reliable integer solutions,
- enabling RTK positioning in urban areas (where signal obstruction is an issue).

Many more investigations based on simulations are to be expected for years to come before the full potential of real modernized GPS and Galileo data may be exploited. The International Association of Geodesy (IAG) has installed the GNSS Working Group 4.5.4 (see under www.gnss.com.au) to account for MCAR methods and applications. Many additional references may be found there.

7.2.3 Search techniques

A standard approach

When processing the data based on double-differences by least-squares adjustment, the ambiguities are estimated as real or floating-point numbers, hence the first double-difference solution is called the float ambiguity solution. The output is the best estimate of the station coordinates as well as double-difference ambiguities. If the baseline is relatively short, say five kilometers, and the observation span relatively long, say one hour, these float ambiguities would typically be very close to integers. Ambiguity resolution in this case will improve the position accuracy. The change in the station coordinates from the float solution to the fixed ambiguity solution should not be large and in the case when ambiguity resolution fails, the float solution is generally a very good alternative.

As the observation span becomes smaller, the float solution will weaken due to loss of information. Ambiguity resolution will then play a more important role, since its effect on the station coordinates will now be significant. If the observation span is further reduced, the success of ambiguity resolution may determine whether or not the user's positioning specifications are met. As this discussion implies, there is a risk associated with a reduction in the observation span. A wrong integer can degrade the position solution significantly.

The search space concept can be generated for the static case by considering the position accuracy of the float ambiguity solution. A conceptually simpler approach, however, is to directly use the estimated accuracies of the float ambiguities to set their search range. For example, if an ambiguity is estimated to be 87 457 341.88 cycles with a standard deviation of 0.30 cycles, all the integer ambiguities that fall within ± 3 standard deviations of that value (for a high statistical probability) might be searched. This would give potential integer ambiguities of 87 457 340 to 87 457 343 by being conservative. This procedure can be repeated for each of the double-difference ambiguities and the result is a set of potential integer ambiguity combinations.

The number of ambiguity sets to be considered depends on the number of satellites tracked and the search range of the double-difference ambiguities. For example, there are five ambiguities if six satellites are tracked and if the range for each ambiguity is three cycles, the number of combinations to test is $3^5 = 243$. If the search range is increased to five cycles, the total number of combinations is 3 125.

7.2 Ambiguity resolution

Once all the potential ambiguity combinations are identified, each one is tested by constraining (fixing) the ambiguities to the selected integer combination and then computing the measurement residuals. The total redundancy is increased in the fixed ambiguity adjustment since only the station coordinates are estimated. However, the residuals are larger than for the float ambiguity solution. King et al. (1987) present a technique by which the influence of various integer ambiguity combinations can be computed from the float ambiguity solution, rather than initiating a new least-squares adjustment for each of the potential ambiguity combinations.

The sum of squared residuals is used as the final measure of the fit of the ambiguity combination. The integer ambiguity solution corresponding to the smallest sum of squared residuals should be the candidate which is selected. Due to reasons stated earlier, however, no candidate may be significantly better than the other to warrant selection. A ratio test is often used to make this decision. For example, if the ratio of the second smallest sum of squared residuals to the smallest sum of squared residuals is 2 or 3 (depending on the algorithm), then a decision to select the smallest sum of squared residuals as the true solution can be made. Otherwise, no integer ambiguity solution can be determined and then the best estimate for the station coordinates is the float ambiguity solution.

An example given in Cannon and Lachapelle (1993) will illustrate this concept. On a 720 m baseline, six satellites were tracked for 10 minutes. Using double-differences with satellite 19 as reference, the least-squares approach yielded for the ambiguities the following values in cycles:

DD SV	Float ambiguity
2 – 19	17 329 426.278
6 – 19	14 178 677.032
11 – 19	11 027 757.713
16 – 19	−1 575 518.876
18 – 19	−15 754 175.795

The abbreviation DD SV indicates double-differences (DD) for the specified space vehicle (SV) numbers. To get integer values, the float solution is simply rounded to the nearest integer values. To check this solution, possible other ambiguity sets are established by varying each ambiguity in a certain range, say by ±2 cycles, so that, apart from the integer solution obtained from the table above, each ambiguity is varied by −2, −1 and +1, +2 cycles. This means that for each ambiguity five cases are checked. Considering the five double-differences, in total $5^5 = 3\,125$ possible integer sets arise which are to be compared with respect to the sum of the squared residuals. Subsequently, the results for the three smallest sums of squared residuals (abbreviated as SSR) are given:

	1st smallest SSR = 0.044	2nd smallest SSR = 0.386	3rd smallest SSR = 0.453
DD SV	Ambiguity	Ambiguity	Ambiguity
2 – 19	17 329 426	17 329 426	17 329 426
6 – 19	14 178 677	14 178 676	14 178 678
11 – 19	11 027 758	11 027 757	11 027 759
16 – 19	–1 575 519	–1 575 518	–1 575 520
18 – 19	–15 754 176	–15 754 176	–15 754 176

The ambiguity set with the smallest sum of squared residuals is likely to represent the correct integers only if its SSR compared to the 2nd smallest SSR is significantly smaller. The ratio, which amounts to 0.386/0.044 = 8.8 in the example above, should be greater than 3, a threshold which has been determined empirically.

To demonstrate a failing of the ratio test, the same example is taken but the data set is reduced to 5 minutes instead of the original 10 minutes. The results for the double-difference float solution are:

DD SV	Float ambiguity
2 – 19	17 329 426.455
6 – 19	14 178 677.192
11 – 19	11 027 757.762
16 – 19	–1 575 518.471
18 – 19	–15 754 175.411

When checking again the same 3 125 possible integer ambiguity sets as before, the following ambiguity sets represent the best solutions in the sense of minimal sum of squared residuals:

	1st smallest SSR = 0.137	2nd smallest SSR = 0.155	3rd smallest SSR = 0.230
DD SV	Ambiguity	Ambiguity	Ambiguity
2 – 19	17 329 425	17 329 426	17 329 426
6 – 19	14 178 675	14 178 677	14 178 675
11 – 19	11 027 757	11 027 758	11 027 756
16 – 19	–1 575 516	–1 575 519	–1 575 518
18 – 19	–15 754 175	–15 754 176	–15 754 175

The ratio test for the smallest and the second smallest yields 0.155/0.137 = 1.1 and, thus, fails. This means that from the statistical point of view with regard to the squared sum of residuals, the correct solution cannot be extracted safely. Note, however, that the solution of the second smallest sum of squared residuals gives the correct integer ambiguities (as compared to the solution of the full 10-minute data set), but from the chosen criterion of the ratio this is not recognizable. This shows that the technique of comparing the sum of squared residuals is certainly not the most advanced technique.

Ambiguity resolution on-the-fly

The notation "on-the-fly" reflects any type of rover motion. The terms AROF (ambiguity resolution on-the-fly), OTF (on-the-fly), and sometimes OTR (on-the-run) are different abbreviations with the same meaning, namely the development of ambiguity resolution techniques for the kinematic case. Numerous techniques have been developed to deal with the kinematic case.

Code ranges are generally used to define the search space for the kinematic case. A relative code range position is used as the best estimate of antenna location, and the associated standard deviations are used to define the size of the search space. This space can be determined in several ways, for example, it can be a cube, a cylinder, or an ellipsoid.

In order to reduce the number of integer ambiguity combinations to be tested, the code solution should be as accurate as possible, which means that receiver selection becomes important. The availability of low noise, narrow correlator-type code ranges is advantageous since they have a resolution in the order of 10 cm as well as improved multipath reduction compared with standard code receivers.

An example is used to show the direct correlation between the code accuracy and the size of the potential ambiguities to be searched. Suppose a standard code receiver is used to define the search cube. The accuracy of the resulting position is approximately 2 m to give a cube size of 4 m on a side. If six satellites are tracked, there are five double-difference ambiguities to consider. The search range for each ambiguity is approximately $4 \, \text{m}/0.2 \, \text{m} = 20$ cycles (where a typical phase wavelength of 0.2 m is considered) to give $20^5 = 3.2$ million total combinations. If, in contrast, a narrow correlator-type receiver is used, the accuracy of the resulting position is approximately 1 m to give a cube of 2 m on a side and a search range of $2 \, \text{m}/0.2 \, \text{m} = 10$ cycles. Under the same six-satellite geometry, the total combinations are reduced to $10^5 = 100\,000$, which is a significant difference.

The importance of the carrier phase wide lane should be mentioned here in the context of the number of potential ambiguity combinations. If the above example is repeated using a wide lane with a wavelength of 86 cm (cf. Table 7.4), the number of potential ambiguities for a narrow correlator-type receiver would be about 35. The advantage of using this observable instead of the original carrier phase is clear as it tremendously reduces the search time. The only disadvantage of using the wide lane is that the measurement is significantly noisier than the single phase. Many OTF implementations use the wide lane to resolve integer ambiguities and then use the resulting position to directly compute the ambiguities on the original carrier phase data, or at least to significantly limit the number of single-phase ambiguities to be considered. The wide lane is also used extensively for fast static applications where the station occupation time is limited.

The OTF techniques have common features like, e.g., the determination of an initial solution; they differ only in how these features are carried out. A summary of the main features is given in Table 7.5 which is closely related to Erickson (1992b). As far as the search technique (domain, space, reduction of trials) is concerned, there are also combinations of several listed characteristics (e.g., Abidin et al. 1992). Illustrative graphic representations of search spaces lead to an easier understanding of the reduction of trials, see Hatch 1991, Erickson 1992a, Frei and Schubernigg 1992, Abidin 1993.

Table 7.5. Characteristics and options for OTF ambiguity resolution techniques

Initial solution	• Code solution for position X, Y, Z and its accuracy σ_X, σ_Y, σ_Z
	• Carrier solution for X, Y, Z and N_j and accuracies σ_X, σ_Y, σ_Z, σ_{N_j}
Search domain	• Test points (three-dimensional space)
	• Ambiguity sets (n-dimensional integer space, where n is the number of ambiguities)
Search space	• $k\sigma_X$, $k\sigma_Y$, $k\sigma_Z$
	• $k\sigma_{N_j}$
Determination of k	• Empirically
	• Statistically
Reduction of trials	• Grid search (fine, coarse)
	• Double-difference plane intersection
	• Statistically (e.g., correlation of ambiguities)
Selection criterion	• Maximum ambiguity function
	• Minimum variance σ_0^2
Acceptance criterion	• Ratio of largest and second largest ambiguity function
	• Ratio of smallest and second smallest variance σ_0^2
Observation period	• Instantaneous
	• Some minutes
Data required	• Single- or dual-frequency
	• Phase only or phase and code

7.2 Ambiguity resolution

The double-difference plane intersection method to reduce the number of trials as mentioned in Table 7.5 requires a brief explanation. Positions are derived from three double-differences with sets of possible ambiguities. Geometrically, each (linearized) double-difference with its trial ambiguity defines a plane in three-dimensional space (Hatch 1990). Thus, the intersection of three planes yields a possible solution position. The grid spacing is the wavelength of the carrier and is equivalent to the grid spacing in the ambiguity search domain.

Minimizing the variance σ_0^2 as selection criterion is in principle the same as minimizing the sum of the squared residuals. If the position of the receiver is eliminated by a mapping function, as proposed by Walsh (1992), the residuals reflect the ambiguities only.

The subsequent paragraphs explain some out of the many OTF techniques that can be used. Examples are the ambiguity function method, the least-squares ambiguity search, the fast ambiguity resolution approach, the fast ambiguity search filter, least-squares ambiguity decorrelation adjustment method, and ambiguity determination with special constraints.

Numerous approaches may be found in publications. Here are some examples: the fast ambiguity resolution using an integer nonlinear programming method (Wei and Schwarz 1995b); a maximum likelihood method based on undifferenced phases (Knight 1994); the fitting of individual epoch residuals for potential ambiguity candidates to low-order polynomials (Borge and Forssell 1994). Additional methods may be found in the review papers by Chen and Lachapelle (1994), Hatch and Euler (1994), Hein (1995).

Ambiguity function method

Counselman and Gourevitch (1981) proposed the principle of the ambiguity function, Remondi (1984, 1990a) and Mader (1990) further investigated this method. The concept will become clear from the following description. Assume the model (6.43) for the single-difference phase represented by

$$\Phi_{AB}^j(t) = \frac{1}{\lambda} \varrho_{AB}^j(t) + N_{AB}^j + f\,\delta_{AB}(t) \tag{7.37}$$

for the receiver sites A and B, and the satellite j. If point A is assumed known and B is a selected candidate from the gridded cube, then the term $\varrho_{AB}^j(t)$ is known and may be shifted to the left side of the equation:

$$\Phi_{AB}^j(t) - \frac{1}{\lambda} \varrho_{AB}^j(t) = N_{AB}^j + f\,\delta_{AB}(t)\,. \tag{7.38}$$

The key is to circumvent the ambiguities N_{AB}^j. A special effect occurs if the term $2\pi N_{AB}^j$ is used as the argument of a cosine or sine function because N_{AB}^j is an

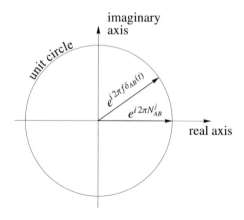

Fig. 7.2. Vector representation in the complex plane

integer. Therefore, the whole expression (7.38) is multiplied by 2π and placed into the complex plane by raising both the left and right side to the power of $e^i = \exp\{i\}$, where $i = \sqrt{-1}$ is the imaginary unit. In detail,

$$\exp\left\{i\left[2\pi \Phi_{AB}^j(t) - \frac{2\pi}{\lambda} \varrho_{AB}^j(t)\right]\right\} = \exp\left\{i[2\pi N_{AB}^j + 2\pi f \delta_{AB}(t)]\right\}, \quad (7.39)$$

where the right side may also be written as

$$\exp\left\{i\, 2\pi N_{AB}^j\right\} \exp\left\{i\, 2\pi f \delta_{AB}(t)\right\}. \quad (7.40)$$

It is illustrative to consider this situation in the complex plane (Fig. 7.2). Note the equivalence

$$\exp\{i\alpha\} = \cos\alpha + i\,\sin\alpha, \quad (7.41)$$

which may be represented as a unit vector with the components $\cos\alpha$ and $\sin\alpha$ if a real axis and an imaginary axis are used. Therefore,

$$\exp\left\{i\, 2\pi N_{AB}^j\right\} = \cos(2\pi N_{AB}^j) + i\,\sin(2\pi N_{AB}^j) = 1 + i\cdot 0 \quad (7.42)$$

results because of the integer nature of N_{AB}^j. Hence, for one epoch and one satellite, (7.39) reduces to

$$\exp\left\{i\left[2\pi \Phi_{AB}^j(t) - \frac{2\pi}{\lambda} \varrho_{AB}^j(t)\right]\right\} = \exp\left\{i\, 2\pi f \delta_{AB}(t)\right\} \quad (7.43)$$

7.2 Ambiguity resolution

by applying (7.40) and (7.42). Considering n_s satellites and forming the sum over these satellites for the epoch t leads to

$$\sum_{j=1}^{n_s} \exp\left\{i\left[2\pi\,\Phi_{AB}^j(t) - \frac{2\pi}{\lambda}\,\varrho_{AB}^j(t)\right]\right\} = n_s \exp\{i\,2\pi\,f\,\delta_{AB}(t)\}\,. \tag{7.44}$$

Considering more than one epoch, the fact that the clock error $\delta_{AB}(t)$ varies with time must be taken into account. Recall that $\exp\{i\,2\pi\,f\,\delta_{AB}(t)\}$ is a unit vector as indicated in Fig. 7.2. Thus, when $\|\exp\{i\,2\pi\,f\,\delta_{AB}(t)\}\| = 1$ is applied to (7.44), the relation

$$\left\|\sum_{j=1}^{n_s} \exp\left\{i\left[2\pi\,\Phi_{AB}^j(t) - \frac{2\pi}{\lambda}\,\varrho_{AB}^j(t)\right]\right\}\right\| = n_s \tag{7.45}$$

is obtained, where the clock error has now vanished.

Take for example four satellites and an error-free situation (i.e., neither measurement errors nor model errors, and correct coordinates for the points A and B). In this case, the evaluation of the left side of (7.45) should yield 4, where $\Phi_{AB}^j(t)$ are the single-differences of measured phases and $\varrho_{AB}^j(t)$ can be calculated from the known points and satellite positions. However, if point B was chosen incorrectly, then the result must be less than 4. In reality, this maximum can probably never be achieved precisely because of measurement errors and incomplete modeling. Thus, the task is restricted to obtaining the maximum of (7.45) by varying B.

With highly stable receiver clocks and close epoch spacing it is theoretically possible to include more than one epoch within the absolute value. Using n_t epochs, the contribution of all epochs may be summed up by

$$\sum_{t=1}^{n_t} \left\|\sum_{j=1}^{n_s} \exp\left\{i\left[2\pi\,\Phi_{AB}^j(t) - \frac{2\pi}{\lambda}\,\varrho_{AB}^j(t)\right]\right\}\right\| = n_t\,n_s\,, \tag{7.46}$$

where for simplicity the same number of satellites at all epochs is assumed. Following Remondi (1984, 1990a), the left side of (7.46) is denoted as an ambiguity function. Analogous to the case with one epoch, the maximum of the ambiguity function must be found. In general it will, as before, be less than the theoretical value $n_t\,n_s$.

The ambiguity function procedure is simple. Assume an approximate solution for point B, e.g., achieved by triple-differences. Then, place this solution into the center of a cube (Fig. 7.3) and partition the cube into grid points. Each grid point is a candidate for the final solution, and the ambiguity function (7.46) is calculated for all single-differences. The grid point yielding the maximum ambiguity function value, which should theoretically be equal to the total number of single-differences

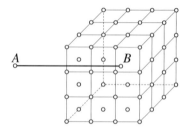

Fig. 7.3. Search space

(i.e., $n_t\, n_s$), is the desired solution. Having found this solution, the ambiguities could be computed using double-differences. Also, an adjustment using double-differences might be performed to verify the position of B and the ambiguities. The computation of point B with fixed ambiguities is the final step.

It is worth noting that the ambiguity function method is completely insensitive to cycle slips. The reason can easily be seen from Eq. (7.42). Even if the ambiguity changes by an arbitrary integer amount ΔN_{AB}^j, then $\exp\{i\, 2\pi\, (N_{AB}^j + \Delta N_{AB}^j)\}$ is still a unit vector and the subsequent equations, therefore, remain unchanged. Other methods require that cycle slips be repaired before computing the ambiguity.

Remondi (1984) shows detailed examples of how to speed up the procedure, how to choose the density of the grid points within the cube, and how to find the correct maximum if there are many relative maxima for the ambiguity function. These considerations are significant, since the computational burden could, otherwise, become overwhelming. For illustrative purposes, assume a $6\,\mathrm{m} \times 6\,\mathrm{m} \times 6\,\mathrm{m}$ cube with a one centimeter grid. Then $(601)^3 \approx 2.17 \cdot 10^8$ possible solutions must be checked with the ambiguity function (7.46).

Least-squares ambiguity search technique

The method described here is investigated in further details in Hatch (1990, 1991). The least-squares ambiguity search technique (LSAST) requires an approximate solution for the position which may be obtained from a code range solution. The search area may be established by surrounding the approximate position by a 3σ region. One of the basic principles of the approach is the separation of the satellites into a primary and a secondary group. The primary group consists of four satellites. Based on these four satellites, which should have a good position dilution of precision (Sect. 7.3.4), the possible ambiguity sets are determined. The remaining secondary satellites are used to eliminate candidates of the possible ambiguity sets.

The set of potential solutions may be found in the following way. Assume the simplified double-difference model (7.14). If the ambiguities are moved to the left

7.2 Ambiguity resolution

side as if they were known, the model reads $\lambda \Phi - N = \varrho$, where all indices have been omitted. For four satellites, three equations of this type may be set up. The three unknown station coordinates contained in the right side of the equation may be solved by linearizing ϱ and inverting the 3×3 design matrix. Specifying and varying the three ambiguities on the left side gives new position solutions, whereas the inverted design matrix remains unchanged. Depending on the variation of the three ambiguities, the set of potential solutions is obtained. Note that Hatch (1990) does not use double-differences but undifferenced phases to avoid any biasing.

From the set of potential solutions, incorrect solutions are removed by taking into account the information of the secondary group of satellites. Sequential least-squares adjustment would be appropriately used for this task. Finally, the sum of the squared residuals may be taken as criterion for the quality indicator of the solution. Ideally, only the true set of ambiguities should remain. If this is not the case, then, as described previously, the solution with the smallest sum of squared residuals should be chosen (after comparing it with the second smallest sum).

Fast ambiguity resolution approach

The development of the fast ambiguity resolution approach (FARA) is given in Frei (1991) and summarized in Frei and Schubernigg (1992). Following the latter publication, the main characteristics are (1) to use statistical information from the initial adjustment to select the search range, (2) to use information of the variance-covariance matrix to reject ambiguity sets that are not acceptable from the statistical point of view, and (3) to apply statistical hypothesis testing to select the correct set of integer ambiguities.

Following Erickson (1992a), the FARA algorithm may be partitioned into four steps: (1) computing the float carrier phase solution, (2) choosing ambiguity sets to be tested, (3) computing a fixed solution for each ambiguity set, and (4) statistically testing the fixed solution with the smallest variance.

In the first step, real values for double-difference ambiguities are estimated based on carrier phase measurements and calculated by an adjustment procedure which also computes the cofactor matrix of the unknown parameters and the a posteriori variance of unit weight (a posteriori variance factor). Based on these results, the variance-covariance matrix of the unknown parameters and the standard deviations of the ambiguities may also be computed.

In the second step, the criteria for the ambiguity ranges to be investigated are based on confidence intervals of the real values of the ambiguities. Therefore, the quality of the initial solution of the first step affects the possible ambiguity ranges. In more detail, if σ_N represents the standard deviation of the ambiguity N, then $\pm k \sigma_N$ is the search range for this ambiguity, where k is derived statistically from Student's t-distribution. This is the first criterion for selecting possible ambiguity sets.

A second criterion is the use of the correlation of the ambiguities. Assuming the double-difference ambiguities N_i and N_j and the difference

$$N_{ij} = N_j - N_i, \tag{7.47}$$

the standard deviation follows from the error propagation law as

$$\sigma_{N_{ij}} = \sqrt{\sigma_{N_i}^2 - 2\sigma_{N_i N_j} + \sigma_{N_j}^2}, \tag{7.48}$$

where $\sigma_{N_i}^2$, $\sigma_{N_i N_j}$, and $\sigma_{N_j}^2$ are contained in the variance-covariance matrix of the parameters. The search range for the ambiguity difference N_{ij} is $k_{ij}\,\sigma_{N_{ij}}$, where k_{ij} is analogous to the search range for individual double-difference ambiguities. This criterion significantly reduces the number of possible integer sets. An even more impressive reduction is achieved if dual-frequency phase measurements are available. Illustrative figures demonstrating this reduction are given in Frei and Schubernigg (1992).

In the third step, least-squares adjustment with fixed ambiguities is performed for each statistically accepted ambiguity set yielding adjusted baseline components and a posteriori variance factors.

In the fourth and final step, the solution with the smallest a posteriori variance is further investigated. The baseline components of this solution are compared with the float solution. If the solution is compatible, it is accepted. As shown in Erickson (1992a), the compatibility may be checked by a χ^2-distribution which tests the compatibility of the a posteriori variance with the a priori variance. Furthermore, another test may be applied to ensure that the second smallest variance is sufficiently less likely than the smallest variance. Note, however, that these two variances are not independent (Teunissen 1996: Sect. 8.2.3).

As seen from the algorithm, FARA only requires data for double-difference phases; thus, in principle, neither code data nor dual-frequency data are required; however, if these data are added, the number of possible ambiguity sets increases dramatically (see the second step of the algorithm).

Euler et al. (1990) present an efficient and rapid search technique, similar to FARA based on the a posteriori variance (resulting from the sum of the squared residual errors). First, an integer set of ambiguities is introduced in the adjustment computation as constraints leading to an initial solution and the corresponding a posteriori variance. The influence of other ambiguity sets on the initial solution and the a posteriori variance is then determined without recomputing the whole adjustment. This influence may be calculated by some simple matrix and vector operations where only a reduced matrix with the dimension of the constraint ambiguities must be inverted. Following Landau and Euler (1992), the computation time for the matrix inversion may be optimized when the Cholesky factorization method is applied which decomposes a symmetric matrix into a product of a lower and an

7.2 Ambiguity resolution

upper triangle matrix. The impact of a changed ambiguity set on the sum of the squared residuals may be reduced by the Cholesky factorization to the computation of an inner product of two vectors. Furthermore, not even the full inner product must be computed in all cases. Based on a threshold, the computation of the inner product for some integer ambiguity sets may be interrupted and the corresponding ambiguity set rejected.

The performance of this method is demonstrated in Landau and Euler (1992). Assuming six satellites and therefore five double-difference ambiguities with a 10-cycle uncertainty each, the total number of possible combinations is 3.2 millions. Using a 486 PC (even if this is pretty old-fashioned today, the ratio of the given results is still a good indicator of the performance), the computation by the Cholesky factorization took 49.1 seconds. Optimizing the Cholesky factorization by introducing the above mentioned threshold for the inner product, the computation time reduces to 0.2 seconds. For a larger search window of ± 50 cycles, the corresponding computations amount to 1.5 days for the Cholesky factorization and 3 seconds for the optimized method. The method may be extended to dual-frequency data. The appropriate formulas are given in Landau and Euler (1992).

The search techniques described so far performed the search in the ambiguity domain. An alternate technique substitutes the position as known and solves for the ambiguities as unknowns. This could be performed in the following way. Eliminate the ambiguities by forming triple-differences and obtain a first estimate for the position and its standard deviation σ by an adjustment. Now center the approximate position within a cube of dimension $\pm 3\sigma$ in each coordinate direction and partition the cube into a regular spatial grid. The cube, thus, contains a matrix of points where the center point is the triple-difference solution (Fig. 7.3). Each of these grid points is considered a candidate for the correct solution. Consequently, one by one, each candidate position is substituted into the observation equation. Then the adjustment (holding the trial position fixed) is performed and the ambiguities are computed. When all points within the cube have been considered, select the solution where the estimated real values of the ambiguities appear as close as possible to integer values. Now, fix the ambiguities to these integer values and compute (holding the ambiguities fixed) the final position which will, in general, be slightly different from the corresponding grid point of the cube.

Fast ambiguity search filter

Following Chen (1994) and Chen and Lachapelle (1994), the fast ambiguity search filter (FASF) algorithm comprises basically three components: (1) a Kalman filter is applied to predict a state vector which is treated as observable, (2) the search of the ambiguities is performed at every epoch until they are fixed, and (3) the search ranges for the ambiguities are computed recursively and are related to each other.

By applying the Kalman filter, information from the initial epoch to the current epoch is taken into account. The state vector of the Kalman filter also contains the ambiguities which are estimated as real numbers if they cannot be fixed. After fixing the ambiguities, the state vector is modified accordingly. The state vector of the Kalman filter is considered an observable and establishes, along with the regular observables (i.e., double-difference phase equations), the design matrix.

The recursively determined search ranges are based on the a priori geometric information and the effect of other (preliminarily) fixed ambiguities. As an example, take the case of four double-difference ambiguities. The first ambiguity is computed without fixing any other ambiguity. The search range for the second ambiguity is computed with the first ambiguity introduced as a known integer quantity (although it may even be the wrong integer number), the search range for the third ambiguity is computed with the first and the second ambiguity introduced as known integer quantities, and the procedure is continued for the fourth ambiguity. According to Chen and Lachapelle (1994), this concept is denoted as recursive computation of the search range.

To avoid large search ranges, a computational threshold is used. Ambiguities which cross this threshold are not fixed but computed as real numbers. Thus, an attempt to fix the ambiguities is only made if the number of potential ambiguity sets is below this threshold. Under normal circumstances, the number of potential ambiguity sets should decrease with accumulating observations. Ideally, there should finally remain a single potential ambiguity set. In practice, however, this will usually not be the case so that, conventionally, a ratio test of the sum of the squared residuals between the minimum and the second-best minimum is calculated. If this ratio fulfills a specified criterion number, the minimum solution is considered to yield the true set of ambiguities.

Once the ambiguities are fixed properly, they are removed from the state vector of the Kalman filter, i.e., from the estimation. Accordingly, the corresponding observation equation is rearranged.

The ranges of the loops for the ambiguities, i.e., the uncertainties, are calculated by using a least-squares approach with parameter elimination. First, the parameters representing the station coordinates are eliminated from the normal equations so that the ambiguities are the only remaining parameters of the model based on double-differences. Furthermore, according to the previous discussion on loops associated with the ambiguities, the ambiguities of the outer loops are constrained as integers. Returning to the example of the four ambiguities, if the range of the third ambiguity is to be determined, the first and the second ambiguity are assumed to be known and introduced as constraints (which is equivalent to removing them from the estimation vector). In fact, this may be done very efficiently as shown in Chen and Lachapelle (1994) where only single rows and columns of the adjustment matrices must be taken into account. As result of this parameter elimination, a float estimation of the corresponding ambiguity and its variance are finally obtained.

7.2 Ambiguity resolution

Multiplying the variance by a scale factor and subtracting and adding this result with respect to the float solution yields the search range for this specific ambiguity.

Note that if the uncertainty ranges are not calculated correctly, the true ambiguity set may not be found.

Least-squares ambiguity decorrelation adjustment method

Teunissen (1993) proposed the idea and further developed the least-squares ambiguity decorrelation adjustment (LAMBDA) method. A fairly detailed description of Teunissen's method is (slightly modified) given here. At present, this method is both theoretically and practically at the top level among the ambiguity determination methods.

The conventional formulation of the adjustment by parameters is

$$\mathbf{v}^T \mathbf{P} \mathbf{v} = \text{minimum!}, \tag{7.49}$$

where $\mathbf{v}$ is the vector of residuals and $\mathbf{P}$ is the weight matrix. This formulation implies that the weighted sum of squared residuals is minimized. As shown in Eq. (7.70), the weight matrix equals the inverse of the cofactor matrix $\mathbf{Q}$ of observations. Consequently,

$$\mathbf{v}^T \mathbf{Q}^{-1} \mathbf{v} = \text{minimum!} \tag{7.50}$$

is an equivalent relation.

Applying the least-squares adjustment methods for, e.g., relative positioning based on double-difference phase observations, the unknowns being determined are coordinate increments for the unknown station and double-difference ambiguities. The values obtained from the adjustment procedures are in the sense of this minimum principle the most likely ones. However, the double-difference ambiguities are obtained as real values but should be integer values. The main objective is, thus, to obtain integer ambiguities which are the most likely ones. Denoting the vector of adjusted float ambiguities by $\hat{\mathbf{N}}$ and the vector of the corresponding integer ambiguities by $\mathbf{N}$, the difference between the two vectors may be regarded as residuals of ambiguities. Consequently, it makes sense to minimize these residuals again by the same principle, i.e., the weighted sum of squared residuals. Following Teunissen et al. (1995),

$$\chi^2(\mathbf{N}) = (\hat{\mathbf{N}} - \mathbf{N})^T \mathbf{Q}_{\hat{\mathbf{N}}}^{-1} (\hat{\mathbf{N}} - \mathbf{N}) = \text{minimum!} \tag{7.51}$$

is obtained, where $\mathbf{Q}_{\hat{\mathbf{N}}}$ is the cofactor matrix of the adjusted float ambiguities. Note that sometimes the covariance matrix is used instead of the cofactor matrix. According to Eq. (7.69), these two matrices only differ by a factor.

The solution of this problem is denoted as the integer least-squares estimate of the ambiguities. Certainly, an approach different from the usual least-squares adjustment calculation must be chosen to account for the integer nature of the still unknown ambiguities **N**.

The following simple example demonstrates the solution principle. Considering two ambiguities and assuming $Q_{\hat{N}}$ as diagonal matrix

$$Q_{\hat{N}} = \begin{bmatrix} q_{\hat{N}_1 \hat{N}_1} & 0 \\ 0 & q_{\hat{N}_2 \hat{N}_2} \end{bmatrix}, \tag{7.52}$$

Eq. (7.51) yields the result

$$\chi^2(\mathbf{N}) = \frac{(\hat{N}_1 - N_1)^2}{q_{\hat{N}_1 \hat{N}_1}} + \frac{(\hat{N}_2 - N_2)^2}{q_{\hat{N}_2 \hat{N}_2}}. \tag{7.53}$$

The minimum is achieved if the N_i are chosen as those integer values being nearest to the real values. In other words, rounding the real value ambiguities to their nearest integer values yields the desired minimum for $\chi^2(\mathbf{N})$.

Since $Q_{\hat{N}}$ was assumed as diagonal matrix, the resulting N_1 and N_2 are still fully decorrelated which is also evident from Eq. (7.53). Geometrically, if two coordinate axes are associated with ambiguities N_1 and N_2, this equation represents an ellipse centered around the ambiguities $\hat{\mathbf{N}}$ and with the semiaxes

$$\begin{aligned} a &= \chi(\mathbf{N}) \sqrt{q_{\hat{N}_1 \hat{N}_1}}, \\ b &= \chi(\mathbf{N}) \sqrt{q_{\hat{N}_2 \hat{N}_2}}, \end{aligned} \tag{7.54}$$

where $\chi(\mathbf{N})$ acts as a scale factor. The axes of the ellipse are parallel to the direction of the coordinate axes. This ellipse is regarded as an ambiguity search space. Mathematically, the two integer ambiguities are contained in the two-dimensional integer space.

In reality, $Q_{\hat{N}}$ will be a fully occupied symmetric matrix. The result is still an ellipse, but its axes are rotated with respect to the coordinate system associated with N_1 and N_2, which implies a correlation of the two ambiguities so that it is more complicated to find the minimum for $\chi^2(\mathbf{N})$. In other words, the rounding to the nearest integer principle no longer works. To return to this convenient feature, the idea is to apply a transformation that decorrelates the ambiguities, which means that the transformed covariance matrix of the ambiguities becomes a diagonal matrix.

Finding a transformation that produces a diagonal matrix for $Q_{\hat{N}}$ seems to be trivial since an eigenvalue decomposition yields a diagonal matrix as output. Explicitly, each symmetric matrix

$$\mathbf{Q} = \begin{bmatrix} q_{11} & q_{12} \\ q_{12} & q_{22} \end{bmatrix} \tag{7.55}$$

7.2 Ambiguity resolution

can be transformed into the diagonal matrix

$$\mathbf{Q}' = \begin{bmatrix} \lambda_1 & 0 \\ 0 & \lambda_2 \end{bmatrix}. \tag{7.56}$$

The eigenvalues are defined by

$$\begin{aligned} \lambda_1 &= \tfrac{1}{2}(q_{11} + q_{22} + w), \\ \lambda_2 &= \tfrac{1}{2}(q_{11} + q_{22} - w), \end{aligned} \tag{7.57}$$

with the auxiliary quantity

$$w = \sqrt{(q_{11} - q_{22})^2 + 4q_{12}^2}. \tag{7.58}$$

The two eigenvectors are orthogonal to each other and are defined by the rotation angle φ, which can be calculated by

$$\tan 2\varphi = \frac{2q_{12}}{q_{11} - q_{22}}. \tag{7.59}$$

The only problem is that the integer ambiguities $\mathbf{N}$ must also be transformed and must preserve their integer nature. Thus, an ordinary eigenvalue decomposition will not work.

Generally, the task may be formulated in the following way. The ambiguities $\mathbf{N}$ and $\hat{\mathbf{N}}$ are reparameterized by matrix $\mathbf{Z}$. Note that Teunissen uses the transposed matrix $\mathbf{Z}^T$, but the principle remains the same. Hence,

$$\begin{aligned} \mathbf{N}' &= \mathbf{Z}\mathbf{N}, \\ \hat{\mathbf{N}}' &= \mathbf{Z}\hat{\mathbf{N}}, \\ \mathbf{Q}_{\hat{\mathbf{N}}'} &= \mathbf{Z}\,\mathbf{Q}_{\hat{\mathbf{N}}}\,\mathbf{Z}^T, \end{aligned} \tag{7.60}$$

where the transformation of the cofactor matrix is obtained by applying the error propagation law. The ambiguities $\mathbf{N}'$ obtained after transformation must remain integer values. That restricts the matrix $\mathbf{Z}$ to a specific class of transformations where three conditions must be fulfilled (Teunissen 1994, 1995b). These conditions are the following (1) the elements of the transformation matrix $\mathbf{Z}$ must be integer values, (2) the transformation must be volume preserving, and (3) the transformation must reduce the product of all ambiguity variances.

Note that the inverse of the transformation matrix $\mathbf{Z}$ must also consist of integer values only, because upon a retransformation of the (determined) integer ambiguities $\mathbf{N}'$, the integer nature of the ambiguities must be kept.

For the two-dimensional example shown, volume preserving reduces to area preserving of the ellipse represented by the two-dimensional cofactor (covariance) matrix.

If the three conditions are fulfilled, the transformed integer ambiguities are again integer values and the cofactor (covariance) matrix of the transformed ambiguities is more diagonal than the cofactor (covariance) matrix of the original ambiguities (Teunissen 1994).

The Gauss transformation is one of the possible candidates and may either be expressed by

$$\mathbf{Z}_1 = \begin{bmatrix} 1 & 0 \\ \alpha_1 & 1 \end{bmatrix}, \qquad \alpha_1 = -\mathrm{INT}\left[q_{\hat{N}_1\hat{N}_2}/q_{\hat{N}_1\hat{N}_1}\right], \qquad (7.61)$$

or by the other form

$$\mathbf{Z}_2 = \begin{bmatrix} 1 & \alpha_2 \\ 0 & 1 \end{bmatrix}, \qquad \alpha_2 = -\mathrm{INT}\left[q_{\hat{N}_1\hat{N}_2}/q_{\hat{N}_2\hat{N}_2}\right], \qquad (7.62)$$

since the roles of the two ambiguities may be interchanged. In the transformation (7.61), the ambiguity $\hat{N}_1$ remains unchanged and $\hat{N}_2$ is transformed. Analogously, $\hat{N}_2$ may be kept unchanged and $\hat{N}_1$ will be transformed as achieved by (7.62). For a better distinction of the two transformations, the subscripts 1 and 2 were introduced. Here, the operator INT performs the rounding to the nearest integer. The theoretical background of the transformation procedure comprises the conditional least-squares estimate (Teunissen 1994).

The transformed ambiguities are obtained from

$$\begin{bmatrix} \hat{N}'_1 \\ \hat{N}'_2 \end{bmatrix} = \begin{bmatrix} 1 & -\mathrm{INT}[q_{\hat{N}_1\hat{N}_2}/q_{\hat{N}_2\hat{N}_2}] \\ 0 & 1 \end{bmatrix} \begin{bmatrix} \hat{N}_1 \\ \hat{N}_2 \end{bmatrix}. \qquad (7.63)$$

For a numerical example, Teunissen (1996: Sect. 8.5.2) assumed that after a least-squares adjustment the ambiguities

$$\hat{\mathbf{N}} = \begin{bmatrix} \hat{N}_1 \\ \hat{N}_2 \end{bmatrix} = \begin{bmatrix} 1.05 \\ 1.30 \end{bmatrix}$$

and

$$\mathbf{Q}_{\hat{\mathbf{N}}} = \begin{bmatrix} q_{\hat{N}_1\hat{N}_1} & q_{\hat{N}_1\hat{N}_2} \\ q_{\hat{N}_1\hat{N}_2} & q_{\hat{N}_2\hat{N}_2} \end{bmatrix} = \begin{bmatrix} 53.4 & 38.4 \\ 38.4 & 28.0 \end{bmatrix}$$

were calculated. Now the transformation is applied to $\mathbf{Q}_{\hat{\mathbf{N}}}$. Translating the matrix elements to variances, the ambiguity $\hat{N}_1$ has a larger variance than $\hat{N}_2$. Hence it is

7.2 Ambiguity resolution

preferable first to change $\hat{N}_1$ and keep $\hat{N}_2$ unchanged, i.e., to apply a transformation based on $\mathbf{Z}_2$. From (7.62),

$$\alpha_2 = -\text{INT}\left[q_{\hat{N}_1\hat{N}_2}/q_{\hat{N}_2\hat{N}_2}\right] = -\text{INT}[38.4/28.0] = -1$$

and

$$\mathbf{Z}_2 = \begin{bmatrix} 1 & -1 \\ 0 & 1 \end{bmatrix}$$

are obtained. The transformation according to (7.60) reads

$$\mathbf{Q}_{\hat{N}'} = \mathbf{Z}_2\, \mathbf{Q}_{\hat{N}}\, \mathbf{Z}_2^T = \begin{bmatrix} 1 & -1 \\ 0 & 1 \end{bmatrix} \begin{bmatrix} 53.4 & 38.4 \\ 38.4 & 28.0 \end{bmatrix} \begin{bmatrix} 1 & 0 \\ -1 & 1 \end{bmatrix}$$

and gives

$$\mathbf{Q}_{\hat{N}'} = \begin{bmatrix} 4.6 & 10.4 \\ 10.4 & 28.0 \end{bmatrix}.$$

The effect of this transformation can be seen best if the ambiguity search space, represented by the standard ellipse (which is centered around the corresponding ambiguities), is considered. The parameters of the standard ellipse follow from (7.55) through (7.59) if $\mathbf{Q}$ is replaced by $\mathbf{Q}_{\hat{N}}$ and $\mathbf{Q}_{\hat{N}'}$ respectively. The eigenvalues of the matrices equal the squared semiaxes of the ellipse and φ defines the direction of the semimajor axis. Explicitly, the data

$$\mathbf{Q}_{\hat{N}} : \quad a = 9.0, \quad b = 0.5, \quad \varphi = 35°,$$
$$\mathbf{Q}_{\hat{N}'} : \quad a = 5.7, \quad b = 0.8, \quad \varphi = 69°$$

are obtained. Graphically, the standard ellipses are shown in Fig. 7.4. The standard ellipse for $\mathbf{Q}_{\hat{N}}$ is centered around the ambiguities $\hat{N}$, i.e., the origin is at $\hat{N}_1 = 1.05$ and $\hat{N}_2 = 1.30$. The standard ellipse for $\mathbf{Q}_{\hat{N}'}$ is centered around the ambiguities $\hat{N}'$, i.e., the origin follows from $\hat{N}' = \mathbf{Z}_2\, \hat{N}$ and amounts to $\hat{N}'_1 = -0.25$ and $\hat{N}'_2 = 1.30$. In Fig. 7.4, search windows are also indicated with sides parallel to the two axes of the two-dimensional integer search space, i.e., two horizontal and two vertical tangents of the ellipse. The "volumes" of the two ellipses are the same because the transformation is volume preserving, but the shape and the orientation of the ellipse has changed. The distance between the two horizontal tangents has not changed because these two tangents bound the search range for the N_2 ambiguity which remained unaltered by the $\mathbf{Z}_2$ transformation, whereas the distance of the two vertical tangents has changed.

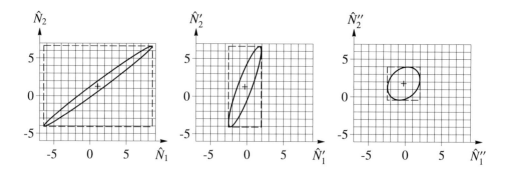

Fig. 7.4. Ambiguity search space for $\mathbf{Q}_{\hat{\mathbf{N}}}$ (left) and transformed ambiguity search spaces for $\mathbf{Q}_{\hat{\mathbf{N}}'}$ (center) and $\mathbf{Q}_{\hat{\mathbf{N}}''}$ (right)

Each grid point represents one pair of ambiguities. Under the assumption that each grid point of the search window must be regarded as a possible candidate to be investigated for a reasonable solution, the advantage of the transformed search space becomes obvious.

From comparing the off-diagonal elements of $\mathbf{Q}_{\hat{\mathbf{N}}}$ and of the transformed $\mathbf{Q}_{\hat{\mathbf{N}}'}$, the decrease of correlation is evident.

Another transformation may now be applied to $\mathbf{Q}_{\hat{\mathbf{N}}'}$. Since ambiguity $\hat{N}'_2$ has a larger variance than $\hat{N}'_1$, it is preferable to change $\hat{N}'_2$ and keep $\hat{N}'_1$ unchanged, i.e., to apply a transformation based on $\mathbf{Z}_1$. First, from (7.61)

$$\alpha_1 = -\text{INT}[q_{\hat{N}_1 \hat{N}_2}/q_{\hat{N}_1 \hat{N}_1}] = -\text{INT}[10.4/4.6] = -2$$

is determined giving

$$\mathbf{Z}_1 = \begin{bmatrix} 1 & 0 \\ -2 & 1 \end{bmatrix}$$

and

$$\mathbf{Q}_{\hat{\mathbf{N}}''} = \mathbf{Z}_1\, \mathbf{Q}_{\hat{\mathbf{N}}'}\, \mathbf{Z}_1^T = \begin{bmatrix} 1 & 0 \\ -2 & 1 \end{bmatrix} \begin{bmatrix} 4.6 & 10.4 \\ 10.4 & 28.0 \end{bmatrix} \begin{bmatrix} 1 & -2 \\ 0 & 1 \end{bmatrix},$$

where the double prime expresses that the transformation is applied on the once transformed matrix. The result is

$$\mathbf{Q}_{\hat{\mathbf{N}}''} = \begin{bmatrix} 4.6 & 1.2 \\ 1.2 & 4.8 \end{bmatrix}.$$

The standard ellipse for $\mathbf{Q}_{\hat{\mathbf{N}}''}$ is given by $a = 2.4$, $b = 1.9$, $\varphi = 47°$ and is shown in Fig. 7.4. The standard ellipse for $\mathbf{Q}_{\hat{\mathbf{N}}''}$ is centered around the ambiguities $\hat{\mathbf{N}}''$, i.e.,

7.2 Ambiguity resolution

the origin follows from $\hat{\mathbf{N}}'' = \mathbf{Z}_1 \hat{\mathbf{N}}'$ and amounts to $\hat{N}_1'' = -0.25$ and $\hat{N}_2'' = 1.80$. As far as the search window is concerned, the effect may easily be seen from the much smaller search area (represented by the window) of $\mathbf{Q}_{\hat{\mathbf{N}}''}$.

Accordingly, the distance between the two vertical tangents has not changed because these two tangents bound the search range for the N_1 ambiguity which remained unaltered by the $\mathbf{Z}_1$ transformation, whereas the distance of the two horizontal tangents has changed.

From comparing the off-diagonal elements of $\mathbf{Q}_{\hat{\mathbf{N}}'}$ and of the transformed $\mathbf{Q}_{\hat{\mathbf{N}}''}$, the decrease of correlation is evident. However, the ambiguities are still not fully decorrelated.

The two transformations may also be combined to a single transformation. Using $\mathbf{Q}_{\hat{\mathbf{N}}''} = \mathbf{Z}_1 \mathbf{Q}_{\hat{\mathbf{N}}'} \mathbf{Z}_1^T$ and substituting $\mathbf{Q}_{\hat{\mathbf{N}}'} = \mathbf{Z}_2 \mathbf{Q}_{\hat{\mathbf{N}}} \mathbf{Z}_2^T$ leads to

$$\mathbf{Q}_{\hat{\mathbf{N}}''} = \underbrace{\mathbf{Z}_1 \mathbf{Z}_2}_{\mathbf{Z}} \mathbf{Q}_{\hat{\mathbf{N}}} \underbrace{\mathbf{Z}_2^T \mathbf{Z}_1^T}_{\mathbf{Z}^T},$$

where

$$\mathbf{Z} = \begin{bmatrix} 1 & 0 \\ -2 & 1 \end{bmatrix} \begin{bmatrix} 1 & -1 \\ 0 & 1 \end{bmatrix} = \begin{bmatrix} 1 & -1 \\ -2 & 3 \end{bmatrix}$$

so that now the single transformation matrix $\mathbf{Z}$ represents the composition of the $\mathbf{Z}_2$ and the $\mathbf{Z}_1$ transformation.

The extension of the reparameterization of the ambiguity search space to higher dimensions is possible. Teunissen (1996: Sect. 8.5.3) gives the decorrelating ambiguity transformation $\mathbf{Z}$ for the three-dimensional case which would apply if double-differences of four satellites are used, and a twelve-dimensional transformation for seven satellites and dual-frequency data. Rizos and Han (1995) propose an iterative procedure to generate the decorrelating ambiguity transformation $\mathbf{Z}$. Note that the ambiguity search space becomes an ellipsoid for the three-dimensional example and an n-dimensional hyperellipsoid for $n > 3$.

After the decorrelation of the ambiguities by the $\mathbf{Z}$ transformation, the task of actually solving ambiguity estimates remains. The search can be carried out efficiently by using the sequential conditional adjustment. This adjustment determines the ambiguities step by step (i.e., sequential) one after the other. For the i-th ambiguity to be estimated, the previously determined $i - 1$ ambiguities are fixed (i.e., conditional). The sequential conditional least-squares adjustment ambiguities are not correlated. This means that the effect of the $\mathbf{Z}$ transformation will not be destroyed. An overview of the procedure is given in Jonge and Tiberius (1995) and some details are covered in Teunissen (1996: Sect. 8.3.2).

Details on the actual discrete search strategy are given in Teunissen (1994), Teunissen et al. (1994), Teunissen (1996: Sects. 8.3.2, 8.5.3).

In summary, Teunissen's LAMBDA method may be separated into the following steps:

1. A conventional least-squares adjustment is carried out to yield the baseline components and float ambiguities.
2. Using the **Z** transformation, the ambiguity search space is reparameterized to decorrelate the float ambiguities.
3. Using the sequential conditional least-squares adjustment together with a discrete search strategy, the integer ambiguities are estimated. By the inverse transformation $\mathbf{Z}^{-1}$, the ambiguities are retransformed to the original ambiguity space where the baseline components are given. Since $\mathbf{Z}^{-1}$ consists only of integer elements, the integer nature of the ambiguities is kept.
4. The integer ambiguities are fixed as known quantities and another conventional least-squares adjustment to determine the final baseline components is performed.

Ambiguity determination with special constraints

Several multiple receiver methods for kinematic applications exist. One common procedure of this technique is to place two or more receivers at fixed locations (usually short distances apart) of the moving object. Since the locations of the antennas are fixed, constraints (e.g., the distance between two antennas) may be formulated which can be used to increase the efficiency of the ambiguity resolution. In principle, the gain by using constraints results in a reduction of the potential ambiguity sets. This is illustrated briefly by two examples.

The first example, taken from Lu and Cannon (1994), concerns attitude determination in a marine environment and employs the distances between the antennas on a ship as constraints for the ambiguity resolution. Here, only the principle of the ambiguity resolution with the constraint of the known distance for a single baseline is described. Referring to the double-difference model (7.14), four satellites yielding three double-differences are considered. Analogous to the procedure for the least-squares ambiguity search technique described earlier, the equations are reformulated as $\lambda \Phi - N = \varrho$, where all indices have been omitted.

Lu and Cannon (1994) and Lu (1995) reduce the search space by introducing the known distance of the baseline. Referring to the system $\lambda \Phi - N = \varrho$, three double-differences are considered and the linearization of ϱ is performed with respect to the reference station of the baseline. Thus, the linearized system may be written as $\mathbf{w} = \mathbf{A}\mathbf{x}$, where $\mathbf{A}$ is a 3×3 design matrix resulting from the linearization, $\mathbf{x}$ contains the unknown baseline components (since the linearization was carried out with respect to the known station), and the left side of the equation contains the residual vector $\mathbf{w}$, which also comprises the ambiguities. Since $\mathbf{x}$ represents the baseline components, the constraint of the length of the baseline, denoted by b, may

7.2 Ambiguity resolution

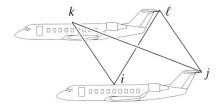

Fig. 7.5. Aircraft-to-aircraft GNSS positioning with four receivers

be introduced by first forming $\mathbf{A}^{-1}\mathbf{w} = \mathbf{x}$ and then $b^2 = \mathbf{x}^T\mathbf{x} = \mathbf{w}^T(\mathbf{A}\mathbf{A}^T)^{-1}\mathbf{w}$. This system may be further simplified by applying to $\mathbf{A}\mathbf{A}^T$ a Cholesky decomposition which reduces $\mathbf{A}$ to a lower triangle matrix. The advantage obtained from this decomposition is that the third ambiguity may be expressed by a quadratic equation containing the other two ambiguities. Thus, introducing search trials for these two other ambiguities yields two solutions for the third ambiguity. Therefore, the constraint significantly reduces the search space. Redundant satellites may be used to further reduce the size of the search space.

The performance of this method can best be seen by means of a simple example. Assuming a 15-cycle uncertainty for the three unknowns would yield (together with the one ambiguity set obtained by rounding the calculated unknowns to their nearest integer values) $31 \times 31 \times 31 = 29\,791$ possible ambiguity sets, whereas taking into account the constraint as described above reduces the set of possible ambiguities to $31 \times 31 \times 2 = 1\,922$.

The second example presented here refers to the introduction of constraints for an aircraft-to-aircraft positioning as proposed in Lachapelle et al. (1994). The situation is shown in Fig. 7.5. Each of the two aircraft is equipped with two receivers. The corresponding distances of the antennas between i and j on one aircraft and k and ℓ on the other aircraft are known and may be introduced as constraints to determine the double-difference ambiguities for each airplane separately, i.e, the double-differences N_{ij} and $N_{k\ell}$ for the available satellites (which are not indicated here by appropriate superscripts). These resolved ambiguities N_{ij} and $N_{k\ell}$ may now be used to interrelate the two aircraft by constraints. As proposed in Lachapelle et al. (1994), three sets of double-difference ambiguity relations are constrained by using, e.g., $N_{ij} = N_{jk} - N_{ik}$, $N_{ij} = N_{j\ell} - N_{i\ell}$, and $N_{k\ell} = N_{\ell i} - N_{ki}$. Thus, for five satellites there are 4×3 double-difference equations of this type which are used to reduce the number of potential ambiguity solutions. Note that these relations are not independent from each other but may still contribute to average out several error sources like carrier phase noise and multipath.

Examples given in Lachapelle et al. (1994) demonstrate that for two aircraft within 1 km typically 4 to 6 minutes of measurements (with a data rate of 1 Hz) are sufficient to obtain a unique solution. The correctness of the ambiguities may roughly be checked by the double-difference phase residuals which must not show a significant drift over time. A drift would be an indication of wrong ambiguities. The root mean square error of the double-difference phase residuals was in the amount of 0.8 cm.

Based on the given data set, several trials were performed by shifting the initial epoch from one trial to the next by 90 seconds. Of these trials, some 50% yielded the same ambiguities.

7.2.4 Ambiguity validation

After the determination of the integer ambiguities, it is of interest to validate the quality of the obtained quantities (Wang 1999). Therefore, the uncertainty of the estimated integer ambiguities is to be determined. As pointed out in Joosten and Tiberius (2000), the distribution of the estimated integer ambiguities will be a probability mass function. For a probabilistic measure, the ambiguity success rate is defined, which quantifies the probability that the integer ambiguities are correctly estimated. The ambiguity success rate equals the integral of the probability density function of the float ambiguities. The integral extends over the so-called pull-in region (Teunissen 1999a, Verhagen 2005: Chap. 3), in which all float solutions are pulled by the integer least-squares criterion to the correct integer ambiguity solution. Due to its definition as probability measure, the success rate is only a single number between 0 and 1 (which may also be expressed as percentage between 0% and 100%).

As mentioned earlier, the ambiguity success rate depends on the functional model, the stochastic model, and the chosen method of integer estimation. Similar to the dilution of precision (DOP) computations, the success rate may be calculated without actual measurements if the functional and the stochastic model are known. In the context of ambiguity validation, a theoretical problem, frequently neglected, should be mentioned. Consider a double-difference model; if integer least-squares adjustment is applied, implicitly a stochastic behavior of the observation vector is assumed. As a result, the fixed ambiguities resulting from this adjustment procedure will also be stochastic. This should be taken into account when validating the ambiguities; this is frequently neglected – however, sometimes justified (Verhagen 2004).

With respect to the integer estimation method, Teunissen (1999a, b) has proven that the LAMBDA method delivers the optimum success rate of all admissible integer estimators. A proper choice of the weight matrix is also important for the ambiguity resolution. Either a too optimistic or too pessimistic precision description will result in a less than optimal ambiguity success rate. Jonkman (1998) and Teunissen et al. (1998) demonstrate examples of an increased ambiguity success

7.2 Ambiguity resolution

rate by improving the stochastic modeling.

Several methods exist for the computation of the success rate. Joosten and Tiberius (2000) describe a simulation procedure based on a random number generator and somewhere between 100 000 and 1 million samples and achieve a success rate of 99.9 percent. Another method mentioned in Joosten and Tiberius (2000) is the computation of a "sharp lower bound" of the probability of correct integer least-squares estimation using conditional standard deviations of the ambiguities which follow directly from the triangular decomposition of the float ambiguity variance-covariance matrix. Applying the LAMBDA method, this decomposition is available without additional computational effort.

Joosten et al. (1999) stress that the success rate should be considered as the measure for judging the success of ambiguity resolution. When using the standard deviations of the ambiguities, this may yield misleading results for two reasons: (1) the correlations are neglected when using only the standard deviations, (2) ambiguity transformations change the standard deviations. In contrast to this, the success rate as defined previously is invariant for any ambiguity transformation.

Verhagen (2004) compares several of the integer validation methods proposed in literature systematically with the assumption that integer least-squares adjustment is used for ambiguity resolution. As known from Teunissen (1999a), this will result in the optimal solution in the sense that the probability of correct integer estimation is maximized. Verhagen (2004) shows that only the best and the second-best integer candidate solutions must be validated. This leads to the ratio test, one of the earliest and most popular ways to validate the integer ambiguity solution (Teunissen and Verhagen 2004). The ratio is formed by the squared norm of the second-best ambiguity residual vector and the squared norm of the best ambiguity residual vector. This ratio is compared against a certain threshold, the critical value. This critical value plays a key role since it is the indicator if the two compared solutions are considered to be discriminated with sufficient confidence.

The choice of the critical value may be regarded as a kind of question mark. Euler and Schaffrin (1990) propose a critical value between 5 and 10 depending on the degrees of freedom. Wei and Schwarz (1995b) choose 2, Han and Rizos (1996) propose 1.5 if elevation-dependent weights are used. Leick (2004: Eq. 7.207) states that many softwares simply use a fixed critical value, for example, 3.

Since a rigorous probabilistic theory for the validation of the integer ambiguities was missing, Teunissen (2003, 2004) developed the theory of integer aperture inference and, as a consequence, the optimal integer aperture estimation.

7.3 Adjustment, filtering, and quality measures

7.3.1 Theoretical considerations

The computation of the position is based on the evaluation of Eq. (6.2), which relates the range observations to the receiver position with respect to the satellite position:

$$\varrho_r^s(t) \equiv f(X_r, Y_r, Z_r). \tag{7.64}$$

The nonlinear equation can be solved applying closed-form algorithms (Kleusberg 1994, Lichtenegger 1995, Grafarend and Shan 2002). Linearizing the equations, in contrast, simplifies the algorithm and allows to implement adjustment algorithms.

Standard least-squares adjustment

In a simplified and linear form, Eq. (7.64) reads

$$\boldsymbol{\ell} = \mathbf{A}\mathbf{x}, \tag{7.65}$$

where

$\boldsymbol{\ell}$ $[n \times 1]$ ... vector of observations,
$\mathbf{A}$ $[n \times u]$ ... design matrix,
$\mathbf{x}$ $[u \times 1]$ ... vector of unknowns (parameter vector).

The design matrix maps the parameter vector onto the observations. The dimension of the vectors and matrices is expressed by $[rows \times columns]$. Equation (7.65) describes the complex physical context in a simplified and limited mathematical model. In a nonlinear situation, like (7.64), a Taylor series expansion is applied and the series is truncated after the term of first order to obtain a linear function with respect to the unknowns (cf. Sect. 7.3.2). Considering a consistent equation, the solution of (7.65) with respect to the parameters corresponds to

$$\mathbf{x} = \mathbf{A}^{-1}\boldsymbol{\ell}. \tag{7.66}$$

This requires that the number of unknown parameters u and the number of observations n are identical. Furthermore the observations have to be mathematically independent, otherwise $\mathbf{A}^{-1}$, the inverse matrix, will be singular, thus considered to be rank deficient. A number of independent observations greater than the number of unknowns leads to an overdetermined problem. Thereby, $n-u$ describes the degree of redundancy. Due to observation noise, Eq. (7.65) becomes inconsistent. Redundant problems are commonly solved using least-squares estimation techniques, to

7.3 Adjustment, filtering, and quality measures

optimally account for the complex physical relations despite of the limited mathematical models, thus, increasing the precision of the solution while at the same time getting a quality control. For a detailed discussion of the adjustment theory of linear models refer to, e.g., Mikhail and Gracie (1981). A geometric interpretation of the least-squares adjustment is given in Perović (2005: Chap. 22). The following discussions concentrate on adjustment with parameters using the observation model. Other models are discussed in, e.g., Leick (2004: Chap. 4).

Observation noise and uncertainty are assumed to be Gaussian normally distributed with zero mean. These are two fundamental assumptions described by $\mathbf{v} \sim N(\mathbf{0}, \mathbf{\Sigma_v})$, where $\mathbf{v}$ is the noise or residual vector and $\mathbf{\Sigma_v}$ denotes its (variance-) covariance matrix. The observation equation therefore reads

$$\boldsymbol{\ell} = \mathbf{A}\mathbf{x} + \mathbf{v} \tag{7.67}$$

and corresponds to a general form of a Gauss–Markov model. The model complements the deterministic mathematical model also denoted as functional model

$$E[\boldsymbol{\ell}] = \mathbf{A}\mathbf{x}, \qquad E[\mathbf{v}] = \mathbf{0} \tag{7.68}$$

by the stochastic model

$$D[\boldsymbol{\ell}] = \mathbf{\Sigma}_\ell = \sigma_0^2\, \mathbf{Q}_\ell, \tag{7.69}$$

where $E[\ ]$ denotes the expectation operator and $D[\boldsymbol{\ell}]$ describes the dispersion matrix of the observations that defines the covariance matrix $\mathbf{\Sigma}_\ell$. The factor σ_0^2 is denoted as a priori variance of unit weight, often assumed to be 1, and $\mathbf{Q}_\ell$ corresponds to the cofactor matrix. Its inverse is commonly used as weight matrix

$$\mathbf{P} = \mathbf{Q}_\ell^{-1}. \tag{7.70}$$

Considering uncorrelated observations, then the cofactor matrix degenerates to a diagonal matrix, and in the special case of equal accuracy to the unit matrix $\mathbf{Q}_\ell = \mathbf{I}$.

Off-diagonal elements of $\mathbf{\Sigma}_\ell$ and $\mathbf{Q}_\ell$, respectively, express the correlation between the measurements. These correlations are induced either mathematically (e.g., differentiation operation to compute double- or triple-differences) or by physical relations.

Equation (7.67) is consistent again but ambiguous without any additional constraint with respect to $\mathbf{v}$. The strategy to solve (7.67) is to minimize the sum of the squares of the residuals, thus the denotation least-squares adjustment. The minimization criteria reads

$$\mathbf{v}^T \mathbf{P} \mathbf{v} = (\boldsymbol{\ell} - \mathbf{A}\mathbf{x})^T\, \mathbf{P}\, (\boldsymbol{\ell} - \mathbf{A}\mathbf{x}) = \text{minimum!} \tag{7.71}$$

and the minimization process yields

$$\frac{d}{d\hat{\mathbf{x}}}\left(\mathbf{v}^T\mathbf{P}\mathbf{v}\right) = \mathbf{A}^T\mathbf{P}\mathbf{A}\hat{\mathbf{x}} - \mathbf{A}^T\mathbf{P}\boldsymbol{\ell} = \mathbf{0}, \tag{7.72}$$

where the parameter vector $\mathbf{x}$ has been replaced by the estimated vector $\hat{\mathbf{x}}$, assuming that

$$E\left[\hat{\mathbf{x}}\right] = \mathbf{x}. \tag{7.73}$$

Finally, a reordering of Eq. (7.72) yields

$$\hat{\mathbf{x}} = \left(\mathbf{A}^T\mathbf{P}\mathbf{A}\right)^{-1}\mathbf{A}^T\mathbf{P}\boldsymbol{\ell} = \mathbf{N}^{-1}\mathbf{g}, \tag{7.74}$$

where the matrix product $\mathbf{A}^T\mathbf{P}\mathbf{A}$ is denoted as the normal equation matrix $\mathbf{N}$. The vector $\mathbf{g}$ is defined by $\mathbf{A}^T\mathbf{P}\boldsymbol{\ell}$. The matrix $\mathbf{N}$ is a symmetric matrix, which has to be nonsingular in order to solve (7.74).

The cofactor matrix of the estimated parameter vector, $\mathbf{Q}_{\hat{\mathbf{x}}}$, follows from (7.74) by the covariance propagation law as

$$\mathbf{Q}_{\hat{\mathbf{x}}} = \left(\mathbf{N}^{-1}\mathbf{A}^T\mathbf{P}\right)\mathbf{Q}_{\ell}\left(\mathbf{N}^{-1}\mathbf{A}^T\mathbf{P}\right)^T \tag{7.75}$$

and considering (7.70) it reduces to

$$\mathbf{Q}_{\hat{\mathbf{x}}} = \mathbf{N}^{-1}. \tag{7.76}$$

As mentioned previously, least-squares adjustment relies on the fundamental assumptions of unbiased and normally distributed statistical elements. The mathematical and stochastic models are not conceived to account for biases and outliers in the observations. Therefore any bias or outlier will deteriorate the models and result into faulty parameters and also faulty statistical values. A small residual vector, thus, does not necessarily indicate the absence of outliers. Blunders may even avoid a convergence of iteratively processed linearized equations of the least-squares estimation. A major task in advance to the adjustment is the detection and exclusion of outliers. Stochastic formulations that are insensitive to biases of certain degrees are denoted as robust and are discussed, e.g., in Wieser (2001).

The estimated parameters $\hat{\mathbf{x}}$ are used to estimate the residuals

$$\hat{\mathbf{v}} = \boldsymbol{\ell} - \mathbf{A}\hat{\mathbf{x}}, \tag{7.77}$$

which are used in turn to estimate the a posteriori variance of unit weight

$$\hat{\sigma}_0^2 = \frac{\hat{\mathbf{v}}^T\mathbf{P}\hat{\mathbf{v}}}{n-u}. \tag{7.78}$$

7.3 Adjustment, filtering, and quality measures

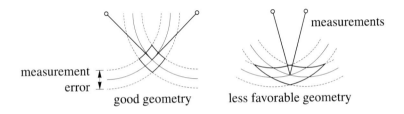

Fig. 7.6. Position error as a function of geometry and measurement error

Although the a posteriori variance does not have to be numerically equal to the a priori variance of unit weight, their statistical equivalence has to be proven by a χ^2-test to guarantee correctness of the adjustment (Leick 2004: p. 136).

The covariance matrix of the estimated parameters, thus, reads

$$\Sigma_{\hat{x}} = \hat{\sigma}_0^2 \, \mathbf{Q}_{\hat{x}} = \hat{\sigma}_0^2 \left(\mathbf{A}^T \mathbf{Q}_\ell^{-1} \mathbf{A} \right)^{-1} . \tag{7.79}$$

In terms of satellite navigation, $\Sigma_{\hat{x}}$ contains as diagonal elements the position error in the three coordinate directions, i.e., σ_X^2, σ_Y^2, and σ_Z^2. The off-diagonal elements express the correlation between the position errors in the three coordinate directions.

The covariance matrix Σ_ℓ contains the variances of the range measurements, which are composed of the satellite clock uncertainty (σ_{sc}^2), the ephemerides uncertainty (σ_{eph}^2), atmospheric uncertainties (σ_{iono}^2, σ_{trop}^2), the multipath error (σ_{mp}^2), receiver clock error (σ_{rc}^2), and a white noise error (σ_{noise}^2). A further extension into various other errors, e.g., interchannel biases, is possible. The overall error budget is estimated by the user equivalent range error (UERE)

$$\sigma_{\text{UERE}} = \sqrt{\sigma_{sc}^2 + \sigma_{eph}^2 + \sigma_{iono}^2 + \sigma_{trop}^2 + \sigma_{mp}^2 + \sigma_{rc}^2 + \sigma_{noise}^2} . \tag{7.80}$$

This measurement error is mapped onto the position error by the receiver–satellite geometry (cf. Sect. 7.3.4). The combination of both, as Fig. 7.6 illustrates, determines the quality of the position solution.

Recursive least-squares adjustment

The recursive application of least-squares adjustment algorithms onto two or several sets of observations decreases the processing load while increasing the number of observations. Another motivation for introducing recursive least-squares adjustment becomes obvious when discussing the concept of Kalman filtering.

Assume a partitioning of the observation model of (7.67) into two subsets indicated by subscripts 1,2:

$$\begin{bmatrix} \ell_1 \\ \ell_2 \end{bmatrix} = \begin{bmatrix} \mathbf{A}_1 \\ \mathbf{A}_2 \end{bmatrix} \mathbf{x} + \begin{bmatrix} \mathbf{v}_1 \\ \mathbf{v}_2 \end{bmatrix} . \tag{7.81}$$

The measurement vector ℓ and the corresponding residual vector $\mathbf{v}$ are composed of two vectors of dimension $[n_1 \times 1]$ and $[n_2 \times 1]$ each. The matrix $\mathbf{A}$ splits into two submatrices of dimension $[n_1 \times u]$ and $[n_2 \times u]$. The parameter vector $\mathbf{x}$ remains unchanged. Splitting up the parameter vector into two subvectors, e.g., a position coordinates and an ambiguity vector, will result into the blockwise least-squares adjustment as discussed by, e.g., Xu (2003: Chap. 7.5). Assuming that there is no correlation between the two measurement groups ℓ_1 and ℓ_2, the cofactor matrix $\mathbf{Q}_\ell$ has block-diagonal structure

$$\mathbf{Q}_\ell = \begin{bmatrix} \mathbf{Q}_{\ell,1} & 0 \\ 0 & \mathbf{Q}_{\ell,2} \end{bmatrix} = \mathbf{P}^{-1} = \begin{bmatrix} \mathbf{P}_1^{-1} & 0 \\ 0 & \mathbf{P}_2^{-1} \end{bmatrix}, \quad (7.82)$$

where $\mathbf{Q}_{\ell,1}$ has the dimension $[n_1 \times n_1]$, and $\mathbf{Q}_{\ell,2}$ the dimension $[n_2 \times n_2]$. Assuming $n_1 \geq u$ and considering (7.74) and (7.76), the solution of the parameters using the first subset of observations leads to

$$\hat{\mathbf{x}}_{(1)} = \left(\mathbf{A}_1^T \mathbf{P}_1 \mathbf{A}_1\right)^{-1} \mathbf{A}_1^T \mathbf{P}_1 \ell_1, \quad (7.83)$$

$$\mathbf{Q}_{\hat{\mathbf{x}},(1)} = \left(\mathbf{A}_1^T \mathbf{P}_1 \mathbf{A}_1\right)^{-1} = \mathbf{N}_1^{-1}. \quad (7.84)$$

The subscript (1) indicates a first estimate. If the change of the first estimate of parameters due to the additional observation set ℓ_2 is denoted as $\Delta\mathbf{x}$, then

$$\hat{\mathbf{x}}_{(2)} = \hat{\mathbf{x}}_{(1)} + \Delta\mathbf{x}_{(2)}. \quad (7.85)$$

The matrix $\mathbf{N}$ and the vector $\mathbf{g}$ for the adjustment of the full set of observations result from adding the corresponding matrices and vectors of the two subsets

$$\mathbf{N} = \mathbf{A}^T \mathbf{P} \mathbf{A} = \left(\mathbf{A}_1^T \mathbf{P}_1 \mathbf{A}_1 + \mathbf{A}_2^T \mathbf{P}_2 \mathbf{A}_2\right) = \mathbf{N}_1 + \mathbf{N}_2, \quad (7.86)$$

$$\mathbf{g} = \mathbf{A}^T \mathbf{P} \ell = \left(\mathbf{A}_1^T \mathbf{P}_1 \ell_1 + \mathbf{A}_2^T \mathbf{P}_2 \ell_2\right) = \mathbf{g}_1 + \mathbf{g}_2. \quad (7.87)$$

The second estimate of the parameters $\hat{\mathbf{x}}_{(2)}$ could be solved by using all observations in a common adjustment computation, i.e., $\hat{\mathbf{x}}_{(2)} = \mathbf{N}^{-1} \mathbf{g}$. Thus, combining Eqs. (7.85) through (7.87) yields

$$(\mathbf{N}_1 + \mathbf{N}_2)(\hat{\mathbf{x}}_{(1)} + \Delta\mathbf{x}_{(2)}) = \mathbf{g}_1 + \mathbf{g}_2. \quad (7.88)$$

The equation slightly rearranged reads

$$(\mathbf{N}_1 + \mathbf{N}_2) \Delta\mathbf{x}_{(2)} = \mathbf{g}_1 + \mathbf{g}_2 - (\mathbf{N}_1 + \mathbf{N}_2) \hat{\mathbf{x}}_{(1)}, \quad (7.89)$$

where the right side simplifies due to $\mathbf{g}_1 - \mathbf{N}_1 \hat{\mathbf{x}}_{(1)} = \mathbf{0}$ so that

$$(\mathbf{N}_1 + \mathbf{N}_2) \Delta\mathbf{x}_{(2)} = \mathbf{g}_2 - \mathbf{N}_2 \hat{\mathbf{x}}_{(1)} \quad (7.90)$$

7.3 Adjustment, filtering, and quality measures

results. Resubstituting $\mathbf{g}_2$ and $\mathbf{N}_2$ from Eqs. (7.86) and (7.87) yields

$$\Delta \mathbf{x}_{(2)} = (\mathbf{N}_1 + \mathbf{N}_2)^{-1} \mathbf{A}_2^T \mathbf{P}_2 (\boldsymbol{\ell}_2 - \mathbf{A}_2 \hat{\mathbf{x}}_{(1)}) \qquad (7.91)$$

or finally

$$\Delta \mathbf{x}_{(2)} = \mathbf{K}_2 (\boldsymbol{\ell}_2 - \mathbf{A}_2 \hat{\mathbf{x}}_{(1)}) , \qquad (7.92)$$

where

$$\mathbf{K}_2 = (\mathbf{N}_1 + \mathbf{N}_2)^{-1} \mathbf{A}_2^T \mathbf{P}_2 . \qquad (7.93)$$

The term $\mathbf{A}_2 \hat{\mathbf{x}}_{(1)}$ in (7.92) can formally be considered as prediction for the observations $\boldsymbol{\ell}_2$.

The change $\Delta \mathbf{Q}_{\hat{\mathbf{x}},(2)}$ with respect to the preliminary cofactor matrix $\mathbf{Q}_{\hat{\mathbf{x}},(1)}$ is obtained from the relation

$$\mathbf{N} \mathbf{Q}_{\hat{\mathbf{x}},(2)} = (\mathbf{N}_1 + \mathbf{N}_2) (\mathbf{Q}_{\hat{\mathbf{x}},(1)} + \Delta \mathbf{Q}_{\hat{\mathbf{x}},(2)}) = \mathbf{I} , \qquad (7.94)$$

where $\mathbf{I}$ denotes the unit matrix. Taking into account $\mathbf{N}_1 \mathbf{Q}_{\hat{\mathbf{x}},(1)} = \mathbf{I}$, this equation is reformulated to

$$\Delta \mathbf{Q}_{\hat{\mathbf{x}},(2)} = - (\mathbf{N}_1 + \mathbf{N}_2)^{-1} \mathbf{N}_2 \mathbf{Q}_{\hat{\mathbf{x}},(1)} \qquad (7.95)$$

and, by resubstituting $\mathbf{N}_2$ from (7.86), the relation

$$\Delta \mathbf{Q}_{\hat{\mathbf{x}},(2)} = - (\mathbf{N}_1 + \mathbf{N}_2)^{-1} \mathbf{A}_2^T \mathbf{P}_2 \mathbf{A}_2 \mathbf{Q}_{\hat{\mathbf{x}},(1)} \qquad (7.96)$$

follows. Comparing this equation with (7.93), $\mathbf{K}_2$ may be substituted and

$$\Delta \mathbf{Q}_{\hat{\mathbf{x}},(2)} = -\mathbf{K}_2 \mathbf{A}_2 \mathbf{Q}_{\hat{\mathbf{x}},(1)} \qquad (7.97)$$

results. Thus, the cofactor matrix of the new estimation decreases, expressed by the minus, with the availability of new measurements. Matrix $\mathbf{K}$, which is denoted as gain matrix, satisfies the relation

$$\mathbf{K}_2 = (\mathbf{N}_1 + \mathbf{N}_2)^{-1} \mathbf{A}_2^T \mathbf{P}_2 = \mathbf{N}_1^{-1} \mathbf{A}_2^T \left(\mathbf{P}_2^{-1} + \mathbf{A}_2 \mathbf{N}_1^{-1} \mathbf{A}_2^T \right)^{-1} , \qquad (7.98)$$

which is based on an equation found by Bennet (1965). The identity of Eq. (7.98) may be proved by multiplying from right by $\left(\mathbf{P}_2^{-1} + \mathbf{A}_2 \mathbf{N}_1^{-1} \mathbf{A}_2^T \right)$ and from left by $(\mathbf{N}_1 + \mathbf{N}_2)$. For additional information refer to Moritz (1980: p. 146). The point of this equation is its application to the inversion of modified matrices of the type $(\mathbf{C} + \mathbf{D})$, where $\mathbf{C}^{-1}$ is known a priori. It is essential to learn from Eq. (7.98) that the first form for $\mathbf{K}_2$ implies the inversion of a $[u \times u]$ matrix, whereas for the second form an inversion of an $[n_2 \times n_2]$ matrix is necessary. Therefore, the second form is advantageous as long as $n_2 < u$. Note that the second form of (7.98) in different notation reads

$$\mathbf{K}_2 = \mathbf{Q}_{\hat{\mathbf{x}},(1)} \mathbf{A}_2^T \left(\mathbf{Q}_{\ell,(2)} + \mathbf{A}_2 \mathbf{Q}_{\hat{\mathbf{x}},(1)} \mathbf{A}_2^T \right)^{-1}. \tag{7.99}$$

Assuming a general recursion step with subscripts $k-1$ and k instead of subscripts 1 and 2, the formalism of recursive least-squares adjustment yields

$$\mathbf{K}_k = \mathbf{N}_{k-1}^{-1} \mathbf{A}_k^T \left(\mathbf{P}_k^{-1} + \mathbf{A}_k \mathbf{N}_{k-1}^{-1} \mathbf{A}_k^T \right)^{-1}, \tag{7.100}$$

$$\hat{\mathbf{x}}_{(k)} = \hat{\mathbf{x}}_{(k-1)} + \mathbf{K}_k \left(\ell_k - \mathbf{A}_k \hat{\mathbf{x}}_{(k-1)} \right), \tag{7.101}$$

$$\mathbf{Q}_{\hat{\mathbf{x}},(k)} = \left(\mathbf{I} - \mathbf{K}_k \mathbf{A}_k \right) \mathbf{Q}_{\hat{\mathbf{x}},(k-1)}. \tag{7.102}$$

The recursive process starts with an initial estimate of the parameter vector $\hat{\mathbf{x}}_{(1)}$ and its cofactor matrix $\mathbf{Q}_{\hat{\mathbf{x}},(1)}$. The formalism works only for $k > 1$.

Theoretically and mathematically the recursive adjustment leads to the same solutions as the conventional adjustment. However, the solution suffers from the limited resolution of the numerical processes. Xu (2003: p. 124) mentions that the inaccuracy of numerical computations will accumulate with an increasing number of recursive steps. As a consequence, the position estimate tends to drift over time.

Discrete Kalman filter

A moving object is characterized by its state vector that assembles the nonstationary position and the velocity. The state vector, thus, is a function of time. Its computation is based on the formulation of Kalman filtering (Kalman 1960), which represents a general form of a recursive least-squares adjustment implementing time updates (dynamic model) of the state vector and its variance-covariance matrix. These time updates are based on the prediction of the present into the future state. The predicted values may be considered as approximate solution of nonlinear relations between new observables and the new state vector. In this way, also nonstationary, nonlinear relations might be described by linearized formulations. Strong nonlinearities can still be modeled using linearized but iteratively applied algorithms. The extended Kalman filter (EKF) concept, in contrast, integrates nonlinear observation equations and nonlinear dynamic models in the Kalman filter algorithm (Grewal and Andrews 2001: Chap. 5.7).

Mathematical deductions

Starting from the continuous case, the discrete Kalman filter is derived. The time-dependent state vector $\mathbf{x}(t)$ comprising the unknown parameters of the dynamic system may be modeled by a system of inhomogeneous differential equations of the first order as

$$\dot{\mathbf{x}}(t) = \mathbf{F}(t)\,\mathbf{x}(t) + \mathbf{e}(t), \tag{7.103}$$

7.3 Adjustment, filtering, and quality measures

where

$\dot{\mathbf{x}}(t)$ ... time derivative of the state vector,
$\mathbf{F}(t)$ ... dynamics matrix,
$\mathbf{e}(t)$ ... driving noise (dynamic disturbance).

For the following, at the initial epoch t_0, the state vector $\mathbf{x}(t_0)$ and its cofactor matrix $\mathbf{Q}_{\mathbf{x},0}$ are assumed to be known. A general solution for the system equation (7.103) only exists if the matrix $\mathbf{F}(t)$ contains periodic or constant coefficients. For the later case, this solution can be written as

$$\mathbf{x}(t) = \mathbf{T}(t, t_0)\,\mathbf{x}(t_0) + \int_{t_0}^{t} \mathbf{T}(t, \tau)\,\mathbf{e}(\tau)\,d\tau = \mathbf{T}(t, t_0)\,\mathbf{x}(t_0) + \mathbf{w}(t), \qquad (7.104)$$

where $\mathbf{w}(t)$ denotes system noise. To express the transition matrix $\mathbf{T}$ as function of the dynamics matrix $\mathbf{F}$, the state vector at epoch t is expressed by a Taylor series expansion. Thus,

$$\mathbf{x}(t) = \mathbf{x}(t_0) + \dot{\mathbf{x}}(t_0)\,(t - t_0) + \tfrac{1}{2}\,\ddot{\mathbf{x}}(t_0)\,(t - t_0)^2 + \ldots \qquad (7.105)$$

is obtained. Substituting (7.103), assuming the dynamics matrix to be constant, and neglecting the driving noise $\mathbf{e}(t)$ gives

$$\mathbf{x}(t) = \mathbf{x}(t_0) + \mathbf{F}(t_0)\,\mathbf{x}(t_0)\,(t - t_0) + \tfrac{1}{2}\,\mathbf{F}(t_0)^2\,\mathbf{x}(t_0)\,(t - t_0)^2 + \ldots \,. \qquad (7.106)$$

Consequently, comparing (7.104) and (7.106) and by introducing $\Delta t = t - t_0$, the transition matrix is obtained as an infinite series with respect to $\mathbf{F}$ by

$$\mathbf{T}(t, t_0) = \mathbf{I} + \mathbf{F}(t_0)\,\Delta t + \tfrac{1}{2}\,\mathbf{F}(t_0)^2\,\Delta t^2 + \ldots = \sum_{n=0}^{\infty} \frac{1}{n!}\,\mathbf{F}(t_0)^n\,\Delta t^n. \qquad (7.107)$$

The notation $\mathbf{x}(t_k) = \mathbf{x}_k$ and analogously for other quantities is introduced to indicate that the following formulations are considered at discrete time epochs t_k. The state transition equation for the discrete case, thus, reads

$$\mathbf{x}_k = \mathbf{T}_{k-1}\,\mathbf{x}_{k-1} + \mathbf{w}_k \qquad (7.108)$$

thereby relating two consecutive state vectors (dynamic model). The system noise $\mathbf{w}$ is assumed to follow a Gaussian distribution with zero mean and an $[n \times n]$ covariance matrix $\mathbf{Q}_\mathbf{w}$, briefly $\mathbf{w} \sim N(\mathbf{0}, \mathbf{Q}_\mathbf{w})$. The system noise describes the uncertainties of modeling the dynamic system behavior. The cofactor matrix of the state vector is computed by applying the law of covariance propagation onto (7.108) yielding

$$\mathbf{Q}_{\mathbf{x},k} = \mathbf{T}_{k-1}\,\mathbf{Q}_{\mathbf{x},k-1}\,\mathbf{T}_{k-1}^\mathrm{T} + \mathbf{Q}_{\mathbf{w},k}\,. \qquad (7.109)$$

The state vector of epoch $k-1$ and the system noise of epoch k are assumed to be uncorrelated.

The derivation of the Kalman filter algorithms closely follows the one of the recursive least-squares adjustment. A more detailed derivation and discussion is found, e.g., in Grewal and Andrews (2001) or Brown and Hwang (1997).

The state transition equation (7.108) results in a first estimate to compute the state vector at time epoch t_k. Applying now the principles of recursive least-squares adjustment, the observations $\boldsymbol{\ell}_k$ of time epoch t_k are used to correct the predicted state vector (measurement update). The observation equation in analogy to (7.81) reads

$$\begin{bmatrix} \mathbf{T}_{k-1}\mathbf{x}_{k-1} \\ \boldsymbol{\ell}_k \end{bmatrix} = \begin{bmatrix} \mathbf{I} \\ \mathbf{A}_k \end{bmatrix} \mathbf{x}_k + \begin{bmatrix} -\mathbf{w}_k \\ \mathbf{v}_k \end{bmatrix}. \tag{7.110}$$

The vectors $\mathbf{w}$ and $\mathbf{v}$ are considered to be uncorrelated, both with zero mean and normally distributed. Equation (7.110) is solved by applying least-squares adjustment algorithms, using cofactor matrices instead of the weight notation and introducing $\hat{\mathbf{x}}$ to indicate estimated parameters:

$$\hat{\mathbf{x}}_k = \left(\begin{bmatrix} \mathbf{I} & \mathbf{A}_k^T \end{bmatrix} \mathbf{Q}_k^{-1} \begin{bmatrix} \mathbf{I} \\ \mathbf{A}_k \end{bmatrix} \right)^{-1} \begin{bmatrix} \mathbf{I} & \mathbf{A}_k^T \end{bmatrix} \mathbf{Q}_k^{-1} \begin{bmatrix} \mathbf{T}_{k-1}\hat{\mathbf{x}}_{k-1} \\ \boldsymbol{\ell}_k \end{bmatrix}. \tag{7.111}$$

The stochastic model is described by the block-diagonal cofactor matrix

$$\mathbf{Q}_k = \begin{bmatrix} \tilde{\mathbf{Q}}_{\hat{\mathbf{x}},k} & \mathbf{0} \\ \mathbf{0} & \mathbf{Q}_{\ell,k} \end{bmatrix}. \tag{7.112}$$

Introducing (7.112) into (7.111) and multiplying the matrices reveals

$$\hat{\mathbf{x}}_k = \left(\tilde{\mathbf{Q}}_{\hat{\mathbf{x}},k}^{-1} + \mathbf{A}_k^T \mathbf{Q}_{\ell,k}^{-1} \mathbf{A}_k \right)^{-1} \left(\tilde{\mathbf{Q}}_{\hat{\mathbf{x}},k}^{-1} \mathbf{T}_{k-1}\hat{\mathbf{x}}_{k-1} + \mathbf{A}_k^T \mathbf{Q}_{\ell,k}^{-1} \boldsymbol{\ell}_k \right). \tag{7.113}$$

Adding the zero term

$$\mathbf{0} = \mathbf{A}_k^T \mathbf{Q}_{\ell,k}^{-1} \mathbf{A}_k \mathbf{T}_{k-1}\hat{\mathbf{x}}_{k-1} - \mathbf{A}_k^T \mathbf{Q}_{\ell,k}^{-1} \mathbf{A}_k \mathbf{T}_{k-1}\hat{\mathbf{x}}_{k-1} \tag{7.114}$$

to the second parenthesis in Eq. (7.113) and rearranging the equation yields

$$\begin{aligned}\hat{\mathbf{x}}_k = &\left(\tilde{\mathbf{Q}}_{\hat{\mathbf{x}},k}^{-1} + \mathbf{A}_k^T \mathbf{Q}_{\ell,k}^{-1} \mathbf{A}_k \right)^{-1} \cdot \\ &\begin{pmatrix} \left(\tilde{\mathbf{Q}}_{\hat{\mathbf{x}},k}^{-1} + \mathbf{A}_k^T \mathbf{Q}_{\ell,k}^{-1} \mathbf{A}_k \right) \mathbf{T}_{k-1}\hat{\mathbf{x}}_{k-1} \\ + \mathbf{A}_k^T \mathbf{Q}_{\ell,k}^{-1} (\boldsymbol{\ell}_k - \mathbf{A}_k \mathbf{T}_{k-1}\hat{\mathbf{x}}_{k-1}) \end{pmatrix}.\end{aligned} \tag{7.115}$$

Introducing the gain matrix $\mathbf{K}$ finally reveals

$$\hat{\mathbf{x}}_k = \mathbf{T}_{k-1}\hat{\mathbf{x}}_{k-1} + \mathbf{K}_k (\boldsymbol{\ell}_k - \mathbf{A}_k \mathbf{T}_{k-1}\hat{\mathbf{x}}_{k-1}). \tag{7.116}$$

7.3 Adjustment, filtering, and quality measures

The gain matrix, here also denoted as Kalman weight, was gained after some reordering of the matrices (see also (7.93)).

The system noise $\mathbf{w}$ is generally unknown, therefore the state vector at epoch t_k is predicted using solely the state vector $\hat{\mathbf{x}}_{k-1}$ and the transition matrix $\mathbf{T}_{k-1}$:

$$\tilde{\mathbf{x}}_k = \mathbf{T}_{k-1}\,\hat{\mathbf{x}}_{k-1}\,. \tag{7.117}$$

Here $\mathbf{x}_{k-1}$ has been replaced by $\hat{\mathbf{x}}_{k-1}$ to indicate that the vector has been estimated during the preceding measurement update (t_{k-1}). The notation $\tilde{\mathbf{x}}_k$ denotes predicted parameters.

The Kalman filter process, finally, is accomplished by a recursive application of a three-step concept:

Step 1: Gain computation (Kalman weight)

$$\mathbf{K}_k = \tilde{\mathbf{Q}}_{\hat{\mathbf{x}},k}\,\mathbf{A}_k^T\left(\mathbf{Q}_{\ell,k} + \mathbf{A}_k\tilde{\mathbf{Q}}_{\hat{\mathbf{x}},k}\,\mathbf{A}_k^T\right)^{-1}\,. \tag{7.118}$$

Step 2: Measurement update (correction)

$$\hat{\mathbf{x}}_k = \tilde{\mathbf{x}}_k + \mathbf{K}_k\,(\boldsymbol{\ell}_k - \mathbf{A}_k\,\tilde{\mathbf{x}}_k)\,, \tag{7.119}$$

$$\mathbf{Q}_{\hat{\mathbf{x}},k} = (\mathbf{I} - \mathbf{K}_k\,\mathbf{A}_k)\,\tilde{\mathbf{Q}}_{\hat{\mathbf{x}},k}\,. \tag{7.120}$$

Step 3: Time update (prediction)

$$\tilde{\mathbf{x}}_{k+1} = \mathbf{T}_k\,\hat{\mathbf{x}}_k\,, \tag{7.121}$$

$$\tilde{\mathbf{Q}}_{\hat{\mathbf{x}},k+1} = \mathbf{T}_k\,\mathbf{Q}_{\hat{\mathbf{x}},k}\,\mathbf{T}_k^T + \mathbf{Q}_{\mathbf{w},k+1}\,. \tag{7.122}$$

To avoid lengthy inversion algorithms, the measurement update can be performed for every independent measurement, what at the same time however favors numerical inaccuracy effects.

The term $\mathbf{A}_k\,\tilde{\mathbf{x}}_k$ in (7.119) corresponds to the estimation of the new observations. The difference $(\boldsymbol{\ell}_k - \mathbf{A}_k\,\tilde{\mathbf{x}}_k)$ quantifies how much information (innovation) the new observations add to the state vector estimate in relation to the old estimation. The Kalman filter in this sense weights the additional information according to the measurement variance and transforms the observation into the state vector domain. If $\mathbf{K} = \mathbf{0}$, no information is added by the observations.

Studying Eqs. (7.118) through (7.122), one may conclude that the essential issues of Kalman filtering are the definition of the transition matrix $\mathbf{T}$ or the dynamics matrix $\mathbf{F}$, respectively, and of the cofactor matrix $\mathbf{Q_w}$.

Figure 7.7 illustrates the three steps of the recursive filter strategy (Hofmann-Wellenhof et al. 2003: p. 54). In addition, three sources of external information are included into the schematic, i.e., the measurement sensor, the dynamic model, and an initialization option, which kicks off or resets the filtering process. In order not

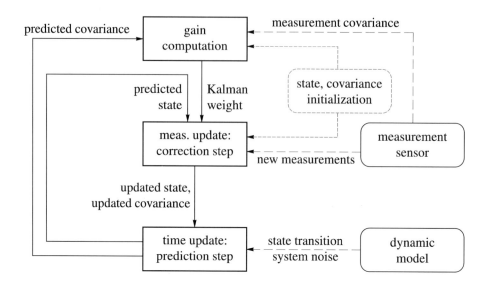

Fig. 7.7. Principle of Kalman filtering

to overload the figure, the possibility of iterative gain computation together with an iterative measurement update has not been included. Also the recursive application of time updates, if no new measurements are available, has not been plotted.

Huddle and Brown (1997: p. 75) emphasize that the Kalman filter in its basic form is a model-dependent filter and it is not adaptive. Consequently, the Kalman filter will provide poor results if the dynamic, mathematical, or stochastic models do not fit the physical truth.

Example
Consider a vehicle moving on a straight line with constant velocity v where the motion is affected by the random acceleration a. Also, assume that the (one-dimensional) position $p(t_0)$ and the velocity $v(t_0)$ as well as the corresponding variances σ_p^2, σ_v^2 and that of the noise, σ_a^2, are known at the initial epoch t_0. Furthermore, it is assumed that the position of the vehicle is observed at an epoch $t = t_0 + \Delta t$ and that the observation has the variance σ_ℓ^2 at this epoch.

The state vector consists of the position and the velocity of the vehicle. Thus, for the initial epoch,

$$\mathbf{x}(t_0) = \begin{bmatrix} p(t_0) \\ v(t_0) \end{bmatrix}, \quad \dot{\mathbf{x}}(t_0) = \begin{bmatrix} \dot{p}(t_0) \\ \dot{v}(t_0) \end{bmatrix} = \begin{bmatrix} v(t_0) \\ 0 \end{bmatrix} \quad (7.123)$$

are obtained. The substitution of these vectors and of the random acceleration a

7.3 Adjustment, filtering, and quality measures

into (7.103) yields the dynamics matrix and the driving noise vector for epoch t_0:

$$\mathbf{F}(t_0) = \begin{bmatrix} 0 & 1 \\ 0 & 0 \end{bmatrix}, \quad \mathbf{e}(t_0) = a \begin{bmatrix} 0 \\ 1 \end{bmatrix}. \tag{7.124}$$

The transition matrix according to (7.107) reads

$$\mathbf{T}(t, t_0) = \begin{bmatrix} 1 & \Delta t \\ 0 & 1 \end{bmatrix}, \tag{7.125}$$

where the infinite series has been truncated after the linear term. Assuming a constant acceleration during the integration interval Δt, the noise vector and its covariance matrix are obtained from (7.104) as

$$\mathbf{w}(t) = a \begin{bmatrix} \frac{1}{2}\Delta t^2 \\ \Delta t \end{bmatrix},$$

$$\mathbf{Q_w} = \frac{1}{\sigma_0^2} D[\mathbf{w}] = \frac{1}{\sigma_0^2} E[\mathbf{w}\mathbf{w}^T] = \frac{1}{\sigma_0^2} \begin{bmatrix} \frac{1}{4}\Delta t^4 \sigma_a^2 & \frac{1}{2}\Delta t^3 \sigma_a^2 \\ \frac{1}{2}\Delta t^3 \sigma_a^2 & \Delta t^2 \sigma_a^2 \end{bmatrix}. \tag{7.126}$$

Under the present assumptions the elements of the predicted state vector $\mathbf{x}(t)$ would also result from the formulas of accelerated motion.

The cofactor matrix $\tilde{\mathbf{Q}}_{\hat{x}}$ of the predicted state vector $\tilde{\mathbf{x}}(t)$ follows from (7.122) as

$$\tilde{\mathbf{Q}}_{\hat{x}} = \frac{1}{\sigma_0^2} \begin{bmatrix} \sigma_p^2 + \Delta t^2 \sigma_v^2 + \frac{1}{4}\Delta t^4 \sigma_a^2 & \Delta t \sigma_v^2 + \frac{1}{2}\Delta t^3 \sigma_a^2 \\ \Delta t \sigma_v^2 + \frac{1}{2}\Delta t^3 \sigma_a^2 & \sigma_v^2 + \Delta t^2 \sigma_a^2 \end{bmatrix}$$

$$= \begin{bmatrix} \tilde{q}_{11} & \tilde{q}_{12} \\ \tilde{q}_{12} & \tilde{q}_{22} \end{bmatrix}. \tag{7.127}$$

Since the inconsistent observation equation is $\ell(t) = p(t)$, the matrix $\mathbf{A}$ in (7.110) shrinks to the row vector

$$\mathbf{A} = \begin{bmatrix} 1 & 0 \end{bmatrix} \tag{7.128}$$

and the gain matrix, assuming $\sigma_0^2 = 1$, reduces to the column vector

$$\mathbf{K} = \frac{1}{\tilde{q}_{11} + \sigma_\ell^2} \begin{bmatrix} \tilde{q}_{11} \\ \tilde{q}_{12} \end{bmatrix}. \tag{7.129}$$

Now, the updated state vector $\hat{\mathbf{x}}(t)$ and the corresponding cofactor matrix $\mathbf{Q}_{\hat{x}}$ can be calculated by Eqs. (7.119) and (7.120).

Smoothing

The process of improving previous estimates of the state vector by a new measurement is called smoothing. Jekeli (2001: p. 216) distinguishes between three different smoothing types. Fixed-interval smoothing is performed backwards in time, thus, contrary to the real-time Kalman filtering, it is a post-mission process. Fixed-lag smoothing algorithms, in contrast, estimate the system state at a time delay to the current state, therefore it is a near real-time system. Fixed-point smoothing estimates the state vector at a fixed point with an increasing number of observations. The smoothing techniques are described in more detail in Brown and Hwang (1997: Chap. 8). The fixed-interval smoothing technique presented here is denoted by Grewal and Andrews (2001: Chap. 4.13) as Rauch-Tung-Striebel two-pass smoother. Using the notations $\tilde{\mathbf{x}}_k$ for the predicted state vector, $\hat{\mathbf{x}}_k$ for the updated state vector, and $\mathring{\mathbf{x}}_k$ for the smoothed state vector, then the equation for optimal smoothing reads

$$\mathring{\mathbf{x}}_k = \hat{\mathbf{x}}_k + \mathbf{D}_k \left[\mathring{\mathbf{x}}_{k+1} - \tilde{\mathbf{x}}_{k+1} \right], \tag{7.130}$$

where the gain matrix is

$$\mathbf{D}_k = \mathbf{Q}_{\hat{\mathbf{x}},k} \, \mathbf{T}_k^T \, \tilde{\mathbf{Q}}_{\hat{\mathbf{x}},k+1}^{-1}. \tag{7.131}$$

At the epoch of the last update measurement, the updated state vector is set equal to the smoothed one and the backwards algorithm can be started. From Eq. (7.130) one may conclude that the process requires the predicted and updated vectors and the cofactor matrices at the update epochs as well as the transition matrices between the updates. This implies, in general, a large amount of data.

7.3.2 Linearization of mathematical models

Considering the models of Chap. 6, the only term comprising unknowns in nonlinear form is ϱ, the geometric range between the observing receiver site r and the satellite s. Linearizing ϱ allows to use the adjustment algorithms of Sect. 7.3.1. The basic formula from Eq. (6.2),

$$\begin{aligned} \varrho_r^s(t) &= \sqrt{(X^s(t) - X_r)^2 + (Y^s(t) - Y_r)^2 + (Z^s(t) - Z_r)^2} \\ &\equiv f(X_r, Y_r, Z_r), \end{aligned} \tag{7.132}$$

shows the range ϱ as a function of the unknown point $\mathbf{X}_r = [X_r, Y_r, Z_r]$. Assuming approximate values $\mathbf{X}_{r0} = [X_{r0}, Y_{r0}, Z_{r0}]$ for the unknowns, an approximate range $\varrho_{r0}^s(t)$ is calculated by

$$\begin{aligned} \varrho_{r0}^s(t) &= \sqrt{(X^s(t) - X_{r0})^2 + (Y^s(t) - Y_{r0})^2 + (Z^s(t) - Z_{r0})^2} \\ &\equiv f(X_{r0}, Y_{r0}, Z_{r0}). \end{aligned} \tag{7.133}$$

7.3 Adjustment, filtering, and quality measures

Using the approximate values, the unknowns X_r, Y_r, Z_r are decomposed to

$$X_r = X_{r0} + \Delta X_r ,$$
$$Y_r = Y_{r0} + \Delta Y_r , \quad (7.134)$$
$$Z_r = Z_{r0} + \Delta Z_r ,$$

where now ΔX_r, ΔY_r, ΔZ_r are the new unknowns, whereas X_{r0}, Y_{r0}, Z_{r0} are known. The advantage of this split-up is that the function $f(X_r, Y_r, Z_r)$ is replaced by an equivalent function $f(X_{r0} + \Delta X_r, Y_{r0} + \Delta Y_r, Z_{r0} + \Delta Z_r)$ which now can be expanded into a Taylor series with respect to the approximate position. This leads to

$$\begin{aligned} f(X_r, Y_r, Z_r) &\equiv f(X_{r0} + \Delta X_r, Y_{r0} + \Delta Y_r, Z_{r0} + \Delta Z_r) \\ &= f(X_{r0}, Y_{r0}, Z_{r0}) + \left. \frac{\partial f(X_r, Y_r, Z_r)}{\partial X_r} \right|_{\mathbf{X}_r = \mathbf{X}_{r0}} \Delta X_r \\ &\quad + \left. \frac{\partial f(X_r, Y_r, Z_r)}{\partial Y_r} \right|_{\mathbf{X}_r = \mathbf{X}_{r0}} \Delta Y_r + \left. \frac{\partial f(X_r, Y_r, Z_r)}{\partial Z_r} \right|_{\mathbf{X}_r = \mathbf{X}_{r0}} \Delta Z_r + \dots , \end{aligned}$$
$$(7.135)$$

where the expansion is truncated after the linear term. The higher terms are assumed to be negligibly small, otherwise the adjustment process has to be iteratively repeated, which means the result of the adjustment is used as the new approximate position. The partial derivatives evaluated at the approximate position $\mathbf{X}_{r0}$ are obtained using (7.133) as

$$\left. \frac{\partial f(X_r, Y_r, Z_r)}{\partial X_r} \right|_{\mathbf{X}_r = \mathbf{X}_{r0}} = -\frac{X^s(t) - X_{r0}}{\varrho_{r0}^s(t)} ,$$

$$\left. \frac{\partial f(X_r, Y_r, Z_r)}{\partial Y_r} \right|_{\mathbf{X}_r = \mathbf{X}_{r0}} = -\frac{Y^s(t) - Y_{r0}}{\varrho_{r0}^s(t)} , \quad (7.136)$$

$$\left. \frac{\partial f(X_r, Y_r, Z_r)}{\partial Z_r} \right|_{\mathbf{X}_r = \mathbf{X}_{r0}} = -\frac{Z^s(t) - Z_{r0}}{\varrho_{r0}^s(t)} ,$$

which correspond to the components of the unit vector pointing from the satellites towards the approximate receiver site r. The substitution of Eqs. (7.133) and (7.136) into Eq. (7.135) gives

$$\varrho_r^s(t) = \varrho_{r0}^s(t) - \frac{X^s(t) - X_{r0}}{\varrho_{r0}^s(t)} \Delta X_r - \frac{Y^s(t) - Y_{r0}}{\varrho_{r0}^s(t)} \Delta Y_r - \frac{Z^s(t) - Z_{r0}}{\varrho_{r0}^s(t)} \Delta Z_r ,$$
$$(7.137)$$

where the equivalence of $f(X_r, Y_r, Z_r)$ with $\varrho_r^s(t)$ has been used. This equation is now linear with respect to the unknowns ΔX_r, ΔY_r, ΔZ_r.

Linear model for point positioning with code ranges

The model is given only in its elementary form and, thus, apart from the geometry, only the clocks are modeled. The ionosphere, troposphere, and other effects are neglected for the moment. According to Eq. (6.6), the model for point positioning with code ranges is given by

$$R_r^s(t) = \varrho_r^s(t) + c\,\delta_r(t) - c\,\delta^s(t), \tag{7.138}$$

which is linearized by substituting (7.137):

$$R_r^s(t) = \varrho_{r0}^s(t) - \frac{X^s(t) - X_{r0}}{\varrho_{r0}^s(t)}\Delta X_r - \frac{Y^s(t) - Y_{r0}}{\varrho_{r0}^s(t)}\Delta Y_r \\ - \frac{Z^s(t) - Z_{r0}}{\varrho_{r0}^s(t)}\Delta Z_r + c\,\delta_r(t) - c\,\delta^s(t). \tag{7.139}$$

Leaving the terms containing unknowns on the right side, the equation is rewritten as

$$R_r^s(t) - \varrho_{r0}^s(t) + c\,\delta^s(t) = -\frac{X^s(t) - X_{r0}}{\varrho_{r0}^s(t)}\Delta X_r - \frac{Y^s(t) - Y_{r0}}{\varrho_{r0}^s(t)}\Delta Y_r \\ - \frac{Z^s(t) - Z_{r0}}{\varrho_{r0}^s(t)}\Delta Z_r + c\,\delta_r(t), \tag{7.140}$$

where the satellite clock bias $\delta^s(t)$ is assumed to be known because satellite clock corrections are received within the navigation message. Model (7.140) comprises four unknowns, namely ΔX_r, ΔY_r, ΔZ_r, and $\delta_r(t)$. Consequently, four satellites are needed to solve the problem. The receiver clock unknown is sometimes introduced by $c\,\delta_r(t)$ for a better numerical stability in the matrix operations. The shorthand notations

$$\begin{aligned} \ell^s &= R_r^s(t) - \varrho_{r0}^s(t) + c\,\delta^s(t), \\ a_{X_r}^s &= -\frac{X^s(t) - X_{r0}}{\varrho_{r0}^s(t)}, \\ a_{Y_r}^s &= -\frac{Y^s(t) - Y_{r0}}{\varrho_{r0}^s(t)}, \\ a_{Z_r}^s &= -\frac{Z^s(t) - Z_{r0}}{\varrho_{r0}^s(t)} \end{aligned} \tag{7.141}$$

help to simplify the representation of the system of equations. Although ℓ^s and a^s are time-dependent, this dependency has not been introduced here explicitly for

7.3 Adjustment, filtering, and quality measures

reasons of simplification. Assuming now four satellites numbered from 1 to 4, then

$$\begin{aligned}
\ell^1 &= a_{X_r}^1 \Delta X_r + a_{Y_r}^1 \Delta Y_r + a_{Z_r}^1 \Delta Z_r + c\,\delta_r(t)\,, \\
\ell^2 &= a_{X_r}^2 \Delta X_r + a_{Y_r}^2 \Delta Y_r + a_{Z_r}^2 \Delta Z_r + c\,\delta_r(t)\,, \\
\ell^3 &= a_{X_r}^3 \Delta X_r + a_{Y_r}^3 \Delta Y_r + a_{Z_r}^3 \Delta Z_r + c\,\delta_r(t)\,, \\
\ell^4 &= a_{X_r}^4 \Delta X_r + a_{Y_r}^4 \Delta Y_r + a_{Z_r}^4 \Delta Z_r + c\,\delta_r(t)
\end{aligned} \quad (7.142)$$

is the appropriate system of equations. Note that the superscripts are the satellite numbers and not exponents. Introducing

$$\boldsymbol{\ell} = \begin{bmatrix} \ell^1 \\ \ell^2 \\ \ell^3 \\ \ell^4 \end{bmatrix}, \quad \mathbf{A} = \begin{bmatrix} a_{X_r}^1 & a_{Y_r}^1 & a_{Z_r}^1 & c \\ a_{X_r}^2 & a_{Y_r}^2 & a_{Z_r}^2 & c \\ a_{X_r}^3 & a_{Y_r}^3 & a_{Z_r}^3 & c \\ a_{X_r}^4 & a_{Y_r}^4 & a_{Z_r}^4 & c \end{bmatrix}, \quad \mathbf{x} = \begin{bmatrix} \Delta X_r \\ \Delta Y_r \\ \Delta Z_r \\ \delta_r(t) \end{bmatrix}, \quad (7.143)$$

the set of linear equations can be written in the matrix-vector form

$$\boldsymbol{\ell} = \mathbf{A}\,\mathbf{x}\,. \quad (7.144)$$

The resubstitution of the matrix $\mathbf{A}$ using (7.141) is given explicitly for one epoch t:

$$\mathbf{A} = \begin{bmatrix}
-\dfrac{X^1(t)-X_{r0}}{\varrho_{r0}^1(t)} & -\dfrac{Y^1(t)-Y_{r0}}{\varrho_{r0}^1(t)} & -\dfrac{Z^1(t)-Z_{r0}}{\varrho_{r0}^1(t)} & c \\
-\dfrac{X^2(t)-X_{r0}}{\varrho_{r0}^2(t)} & -\dfrac{Y^2(t)-Y_{r0}}{\varrho_{r0}^2(t)} & -\dfrac{Z^2(t)-Z_{r0}}{\varrho_{r0}^2(t)} & c \\
-\dfrac{X^3(t)-X_{r0}}{\varrho_{r0}^3(t)} & -\dfrac{Y^3(t)-Y_{r0}}{\varrho_{r0}^3(t)} & -\dfrac{Z^3(t)-Z_{r0}}{\varrho_{r0}^3(t)} & c \\
-\dfrac{X^4(t)-X_{r0}}{\varrho_{r0}^4(t)} & -\dfrac{Y^4(t)-Y_{r0}}{\varrho_{r0}^4(t)} & -\dfrac{Z^4(t)-Z_{r0}}{\varrho_{r0}^4(t)} & c
\end{bmatrix}. \quad (7.145)$$

From the linear system of equations, the coordinate differences ΔX_r, ΔY_r, ΔZ_r and the receiver clock error $\delta_r(t)$ for epoch t are obtained by inversion according to (7.66). The desired point coordinates are finally obtained by (7.134). Recall that the selection of the approximate values for the coordinates was completely arbitrary, they could even be set equal to zero. However, depending on the quality of the approximate values, the equations have to be applied iteratively.

Point positioning with code ranges is applicable for each epoch separately. Therefore, this model may also be used in kinematic applications. The equations only change by introducing a time dependency of the unknown point coordinates $\mathbf{X}_r(t)$.

Linear model for point positioning with carrier phases

The procedure is the same as in the previous section. Using Eq. (6.11), the linearization is performed for $\varrho_r^s(t)$ and known terms are shifted to the left side. Multiplying the equation by λ and using $c = \lambda f$ yields

$$\lambda \Phi_r^s(t) - \varrho_{r0}^s(t) + c\,\delta^s(t) = -\frac{X^s(t) - X_{r0}}{\varrho_{r0}^s(t)} \Delta X_r \\ - \frac{Y^s(t) - Y_{r0}}{\varrho_{r0}^s(t)} \Delta Y_r - \frac{Z^s(t) - Z_{r0}}{\varrho_{r0}^s(t)} \Delta Z_r + \lambda N_r^s + c\,\delta_r(t), \quad (7.146)$$

where compared to point positioning with code ranges the number of unknowns is now increased by the ambiguities. Considering again four satellites, the system is given in matrix-vector form $\ell = \mathbf{A}\,\mathbf{x}$, where

$$\ell = \begin{bmatrix} \lambda \Phi_r^1(t) - \varrho_{r0}^1(t) + c\,\delta^1(t) \\ \lambda \Phi_r^2(t) - \varrho_{r0}^2(t) + c\,\delta^2(t) \\ \lambda \Phi_r^3(t) - \varrho_{r0}^3(t) + c\,\delta^3(t) \\ \lambda \Phi_r^4(t) - \varrho_{r0}^4(t) + c\,\delta^4(t) \end{bmatrix},$$

$$\mathbf{A} = \begin{bmatrix} a_{X_r}^1(t) & a_{Y_r}^1(t) & a_{Z_r}^1(t) & \lambda & 0 & 0 & 0 & c \\ a_{X_r}^2(t) & a_{Y_r}^2(t) & a_{Z_r}^2(t) & 0 & \lambda & 0 & 0 & c \\ a_{X_r}^3(t) & a_{Y_r}^3(t) & a_{Z_r}^3(t) & 0 & 0 & \lambda & 0 & c \\ a_{X_r}^4(t) & a_{Y_r}^4(t) & a_{Z_r}^4(t) & 0 & 0 & 0 & \lambda & c \end{bmatrix}, \quad (7.147)$$

$$\mathbf{x} = \begin{bmatrix} \Delta X_r & \Delta Y_r & \Delta Z_r & N_r^1 & N_r^2 & N_r^3 & N_r^4 & \delta_r(t) \end{bmatrix}^T$$

and where the coefficients of the coordinate increments, cf. (7.141), are supplemented with the time parameter t. Obviously, the four equations are inadequate to determine the eight unknowns. This reflects the fact that point positioning with phases in this form cannot be solved epoch by epoch. Each additional epoch increases the number of unknowns by a new clock term. Thus, for two epochs there are eight equations and nine unknowns (still an underdetermined problem). For three epochs (t_1, t_2, t_3) there are 12 equations and 10 unknowns, thus, a slightly overdetermined problem. The 10 unknowns in the latter example are the coordinate increments ΔX_r, ΔY_r, ΔZ_r for the unknown point, the integer ambiguities N_r^1, N_r^2, N_r^3, N_r^4 for the four satellites, and the receiver clock biases $\delta_r(t_1)$, $\delta_r(t_2)$, $\delta_r(t_3)$ for the three epochs. The design matrix $\mathbf{A}$ has the dimension [12×10]. The solution of this redundant system is performed by least-squares adjustment.

7.3 Adjustment, filtering, and quality measures

Linear model for relative positioning

The previous sections have shown linear models for both code ranges and carrier phases. For the case of relative positioning, the investigation is restricted to carrier phases, since on the one side relative positioning aims at higher accuracies achievable only with carrier phases and on the other side it should be obvious how to change from the more expanded model of phases to a code model. Furthermore, the linearization and setup of the linear equation system remains, in principle, the same for phases and phase combinations and could be performed analogously for each model. Therefore, the double-difference is selected for treatment in detail. The model for the double-difference of Eq. (6.45), multiplied by λ, is

$$\lambda \, \Phi_{AB}^{jk}(t) = \varrho_{AB}^{jk}(t) + \lambda \, N_{AB}^{jk}, \tag{7.148}$$

where the term ϱ_{AB}^{jk} containing the geometry is composed of

$$\varrho_{AB}^{jk}(t) = \varrho_B^k(t) - \varrho_B^j(t) - \varrho_A^k(t) + \varrho_A^j(t), \tag{7.149}$$

which reflects the fact of four measurement quantities for a double-difference. Each of the four terms must be linearized according to (7.137) yielding

$$\begin{aligned}
\varrho_{AB}^{jk}(t) =\ & \varrho_{B0}^k(t) - \frac{X^k(t) - X_{B0}}{\varrho_{B0}^k(t)} \Delta X_B - \frac{Y^k(t) - Y_{B0}}{\varrho_{B0}^k(t)} \Delta Y_B \\
& - \frac{Z^k(t) - Z_{B0}}{\varrho_{B0}^k(t)} \Delta Z_B \\
& - \varrho_{B0}^j(t) + \frac{X^j(t) - X_{B0}}{\varrho_{B0}^j(t)} \Delta X_B + \frac{Y^j(t) - Y_{B0}}{\varrho_{B0}^j(t)} \Delta Y_B \\
& + \frac{Z^j(t) - Z_{B0}}{\varrho_{B0}^j(t)} \Delta Z_B \\
& - \varrho_{A0}^k(t) + \frac{X^k(t) - X_{A0}}{\varrho_{A0}^k(t)} \Delta X_A + \frac{Y^k(t) - Y_{A0}}{\varrho_{A0}^k(t)} \Delta Y_A \\
& + \frac{Z^k(t) - Z_{A0}}{\varrho_{A0}^k(t)} \Delta Z_A \\
& + \varrho_{A0}^j(t) - \frac{X^j(t) - X_{A0}}{\varrho_{A0}^j(t)} \Delta X_A - \frac{Y^j(t) - Y_{A0}}{\varrho_{A0}^j(t)} \Delta Y_A \\
& - \frac{Z^j(t) - Z_{A0}}{\varrho_{A0}^j(t)} \Delta Z_A \, .
\end{aligned} \tag{7.150}$$

Substituting (7.150) into (7.148) and rearranging leads to the linear observation equation

$$\ell_{AB}^{jk}(t) = a_{X_A}^{jk}(t)\,\Delta X_A + a_{Y_A}^{jk}(t)\,\Delta Y_A + a_{Z_A}^{jk}(t)\,\Delta Z_A \\ + a_{X_B}^{jk}(t)\,\Delta X_B + a_{Y_B}^{jk}(t)\,\Delta Y_B + a_{Z_B}^{jk}(t)\,\Delta Z_B + \lambda\,N_{AB}^{jk}\,, \tag{7.151}$$

where the left side

$$\ell_{AB}^{jk}(t) = \lambda\,\Phi_{AB}^{jk}(t) - \varrho_{B0}^{k}(t) + \varrho_{B0}^{j}(t) + \varrho_{A0}^{k}(t) - \varrho_{A0}^{j}(t) \tag{7.152}$$

comprises both the measurement quantities and all terms computed from the approximate values. On the right side of (7.151), the abbreviations

$$a_{X_A}^{jk}(t) = +\frac{X^k(t) - X_{A0}}{\varrho_{A0}^{k}(t)} - \frac{X^j(t) - X_{A0}}{\varrho_{A0}^{j}(t)}\,,$$

$$a_{Y_A}^{jk}(t) = +\frac{Y^k(t) - Y_{A0}}{\varrho_{A0}^{k}(t)} - \frac{Y^j(t) - Y_{A0}}{\varrho_{A0}^{j}(t)}\,,$$

$$a_{Z_A}^{jk}(t) = +\frac{Z^k(t) - Z_{A0}}{\varrho_{A0}^{k}(t)} - \frac{Z^j(t) - Z_{A0}}{\varrho_{A0}^{j}(t)}\,, \tag{7.153}$$

$$a_{X_B}^{jk}(t) = -\frac{X^k(t) - X_{B0}}{\varrho_{B0}^{k}(t)} + \frac{X^j(t) - X_{B0}}{\varrho_{B0}^{j}(t)}\,,$$

$$a_{Y_B}^{jk}(t) = -\frac{Y^k(t) - Y_{B0}}{\varrho_{B0}^{k}(t)} + \frac{Y^j(t) - Y_{B0}}{\varrho_{B0}^{j}(t)}\,,$$

$$a_{Z_B}^{jk}(t) = -\frac{Z^k(t) - Z_{B0}}{\varrho_{B0}^{k}(t)} + \frac{Z^j(t) - Z_{B0}}{\varrho_{B0}^{j}(t)}$$

have been used. The coordinates of one point (e.g., A) must be known for relative positioning. More specifically, the known point A reduces the number of unknowns by three because of

$$\Delta X_A = \Delta Y_A = \Delta Z_A = 0 \tag{7.154}$$

and leads to a slight change in the left-side term

$$\ell_{AB}^{jk}(t) = \lambda\,\Phi_{AB}^{jk}(t) - \varrho_{B0}^{k}(t) + \varrho_{B0}^{j}(t) + \varrho_{A}^{k}(t) - \varrho_{A}^{j}(t)\,. \tag{7.155}$$

7.3 Adjustment, filtering, and quality measures

Assuming now four satellites j, k, l, m and two epochs t_1, t_2, the matrix-vector system

$$\ell = \begin{bmatrix} \ell_{AB}^{jk}(t_1) \\ \ell_{AB}^{jl}(t_1) \\ \ell_{AB}^{jm}(t_1) \\ \ell_{AB}^{jk}(t_2) \\ \ell_{AB}^{jl}(t_2) \\ \ell_{AB}^{jm}(t_2) \end{bmatrix} \quad x = \begin{bmatrix} \Delta X_B \\ \Delta Y_B \\ \Delta Z_B \\ N_{AB}^{jk} \\ N_{AB}^{jl} \\ N_{AB}^{jm} \end{bmatrix}$$

$$A = \begin{bmatrix} a_{X_B}^{jk}(t_1) & a_{Y_B}^{jk}(t_1) & a_{Z_B}^{jk}(t_1) & \lambda & 0 & 0 \\ a_{X_B}^{jl}(t_1) & a_{Y_B}^{jl}(t_1) & a_{Z_B}^{jl}(t_1) & 0 & \lambda & 0 \\ a_{X_B}^{jm}(t_1) & a_{Y_B}^{jm}(t_1) & a_{Z_B}^{jm}(t_1) & 0 & 0 & \lambda \\ a_{X_B}^{jk}(t_2) & a_{Y_B}^{jk}(t_2) & a_{Z_B}^{jk}(t_2) & \lambda & 0 & 0 \\ a_{X_B}^{jl}(t_2) & a_{Y_B}^{jl}(t_2) & a_{Z_B}^{jl}(t_2) & 0 & \lambda & 0 \\ a_{X_B}^{jm}(t_2) & a_{Y_B}^{jm}(t_2) & a_{Z_B}^{jm}(t_2) & 0 & 0 & \lambda \end{bmatrix}$$

(7.156)

is obtained which represents a determined and, thus, solvable system. For only one epoch, the system would have more unknowns than observation equations.

7.3.3 Network adjustment

The previous sections described the linearization of the observation equations. The adjustment itself, i.e., the solution of the system of linear equations, is a purely mathematical task to be solved.

Single-baseline solution

The adjustment principle $\mathbf{v}^T \mathbf{P} \mathbf{v}$ = minimum! requires the implementation of the weight matrix $\mathbf{P}$. The off-diagonal elements of $\mathbf{P}$ express the correlation between the measurements. As shown in Sect. 6.3.3, phases and single-differences are uncorrelated, whereas double- and triple-differences are mathematically correlated. The implementation of double-difference correlations can be easily accomplished. Alternatively, the double-differences can be decorrelated by using a Gram–Schmidt orthogonalization (Remondi 1984). The implementation of the correlation of the triple-differences is more difficult but worth to be implemented (Remondi 1984: Table 7.1).

In the case of an observed network, the use of the single baseline method usually implies a baseline-by-baseline computation for all possible combinations. If n_r denotes the number of observing sites, then $n_r \, (n_r-1)/2$ baselines can be calculated. Only $n_r - 1$ of them are theoretically independent. The redundant baselines can be used for misclosure checks. Alternatively all possible baselines are computed for the different measurement sessions. The resulting vectors of all sessions are subject to a common adjustment.

The disadvantage of the simple single-baseline solution from the theoretical point of view is that it disregards the correlation of simultaneously observed baselines. By solving baseline by baseline, this correlation is ignored.

Multipoint solution

In the multipoint approach, the correlations between the baselines are taken into account by considering all points in the network at once.

The principal correlations have been shown in Sect. 6.3.3. The same theoretical aspects also apply to the extended case of a network.

Single-difference example for a network

When three points A, B, C and a single satellite j at a single epoch t are considered, two independent baselines can be defined. Taking A as reference site, for the two baselines $A-B$ and $A-C$ the two single-differences

$$\Phi^j_{AB}(t) = \Phi^j_B(t) - \Phi^j_A(t),$$
$$\Phi^j_{AC}(t) = \Phi^j_C(t) - \Phi^j_A(t) \tag{7.157}$$

are set up for the one satellite j at epoch t. The two single-differences and the phases are related by

$$\mathbf{S} = \mathbf{C}\,\mathbf{\Phi}, \tag{7.158}$$

where

$$\mathbf{S} = \begin{bmatrix} \Phi^j_{AB}(t) \\ \Phi^j_{AC}(t) \end{bmatrix}, \quad \mathbf{C} = \begin{bmatrix} -1 & 1 & 0 \\ -1 & 0 & 1 \end{bmatrix}, \quad \mathbf{\Phi} = \begin{bmatrix} \Phi^j_A(t) \\ \Phi^j_B(t) \\ \Phi^j_C(t) \end{bmatrix}. \tag{7.159}$$

To find the correlation, the covariance propagation law is applied by $\mathbf{\Sigma_S} = \mathbf{C}\,\mathbf{\Sigma_\Phi}\,\mathbf{C}^T$ leading to

$$\mathbf{\Sigma_S} = \sigma^2\,\mathbf{C}\,\mathbf{C}^T \tag{7.160}$$

7.3 Adjustment, filtering, and quality measures

because of $\Sigma_\Phi = \sigma^2 \mathbf{I}$, cf. Eq. (6.63). Substituting matrix $\mathbf{C}$ from (7.159) and evaluating the matrix operation yields

$$\Sigma_S = \sigma^2 \begin{bmatrix} 2 & 1 \\ 1 & 2 \end{bmatrix}, \tag{7.161}$$

which shows, as is to be expected, a correlation of the single-differences of the two baselines with a common point. Recall that single-differences of a single baseline are uncorrelated as pointed out in Sect. 6.3.3.

Double-difference example for a network
Since double-differences are already correlated for a single baseline, a correlation must be expected for the network too. Nevertheless, the subsequent slightly larger example will demonstrate the increasing complexity. Assume again three points A, B, C with A as reference site for the two baselines $A-B$ and $A-C$. Consider a single epoch t for four satellites j, k, ℓ, m, where j is taken as the reference satellite for the double-differences.

There are $(n_r - 1)(n_s - 1)$ independent double-differences for n_r points and n_s satellites. For the given example, $n_r = 3$ and $n_s = 4$ and, thus, 6 double-differences can be identified. Based on Eq. (6.47), these are

$$\begin{aligned}
\Phi_{AB}^{jk}(t) &= \Phi_B^k(t) - \Phi_B^j(t) - \Phi_A^k(t) + \Phi_A^j(t), \\
\Phi_{AB}^{j\ell}(t) &= \Phi_B^\ell(t) - \Phi_B^j(t) - \Phi_A^\ell(t) + \Phi_A^j(t), \\
\Phi_{AB}^{jm}(t) &= \Phi_B^m(t) - \Phi_B^j(t) - \Phi_A^m(t) + \Phi_A^j(t), \\
\Phi_{AC}^{jk}(t) &= \Phi_C^k(t) - \Phi_C^j(t) - \Phi_A^k(t) + \Phi_A^j(t), \\
\Phi_{AC}^{j\ell}(t) &= \Phi_C^\ell(t) - \Phi_C^j(t) - \Phi_A^\ell(t) + \Phi_A^j(t), \\
\Phi_{AC}^{jm}(t) &= \Phi_C^m(t) - \Phi_C^j(t) - \Phi_A^m(t) + \Phi_A^j(t)
\end{aligned} \tag{7.162}$$

for the assumptions made. As in the previous example, a matrix-vector relation is desired. By introducing for the matrix $\mathbf{C}$:

$$\mathbf{C} = \begin{bmatrix}
1 & -1 & 0 & 0 & -1 & 1 & 0 & 0 & 0 & 0 & 0 & 0 \\
1 & 0 & -1 & 0 & -1 & 0 & 1 & 0 & 0 & 0 & 0 & 0 \\
1 & 0 & 0 & -1 & -1 & 0 & 0 & 1 & 0 & 0 & 0 & 0 \\
1 & -1 & 0 & 0 & 0 & 0 & 0 & 0 & -1 & 1 & 0 & 0 \\
1 & 0 & -1 & 0 & 0 & 0 & 0 & 0 & -1 & 0 & 1 & 0 \\
1 & 0 & 0 & -1 & 0 & 0 & 0 & 0 & -1 & 0 & 0 & 1
\end{bmatrix}, \tag{7.163}$$

and for the vectors **D** and **Φ**:

$$\mathbf{D} = \begin{bmatrix} \Phi_{AB}^{jk}(t) \\ \Phi_{AB}^{j\ell}(t) \\ \Phi_{AB}^{jm}(t) \\ \Phi_{AC}^{jk}(t) \\ \Phi_{AC}^{j\ell}(t) \\ \Phi_{AC}^{jm}(t) \end{bmatrix}, \qquad \mathbf{\Phi} = \begin{bmatrix} \Phi_{A}^{j}(t) \\ \Phi_{A}^{k}(t) \\ \Phi_{A}^{\ell}(t) \\ \Phi_{A}^{m}(t) \\ \Phi_{B}^{j}(t) \\ \vdots \end{bmatrix}, \qquad (7.164)$$

the relation

$$\mathbf{D} = \mathbf{C}\,\mathbf{\Phi} \qquad (7.165)$$

is valid. The covariance follows by

$$\mathbf{\Sigma_D} = \mathbf{C}\,\mathbf{\Sigma_\Phi}\,\mathbf{C}^T \qquad (7.166)$$

which reduces to

$$\mathbf{\Sigma_D} = \sigma^2 \mathbf{C}\,\mathbf{C}^T \qquad (7.167)$$

because of the uncorrelated phases. Explicitly, the matrix product

$$\mathbf{C}\,\mathbf{C}^T = \begin{bmatrix} 4 & 2 & 2 & 2 & 1 & 1 \\ 2 & 4 & 2 & 1 & 2 & 1 \\ 2 & 2 & 4 & 1 & 1 & 2 \\ 2 & 1 & 1 & 4 & 2 & 2 \\ 1 & 2 & 1 & 2 & 4 & 2 \\ 1 & 1 & 2 & 2 & 2 & 4 \end{bmatrix} \qquad (7.168)$$

is a full matrix as expected. Finally, the weight matrix results from the inverse of (7.168): $\mathbf{P} = (\mathbf{C}\,\mathbf{C}^T)^{-1}$.

Beutler et al. (1986) show some results of network campaigns where the correlations have either been totally neglected, introduced in a single-baseline mode, or calculated correctly (Beutler et al. 1987). For small networks with baselines not exceeding 10 km, the differences of the three methods are in the range of a few millimeters. Clearly, the solution without any correlation deviates from the theoretically correct values by a greater amount. It is estimated that the single-baseline method deviates from the multibaseline (correlated) solution by a maximum of 2σ.

7.3 Adjustment, filtering, and quality measures

Single-baseline versus multipoint solution

The implementation of the single-baseline method is much simpler; and it takes less effort to detect and eliminate faulty measurements. The multipoint solution, in contrast, accounts for the correlation and more easily detects and repairs cycle slips.

The economic implementation of the full correlation for a multipoint solution only works properly for networks with the same observation pattern at each receiver site. In the event of numerous data outages it is better to recalculate the covariance matrix.

Even in the case of the multipoint approach, it becomes questionable whether the correlations can be modeled properly. An illustrative example is given in Beutler et al. (1990) where single- and dual-frequency receivers are combined in a network. For the dual-frequency receivers, the ionosphere-free combination is formed from dual-frequency measurements and processed together with the data of the single-frequency receivers. Thus, a correlation is introduced because of the single-frequency data. A proper modeling of the correlation biases the ionosphere-free baseline by the ionosphere of the single-frequency baseline, an effect which is definitely undesirable.

Least-squares adjustment of baselines

Considering networks, usually the number of measured baselines will exceed the minimum amount necessary, e.g., due to several measurement campaigns. In this case, redundant information is available and the determination of the coordinates of the network points may be carried out by a least-squares adjustment.

The baseline vectors $\mathbf{X}_{ij}$ between the unknown network points $\mathbf{X}_j$ and $\mathbf{X}_i$ are introduced as observables into the least-squares adjustment. Consequently, the linear expression

$$\mathbf{X}_{ij} = \mathbf{X}_j - \mathbf{X}_i \tag{7.169}$$

referring the observables to the unknowns is formulated. Linearizing the equation (i.e., $\mathbf{X}_{ij} = \mathbf{X}_{ij0} + \Delta \mathbf{X}_{ij}$) is not necessary but may be applied for numerical stability reasons. In case of redundancy, residuals $\mathbf{v}_{ij}$ are added to the observables to assure consistency, cf. (7.67), thus

$$\ell_{ij} = \mathbf{X}_{ij} = \mathbf{X}_j - \mathbf{X}_i + \mathbf{v}_{ij}. \tag{7.170}$$

The components $\mathbf{X}_{ij} = [X_{ij}, Y_{ij}, Z_{ij}]$ of the baseline vector are considered as observables. The elements of the design matrix are the coefficients of the unknowns $\mathbf{X}_i$ and $\mathbf{X}_j$ and amount to 0, +1, or −1. The relation (7.170) is solved by applying the least-squares principle $\mathbf{v}^T \mathbf{v}$ = minimum! if equal weights for the baselines are assumed. Otherwise, a weight matrix **P** must be taken into account according to Eq. (7.71).

Introducing the coordinate differences X_{ij}, Y_{ij}, Z_{ij} as the only observables, then absolute coordinates cannot be derived because the matrix of normal equations becomes singular. Considering the most general case, the rank deficiency of a three-dimensional network amounts to seven, corresponding to the seven degrees of freedom of a three-dimensional network or the seven parameters of a similarity transformation in space (three translations, three rotations, and one scale factor).

For relative positioning, orientation and scale of the network of baseline vectors (and also its shape) are determined due to the definition of the satellite orbits. This means that four, i.e., three rotations and the scale, of seven parameters are determined. The three translations of the whole network, i.e., a shift vector, are still undetermined so that the rank deficiency of the normal equations equals three. Selecting and fixing a single point of the network (i.e., considering its coordinates as known) solves the problem of the shift vector and leads to the minimally constrained solution. Note, however, that the fixing of coordinates should be restricted to one point. Otherwise, constraints would be induced into the network which could obviate the strong geometry and result in network distortions.

Previously, the problem of correlations was discussed. For single-baseline solutions, only correlations between vector components are computed. For multipoint baseline solutions, correlations between the various baselines are computed. In the first case, the correlations may be disregarded, whereas in the latter case, the correlations should be taken into account.

The adjustment described for three-dimensional Cartesian coordinates may also be performed in ellipsoidal coordinates. The principle of formulating the observation equations is precisely the same. However, the expression which relates the observables to the unknowns is more sophisticated.

7.3.4 Dilution of precision

The geometry of the visible satellites is an important factor in achieving high-quality results especially for point positioning and kinematic surveying (Fig. 7.6). Visibility, thereby, is characterized by the unobstructed line of sight between receiver and satellite. The geometry changes with time due to the relative motion of the user and satellites. A measure of the instantaneous geometry is the dilution of precision (DOP) factor.

First, the specific case of four satellites is considered. The linearized observation equations for the point positioning model with code ranges are given by Eq. (7.144), and the solution for the (four) unknowns follows from the inverse relation $\mathbf{x} = \mathbf{A}^{-1}\boldsymbol{\ell}$. The design matrix **A** is given by Eq. (7.145), where the first three elements in each row are the components of the unit vectors ϱ_r^s, $s = 1, 2, 3, 4$,

7.3 Adjustment, filtering, and quality measures

pointing from the four satellites to the observing site r. The solution fails if the design matrix is singular or, equivalently, if its determinant becomes zero. The determinant is proportional to the scalar triple product

$$\left((\varrho_r^4 - \varrho_r^1),\ (\varrho_r^3 - \varrho_r^1),\ (\varrho_r^2 - \varrho_r^1)\right) \tag{7.171}$$

which can geometrically be interpreted by the volume of a tetrahedron. This body is formed by the intersection points of the site–satellite vectors with the unit sphere centered at the observing site. The larger the volume of this body, the better the satellite geometry. Since good geometry should mirror a low DOP value, the reciprocal value of the volume of the geometric body is directly proportional to DOP. The critical configuration is given when the body degenerates to a plane. This is the case when the unit vectors ϱ_r^s form a cone with the observing site as apex (Wunderlich 1992, Grafarend and Shan 2002). As Leick (2004: p. 182) emphasizes, the relative motion of the satellites, however, avoids critical configurations over long periods. The size of the tetrahedron body will become maximal if one satellite is in zenith, whereas all other satellites are evenly distributed in azimuth at the horizon. Considering geometries of satellites below the horizon is impractical for receivers on earth but does make sense for spaceborne receivers. The satellite constellation of a GNSS poses constraints on the minimal and maximal DOP values.

More generally, DOP can be calculated from the inverse of the normal equation matrix of (7.74). The cofactor matrix of the parameters $\mathbf{Q_X}$ follows from

$$\mathbf{Q_X} = (\mathbf{A}^T \mathbf{A})^{-1}. \tag{7.172}$$

Capital $\mathbf{X}$ is used here as an indication of coordinates of an ECEF system. In this case, the weight matrix has been assumed to be a unit matrix, otherwise

$$\mathbf{Q_X} = (\mathbf{A}^T \mathbf{P} \mathbf{A})^{-1}. \tag{7.173}$$

The cofactor matrix $\mathbf{Q_X}$ is a [4×4] matrix, where three components are contributed by the site position X, Y, Z and one component by the receiver clock. Denoting the elements of the cofactor matrix as

$$\mathbf{Q_X} = \begin{bmatrix} q_{XX} & q_{XY} & q_{XZ} & q_{Xt} \\ q_{XY} & q_{YY} & q_{YZ} & q_{Yt} \\ q_{XZ} & q_{YZ} & q_{ZZ} & q_{Zt} \\ q_{Xt} & q_{Yt} & q_{Zt} & q_{tt} \end{bmatrix}, \tag{7.174}$$

the diagonal elements are used for the following DOP definitions:

$$\begin{aligned}
\text{GDOP} &= \sqrt{q_{XX} + q_{YY} + q_{ZZ} + q_{tt}} & \ldots & \quad \text{geometric dilution of precision;} \\
\text{PDOP} &= \sqrt{q_{XX} + q_{YY} + q_{ZZ}} & \ldots & \quad \text{position dilution of precision;} \\
\text{TDOP} &= \sqrt{q_{tt}} & \ldots & \quad \text{time dilution of precision.}
\end{aligned} \qquad (7.175)$$

It should be noted that the previous DOP explanation using the geometric body refers to GDOP. The derivation and the definition of the DOP values highlights that the dilution of precision is only a function of the geometry between user and satellite. No observations are needed to compute DOP values, thus, DOP values may be computed in advance to measurement campaigns by using almanac data or other orbit information. As shown in Fig. 7.8, the DOP values are a function of satellite constellation, and thus a function of time. The figures are based on a simulated constellation of 27 satellites equally distributed in 3 orbital planes, an orbit inclination of 56°, and an orbital height of 23 200 km.

The definitions in (7.175) deserve brief explanations in order to avoid confusion. Quite often the elements under the square root are displayed as quadratic terms. This depends on the designation of the elements of the cofactor matrix. Here,

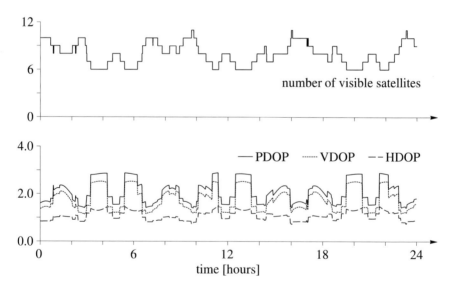

Fig. 7.8. Satellite visibility and dilution of precision for Graz, Austria (47.1°N, 15.5°E, 400 m; mask angle = 10°, i.e., satellites with an elevation of less than 10° are omitted)

7.3 Adjustment, filtering, and quality measures

the diagonal elements of the cofactor matrix have been denoted as q_{XX}, q_{YY}, q_{ZZ}, and q_{tt}, thus no superscripts appear in the DOP definitions. If the diagonal elements are denoted as, e.g., q_X^2, q_Y^2, q_Z^2, q_t^2, then of course the superscripts also appear in the DOP definitions. The general rule used here is the following: when computing DOP values, the elements of the cofactor matrix are not squared, e.g., for GDOP the square root of the trace of the cofactor matrix must be calculated.

The DOP values in (7.175) are expressed in the equatorial system. When the topocentric local coordinate system with its axes along the local north, east, and up (i.e., vertical) is used, the global cofactor matrix $\mathbf{Q_X}$ must be transformed into the local cofactor matrix $\mathbf{Q_x}$ by the law of covariance propagation. Denoting now as $\mathbf{Q_X}$ that part of the cofactor matrix that contains the geometrical components (disregarding the time-correlated components), the transformation reads

$$\mathbf{Q_x} = \mathbf{R}\,\mathbf{Q_X}\,\mathbf{R}^T = \begin{bmatrix} q_{nn} & q_{ne} & q_{nu} \\ q_{ne} & q_{ee} & q_{eu} \\ q_{nu} & q_{eu} & q_{uu} \end{bmatrix}, \tag{7.176}$$

where the rotation matrix $\mathbf{R}^T = [\mathbf{n}\ \mathbf{e}\ \mathbf{u}]$ contains the axes of the local coordinate system as given in Sect. 8.2.2. Because of the invariance of the trace of a matrix with respect to rotation, the PDOP value in the local system is identical to the value in the global system. In addition to the PDOP, two further DOP definitions are given. HDOP, the dilution of precision in the horizontal position, and VDOP, denoting the corresponding value for the vertical component, the height:

$$\begin{aligned} \text{HDOP} &= \sqrt{q_{nn} + q_{ee}}\,, \\ \text{VDOP} &= \sqrt{q_{uu}}\,. \end{aligned} \tag{7.177}$$

Satellites below the horizon are not visible, consequently, they do not contribute to the geometry. Vertical DOP values, thus, are generally higher, i.e., worse than the horizontal DOP (cf. Fig. 7.8). In contrast, spaceborne applications can benefit from satellites below the horizon. Using the previously mentioned satellite constellation as example to compute DOP values all across the globe (mask angle = $0°$), then HDOP values between 0.6 and 1.3, VDOP values between 0.8 and 2.6, and PDOP values between 1.1 and 2.9 are computed for users on ground. The number of visible satellites thereby varies between 6 and 12. Generally speaking, the lower the DOP values the better the constellation. Good geometry is available if the PDOP values are less than 3 and HDOP values are less than 2.

DOP computations are not restricted to point positioning but can also be applied to relative positioning. Starting with the design matrix for a baseline vector determination, the cofactor matrix is computed. These DOP values are considered as relative DOP values.

The purposes of DOP are threefold. First, it is useful in planning a survey and, second, it may be helpful in interpreting processed baseline vectors. For example, data with poor DOP could possibly be omitted. Finally, the DOP projects the errors from the measurement domain onto the errors in the position domain. A poor geometry amplifies the range errors (cf. Fig. 7.6). Denoting the measurement accuracy by σ_{range} (i.e., the standard deviation) and assuming equal accuracy of and no correlation between range measurements, then the positioning accuracy follows from the product of DOP and the measurement accuracy. Applied to the specific DOP definitions,

$$\text{GDOP}\, \sigma_{\text{range}} \quad \ldots \quad \text{geometric accuracy in position and time,}$$
$$\text{PDOP}\, \sigma_{\text{range}} \quad \ldots \quad \text{accuracy in position,}$$
$$\text{TDOP}\, \sigma_{\text{range}} \quad \ldots \quad \text{accuracy in time,}$$
$$\text{HDOP}\, \sigma_{\text{range}} \quad \ldots \quad \text{accuracy in the horizontal position,}$$
$$\text{VDOP}\, \sigma_{\text{range}} \quad \ldots \quad \text{accuracy in vertical direction}$$

are obtained (Wells et al. 1987: Chap. 4). The list of DOP definitions is not restricted to those given here. Since the measurement accuracy σ_{range} is unknown, it is estimated by σ_{UERE}.

7.3.5 Quality parameters

The performance of a navigation system is characterized by a number of statistical quality parameters. The definitions are given in alphabetical order and are based, if not otherwise stated, on the 2005 US Federal Radionavigation Plan (Department of Defense et al. 2005: Appendix A). Variations of the definitions exist depending on the application and navigation system used. The parameters are commonly not individually defined, but are correlated to each other. In a document submitted to the International Maritime Organization (IMO), the European Commission (2003a) plots schematically a pyramid with the accuracy as basis (Fig. 7.9). Integrity, in this ordering, is solely a function of accuracy.

The satellite geometry continuously changes over time. Satellites, furthermore, are temporarily removed from service for routine maintenance operations. Therefore, the quality parameters generally refer to a certain instant of time, a time interval, or an average over time.

The assessment of the parameters is not given here, since most application domains have different certification procedures. For a simplified assessment assume a position log over a long time interval at a known position or trajectory. One may assume stationarity and even ergodicity to simplify parameter determination. The measurement campaign and the time interval have to be designed carefully to guar-

7.3 Adjustment, filtering, and quality measures

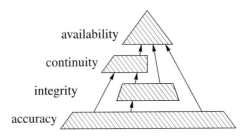

Fig. 7.9. Hierarchy of quality parameters

antee a statistically significant distribution of the measurements (Institute of Navigation 1997).

Accuracy

Accuracy is a statistical measure that provides the degree of conformance between the estimated or measured parameter (e.g., position and/or velocity) of an object at a given time and the true parameter. The accuracy of a navigation system is usually presented as a statistical measure of the system error together with a confidence level (e.g., 95%) reflecting the probability of the given value. Three types of accuracy are distinguished:

- Predictable (or absolute) accuracy quantifies the degree of conformance between the position solution of the navigation system with respect to the true (i.e., charted) position.
- Repeatable accuracy (or precision) describes the accuracy with which a user can return to a position whose coordinates have been determined at a previous time with the same navigation system. The statistical measure of the repeatable accuracy does not account for the true position.
- Relative accuracy denotes the accuracy with which a user can determine his/her position relative to that of another user of the same navigation system at the same instant of time, regardless of any error in the true position.

The visual comparison of precision and accuracy is schematically given in Fig. 7.10. For the position computation, GNSS measurements are commonly assumed to be unbiased and Gaussian normally distributed that would result in accurate position solutions. In practice, thus, the accuracy notation is mistakenly used to paraphrase the definitions of accuracy in combination with precision. Considering atmospheric effects or multipath, these assumptions however are not valid.

The accuracy of a system is decomposed into the UERE and the geometric dilution of precision. Accuracy and precision are described at various confidence levels as specified in Sect. 7.3.6.

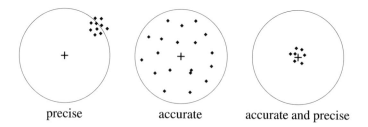

Fig. 7.10. Accuracy and precision of measurement

Availability

The availability of a navigation system is a measure of the percentage of time, during which the system performs within its coverage area under specified conditions, e.g., the receiver provides the required levels of accuracy, integrity, and/or continuity. Availability, thus, is an indication of the ability of the system to provide a usable service within the specified coverage area. Signal availability, furthermore, corresponds to the percentage of time that the signals transmitted from satellites are available for use. Signal availability is a function of both the physical characteristics of the environment and the technical capabilities of the signal generation and transmission facilities.

Capacity

Capacity is the number of users which can use a navigation system simultaneously.

Continuity

Continuity is the ability of a system to perform a function without nonscheduled interruptions during an intended operation. More specifically, continuity quantifies the probability that the specified system performance will be maintained for the duration of a phase of operation, presuming that the system was available at the beginning of that phase of operation. The landing phase of an aircraft is the best example. Continuity risk, in contrast, describes the probability that the specified system performance will not be maintained throughout the phase of operation. Satellite outages announced well in advance do not contribute to continuity risk.

Coverage

The coverage of a navigation system is that surface area or space volume where the performance of the system is adequate to permit the user to determine a position to a specified level of accuracy. In case of radionavigation, the coverage is influenced by system geometry, signal power levels, receiver sensitivity, atmospheric noise

7.3 Adjustment, filtering, and quality measures

conditions, blockage, jamming, and other factors that affect signal availability. The nominal GNSS coverage is global. This kind of coverage refers to the availability of signal in space, neglecting any obstructions of topography, buildings, etc.

Integrity

Definitions
Integrity is the measure of the trust that can be placed in the correctness of the information provided by a navigation system. Integrity is the ability of a navigation system to provide timely warnings to the users when the system should not be used. Integrity involves several subparameters such as the time to alarm or the alarm limit, the latter reflects the maximum tolerable error of the system.

Integrity risk, in contrast, is defined as the probability of providing incorrect information without warning the user in the given period of time. In terms of position, the position error exceeds an alarm limit (AL) although the user is not notified in the specified time to alarm/alert (TTA). The specification of the AL is correlated to the accuracy and integrity risk levels chosen.

Two major approaches of integrity computation are distinguished. The first computes the integrity risk at the given AL. The second estimates the error magnitude with respect to the given integrity risk. Based on the second approach, the aviation community developed a mathematical model to describe the integrity and integrity risk using protection levels (PL).

The true instantaneous error, which is commonly unknown, is denoted as position error (PE). The alarm limit, as defined before, specifies the maximum allowable PE before an alarm shall be raised. PL is the estimate of the navigation system error that shall bound the PE. Integrity is guaranteed as long as PL > PE. If PL > AL, the integrity is specified to be lost, an alarm is triggered, and, at the same time, continuity and availability are lost too.

The vertical PL (VPL) and the horizontal PL (HPL) are generally considered as independent levels, therefore the error figure and, thus, the integrity figure degenerates from an ellipsoid to a cylinder – a concept having its roots in civil aviation. DeCleene (2000) emphasizes that the normal distribution with a known variance and zero mean allows for straightforward computation of the protection levels and therefore of the integrity risk. In the absence of any biases, the protection levels are defined by

$$\text{VPL} = \kappa_{\text{VPL}} \, \sigma_0 \, \sqrt{q_{uu}}, \tag{7.178}$$

$$\text{HPL} = \kappa_{\text{HPL}} \, \sigma_0 \, \sqrt{\frac{q_{nn} + q_{ee}}{2} + \sqrt{\left(\frac{q_{nn} - q_{ee}}{2}\right)^2 + q_{ne}^2}}. \tag{7.179}$$

The elements q_{nn}, q_{ee}, q_{ne}, and q_{uu} are the ones of the cofactor matrix of the estimated parameters in the local coordinate system (7.176). These equations, as the

Radio Technical Commission for Aeronautical Services (2006) states, are valid if the measurement accuracies have been introduced as weights into the adjustment process, thus $\mathbf{Q_x} = \mathbf{R}(\mathbf{A^T P A})^{-1}\mathbf{R^T}$. The κ factor in (7.178) and (7.179), as Roturier et al. (2001) mention, scales the position domain variances to a level compatible with the integrity requirement, taking into account the error distribution function with respect to the dimension and any biases (e.g., multipath), and the integrity risk per independent sample. For example, for precision approach applications in the aviation domain, the κ values for horizontal and vertical protection levels are $\kappa_{VPL} = 5.33$, $\kappa_{HPL} = 6.0$ assuming a normal distribution function (Radio Technical Commission for Aeronautical Services 2006).

Figure 7.11 contrasts the position error to the estimated protection level and differentiates between six conditions with regard to PE, PL, and AL. If the protection levels are too conservative, PL will always bound the PE but will frequently exceed the AL that causes a low continuity and availability. If the chosen protection level is too optimistic, in contrast, the probability of a hazardously misleading information (HMI) increases. The PL computation has been chosen perfectly if the cloud of points as indicated in Fig. 7.11 is just above the diagonal, hence, neither too optimistic nor too conservative. Plotting the PE against PL is commonly denoted as Stanford diagram. Variations of it are introduced by Tossaint et al. (2006).

Equations (7.178) and (7.179) are justifiable for Gaussian normal distribution and zero mean. The augmentation information, as described in Sect. 12.4, mitigates most of the system biases and biases induced by atmosphere; however, the local effects cannot be accounted for. As a consequence, as Roturier et al. (2001) emphasize, the distribution of the pseudorange residual errors may not meet Gaussian normal distribution and zero mean. Concepts have been introduced to compute the protection levels, even in the condition of a faulty measurement (e.g., DeCleene 2000), by assuming a faulty pseudorange bias error associated to one satellite.

Integrity monitoring

The objective of integrity monitoring is to determine whether the system or individual measurements, respectively, meet the navigation performance requirements. One option for integrity monitoring is to integrate dissimilar or complementary navigation systems and techniques (Hofmann-Wellenhof et al. 2003: Chap. 13). The second option considers dedicated monitoring stations to continuously monitor the satellite signals and forward information about any erroneous behavior to the user. Finally, the receiver autonomous integrity monitoring (RAIM) technique exploits measurement redundancy to provide a high level of integrity and safety. Common to all systems is that an alarm will be triggered if the integrity goes unmonitored.

In the case of integrity monitoring services, a network of monitoring stations analyzes the satellite signals with regard to their integrity and forwards the infor-

7.3 Adjustment, filtering, and quality measures

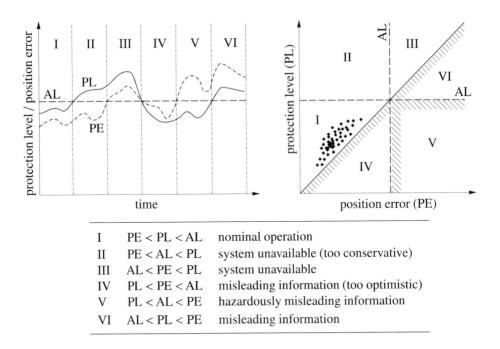

I	PE < PL < AL	nominal operation
II	PE < AL < PL	system unavailable (too conservative)
III	AL < PE < PL	system unavailable
IV	PL < PE < AL	misleading information (too optimistic)
V	PL < AL < PE	hazardously misleading information
VI	AL < PL < PE	misleading information

Fig. 7.11. Protection levels

mation to the users, either using the data message of the navigation signal or a separate means of communication. The implementation can be based on a simple integrity flag or on enhanced accuracy estimates that combine the different satellite clock and orbit biases into accuracy measures. The estimated signal-in-space accuracy parameter thereby shall overbound the true signal-in-space error. Integrity monitoring services cannot take into account local effects, e.g, multipath. Consequently, the information of monitoring services is generally combined with RAIM techniques to provide a high level of integrity.

The RAIM relies on the redundancy of navigation signals to monitor the integrity by autonomously processing the signals independently from external integrity sources. Ober (2003) emphasizes that the receiver detects, excludes, and/or isolates failures which cannot be handled by a robust position estimator. The algorithms are based either on a snapshot concept or on continuous, filtered approaches. The snapshot algorithms, as Young and McGraw (2003) state, are commonly based on least-squares methods, whereas the filtered RAIM techniques utilize Kalman filter concepts. Both approaches require redundancy to compare and statistically evaluate different solutions. Unavailability of redundancy will allow any bias to distort the position solution. If the redundancy equals 1, the RAIM will, depending on the size of the bias, detect a bias in the measurements. Nevertheless, the biased measurement cannot be identified. The redundancy has to be greater than

1 to detect an error and identify and exclude the concerning measurement. Beside the redundancy, as Hewitson and Wang (2006) state, also the geometry influences the correlation of the test statistics and therefore the performance of the RAIM procedures.

There are numerous different RAIM algorithms to detect and isolate any measurement error. The position domain concept, as an example, evaluates different position solutions using alternating combinations of observables out of the population of measurements. The position solutions with respect to the faulty measurement will deviate from all other solutions. As a consequence, this measurement can be excluded from the position computation. This concept becomes lengthy when processing measurements of several GNSS in a combined solution.

Reliability

The reliability of a navigation system quantifies the probability of performing a specified function without failure under given conditions for a specified period of time. The higher the reliability the more seldom integrity events occur. The reliability of a repairable system is closely related to the mean time between failures and the mean time to repair.

Risk

Risk, as defined by the European Space Agency (2004), describes undesirable situations or circumstances that have both a likelihood of occurrence and a potential negative consequence. The US Department of Defense (2000) defines the hazard probability as the aggregate probability of occurrence of the individual events that create a specified hazard. Four levels are distinguished: frequent, occasionally, improbable, and impossible. The severity is outlined by again four different hazard levels: catastrophic, critical, marginal, and negligible, where marginal may be additionally divided into major and minor.

7.3.6 Accuracy measures

Accuracy and precision are described at various percentile levels (cf. Fig. 7.12). Note that an accuracy probability level of $\varepsilon\%$ always implies that $(100 - \varepsilon)\%$ of the position solution exceed this limit.

The definition of statistical terms follows Kreyszig (2006: Chaps. 24, 25). Let x be a random or stochastic variable and $f(x)$ the density function of its distribution, then the integral

$$P(a < x < b) = \int_a^b f(x)\,dx \qquad (7.180)$$

7.3 Adjustment, filtering, and quality measures

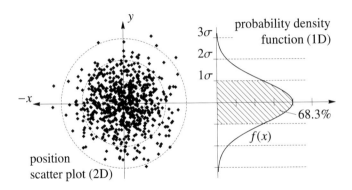

Fig. 7.12. Confidence level

defines the probability that the variable x assumes any value in the interval $a < x < b$. The mean or expectation μ of x is given by

$$\mu = \int_{-\infty}^{+\infty} x f(x) \, dx \tag{7.181}$$

and the variance σ^2 of x follows from

$$\sigma^2 = \int_{-\infty}^{+\infty} (x - \mu)^2 f(x) \, dx. \tag{7.182}$$

The square root of the variance is denoted as standard deviation σ.

A stochastic variable x with mean μ and variance σ^2 may be converted (or standardized) into the variable $(x - \mu)/\sigma$. This standardized variable has zero mean and unit variance.

Distribution functions

Normal distribution
The density function of the most frequently used Gaussian or normal distribution is defined by

$$f(x) = \frac{1}{\sigma \sqrt{2\pi}} e^{-(x-\mu)^2/(2\sigma^2)} \tag{7.183}$$

and reduces to

$$f(z) = \frac{1}{\sqrt{2\pi}} e^{-z^2/2} \tag{7.184}$$

for a standardized variable z.

Chi-square distribution

Let x be the sum of the squares of n independent and standardized stochastic variables with Gaussian distribution. The density of the associated Chi-square or briefly χ^2-distribution (Bronstein et al. 2005: p. 785) is defined by

$$f_n(x) = \frac{1}{2^{n/2}\,\Gamma(n/2)} x^{(n/2-1)} e^{-x/2}, \qquad x > 0, \tag{7.185}$$

where n is denoted as degree of freedom. The Gamma-function $\Gamma(n/2)$ for $n > 0$ is given by

$$\Gamma(n/2) = \int_0^{+\infty} t^{(n/2-1)} e^{-t} dt \tag{7.186}$$

and has the numerical values $\Gamma(1/2) = \sqrt{\pi}$, $\Gamma(2/2) = 1$, $\Gamma(3/2) = \sqrt{\pi}/2$ for the numbers $n = 1, 2, 3$ (Bronstein et al. 2005: p. 479).

Specifications

One-dimensional accuracy measures

The one-dimensional case is specified by $n = 1$ and $x = z^2$. The χ^2-distribution converts to the Gaussian distribution with zero mean and unit variance, which may be proved by transforming the probability $P_1 = \int f_1(x)\,dx$ by substituting $x = z^2$ and $dx = 2z\,dz$. The associated probability follows as

$$P_1(-\alpha < z < \alpha) = \frac{1}{\sqrt{2\pi}} \int_{-\alpha}^{+\alpha} e^{-z^2/2}\,dz. \tag{7.187}$$

For a nonstandardized random variable x, the probability P_1 corresponds to the symmetric interval $(\mu - \alpha\sigma, \mu + \alpha\sigma)$. Numerical values for the probability P_1 of the normal distribution are extensively tabulated in most textbooks on statistics. Some examples are given in Table 7.6. Note that a synonym for probability is the confidence level.

Table 7.6. One-dimensional accuracy measures

α	Probability [%]	Notation
0.67	50.0	Linear error probable (LEP)
1.00	68.3	1σ level or root mean square (RMS)
1.96	95.0	95% confidence level
2.00	95.4	2σ level
3.00	99.7	3σ level

7.3 Adjustment, filtering, and quality measures

Table 7.7. Two-dimensional accuracy measures

$\sqrt{\alpha}$	Probability [%]	Notation
1.00	39.3	1σ or standard ellipse
1.18	50.0	Circular error probable (CEP)
$\sqrt{2}$	63.2	Distance RMS (DRMS)
2.00	86.5	2σ ellipse
2.45	95.0	95% confidence level
$2\sqrt{2}$	98.2	Twice distance RMS (2DRMS)
3.00	98.9	3σ ellipse

Two-dimensional accuracy measures
The two-dimensional case is specified by $n = 2$ and $x = z_1^2 + z_2^2$. The density function of the χ^2-distribution becomes $f_2(x) = e^{-x/2}/2$. The associated probability reads

$$P_2(0 < x < \alpha) = 1 - e^{-\alpha/2}, \tag{7.188}$$

where the subscript of P_2 indicates the degree of freedom. The χ^2-distribution function is no longer symmetric due to the squared sum of the random variables z_1, z_2.

When z_1 and z_2 are identified with the semimajor axes a, b of the standard error ellipse (i.e., ellipse of constant probability scaled with 1σ), then x corresponds to the squared mean position error σ_P. The (standard) ellipse converts to a circle if both semimajor axes have the same length. The radius of the circle scaled with 1.18σ defines the circular error probable (CEP). The radius of the circle scaled with $\sqrt{2}\sigma$ is denoted distance root mean square (DRMS) or, simply, mean radial error. Numerical values for the probability P_2, obtained from (7.188), are given in Table 7.7, where, for geometric reasons, $\sqrt{\alpha}$ is chosen as an entry. The given probability levels are valid for circular and near circular error distributions. With increasing deviation of the circular distribution also the probability levels change. Lachapelle (1998) emphasizes that for $\sigma_x = 10\sigma_y$ the probability level of DRMS increases to 68.2%. Following Mikhail (1976: Sect. 2.6), the CEP may be approximated by CEP $\cong 0.589(\sigma_x + \sigma_y)$ if $0.2 \leq \sigma_{min}/\sigma_{max}$, where σ_{min} denotes the minimum value of σ_x, σ_y, and σ_{max} correspondingly the maximum value.

Three-dimensional accuracy measures
The three-dimensional case is specified by $n = 3$ and $x = z_1^2 + z_2^2 + z_3^2$. The density function of the χ^2-distribution becomes more complicated and the associated probability is given by

$$P_3(0 < x < \alpha) = \frac{1}{\sqrt{2\pi}} \int_0^{+\alpha} \sqrt{x}\, e^{-x/2}\, dx. \tag{7.189}$$

Table 7.8. Three-dimensional accuracy measures

$\sqrt{\alpha}$	Probability [%]	Notation
1.00	19.9	1σ or standard ellipsoid
1.54	50.0	Spherical error probable (SEP)
$\sqrt{3}$	60.8	Mean radial spherical error (MRSE)
2.00	73.9	2σ ellipsoid
2.80	95.0	95% confidence level
3.00	97.1	3σ ellipsoid

Numerical values for the probability P_3 can be taken from statistical and mathematical handbooks (e.g., Hartung et al. 2005). Some examples are given in Table 7.8, where $\sqrt{\alpha}$ is used as an entry again. The variable x can be interpreted as the sum of the squares of the semimajor axis of a standard error ellipsoid. The ellipsoid converts to a sphere if the semiaxes of the ellipsoid have same lengths. The radius of the sphere scaled with $\sqrt{3}\,\sigma$ is denoted as mean radial spherical error (MRSE), which is a single measure for the three-dimensional case. Again following Mikhail (1976: Sect. 2.6), the spherical error probable (SEP) may be approximated by $\text{SEP} \cong 0.513(\sigma_x + \sigma_y + \sigma_z)$ if $0.35 \leq \sigma_{min}/\sigma_{max}$.

Interrelation of accuracy measures

The equivalence of, e.g., LEP and RMS in Table 7.6 is given by $\text{LEP}/0.67 = \text{RMS}/1.00$.

According to Diggelen (1998), Table 7.9 relates the accuracy measures of the different dimensions based on a specific GNSS constellation. More details on the presuppositions and an extended table are given in Diggelen (1998). Interpreting Table 7.9 by an example, $\text{SEP} = 2.0\,\text{CEP}$, thus, $\text{CEP} = 2.5\,\text{m}$ is equivalent to $\text{SEP} = 5\,\text{m}$. Further GNSS accuracy equivalences are discussed in Diggelen (2007).

Table 7.9. Interrelation of accuracy measures

RMS	CEP	2DRMS	SEP	
1	0.44	1.1	0.88	RMS (1D)
	1	2.4	2.0	CEP (2D)
		1	0.85	2DRMS (2D)
			1	SEP (3D)

8 Data transformation

8.1 Introduction

From the user's point of view, one official reference frame of GNSS would be desirable. For several reasons, the reality is far from this idealization: the reference frame of GPS is WGS-84, for GLONASS it is PE-90, and also Galileo with GTRF will have its own reference frame (see Sects. 9.2, 10.2, and 11.2, respectively). Nevertheless, the main property of these reference frames is the same; they are realized by a geocentric Cartesian coordinate system. Therefore, when using GNSS, the coordinates of terrestrial sites are obtained in the respective reference frame. The surveyor is not, usually, interested in coordinates of the terrestrial points referring to a global frame; rather, the results are preferred in a local coordinate frame either as geodetic (i.e., ellipsoidal) coordinates, as plane coordinates, or as vectors combined with other terrestrial data. Since the realization of the GNSS reference frame (WGS-84, PE-90, GTRF) is a geocentric system and the local system usually is not, certain transformations are required. The subsequent sections deal with the transformations most frequently used.

8.2 Coordinate transformations

8.2.1 Cartesian coordinates and ellipsoidal coordinates

Denoting the (global) Cartesian coordinates of a point in space by X, Y, Z and assuming an ellipsoid of revolution with the same origin as the Cartesian coordinate system, the point can also be expressed by the ellipsoidal coordinates φ, λ, h (Fig. 8.1). According to Hofmann-Wellenhof and Moritz (2006: p. 195), the relation between the Cartesian coordinates and the ellipsoidal coordinates is given by

$$\begin{aligned} X &= (N + h) \cos\varphi \cos\lambda\,, \\ Y &= (N + h) \cos\varphi \sin\lambda\,, \\ Z &= \left(\frac{b^2}{a^2} N + h\right) \sin\varphi\,, \end{aligned} \qquad (8.1)$$

where N is the radius of curvature in the prime vertical which is obtained by

$$N = \frac{a^2}{\sqrt{a^2 \cos^2\varphi + b^2 \sin^2\varphi}}\,, \qquad (8.2)$$

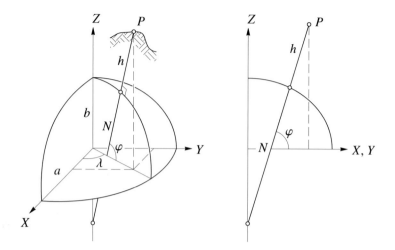

Fig. 8.1. Cartesian coordinates X, Y, Z and ellipsoidal coordinates φ, λ, h (left) and N, the radius of curvature in the prime vertical (right)

and a, b are the semiaxes of the ellipsoid. Recall that the Cartesian coordinates related to the geocentric system are also denoted ECEF coordinates and that the origin is the geocenter.

The formulas (8.1) transform ellipsoidal coordinates φ, λ, h into Cartesian coordinates X, Y, Z. For GNSS applications, the inverse transformation is more important since the Cartesian coordinates are given and the ellipsoidal coordinates sought. Thus, the task is now to compute the ellipsoidal coordinates φ, λ, h from the Cartesian coordinates X, Y, Z. Usually, this problem is solved iteratively, although a solution in closed form is possible. From X and Y, the radius of a parallel,

$$p = \sqrt{X^2 + Y^2} = (N + h) \cos \varphi, \tag{8.3}$$

can be computed. This equation is rearranged as

$$h = \frac{p}{\cos \varphi} - N \tag{8.4}$$

so that the ellipsoidal height appears explicitly. Introducing by

$$e^2 = \frac{a^2 - b^2}{a^2} \tag{8.5}$$

the first numerical eccentricity, the relation $b^2/a^2 = 1 - e^2$ follows. This can be substituted into the equation for Z in (8.1). The result

$$Z = (N + h - e^2 N) \sin \varphi \tag{8.6}$$

8.2 Coordinate transformations

can be rearranged as

$$Z = (N+h)\left(1 - e^2 \frac{N}{N+h}\right) \sin \varphi. \tag{8.7}$$

Dividing this expression by Eq. (8.3) gives

$$\frac{Z}{p} = \left(1 - e^2 \frac{N}{N+h}\right) \tan \varphi, \tag{8.8}$$

which yields

$$\tan \varphi = \frac{Z}{p}\left(1 - e^2 \frac{N}{N+h}\right)^{-1}. \tag{8.9}$$

For the longitude λ, the equation

$$\tan \lambda = \frac{Y}{X} \tag{8.10}$$

is obtained from Eq. (8.1) by dividing the first and the second equation.

The longitude λ can be directly computed from Eq. (8.10). The height h and the latitude φ are determined by Eqs. (8.4) and (8.9). The problem is that both equations depend on the latitude and the height. A solution can be found iteratively by the following steps:

1. Compute $p = \sqrt{X^2 + Y^2}$.
2. Compute an approximate value $\varphi_{(0)}$ from
$$\tan \varphi_{(0)} = \frac{Z}{p}(1 - e^2)^{-1}.$$
3. Compute an approximate value $N_{(0)}$ from
$$N_{(0)} = \frac{a^2}{\sqrt{a^2 \cos^2 \varphi_{(0)} + b^2 \sin^2 \varphi_{(0)}}}.$$
4. Compute the ellipsoidal height by
$$h = \frac{p}{\cos \varphi_{(0)}} - N_{(0)}.$$
5. Compute an improved value for the latitude by
$$\tan \varphi = \frac{Z}{p}\left(1 - e^2 \frac{N_{(0)}}{N_{(0)} + h}\right)^{-1}.$$
6. Check for another iteration step: if $\varphi = \varphi_{(0)}$, then the iteration is completed; otherwise set $\varphi_{(0)} = \varphi$ and continue with step 3.

Many other computation methods have been devised. One example for the transformation of X, Y, Z into φ, λ, h without iteration but with an inherent approximation is

$$\varphi = \arctan \frac{Z + e'^2 b \sin^3 \theta}{p - e^2 a \cos^3 \theta},$$

$$\lambda = \arctan \frac{Y}{X}, \qquad (8.11)$$

$$h = \frac{p}{\cos \varphi} - N,$$

where

$$\theta = \arctan \frac{Z a}{p b} \qquad (8.12)$$

is an auxiliary quantity and

$$e'^2 = \frac{a^2 - b^2}{b^2} \qquad (8.13)$$

is the second numerical eccentricity. Actually, there is no reason why these formulas are less popular than the iterative procedure since there results no significant difference between the two methods. A computation method with neither iteration nor approximation is, e.g., given by Zhu (1993).

For a numerical example, consider a point with $\varphi = 47°$, $\lambda = 15°$, and $h = 2\,000$ m referring to the Geodetic Reference System 1980 (GRS-80). Using the semiaxes of GRS-80, $a = 6\,378\,137.0000$ m and $b = 6\,356\,752.3141$ m (Hofmann-Wellenhof and Moritz 2006: Tables 2.1, 2.2), the desired ECEF coordinates are $X = 4\,210\,520.621$ m, $Y = 1\,128\,205.600$ m, $Z = 4\,643\,227.496$ m as computed from (8.1). It is recommended that the inverse transformation be performed as a check.

Note that the values of the GRS-80 semiaxes are identical with the respective values of the WGS-84; the latter may be regarded as a descendant of the GRS-80.

8.2.2 Global coordinates and local-level coordinates

Baseline vectors

The global coordinates are identical with the Cartesian coordinates of the previous section; however, instead of using the components X, Y, Z, the vector notation $\mathbf{X}$ is preferred. Thus, the vectors $\mathbf{X}_i$ and $\mathbf{X}_j$ represent two terrestrial points P_i and P_j. Defining the baseline vector between these two points in the global coordinate system by $\mathbf{X}_{ij} = \mathbf{X}_j - \mathbf{X}_i$, this vector may also be defined in the local-level system referenced to the tangent plane at P_i and introducing the notation $\mathbf{x}_{ij}$.

8.2 Coordinate transformations

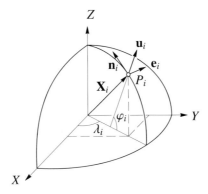

Fig. 8.2. Global and local-level coordinates

The axes $\mathbf{n}_i$, $\mathbf{e}_i$, $\mathbf{u}_i$ of the local (tangent plane) coordinate system at P_i corresponding to the north, east, up direction are represented in the global system by

$$\mathbf{n}_i = \begin{bmatrix} -\sin\varphi_i \cos\lambda_i \\ -\sin\varphi_i \sin\lambda_i \\ \cos\varphi_i \end{bmatrix}, \quad \mathbf{e}_i = \begin{bmatrix} -\sin\lambda_i \\ \cos\lambda_i \\ 0 \end{bmatrix}, \quad \mathbf{u}_i = \begin{bmatrix} \cos\varphi_i \cos\lambda_i \\ \cos\varphi_i \sin\lambda_i \\ \sin\varphi_i \end{bmatrix}, \quad (8.14)$$

where the vectors $\mathbf{n}_i$ and $\mathbf{e}_i$ span the tangent plane at P_i (Fig. 8.2). The third coordinate axis of the local-level system, i.e., the vector $\mathbf{u}_i$, is orthogonal to the tangent plane and coincides with the ellipsoidal normal.

The local-level system refers to the (natural) plumb line at P_i (more precisely: to the tangent at P_i of the slightly curved plumb line) if the ellipsoidal coordinates φ_i, λ_i in (8.14) are replaced by the astronomical latitude and longitude.

Now the components n_{ij}, e_{ij}, u_{ij} of the vector $\mathbf{x}_{ij}$ in the local-level system are introduced. These coordinates are sometimes denoted as ENU (east, north, up) coordinates. Considering Fig. 8.3, these components are obtained by a projection of vector $\mathbf{X}_{ij}$ onto the local-level axes $\mathbf{n}_i$, $\mathbf{e}_i$, $\mathbf{u}_i$. Analytically, this is achieved by inner products (also denoted as scalar or dot products). Therefore,

$$\mathbf{x}_{ij} = \begin{bmatrix} n_{ij} \\ e_{ij} \\ u_{ij} \end{bmatrix} = \begin{bmatrix} \mathbf{n}_i \cdot \mathbf{X}_{ij} \\ \mathbf{e}_i \cdot \mathbf{X}_{ij} \\ \mathbf{u}_i \cdot \mathbf{X}_{ij} \end{bmatrix} \quad (8.15)$$

is obtained. Assembling the vectors $\mathbf{n}_i$, $\mathbf{e}_i$, $\mathbf{u}_i$ of the local-level system as columns in a matrix $\mathbf{R}_i$, i.e.,

$$\mathbf{R}_i = \begin{bmatrix} -\sin\varphi_i \cos\lambda_i & -\sin\lambda_i & \cos\varphi_i \cos\lambda_i \\ -\sin\varphi_i \sin\lambda_i & \cos\lambda_i & \cos\varphi_i \sin\lambda_i \\ \cos\varphi_i & 0 & \sin\varphi_i \end{bmatrix}, \quad (8.16)$$

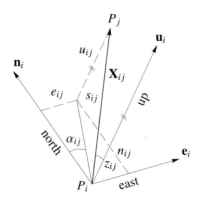

Fig. 8.3. Measurement quantities in the local-level system

relation (8.15) may be simplified to

$$\mathbf{x}_{ij} = \mathbf{R}_i^T \mathbf{X}_{ij}. \tag{8.17}$$

The components of $\mathbf{x}_{ij}$ may also be expressed by the spatial distance s_{ij}, the azimuth α_{ij}, and the zenith angle z_{ij}, which is assumed to be corrected for refraction. The appropriate relation is

$$\mathbf{x}_{ij} = \begin{bmatrix} n_{ij} \\ e_{ij} \\ u_{ij} \end{bmatrix} = \begin{bmatrix} s_{ij} \sin z_{ij} \cos \alpha_{ij} \\ s_{ij} \sin z_{ij} \sin \alpha_{ij} \\ s_{ij} \cos z_{ij} \end{bmatrix}, \tag{8.18}$$

where the terrestrial measurement quantities s_{ij}, α_{ij}, z_{ij} refer to P_i, i.e., the measurements were taken at P_i. Inverting (8.18) gives the measurement quantities explicitly:

$$\begin{aligned} s_{ij} &= \sqrt{n_{ij}^2 + e_{ij}^2 + u_{ij}^2}, \\ \tan \alpha_{ij} &= \frac{e_{ij}}{n_{ij}}, \\ \cos z_{ij} &= \frac{u_{ij}}{\sqrt{n_{ij}^2 + e_{ij}^2 + u_{ij}^2}}. \end{aligned} \tag{8.19}$$

Substituting (8.15) for n_{ij}, e_{ij} and u_{ij}, the measurement quantities may be expressed by the components of the vector $\mathbf{X}_{ij}$ in the global system.

Line-of-sight vectors

If the position vector of the terrestrial site $\mathbf{X}_i$ is known (at least approximately, i.e., coordinates from a map are sufficient) and if the vector $\mathbf{X}_j$ is replaced by the posi-

8.2 Coordinate transformations

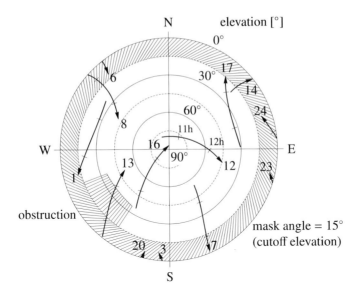

Fig. 8.4. Sky plot

tion vector of the satellite (known from the satellite ephemerides), then Eq. (8.19) yields apart from the spatial distance the azimuth and zenith angle of the satellite with respect to the observing site.

These data form the input for visibility charts or sky plots which are proper means for GNSS survey planning. The sky plots are polar orthogonal plots showing the satellite paths as a function of elevation and azimuth. Such sky plots are often supplemented by time tags and by the image of the local horizon. Thus, sky plots give also an imagination on the geometry of satellite constellation (Fig. 8.4) which mathematically is expressed by the geometric dilution of precision (GDOP) factor, cf. Sect. 7.3.4.

8.2.3 Ellipsoidal coordinates and plane coordinates

General remarks

In contrast to the previous section, only points on the ellipsoid are considered. Thus, ellipsoidal latitude φ and longitude λ are of interest here. The objective is to map a point φ, λ on the ellipsoid into a point x, y in a plane. Map projections may be expressed by

$$x = x(\varphi, \lambda; a, b), \qquad y = y(\varphi, \lambda; a, b) \tag{8.20}$$

in general form. Geodetic applications require conformal mapping. Conformality means that an angle on the ellipsoid is preserved after mapping it into the plane.

More precisely, the angle included by two geodesics on the ellipsoid is preserved if the two geodesics are conformally mapped into the plane.

Some of the most important conformal mappings of the ellipsoid into a plane are subsequently interpreted geometrically (see also Fig. 8.5), although they are defined analytically. Detailed formulas for many conformal mappings can be found, e.g., in Richardus and Adler (1972) or in Hofmann-Wellenhof et al. (1994).

- Conical projection. Consider a cone which is tangent to the ellipsoid at a selected (standard) parallel. After development of the conical surface, the meridians are straight lines converging at a point called the apex. This point is also the center of circles which represent the projected parallels. The standard parallel is mapped without distortion. One example of this projection is the conformal Lambert projection.
- Cylindrical projection. This is a special case of the conical projection if the apex is moved to infinity so that the cone becomes a cylinder which is tangent at the equator. In the transverse position, the cylinder is tangent at a standard meridian. After development of the cylindrical surface, the standard meridian is mapped without distortion. Two examples of this projection are the transverse Mercator projection and the universal transverse Mercator (UTM) projection. Because both methods are widely used, more details are given below.
- Azimuthal projection. This is also a special case of the conical projection if the apex is moved to the pole so that the cone becomes a plane which is tangent at the pole. The pole is the center of circles representing the parallels and of straight lines representing the meridians. More generally, the projection plane can be defined as plane tangent at any point on the ellipsoid. One example of this projection is the stereographic projection.

Equivalent projection is another type of mapping where the characteristic feature is that areas remain undistorted. Three examples of this projection are given in Fig. 8.5.

Transverse Mercator projection

This method is also referred to as Gauss–Krüger projection. The ellipsoid is partitioned into 120 zones of 3° longitude each where the central meridian with longitude λ_0 is in the center of each zone. The central meridian is mapped onto the plane without scale distortion and represents the y-axis (north direction). The x-axis is the mapping of the equator. The central meridian of the zone and the equator are special cases since all other meridians and parallels are mapped as curved lines. Due to the property of conformality, the mapped images of the meridians and parallels are orthogonal to each other.

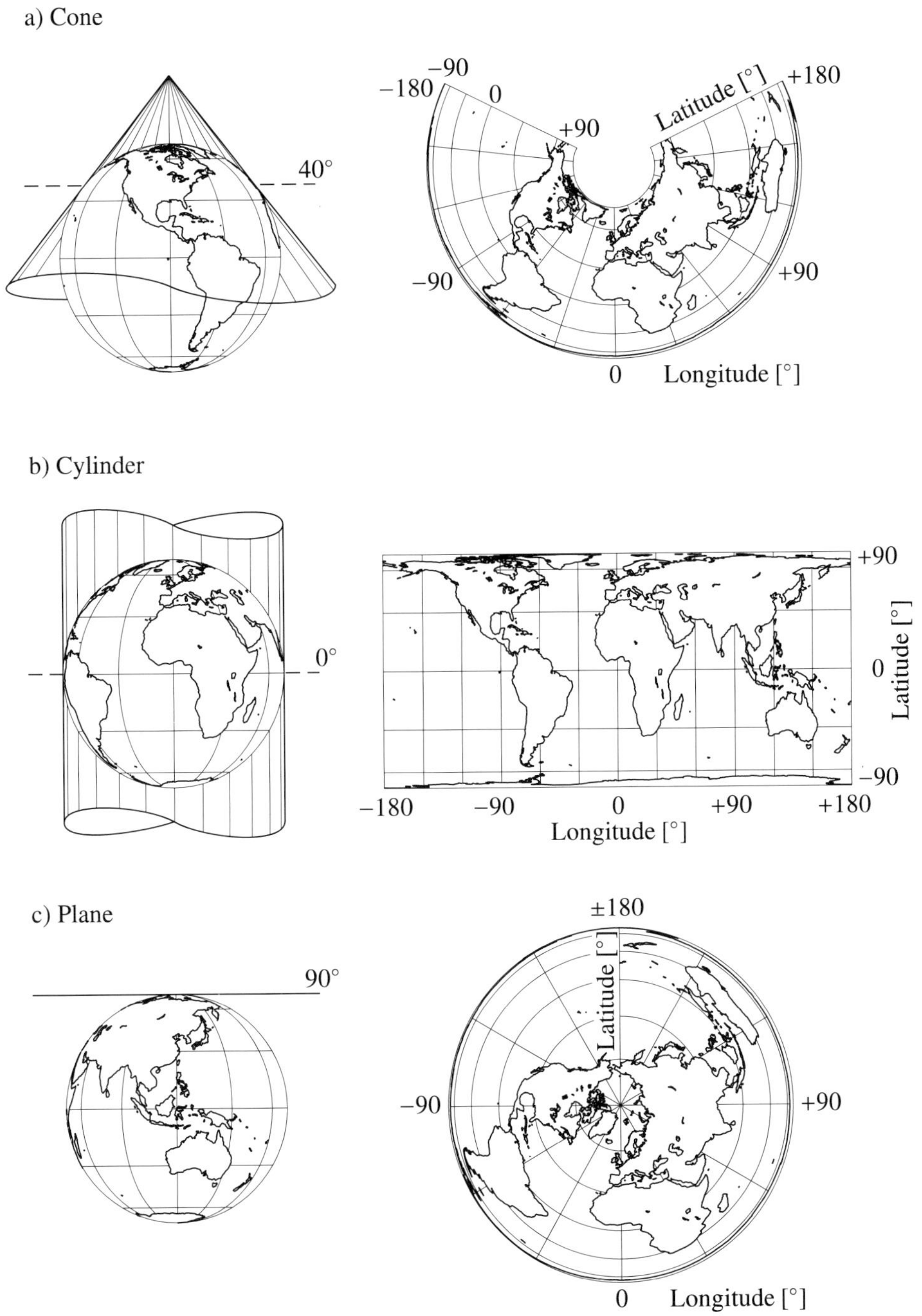

8.2 Coordinate transformations

a) Cone

b) Cylinder

c) Plane

Fig. 8.5. Conical, cylindrical, and azimuthal projection
(left: parallel projection, right: equivalent projection)

The numbering of the zones is related to Greenwich. In some cases, the numbering is related to Ferro. By definition, the latter is situated 17°40′ west of Greenwich.

For the transverse Mercator projection, the ellipsoidal point φ, λ being mapped into a point y, x of the plane is given by the series expansions

$$y = B(\varphi) + \frac{t}{2} N \cos^2\varphi \, \ell^2 + \frac{t}{24} N \cos^4\varphi \, (5 - t^2 + 9\eta^2 + 4\eta^4) \, \ell^4$$

$$+ \frac{t}{720} N \cos^6\varphi \, (61 - 58 t^2 + t^4 + 270 \eta^2 - 330 t^2\eta^2) \, \ell^6$$

$$+ \frac{t}{40\,320} N \cos^8\varphi \, (1385 - 3111 t^2 + 543 t^4 - t^6) \, \ell^8 + \ldots, \tag{8.21}$$

$$x = N \cos\varphi \, \ell + \frac{1}{6} N \cos^3\varphi \, (1 - t^2 + \eta^2) \, \ell^3$$

$$+ \frac{1}{120} N \cos^5\varphi \, (5 - 18 t^2 + t^4 + 14 \eta^2 - 58 t^2\eta^2) \, \ell^5$$

$$+ \frac{1}{5040} N \cos^7\varphi \, (61 - 479 t^2 + 179 t^4 - t^6) \, \ell^7 + \ldots,$$

where

$B(\varphi)$	…	arc length of meridian, see (8.22),
N	…	radius of curvature in prime vertical, see Fig. 8.1 and relation (8.2),
$\eta = \dfrac{\cos\varphi}{b} \sqrt{a^2 - b^2}$	…	auxiliary quantity,
$t = \tan\varphi$	…	auxiliary quantity,
$\ell = \lambda - \lambda_0$	…	longitude difference,
λ_0	…	longitude of the central meridian

are used. By convention, y is given first followed by x because the pair of coordinates (y, x) corresponds to (φ, λ). The arc length of meridian $B(\varphi)$ is the ellipsoidal distance from the equator to the point to be mapped and can be computed by the series expansion

$$B(\varphi) = \alpha \, [\varphi + \beta \sin 2\varphi + \gamma \sin 4\varphi + \delta \sin 6\varphi + \varepsilon \sin 8\varphi + \ldots], \tag{8.22}$$

8.2 Coordinate transformations

where

$$\alpha = \frac{a+b}{2}\left(1 + \frac{1}{4}n^2 + \frac{1}{64}n^4 + \ldots\right),$$

$$\beta = -\frac{3}{2}n + \frac{9}{16}n^3 - \frac{3}{32}n^5 + \ldots,$$

$$\gamma = \frac{15}{16}n^2 - \frac{15}{32}n^4 + \ldots, \qquad (8.23)$$

$$\delta = -\frac{35}{48}n^3 + \frac{105}{256}n^5 - \ldots,$$

$$\varepsilon = \frac{315}{512}n^4 + \ldots,$$

and

$$n = \frac{a-b}{a+b}. \qquad (8.24)$$

For a numerical example, consider a point on the GRS-80 ellipsoid with $\varphi = 47°$ N and $\lambda = 16°$ E and compute the Gauss–Krüger coordinates. First, from (8.23) and (8.24), the values

$$\alpha = 6\,367\,449.1458 \text{ m},$$
$$\beta = -2.518\,827\,93 \cdot 10^{-3},$$
$$\gamma = 2.643\,54 \cdot 10^{-6}, \qquad (8.25)$$
$$\delta = -3.45 \cdot 10^{-9},$$
$$\varepsilon = 5 \cdot 10^{-12}$$

are obtained. Then, using (8.22), $B(\varphi) = 5\,207\,247.009$ m results. Finally, with $\lambda_0 = 15°$, there follows from (8.21) $y = 5\,207\,732.441$ m, $x = 76\,055.734$ m.

The inverse Gauss–Krüger projection involves the mapping of a point y, x in the plane to a point φ, λ on the ellipsoid. The formulas are given by the series expansions

$$\varphi = \varphi_f + \frac{t_f}{2N_f^2}(-1-\eta_f^2)x^2$$

$$+ \frac{t_f}{24 N_f^4}(5 + 3t_f^2 + 6\eta_f^2 - 6t_f^2\eta_f^2 - 3\eta_f^4 - 9t_f^2\eta_f^4)x^4$$

$$+ \frac{t_f}{720 N_f^6}(-61 - 90 t_f^2 - 45 t_f^4 - 107 \eta_f^2 + 162 t_f^2\eta_f^2 + 45 t_f^4\eta_f^2)x^6$$

$$+ \frac{t_f}{40\,320\, N_f^8}(1385 + 3633 t_f^2 + 4095 t_f^4 + 1575 t_f^6)x^8 + \ldots,$$

$$\lambda = \lambda_0 + \frac{1}{N_f \cos\varphi_f}x + \frac{1}{6 N_f^3 \cos\varphi_f}(-1 - 2t_f^2 - \eta_f^2)x^3$$

$$+ \frac{1}{120 N_f^5 \cos\varphi_f}(5 + 28 t_f^2 + 24 t_f^4 + 6\eta_f^2 + 8t_f^2\eta_f^2)x^5$$

$$+ \frac{1}{5040 N_f^7 \cos\varphi_f}(-61 - 662 t_f^2 - 1320 t_f^4 - 720 t_f^6)x^7 + \ldots,$$

(8.26)

where the terms with the subscript f must be calculated based on the footpoint latitude φ_f. For the footpoint latitude, the series expansion is given by

$$\varphi_f = \bar{y} + \bar{\beta}\sin 2\bar{y} + \bar{\gamma}\sin 4\bar{y} + \bar{\delta}\sin 6\bar{y} + \bar{\varepsilon}\sin 8\bar{y} + \ldots, \qquad (8.27)$$

where

$$\bar{\alpha} = \frac{a+b}{2}\left(1 + \frac{1}{4}n^2 + \frac{1}{64}n^4 + \ldots\right),$$

$$\bar{\beta} = \frac{3}{2}n - \frac{27}{32}n^3 + \frac{269}{512}n^5 + \ldots,$$

$$\bar{\gamma} = \frac{21}{16}n^2 - \frac{55}{32}n^4 + \ldots, \qquad (8.28)$$

$$\bar{\delta} = \frac{151}{96}n^3 - \frac{417}{128}n^5 + \ldots,$$

$$\bar{\varepsilon} = \frac{1097}{512}n^4 + \ldots,$$

and the relation

$$\bar{y} = \frac{y}{\bar{\alpha}} \qquad (8.29)$$

8.2 Coordinate transformations

are used. Note that the coefficient $\bar{\alpha}$ is identical to α in (8.23).

For a numerical example, consider the task inverse to the previous one. Given are the Gauss–Krüger coordinates $y = 5\,207\,732.441$ m, $x = 76\,055.734$ m and the ellipsoidal coordinates related to the GRS-80 ellipsoid are to be calculated. First, from (8.28) and (8.29, the values

$$\bar{\alpha} = 6\,367\,449.1458 \text{ m},$$
$$\bar{\beta} = 2.518\,826\,60 \cdot 10^{-3},$$
$$\bar{\gamma} = 3.700\,95 \cdot 10^{-6}, \qquad (8.30)$$
$$\bar{\delta} = 7.45 \cdot 10^{-9}$$
$$\bar{\varepsilon} = 17 \cdot 10^{-12}$$

are obtained. Then, using (8.27), the footpoint latitude $\varphi_f = 47.004\,366\,54°$ results. Finally, from (8.26) $\varphi = 47°$ N and $\lambda = 16°$ E are obtained.

Universal transverse Mercator system

The universal transverse Mercator (UTM) system is a modification of the transverse Mercator system. First, the ellipsoid is partitioned into 60 zones with a width of $6°$ longitude each. Second, a scale factor of 0.9996 is applied to the conformal coordinates in the plane. The reason for this factor is to avoid fairly large distortions in the outer areas of a zone.

The zone numbering starts with M1 for the central meridian $\lambda_0 = 177°$ W and continues with M2 for the central meridian $\lambda_0 = 171°$ W. An adequate formula to calculate the zone number is given by

$$\text{INT}\left(\frac{180 \pm \lambda}{6}\right) + 1, \qquad (8.31)$$

where the plus sign must be used for eastern longitudes and the minus sign for western longitudes. The INT operator denotes the integer part of a real number, i.e., the decimal places are omitted.

For a numerical example, consider a point on the GRS-80 ellipsoid with $\varphi = 47°$ N and $\lambda = 16°$ E and compute the UTM coordinates. The zone is M33 according to Eq. (8.31) and the central meridian becomes $\lambda_0 = 15°$ E. The final result $y = 5\,205\,649.348$ m, $x = 76\,025.312$ m is obtained by the formulas (8.21) after multiplication by the scale factor. It is recommended that the inverse mapping be performed to demonstrate that millimeter accuracy is achieved.

8.2.4 Height transformation

Definitions

In the previous section, a point on the ellipsoid was mapped onto a plane and vice versa. The ellipsoidal height could be completely ignored. In this section, the primary interest is the height.

The ellipsoidal height h results from the projection of a point in space onto the chosen ellipsoid. The orthometric height H is the separation of a point from the geoid, measured along the (slightly curved) plumb line. The formula

$$h = H + N \tag{8.32}$$

is the relationship between the ellipsoid and the geoid, where N is the geoidal height (undulation).

As shown in Fig. 8.6, this formula is an approximation but is sufficiently accurate for all practical purposes. The angle ε expresses the deflection of the vertical between the slightly curved plumb line and the ellipsoidal normal. This angle does not exceed 30 arcseconds in most areas.

Positioning with GNSS results in X, Y, Z coordinates. After applying the transformation (8.11), ellipsoidal heights are obtained. If, additionally, one of the two remaining terms in (8.32) is given, the other one can be calculated. Thus, if the geoid is known, orthometric heights can be derived, or, if orthometric heights are known, geoidal heights can be derived.

As long as geodesists were performing horizontal surveys (e.g., triangulation) separate from vertical surveys (e.g., leveling), the influence of the difference between the ellipsoid and the geoid was not significant. This changed with the advent of satellite geodesy which yields both horizontal and vertical information. Now, the geometrically defined (i.e., ellipsoidal) heights obtained from satellite geodesy must be combined with physically defined (i.e., orthometric) heights. Thus, a short explanation is given below for readers not familiar with the ellipsoid and the geoid.

Ellipsoid

An ellipsoid of revolution is an approximation of the surface of the earth. This surface is formed by choosing a proper sized ellipse and rotating it about its minor axis. The ellipsoid is a convenient mathematical surface which has been subdivided by latitude and longitude to form a coordinate system. Distances and azimuths can be computed on this ellipsoid to millimeter accuracy and hundredths of an arcsecond.

It should be noted that there is a distinction between locally best-fitting (non-geocentric) ellipsoids and a global (geocentric) ellipsoid. Different local ellipsoids are primarily used because they fit a particular portion of the earth better than a

8.2 Coordinate transformations

global ellipsoid. In fact, the ellipsoid chosen for an area – for example, the Clarke ellipsoid for the former North American Datum 1927 (NAD-27) – was picked to minimize the difference between that ellipsoid and the geoid in that area. The center of a local ellipsoid does not coincide with the center of the true earth, but is displaced from the true center by up to some 100 meters.

The Geodetic Reference System 1980 (GRS-80) is an example of a global ellipsoid. The WGS-84 used for GPS is virtually identical with the GRS-80. The center of the GRS-80 ellipsoid coincides with the true center of the earth, and its surface provides an average fit of the geoid. The fit of this new surface to the geoid is not an improvement for all places on the earth. For example, the old Clarke ellipsoid surface only differs by small amounts from the geoid surface for the US; whereas, in the eastern US, the GRS-80 ellipsoid is approximately 30 m lower than the geoid. Surveyors using the old NAD-27 datum were accustomed to reduce their distance measurements to sea level when actually they were correcting lengths to map them onto the ellipsoid. Now on the east coast of the US, they must take into account approximately 30 m to obtain the correct height to use for the reduction to the ellipsoid.

Geoid

The geoid is defined physically and is the surface that is used to represent the shape of the earth. Drawings of the geoid show it as a bumpy surface with hills and valleys similar to a topographic model.

The center of the geoid coincides with the true center of the earth and its surface is an equipotential surface. The geoid can be visualized by imagining that the earth were completely covered by water. This water surface would (in theory) be an equipotential surface since the water would flow to compensate for any height difference that would occur. In actuality, the sea level differs slightly from a true equipotential surface due to bumps formed by different ocean currents, salinities, temperatures, etc.

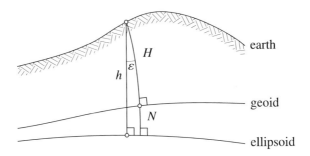

Fig. 8.6. Definition of heights

The geoid is the surface chosen for leveling datums. In many countries, the geoid that most nearly coincides with the average sea level value (measured by tide gauges) is chosen as the zero elevation. Because of this choice, the datum is often referred to as the sea level datum. This does not mean, however, that the zero elevation will coincide with the average level of sea level along a coastline. In fact, a standard deviation of about one meter exists between the zero elevation and the average sea level at any given point. The average sea level at a given point along the coast is affected by many factors so that average sea levels along the coast do not form an equipotential surface.

The geoidal surface is very irregular and it is virtually impossible to realize an exact mathematical model for the geoid. A good approximation of the geoid has been achieved through the use of spherical harmonics (International Association of Geodesy 1995).

Current research results are published by a Working Group of the IAG, see under www.iges.polimi.it. Among other publications, the Bulletins of this Working Group are a representative series. Tziavos and Barzaghi (2002) edited a collection of more than twenty papers both on numerical and theoretical aspects of regional gravity field estimation.

Long-term observations including the respective analysis resulted in global gravity field models as the Earth Gravitational Model 1996 (EGM96) based on a maximum spherical harmonic degree of 360 and essentially developed by the NASA Goddard Space Flight Center (GSFC), the National Imagery and Mapping Agency (NIMA), and the Ohio State University (OSU) (Lemoine et al. 1998). Another example of a global gravity field model is computed from satellite orbit perturbations (Biancale et al. 2000).

Flury and Rummel (2005) cover future aspects. The challenging minisatellite payload (CHAMP) and the gravity recovery and climate experiment (GRACE) satellite missions have tremendously improved the gravity field determination. For the first time, GRACE enabled the observation of temporal gravitational field variations not caused by tides but mainly by seasonal, interannual and long-term mass redistributions in the atmosphere, hydrosphere, cryosphere, and solid earth (Ilk et al. 2005: p. 14).

The gravity field and steady-state ocean circulation explorer (GOCE) mission is a core mission of the ESA Living Planet Programme. The primary objectives of the GOCE mission are to measure the earth's stationary gravity field and to model the geoid with extremely high accuracy. More specifically: to determine the gravity anomalies with an accuracy of 1 mgal, to determine the geoid with an accuracy of 1–2 cm, and to achieve these results at a spatial resolution better than 100 km (www.esa.int/export/esaLP/goce.html, European Space Agency 1999).

However, geoid determination will undoubtedly occupy geodesists for the next years, decades, and centuries.

8.3 Datum transformations

The coordinate transformations in the previous section dealt with the transition from one set of coordinates to another set of coordinates for the same point. Global coordinates X, Y, Z have been transformed into ellipsoidal coordinates φ, λ, h and into local-level coordinates n, e, u. Then, two-dimensional ellipsoidal surface coordinates φ, λ have been transformed into plane coordinates y, x. Finally, the ellipsoidal height h has been transformed either to the orthometric height H or to the geoidal height N.

A (geodetic) datum defines the relationship between a global and a local three-dimensional Cartesian coordinate system; therefore, a datum transformation transforms one coordinate system of a certain type to another coordinate system of the same type. This is one of the primary tasks when combining GNSS data with terrestrial data, i.e., the transformation of geocentric coordinates to local terrestrial coordinates. As mentioned earlier, the terrestrial system uses a locally best-fitting ellipsoid, e.g., the Clarke ellipsoid or the GRS-80 ellipsoid in the US and the Bessel ellipsoid in many parts of Europe. The local ellipsoid is linked to a nongeocentric Cartesian coordinate system where the origin coincides with the center of the ellipsoid. Plane coordinates such as Gauss–Krüger coordinates are obtained by mapping the local ellipsoid into the plane.

In the context of datum transformations, the fiducial point concept must be mentioned, although it was primarily used for orbit improvement in the past (e.g., Ashkenazi et al. 1990). Fiducial points are sites whose positions are accurately known from a (GNSS-independent) method such as VLBI or SLR. The concept of fiducial points is quite simple: during a GNSS campaign, at least three fiducial points in the area of the campaign are also equipped with receivers (apart from the points to be determined). This enables the transformation of GNSS coordinates into the frame of the fiducial points by a three-dimensional similarity transformation. Note that the geometry of the fiducial points with respect to the remaining points has a strong effect on the GNSS accuracy in this region.

8.3.1 Three-dimensional transformation

Consider two sets of three-dimensional Cartesian coordinates forming the vectors $\mathbf{X}$ and $\mathbf{X}_T$ (Fig. 8.7). The similarity transformation, also denoted as Helmert transformation, between the two sets can be formulated by the relation

$$\mathbf{X}_T = \mathbf{c} + \mu\, \mathbf{R}\, \mathbf{X}, \tag{8.33}$$

where $\mathbf{c}$ is the translation (or shift) vector, μ is a scale factor, and $\mathbf{R}$ is a rotation matrix. The components of the shift vector

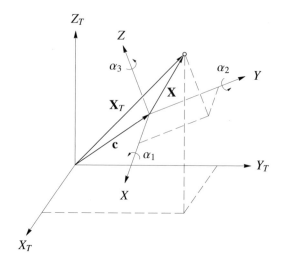

Fig. 8.7. Three-dimensional transformation

$$\mathbf{c} = \begin{bmatrix} c_1 \\ c_2 \\ c_3 \end{bmatrix} \tag{8.34}$$

account for the coordinates of the origin of the $\mathbf{X}$ system in the $\mathbf{X}_T$ system. Note that a single scale factor is considered. More generally (but with GNSS not necessary), three scale factors, one for each axis, could be used. The rotation matrix is an orthonormal matrix which is composed of three successive rotations

$$\mathbf{R} = \mathbf{R}_3\{\alpha_3\}\,\mathbf{R}_2\{\alpha_2\}\,\mathbf{R}_1\{\alpha_1\}, \tag{8.35}$$

where the subscripts indicate the respective rotation axis. Explicitly,

$$\mathbf{R} = \begin{bmatrix} \cos\alpha_2 \cos\alpha_3 & \cos\alpha_1 \sin\alpha_3 + \sin\alpha_1 \sin\alpha_2 \cos\alpha_3 & \sin\alpha_1 \sin\alpha_3 - \cos\alpha_1 \sin\alpha_2 \cos\alpha_3 \\ -\cos\alpha_2 \sin\alpha_3 & \cos\alpha_1 \cos\alpha_3 - \sin\alpha_1 \sin\alpha_2 \sin\alpha_3 & \sin\alpha_1 \cos\alpha_3 + \cos\alpha_1 \sin\alpha_2 \sin\alpha_3 \\ \sin\alpha_2 & -\sin\alpha_1 \cos\alpha_2 & \cos\alpha_1 \cos\alpha_2 \end{bmatrix}$$

$$\tag{8.36}$$

is obtained.

In the case of known transformation parameters $\mathbf{c}$, μ, $\mathbf{R}$, a point from the $\mathbf{X}$ system can be transformed into the $\mathbf{X}_T$ system by (8.33).

8.3 Datum transformations

If the transformation parameters are unknown, they can be determined with the aid of common (identical) points. This means that the coordinates of the same point are given in both systems. Since each common point (given by $\mathbf{X}_T$ and $\mathbf{X}$) yields three equations, two common points and one additional common component (e.g., height) are sufficient to solve for the seven unknown parameters. In practice, redundant common point information is used and the unknown parameters are calculated by least-squares adjustment.

Since the parameters are mixed nonlinearly in Eq. (8.33), a linearization must be performed where approximate values $\mathbf{c}_0$, μ_0, $\mathbf{R}_0$ are required. In the case of GNSS, a datum transformation between the geocentric system (e.g., WGS-84) and a local system, the approximation $\mu_0 = 1$ is appropriate and the relation

$$\mu = \mu_0 + \Delta\mu = 1 + \Delta\mu \tag{8.37}$$

is obtained. Furthermore, the rotation angles α_i in (8.36) are small and may be considered as differential quantities $\Delta\alpha_i$. Introducing these quantities into (8.36), setting $\cos\Delta\alpha_i = 1$ and $\sin\Delta\alpha_i = \Delta\alpha_i$, and considering only first-order terms gives

$$\mathbf{R} = \begin{bmatrix} 1 & \Delta\alpha_3 & -\Delta\alpha_2 \\ -\Delta\alpha_3 & 1 & \Delta\alpha_1 \\ \Delta\alpha_2 & -\Delta\alpha_1 & 1 \end{bmatrix} = \mathbf{I} + \Delta\mathbf{R}, \tag{8.38}$$

where $\mathbf{I}$ is the unit matrix and $\Delta\mathbf{R}$ is a (skewsymmetric) differential rotation matrix. Thus, the approximation $\mathbf{R}_0 = \mathbf{I}$ is appropriate. Finally, the shift vector is split up in the form

$$\mathbf{c} = \mathbf{c}_0 + \Delta\mathbf{c}, \tag{8.39}$$

where the approximate shift vector

$$\mathbf{c}_0 = \mathbf{X}_T - \mathbf{X} \tag{8.40}$$

follows by substituting the approximations for the scale factor and the rotation matrix into Eq. (8.33).

Introducing Eqs. (8.37), (8.38), (8.39) into (8.33) and skipping details which can be found, for example, in Hofmann-Wellenhof et al. (1994) gives the linearized model for a single point i. This model can be written in the form

$$\mathbf{X}_{T_i} - \mathbf{X}_i - \mathbf{c}_0 = \mathbf{A}_i\,\Delta\mathbf{p}, \tag{8.41}$$

where the left side of the equation is known and may formally be considered as an observation. The design matrix $\mathbf{A}_i$ and the vector $\Delta\mathbf{p}$, containing the unknown parameters, are given by

$$\mathbf{A}_i = \begin{bmatrix} 1 & 0 & 0 & X_i & 0 & -Z_i & Y_i \\ 0 & 1 & 0 & Y_i & Z_i & 0 & -X_i \\ 0 & 0 & 1 & Z_i & -Y_i & X_i & 0 \end{bmatrix}, \tag{8.42}$$

$$\Delta \mathbf{p} = [\ \Delta c_1 \quad \Delta c_2 \quad \Delta c_3 \quad \Delta \mu \quad \Delta \alpha_1 \quad \Delta \alpha_2 \quad \Delta \alpha_3 \ ]^\mathrm{T}.$$

Recall that Eq. (8.41) is now a system of linear equations for point i. For n common points, the design matrix $\mathbf{A}$ is

$$\mathbf{A} = \begin{bmatrix} \mathbf{A}_1 \\ \mathbf{A}_2 \\ \vdots \\ \mathbf{A}_n \end{bmatrix}. \tag{8.43}$$

In detail, for three common points the design matrix is

$$\mathbf{A} = \begin{bmatrix} 1 & 0 & 0 & X_1 & 0 & -Z_1 & Y_1 \\ 0 & 1 & 0 & Y_1 & Z_1 & 0 & -X_1 \\ 0 & 0 & 1 & Z_1 & -Y_1 & X_1 & 0 \\ 1 & 0 & 0 & X_2 & 0 & -Z_2 & Y_2 \\ 0 & 1 & 0 & Y_2 & Z_2 & 0 & -X_2 \\ 0 & 0 & 1 & Z_2 & -Y_2 & X_2 & 0 \\ 1 & 0 & 0 & X_3 & 0 & -Z_3 & Y_3 \\ 0 & 1 & 0 & Y_3 & Z_3 & 0 & -X_3 \\ 0 & 0 & 1 & Z_3 & -Y_3 & X_3 & 0 \end{bmatrix}, \tag{8.44}$$

which leads to a slightly redundant system. Least-squares adjustment yields the parameter vector $\Delta \mathbf{p}$ and the adjusted values by (8.37), (8.38), (8.39). Once the seven parameters of the similarity transformation are determined, formula (8.33) can be used to transform other than the common points.

For a specific example, consider the task of transforming GNSS coordinates of a network to (three-dimensional) coordinates of a (nongeocentric) local system. The GNSS coordinates are denoted by $(X, Y, Z)_\mathrm{GNSS}$ and the local system coordinates are the plane coordinates $(y, x)_\mathrm{LS}$ and the ellipsoidal height h_LS. To obtain the transformation parameters, it is assumed that the coordinates of the common points in both systems are available. The solution of the task is obtained by the following algorithm:

1. Transform the plane coordinates $(y, x)_\mathrm{LS}$ of the common points into the ellipsoidal surface coordinates $(\varphi, \lambda)_\mathrm{LS}$ by using the appropriate mapping formulas.

8.3 Datum transformations

2. Transform the ellipsoidal coordinates $(\varphi, \lambda, h)_{LS}$ of the common points into the Cartesian coordinates $(X, Y, Z)_{LS}$ by (8.1).
3. Determine the seven parameters of a Helmert transformation by using the coordinates $(X, Y, Z)_{GNSS}$ and $(X, Y, Z)_{LS}$ of the common points.
4. For network points other than the common points, transform the coordinates $(X, Y, Z)_{GNSS}$ into $(X, Y, Z)_{LS}$ via Eq. (8.33) using the transformation parameters determined in the previous step.
5. Transform the Cartesian coordinates $(X, Y, Z)_{LS}$ computed in the previous step into ellipsoidal coordinates $(\varphi, \lambda, h)_{LS}$ by (8.11) or by the iterative procedure.
6. Map the ellipsoidal surface coordinates $(\varphi, \lambda)_{LS}$ computed in the previous step into plane coordinates $(y, x)_{LS}$ by the appropriate mapping formulas.

Note that in steps 1, 2, 5, and 6 the same ellipsoid, either a local or a global one, must be used.

The advantage of the three-dimensional approach is that no a priori information is required for the seven parameters of the similarity transformation. The disadvantage of the method is that for the common points ellipsoidal heights (and, thus, geoidal heights) are required. However, as reported by Schmitt et al. (1991), the effect of incorrect heights of the common points often has a negligible effect on the plane coordinates $(y, x)_{LS}$. For example, incorrect heights may cause a tilt of a 20 km × 20 km network by an amount of 5 m in space; however, the effect on the plane coordinates is only approximately 1 mm.

For large areas, the height problem can be solved by adopting approximate ellipsoidal heights for the common points and performing a three-dimensional affine transformation instead of the similarity transformation.

8.3.2 Two-dimensional transformation

Two-dimensional coordinates are now considered. The two different sets of plane coordinates are represented by $\mathbf{x}$ and $\mathbf{x}_T$ (Fig. 8.8). The two-dimensional similarity transformation is defined by

$$\mathbf{x}_T = \mathbf{c} + \mu \mathbf{R} \mathbf{x} \tag{8.45}$$

with the shift vector

$$\mathbf{c} = \begin{bmatrix} c_1 \\ c_2 \end{bmatrix}, \tag{8.46}$$

the scale factor μ, and the rotation matrix

$$\mathbf{R} = \begin{bmatrix} \cos \alpha & -\sin \alpha \\ \sin \alpha & \cos \alpha \end{bmatrix}, \tag{8.47}$$

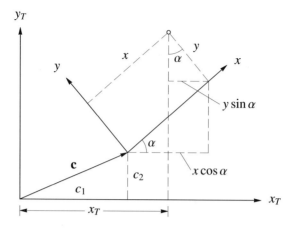

Fig. 8.8. Two-dimensional transformation

which contains a single rotation angle. Hence, Eq. (8.45) comprises four transformation parameters: the two translation components c_1, c_2, the scale factor μ, and the rotation angle α. Substituting (8.46) and (8.47) into (8.45) gives the transformed components explicitly:

$$x_T = c_1 + \mu x \cos\alpha - \mu y \sin\alpha,$$
$$y_T = c_2 + \mu x \sin\alpha + \mu y \cos\alpha. \tag{8.48}$$

These formulas can be verified by the geometry given in Fig. 8.8, where the contributing terms for x_T (without scale factor) are indicated.

In the case of known transformation parameters $\mathbf{c}$, μ, $\mathbf{R}$, a point in the $\mathbf{x}$ system can be transformed into the $\mathbf{x}_T$ system by (8.45).

If the transformation parameters are unknown, they are – analogously to the three-dimensional case – determined using common points. Two common points, each yielding two equations, are sufficient to solve for the four unknown parameters. In practice, redundant common point information is used and the unknown parameters are calculated by least-squares adjustment, where the coordinates x_T and y_T are formally considered as observations.

As seen from (8.48), the unknowns appear again in nonlinear form. However, using the auxiliary unknowns

$$p = \mu \cos\alpha,$$
$$q = \mu \sin\alpha, \tag{8.49}$$

8.3 Datum transformations

linear equations

$$x_T = c_1 + p\,x - q\,y,$$
$$y_T = c_2 + q\,x + p\,y$$
(8.50)

with respect to the unknowns are obtained. The scale factor and the rotation angle are determined from the auxiliary unknowns by

$$\mu = \sqrt{p^2 + q^2},$$
$$\tan \alpha = q/p.$$
(8.51)

For a specific example, consider the task of transforming GNSS coordinates of a network to (two-dimensional) coordinates of a (nongeocentric) local system. The GNSS coordinates are denoted by $(X, Y, Z)_{\text{GNSS}}$ and the plane coordinates in the local system are denoted by $(y, x)_{\text{LS}}$. To get the transformation parameters, it is assumed that for some points (common points) the coordinates in both systems are known. Note that no height information in the local system is available. The solution of the task is obtained by the following algorithm:

1. Transform the Cartesian coordinates $(X, Y, Z)_{\text{GNSS}}$ of all network points into ellipsoidal coordinates $(\varphi, \lambda, h)_{\text{GNSS}}$ by (8.11) or by the iterative procedure. Note that the ellipsoid of the local datum should be used.
2. Disregard for all network points the ellipsoidal heights h_{GNSS} and map the ellipsoidal surface coordinates $(\varphi, \lambda)_{\text{GNSS}}$ computed in the previous step into plane coordinates $(y, x)_{\text{GNSS}}$ by using the appropriate mapping formulas and the ellipsoid of the local datum.
3. Determine the four parameters of a two-dimensional similarity transformation by using the coordinates $(y, x)_{\text{GNSS}}$ and $(y, x)_{\text{LS}}$ of the common points.
4. For network points other than the common points, transform the coordinates $(y, x)_{\text{GNSS}}$ into $(y, x)_{\text{LS}}$ via Eq. (8.45) using the transformation parameters determined in the previous step.

This data transformation works well (even without using height information) for small-sized networks, low elevations, and for areas where geoidal heights do not change. The coordinates $(y, x)_{\text{GNSS}}$, however, depend on the dimension and on the displacement of the local ellipsoid. These dependencies may lead to a distortion of the point clusters which, in principle, is not matched by a similarity transformation (Lichtenegger 1991). A numerical example is given in Schmitt et al. (1991) and shows that the use of the WGS-84 ellipsoid and the two-dimensional similarity transformation leads to discrepancies in the range of 8 mm to 15 mm for a network covering an area of 200 km × 200 km.

One way to overcome the problem is to apply other transformations such as the affine transformation between the two coordinate sets in the plane. Another way is to transform the $(y, x)_{LS}$ coordinates of common points approximately to $(X, Y, Z)_{LS}$ by using approximate ellipsoidal heights. Now the parameters of a three-dimensional similarity transformation can be derived, which enable the approximate transformation of all GNSS coordinates into the local datum. This transformation could also be achieved by only applying an approximate shift vector to the GNSS coordinates. The approximate coordinates are then mapped into the plane, and the final similarity transformation is performed in two-dimensional space.

8.3.3 One-dimensional transformation

General remarks

One of the distinct features of GNSS is that three-dimensional (3D) coordinates are obtained at once. There is no separation between the horizontal coordinates and the height of a point because all three components are calculated together by the same procedure. In classical geodesy, horizontal coordinates and heights were obtained independently.

Now the question arises why in the previous section a two-dimensional (2D) transformation and here a one-dimensional (1D) transformation are discussed. The answer lies in historical data. Many countries have excellent horizontal control networks available but often fairly poor ellipsoidal heights because of lack of geoidal heights. Thus, it is appropriate to use the 2D transformation if height information is not available. Similarly, the 1D transformation can be used to transform heights without detailed knowledge of the geoid.

Symbolically, the 1D transformation is obtained by 3D ⊖ 2D. In more detail, the parameters for the 1D transformation are obtained by "subtracting" the parameters of the 2D transformation from the parameters of the 3D transformation. This looks like

$$\left. \begin{array}{c c c c c c c c} 3D & c_1 & c_2 & c_3 & \mu & \alpha_1 & \alpha_2 & \alpha_3 \\ 2D & c_1 & c_2 & & \mu & & & \alpha_3 \\ \hline 1D & & & c_3 & & \alpha_1 & \alpha_2 & \end{array} \right\} \ominus \qquad (8.52)$$

where for the 2D transformation the rotation angle has been given the corresponding subscript. In other words, the 3D transformation is composed of a 2D transformation for the horizontal coordinates and a 1D transformation for the heights.

Concentrating now on the 1D transformation, from the parameters in (8.52) it can be seen that the transformation consists of a shift along the vertical axis, a tilt (rotation) about the north–south axis, and a tilt about the east–west axis. These three unknowns are determined by using the height information of at least three common points.

8.3 Datum transformations

Transformation using heights

Assume that for some (common) points in a GNSS network the orthometric heights or elevations H_i and the ellipsoidal heights h_i are known. The mathematical model for the one-dimensional transformation is given by

$$H_i = h_i + \Delta h - y_i \Delta \alpha_1 + x_i \Delta \alpha_2, \tag{8.53}$$

where Δh is a vertical shift, and $\Delta \alpha_1$ and $\Delta \alpha_2$ are rotation angles about the x-axis and the y-axis. Formally, Eq. (8.53) corresponds to the third component of the three-dimensional similarity transformation without taking into account a scale factor. Additionally, the equation is now expressed in the local coordinate frame with the position coordinates x_i and y_i. These coordinates are required with only low accuracy and can be taken, e.g., from a map. Geometrically, the model equation (8.53) may be interpreted as the equation of a plane, which enables the interpolation of geoidal heights $N = h - H$ in other than the common points. The interpolation could be extended to a higher-order surface to take into account more irregular geoid structures.

When a geoid model is available, the ellipsoidal heights h_i could be transformed into approximate elevations H_{i0}. Usually, there are discrepancies between the heights H_i and H_{i0} due to the combined effects of the GNSS systematic errors and the errors in the geoid modeling.

The Federal Geodetic Control Subcommittee (FGCS) specifications require for the US that surveys to be included into the national network be tied to a minimum of four benchmarks well-distributed geometrically throughout the project area (e.g., corners of project area). The additional benchmarks (ties) enable a least-squares adjustment of the model equations (8.53) and provide the necessary check on the computation of the rotation of the ellipsoid to the geoid. A good practice is to perform the rotation using three of the elevations and to check the rotated elevation of the fourth point against the true elevation. The two values should agree within a few centimeters under normal conditions. An additional check on the correctness of the transformation is provided by inspecting the magnitude of the two rotation angles computed by the least-squares adjustment. Normally, these angles should be less than a few arcseconds.

Usually, the elevations of points in a network of small size, say 10 km × 10 km, can be determined with an accuracy of about 3 cm. In cases where the generalized model adequately describes the variation of the geoidal heights, much larger areas can be surveyed using this method with comparable accuracies being achieved.

Transformation using height differences

In the preceding paragraph, the importance of geoidal heights has been stressed. The ellipsoidal height determined by GNSS can be transformed to the orthometric height if the geoidal height is known.

There are times, however, when it is only required to measure changes in elevation. For example, when it is desired to measure the subsidence rate of a point (e.g., oil platform). In such cases, the importance of a well-known geoid diminishes because relative heights are considered. For two points,

$$H_1 = h_1 - N_1,$$
$$H_2 = h_2 - N_2$$
(8.54)

are the height relations and

$$H_2 - H_1 = h_2 - h_1 - (N_2 - N_1) \tag{8.55}$$

is the height difference or height change between the points 1 and 2. Here only the difference of the geoidal heights influences the result. Thus, if the geoidal heights are constant in a local area, meaning that the separation between the geoid and the ellipsoid is constant, they can be ignored. Similarly, if the geoid has a constant slope with the ellipsoid, the heights can be computed accurately by rotating the GNSS heights into the geoidal surface as previously described.

8.4 Combining GNSS and terrestrial data

8.4.1 Common coordinate system

So far GNSS and terrestrial networks have been considered separately with respect to the adjustment. The combination, for example, by a datum transformation, was supposed to be performed after individual adjustments. Now the common adjustment of GNSS observations and terrestrial data is investigated. The problem encountered here is that GNSS data refer to a three-dimensional geocentric Cartesian system, whereas terrestrial data refer to the individual local-level (tangent plane) systems at each measurement point referenced to plumb lines. Furthermore, terrestrial data are traditionally separated into position and height, where the position refers to an ellipsoid and the (orthometric) height to the geoid.

For a common adjustment, a common coordinate system to which all observations are transformed, is required. In principle, any convenient system may be introduced as common reference. One possibility is to use two-dimensional (plane) coordinates in the local system as proposed by Daxinger and Stirling (1995). Here, a three-dimensional coordinate system is chosen. The origin of the coordinate system is the center of the ellipsoid adopted for the local system, the Z-axis coincides with the semiminor axis of the ellipsoid, the X-axis is obtained by the intersection of the ellipsoidal Greenwich meridian plane (i.e., $\lambda = 0$) and the ellipsoidal equatorial plane, and the Y-axis completes the right-handed system. Position vectors

8.4 Combining GNSS and terrestrial data

referred to this system are denoted by $\mathbf{X}_{LS}$, where LS indicates the reference to the local system.

After the decision on the common coordinate system, the terrestrial measurements referring to the individual local-level systems at the observing sites must be represented in this common coordinate system. Similarly, GNSS baseline vectors regarded as measurement quantities are to be transformed to this system.

8.4.2 Representation of measurement quantities

Distances

The spatial distance s_{ij} as function of the local-level coordinates is given in (8.19). If n_{ij}, e_{ij}, u_{ij}, the components of $\mathbf{x}_{ij}$, are substituted by (8.15), the relation

$$s_{ij} = \sqrt{n_{ij}^2 + e_{ij}^2 + u_{ij}^2}$$
$$= \sqrt{(X_j - X_i)^2 + (Y_j - Y_i)^2 + (Z_j - Z_i)^2} \tag{8.56}$$

is obtained, where (8.14) has also been taken into account (namely the fact that $\mathbf{n}_i$, $\mathbf{e}_i$, $\mathbf{u}_i$ are unit vectors). The second expression also arises immediately from the Pythagorean theorem. Differentiation of (8.56) yields

$$ds_{ij} = \frac{X_{ij}}{s_{ij}}(dX_j - dX_i) + \frac{Y_{ij}}{s_{ij}}(dY_j - dY_i) + \frac{Z_{ij}}{s_{ij}}(dZ_j - dZ_i), \tag{8.57}$$

where

$$\begin{aligned} X_{ij} &= X_j - X_i, \\ Y_{ij} &= Y_j - Y_i, \\ Z_{ij} &= Z_j - Z_i \end{aligned} \tag{8.58}$$

have been introduced accordingly. The relation (8.57) may also be expressed as

$$\Delta s_{ij} = \frac{X_{ij}}{s_{ij}}(\Delta X_j - \Delta X_i) + \frac{Y_{ij}}{s_{ij}}(\Delta Y_j - \Delta Y_i) + \frac{Z_{ij}}{s_{ij}}(\Delta Z_j - \Delta Z_i) \tag{8.59}$$

if the differentials are replaced by differences.

Azimuths

Again the same principle applies: the measured azimuth α_{ij} as function of the local-level coordinates is given in (8.19). If n_{ij}, e_{ij}, u_{ij}, the components of $\mathbf{x}_{ij}$, are substituted by (8.15), the relation

$$\tan \alpha_{ij} = e_{ij}/n_{ij}$$
$$= \frac{-X_{ij} \sin \lambda_i + Y_{ij} \cos \lambda_i}{-X_{ij} \sin \varphi_i \cos \lambda_i - Y_{ij} \sin \varphi_i \sin \lambda_i + Z_{ij} \cos \varphi_i} \tag{8.60}$$

is obtained. After a lengthy derivation, the relation

$$\Delta\alpha_{ij} = \frac{\sin\varphi_i \cos\lambda_i \sin\alpha_{ij} - \sin\lambda_i \cos\alpha_{ij}}{s_{ij}\sin z_{ij}}(\Delta X_j - \Delta X_i)$$
$$+ \frac{\sin\varphi_i \sin\lambda_i \sin\alpha_{ij} + \cos\lambda_i \cos\alpha_{ij}}{s_{ij}\sin z_{ij}}(\Delta Y_j - \Delta Y_i) \qquad (8.61)$$
$$- \frac{\cos\varphi_i \sin\alpha_{ij}}{s_{ij}\sin z_{ij}}(\Delta Z_j - \Delta Z_i)$$
$$+ \cot z_{ij}\sin\alpha_{ij}\,\Delta\varphi_i + (\sin\varphi_i - \cos\alpha_{ij}\cos\varphi_i \cot z_{ij})\,\Delta\lambda_i$$

is obtained.

Directions
Measured directions R_{ij} are related to azimuths α_{ij} by the orientation unknown o_i. The relation reads

$$R_{ij} = \alpha_{ij} - o_i, \qquad (8.62)$$

and the expression

$$\Delta R_{ij} = \Delta\alpha_{ij} - \Delta o_i \qquad (8.63)$$

is immediately obtained.

Zenith angles
The zenith angle z_{ij} as function of the local-level coordinates is given in (8.19). If n_{ij}, e_{ij}, u_{ij}, the components of $\mathbf{x}_{ij}$, are substituted by (8.15), the relation

$$\cos z_{ij} = u_{ij}/s_{ij}$$
$$= \frac{X_{ij}\cos\varphi_i \cos\lambda_i + Y_{ij}\cos\varphi_i \sin\lambda_i + Z_{ij}\sin\varphi_i}{\sqrt{X_{ij}^2 + Y_{ij}^2 + Z_{ij}^2}} \qquad (8.64)$$

is obtained, where (8.56) and (8.58) have been used. After a lengthy derivation, the relation

$$\Delta z_{ij} = \frac{X_{ij}\cos z_{ij} - s_{ij}\cos\varphi_i \cos\lambda_i}{s_{ij}^2 \sin z_{ij}}(\Delta X_j - \Delta X_i)$$
$$+ \frac{Y_{ij}\cos z_{ij} - s_{ij}\cos\varphi_i \sin\lambda_i}{s_{ij}^2 \sin z_{ij}}(\Delta Y_j - \Delta Y_i) \qquad (8.65)$$
$$+ \frac{Z_{ij}\cos z_{ij} - s_{ij}\sin\varphi_i}{s_{ij}^2 \sin z_{ij}}(\Delta Z_j - \Delta Z_i)$$
$$- \cos\alpha_{ij}\,\Delta\varphi_i - \cos\varphi_i \sin\alpha_{ij}\,\Delta\lambda_i$$

8.4 Combining GNSS and terrestrial data

is obtained.

It is presupposed that the zenith angles are reduced to the chord of the light path. This reduction may be modeled by

$$z_{ij} = z_{ij_{\text{meas}}} + \frac{s_{ij}}{2R_e} k, \tag{8.66}$$

where $z_{ij_{\text{meas}}}$ is the measured zenith angle, R_e is the mean radius of the earth, and k is the coefficient of refraction. For k either a standard value may be substituted or the coefficient of refraction is estimated as additional unknown. In the case of estimation, there are several choices, e.g., one value for k for all zenith angles, or one value for a group of zenith angles, or one value per day.

Ellipsoidal height differences

The "measured" ellipsoidal height difference is represented by

$$h_{ij} = h_j - h_i. \tag{8.67}$$

The heights involved are obtained by transforming the Cartesian coordinates into ellipsoidal coordinates according to (8.11) or by using the iterative procedure given in Sect. 8.2.1. The height difference is approximately (neglecting the curvature of the earth) given by the third component of $\mathbf{x}_{ij}$ in the local-level system. Hence,

$$h_{ij} = \mathbf{u}_i \cdot \mathbf{X}_{ij} \tag{8.68}$$

or, by substituting $\mathbf{u}_i$ according to (8.14), the relation

$$h_{ij} = \cos \varphi_i \cos \lambda_i X_{ij} + \cos \varphi_i \sin \lambda_i Y_{ij} + \sin \varphi_i Z_{ij} \tag{8.69}$$

is obtained. This equation may be differentiated with respect to the Cartesian coordinates. If the differentials are replaced by the corresponding differences,

$$\begin{aligned}\Delta h_{ij} = &\cos \varphi_j \cos \lambda_j \Delta X_j + \cos \varphi_j \sin \lambda_j \Delta Y_j + \sin \varphi_j \Delta Z_j \\ &- \cos \varphi_i \cos \lambda_i \Delta X_i - \cos \varphi_i \sin \lambda_i \Delta Y_i - \sin \varphi_i \Delta Z_i\end{aligned} \tag{8.70}$$

is obtained, where the coordinate differences were decomposed into their individual coordinates.

Baselines

From GNSS measurements, baselines $\mathbf{X}_{ij(\text{GNSS})} = \mathbf{X}_{j(\text{GNSS})} - \mathbf{X}_{i(\text{GNSS})}$ in a geocentric Cartesian coordinate system are obtained. The position vectors $\mathbf{X}_{i(\text{GNSS})}$ and $\mathbf{X}_{j(\text{GNSS})}$

may be transformed by a three-dimensional (7-parameter) similarity transformation to a local system indicated by LS. According to Eq. (8.33), the transformation formula reads

$$\mathbf{X}_{LS} = \mathbf{c} + \mu \mathbf{R} \mathbf{X}_{GNSS}, \tag{8.71}$$

where the meaning of the individual quantities is the following:

$\mathbf{X}_{LS}$ ... position vector in the local system,
$\mathbf{X}_{GNSS}$ ... position vector in the geocentric system,
$\mathbf{c}$ ... shift vector,
$\mathbf{R}$ ... rotation matrix,
μ ... scale factor.

Forming the difference of two position vectors, i.e., yielding the baseline $\mathbf{X}_{ij}$, the shift vector $\mathbf{c}$ is eliminated. Using (8.71), there results

$$\mathbf{X}_{ij(LS)} = \mu \mathbf{R} \mathbf{X}_{ij(GNSS)} \tag{8.72}$$

for the baseline. Similar to (8.41), the linearized form is

$$\mathbf{X}_{ij(LS)} = \mathbf{X}_{ij(GNSS)} + \mathbf{A}_{ij} \Delta \mathbf{p}, \tag{8.73}$$

where now the vector $\Delta \mathbf{p}$ and the design matrix $\mathbf{A}_{ij}$ are given by

$$\Delta \mathbf{p} = [\Delta \mu \quad \Delta \alpha_1 \quad \Delta \alpha_2 \quad \Delta \alpha_3]^T,$$

$$\mathbf{A}_{ij} = \begin{bmatrix} X_{ij} & 0 & -Z_{ij} & Y_{ij} \\ Y_{ij} & Z_{ij} & 0 & -X_{ij} \\ Z_{ij} & -Y_{ij} & X_{ij} & 0 \end{bmatrix}_{GNSS}. \tag{8.74}$$

Note that the rotations $\Delta \alpha_i$ refer to the axes of the system adherent to GNSS. If they should refer to the local system, then the signs of the rotations must be changed, i.e., the signs of the elements of the last three columns of matrix $\mathbf{A}_{ij}$ must be reversed.

The vector $\mathbf{X}_{ij(LS)}$ on the left side of (8.73) contains the points $\mathbf{X}_{i(LS)}$ and $\mathbf{X}_{j(LS)}$ in the local system. If these points are unknown, then they are replaced by known approximate values and unknown increments

$$\begin{aligned} \mathbf{X}_{i(LS)} &= \mathbf{X}_{i0(LS)} + \Delta \mathbf{X}_{i(LS)}, \\ \mathbf{X}_{j(LS)} &= \mathbf{X}_{j0(LS)} + \Delta \mathbf{X}_{j(LS)}, \end{aligned} \tag{8.75}$$

where the coefficients of these unknown increments (+1 or −1) together with matrix $\mathbf{A}_{ij}$ form the design matrix. The vector $\mathbf{X}_{ij(GNSS)}$ in (8.73) is regarded as measurement quantity. Thus, finally,

$$\mathbf{X}_{ij(GNSS)} = \Delta \mathbf{X}_{j(LS)} - \Delta \mathbf{X}_{i(LS)} - \mathbf{A}_{ij} \Delta \mathbf{p} + \mathbf{X}_{j0(LS)} - \mathbf{X}_{i0(LS)} \tag{8.76}$$

8.4 Combining GNSS and terrestrial data

is the linearized observation equation.

In principle, any type of geodetic measurement can be employed if the integrated geodesy adjustment model is used. The basic concept is that any geodetic measurement can be expressed as a function of one or more position vectors $\mathbf{X}$ and of the gravity field W of the earth. The usually nonlinear function must be linearized, where the gravity field W is split into the normal potential U of an ellipsoid and the disturbing potential T, thus, $W = U + T$. Applying a minimum principle leads to the collocation formulas (Moritz 1980: Chap. 11).

Numerous examples integrating GNSS and other data can be found in technical publications. A more or less arbitrary and very brief selection is given here. The combination of GNSS and gravity data is investigated by Hein (1990a). Several contributions are cited in Delikaraoglou and Lahaye (1990) and numerical examples are given, e.g., in Hofmann-Wellenhof et al. (1994: Sect. 3.6.4), Daxinger and Stirling (1995). Schaefer et al. (2000) combine GPS with data of a compass and an inclinometer to a multisensor system. Similarly, Petovello et al. (2001) tackle the challenging task to combine GPS with inertial navigation data. Another sensor combination (GPS, accelerometers, and fiber sensors) for structural deformation monitoring is described in Li (2004). Combining GPS and photogrammetric measurements in a common adjustment is the objective of Ellum and El-Sheimy (2005). Fotopoulos (2005) combines GPS with geoidal data. This list could be continued almost endlessly.

9 GPS

9.1 Introduction

9.1.1 Historical review

The Global Positioning System is the responsibility of the Joint Program Office (JPO), a component of the Space and Missile Center at El Segundo, California. In 1973, the JPO was directed by the US Department of Defense (DoD) to establish, develop, test, acquire, and deploy a spaceborne positioning system. The present navigation system with timing and ranging (NAVSTAR) Global Positioning System (GPS) is the result of this initial directive.

GPS was conceived as a ranging system from known positions of satellites in space to unknown positions on land, at sea, in air and space. Effectively, the satellite signal is continually marked with its (own) transmission time so that when received the signal transit period can be measured with a synchronized receiver. The original objectives of GPS were the instantaneous determination of position and velocity (i.e., navigation), and the precise coordination of time (i.e., time transfer). A detailed definition given by W. Wooden in 1985 reads:

"The NAVSTAR Global Positioning System (GPS) is an all-weather, space-based navigation system under development by the Department of Defense (DoD) to satisfy the requirements for the military forces to accurately determine their position, velocity, and time in a common reference system, anywhere on or near the earth on a continuous basis."

Since the DoD is the initiator of GPS, the primary goals were military ones. But the US Congress, with guidance from the President, directed the DoD to promote its civil use. This was greatly accelerated by the production of a "portable" codeless GPS receiver for geodetic surveying that could measure short baselines to millimeter accuracy and long baselines to one part per million (ppm). This instrument developed by C. Counselman and trade-named the Macrometer Interferometric Surveyor^{TM} was in commercial use at the time the military was still testing navigation receivers so that the first productive application of GPS was to establish high-accuracy geodetic networks.

9.1.2 Project phases

Early conceptual and developmental phases
From the technological point of view, GPS may be considered an "aged" system calling for a facelift. It sounds strange if the initial development, dating back to

1973, may be regarded as GPS history and has almost vanished behind the curtain of the past. Five years later, in 1978, the launch of developmental satellites began. Then it took eleven long years to gather experience and further develop mature satellites with well-proven payload elements. In 1989, the first operational satellite could be launched. Woven into these corner stones of development, important policy decisions were met, among them the development of GPS as a dual-use system, thus deviating from the originally purely military objectives. The dual-use foresees military applications for the US and allied forces and civilian applications for a worldwide use. More on the GPS policy history is given in Sect. 9.1.3.

Operational capabilities

There are two operational capabilities: initial operational capability (IOC) and full operational capability (FOC).

IOC was attained in July 1993, when 24 (Block I/II/IIA) GPS satellites were operating and were available for navigation. Officially, IOC was declared by the DoD on December 8, 1993.

FOC was achieved when 24 Block II/IIA satellites were operational in their assigned orbits and the constellation was tested for operational military performance. Even though 24 Block II and Block IIA satellites were available since March 1994, FOC was not declared before July 17, 1995.

GPS modernization

According to the previous operational capabilities, GPS is in full use since more than one decade. Therefore it is natural that on the one hand further technological advances occurred and on the other hand the demand for even better performance with respect to applicability and accuracy arose. The need for improvement was driven from both the military and civil interests and requests. Beyond these arguments also competition is an issue since systems of other countries like the European Galileo or the Chinese Beidou system showed clearer contours, features, and developmental time schedules.

On January 25, 1999, the GPS modernization program was announced officially, aiming at the objectives to satisfy the requirements mentioned. The modernization impacts the space and the control segment (Sects. 9.4.1, 9.4.2) and, specifically, the GPS signals (Sect. 9.5).

9.1.3 Management and operation

For civilian users, the real impact of the originally military GPS occurred in 1983, when the US President offered free civilian access after the incident of the Korean Airlines Flight 007.

9.1 Introduction

The first US national policy on GPS was released in 1996 under the title "US GPS Policy" and the Interagency GPS Executive Board (IGEB) has been established. Important decisions for civilian users were usually announced by Decision Directives of the White House by the President or the Vice President. Some examples are given below.

- In 1996, the acceptance and integration of GPS into peaceful civil, commercial and scientific applications worldwide were encouraged. Furthermore, the commitment of the US to discontinuing the use of selective availability (SA), see Sect. 9.3.3, by 2006 with an annual assessment was announced.
- On March 30, 1998, the US Vice President announced that a second civilian signal will be provided by the GPS. "This new civilian signal will mean significant improvements in navigation, positioning and timing services to millions of users worldwide – from backpackers and fishermen to farmers, airline pilots, and scientists", the Vice President said.
- On January 25, 1999, the US Vice President announced a new initiative that will modernize GPS and will add two new civil signals to future GPS satellites, significantly enhancing the service provided to civil, commercial, and scientific users worldwide.
- On May 1, 2000, the statement by the US President regarding the US decision to stop degrading GPS accuracy was released. A partial citation reads: "Today, I am pleased to announce that the United States will stop the intentional degradation of the Global Positioning System (GPS) signals available to the public beginning at midnight tonight. We call this degradation feature Selective Availability (SA). This will mean that civilian users of GPS will be able to pinpoint locations up to ten times more accurately than they do now. GPS is a dual-use, satellite-based system that provides accurate location and timing data to users worldwide. My March 1996 Presidential Decision Directive included in the goals for GPS to: 'encourage acceptance and integration of GPS into peaceful civil, commercial and scientific applications worldwide; and to encourage private sector investment in and use of US GPS technologies and services'. To meet these goals, I committed the US to discontinuing the use of SA by 2006 with an annual assessment of its continued use beginning this year." From the civilian point of view, sometimes the switching off of SA is considered the first step towards GPS modernization.

What is the structure behind these decisions of the US President and Vice President? The answer becomes clear when studying the US GPS policy superseding the release of 1996 by a new national policy authorized by the US President on December 8, 2004. This national policy "establishes guidance and implementation actions for space-based positioning, navigation, and timing programs, augmentations, and

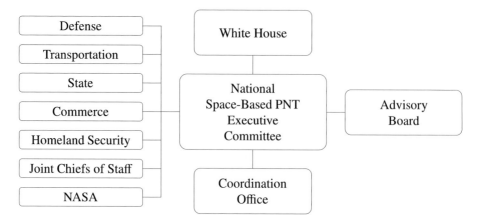

Fig. 9.1. Organizational structure of the US National Space-Based PNT Executive Committee

activities for US national and homeland security, civil, scientific, and commercial purposes".

Also by this Presidential Directive of December 2004, the National Space-Based Positioning, Navigation, and Timing (PNT) Executive Committee was established to advise and coordinate federal departments and agencies on matters concerning GPS and related systems.

On the Web site http://pnt.gov, the main features and tasks of this Executive Committee are explained (given below in shortened form) and the overall organizational structure is shown in Fig. 9.1.

This National Space-Based PNT Executive Committee is chaired jointly by the Deputy Secretaries of the Department of Defense (DoD) and the Department of Transportation (DoT). Its membership includes equivalent-level officials from the Departments of State, Commerce, and Homeland Security, the Joint Chiefs of Staff, and National Aeronautics and Space Administration (NASA). Components of the Executive Office of the President participate as observers to the Executive Committee, and the chairman of the Federal Communications Commission participates as a liaison.

A permanent Coordination Office located in Washington, DC, and activated on November 1, 2005, provides day-to-day staff support to the Executive Committee. It consists of an interagency staff headed by a director. The Coordination Office is a point of contact for inquiries regarding PNT policy.

The US Space-Based PNT Advisory Board will provide independent advice to the Executive Committee through NASA.

The National Space-Based PNT Executive Committee has replaced the IGEB, which has been active from 1996 to 2004. As given in Hothem (2006), the policy objectives are:

- Provide space-based civil PNT services free of direct user fees on a continuous, worldwide basis.
- Commitment to continue the modernization of GPS and its augmentation and, thus, improving global services.
- Ensure that civil requirements are met and civil services exceed (or are at least equivalent to) those of other international civil space-based PNT services.
- Improve resistance to interference for civil, commercial, homeland security, and scientific users.
- Open, free access to information needed to use civil GPS and its augmentations.
- Improve capabilities to deny hostile use of PNT without unduly disrupting civil and commercial access.
- Maintain GPS as a component of US critical infrastructure and investigate plan for backup capabilities and services.
- Ensure that other international PNT systems are interoperable (or at least compatible) with GPS and its augmentations.

To fully understand the last objective, the definitions of "interoperable" and "compatible" are necessary. As given in the Fact Sheet of the US Space-Based PNT Policy (released on December 15, 2004):

- "interoperable" refers to the ability of civil US and foreign space-based positioning, navigation, and timing services to be used together to provide better capabilities at the user level than would be achieved by relying solely on one service or signal;
- "compatible" refers to the ability of US and foreign space-based positioning, navigation, and timing services to be used separately or together without interfering with each individual service or signal, and without adversely affecting navigation warfare.

9.2 Reference systems

9.2.1 Coordinate system

Referring to coordinates, the GPS terrestrial reference system is the World Geodetic System 1984 (WGS-84). This geocentric system was originally realized by the coordinates of about 1 500 terrestrial sites which have been derived from Transit observations. Associated to this frame is a geocentric ellipsoid of revolution, originally defined by the four parameters: semimajor axis a, normalized second-degree zonal gravitational coefficient $\bar{C}_{2,0}$, truncated angular velocity of the earth ω_e, and earth's gravitational constant μ. This frame has been used for GPS since 1987. The gravitational coefficient $\bar{C}_{2,0}$ can be expressed by the flattening parameter f, which is defined by the semiaxes of the ellipsoid: $f = (a - b)/a$.

The comparison of the original WGS-84 and the international terrestrial reference frame (ITRF) revealed remarkable differences (Malys and Slater 1994):

1. The WGS-84 was established through Doppler observations from the Transit satellite system, while ITRF is based on SLR and VLBI observations. The accuracy of the Transit reference stations was estimated to be in the range of 1 to 2 meters, while the accuracy of the ITRF reference stations is at the centimeter level.

2. The numerical values for the original defining parameters differ from those in the ITRF. The only significant difference, however, was in the earth's gravitational constant $d\mu = \mu_{\text{WGS}} - \mu_{\text{ITRF}} = 0.582 \cdot 10^8 \text{ m}^3 \text{ s}^{-2}$, which resulted in measurable differences in the satellite orbits.

Based on this information, the former Defense Mapping Agency (DMA) has proposed to replace the μ-value in the WGS-84 by the standard International Earth Rotation Service (IERS) value and to refine the coordinates of the GPS tracking stations. The revised WGS-84, valid since 1994, has been given the designation WGS-84 (G730), where the number 730 denotes the GPS week number when DMA has implemented the refined system (Bock 1996).

In 1996, the National Imagery and Mapping Agency (NIMA), the successor of DMA, has implemented a revised version of the frame denoted as WGS-84 (G873). The frame is realized by monitor stations with refined coordinates. The associated ellipsoid is now defined by the four parameters listed in Table 9.1, which are slightly different from the respective ITRF values. The refined WGS-84 (G1150) frame was introduced in 2002. With respect to ITRF2005, the current WGS-84 frame shows unsignificant systematic differences in the order of 1 cm. Hence, both frames are virtually identical. For more details on the WGS-84 (G1150) frame, the reader is

Table 9.1. Parameters of the WGS-84 ellipsoid

Parameter and Value	Description
$a = 6\,378\,137.0$ m	Semimajor axis of the ellipsoid
$f = 1/298.257\,223\,563$	Flattening of the ellipsoid
$\omega_e = 7\,292\,115 \cdot 10^{-11}$ rad s^{-1}	Angular velocity of the earth
$\mu = 3\,986\,004.418 \cdot 10^8$ m^3 s^{-2}	Earth's gravitational constant

9.2.2 Time system

Referring to time, the system time of GPS is related to the atomic time system and is referenced to coordinated universal time (UTC) as maintained by the US Naval Observatory (USNO). Nominally the GPS time has a constant offset of 19 seconds with TAI, the international atomic time,

$$\text{TAI} = \text{GPS time} + 19.000^{\text{s}} \tag{9.1}$$

and was coincident with UTC at the GPS standard epoch January $6\overset{d}{.}0$, 1980. According to Eq. (2.22), TAI and UTC differ by an integer number n of seconds. In January 2007, the integer value was $n = 33$ and, thus, GPS time is exactly 14 seconds ahead of UTC.

Starting at the GPS standard epoch, the system time of GPS is counted in terms of GPS weeks and seconds within the current week. For the calculation of the GPS week use the relation

$$\text{WEEK} = \text{INT}[(\text{JD} - 2\,444\,244.5)/7], \tag{9.2}$$

where JD indicates the Julian Date and INT is an integer operator. This formula together with Eq. (2.25) can be used to verify the fact that the epoch J2000.0 (see Table 2.2) corresponds to Saturday in the 1042nd GPS week. Note, however, that at the begin of every 1024th week the week number in the navigation message is set to zero since only 10 bits are reserved for the week number. The first rollover occurred at midnight August 21–22, 1999.

In 2000, "The GPS Toolbox" was opened (Hilla and Jackson 2000). In this toolbox, the source codes for date algorithms are available under the Web site www.ngs.noaa.gov/gps-toolbox.

9.3 GPS services

For point positioning and timing, GPS provides two levels of service: the standard positioning service (SPS) with access for civilian users (Sect. 9.3.1) and the precise positioning service (PPS) with access for authorized users (Sect. 9.3.2).

The SPS may be controlled by the JPO by applying selective availability and anti-spoofing (Sect. 9.3.3) to deny the full system accuracy to nonmilitary users.

GPS information services provide GPS status information, orbital and other data to the civilian users. As described in Sect. 13.4.3, the user may benefit from different GPS information services.

To understand the subsequent sections, basic information on the GPS satellite signal is required. The detailed signal description is given in Sect. 9.5.

The key to the system's accuracy is the fact that all signal components are precisely controlled by atomic clocks. Typical GPS satellite clocks are rubidium or cesium clocks. The long-term frequency stability of these clocks reaches a few parts in 10^{-13} and 10^{-14} over one day. The future hydrogen masers will have an even better stability of 10^{-14} to 10^{-15} over one day. These highly accurate clocks, also denoted as frequency standards, being the heart of GPS satellites produce the fundamental frequency of 10.23 MHz. Coherently derived from this fundamental frequency are mainly (presently) two signals, the L1 and the L2 carrier waves generated by multiplying the fundamental frequency by 154 and 120, respectively, yielding

$$L1 = 1575.42\,\text{MHz},$$
$$L2 = 1227.60\,\text{MHz}.$$

These dual frequencies are essential for eliminating the major source of error, i.e., the ionospheric refraction (Sect. 5.3.2).

The pseudoranges that are derived from measured run times of the signal from each satellite to the receiver use two pseudorandom noise (PRN) codes that are modulated (superimposed) onto the two carriers.

The first code is the coarse/acquisition (C/A) code, which is available for civilian use. The C/A-code, designated as the standard positioning service (SPS), has an effective wavelength of approximately 300 m. The C/A-code is presently modulated upon L1 only and is purposely omitted from L2. This omission allows to deny full system accuracy to nonmilitary users.

The second code is the precision (P) code, which has been reserved for US military and other authorized users. The P-code, designated as the precise positioning service (PPS), has an effective wavelength of approximately 30 m. The P-code is modulated on both carriers L1 and L2. Unlimited access to the P-code was permitted until the system was declared fully operational. Today, the P-code is encrypted to the Y-code to make it available to authorized users only.

In addition to the PRN codes, a data message consisting of status information, satellite clock bias, and satellite ephemerides is modulated onto the carriers. It is worth noting that the present signal structure will be improved in the near future.

9.3.1 Standard positioning service

The SPS is a positioning and timing service. It uses the C/A-code and is provided on the L1 signal only; the L2 signal is not part of the SPS (Department of Defense et al. 2005). The SPS performance refers to the signal in space (SIS). Contributions

9.3 GPS services

Table 9.2. Standard positioning and timing service based on a 95% probability level and SIS only

Accuracy standard	Conditions and constraints
Global average positioning domain accuracy ≤ 13 m horizontal error ≤ 22 m vertical error	• based on a measurement interval of 24 hours *averaged over all points* • all-in-view satellites
Worst site positioning domain accuracy ≤ 36 m horizontal error ≤ 77 m vertical error	• based on a measurement interval of 24 hours *for any point* • all-in-view satellites
Time transfer accuracy ≤ 40 ns time transfer error	• based on a measurement interval of 24 hours *averaged over all points*

of ionosphere, troposphere, receiver, multipath, topography, or interference are not included.

The SPS is freely available to all kinds of users on a continuous and worldwide basis. Adapted from the Department of Defense (2001), the standard positioning and timing service based on a 95% probability level and SIS only is given in Table 9.2. Note the difference between the global average positioning domain accuracy where the measurements are averaged over all points of the 24-hour measurement interval and the individual site positioning domain accuracy to figure out the worst case. The "all-in-view satellites" condition in Table 9.2 implies a 5° mask angle and the removal of the worst two satellites from the constellation.

Note that these official values of the SPS depend on many other factors apart from the given conditions and constraints. Furthermore, the SPS performance is usually much better than the specification. Conley et al. (2006: p. 362) mention average values for a 20-site network of 7.1 m horizontal error and 11.4 m vertical error but stress the large number of possible GPS receiver configurations and integrations and the various environmental conditions.

It is interesting to ask for the deterioration of the errors if the probability level is increased. Referring again to Conley et al. (2006: p. 362), for 99.99% probability level the values for the horizontal and vertical errors "have generally been below 50 m with a few notable exceptions". Those troubles may be avoided by excluding in any case the statistically worst satellites, as it is done for the computation of the standardized SPS.

9.3.2 Precise positioning service

The PPS uses the P-code (Y-code, respectively) on the L1 and the L2 signal. The use of the PPS is restricted to US armed forces, US federal agencies, and some selected allied armed forces and governments. As discussed in the previous section, it is problematic to compare the performance purely by the horizontal and the vertical errors. A wider perspective should be considered in order not to misinterpret the results. Therefore, it is probably most reasonable to state that the accuracy available to PPS users is similar to that of SPS users (Seeber 2003: p. 230) if no intentional degradation of SPS like selective availability is activated. This statement gains attraction by avoiding confusion which might arise if different sources are considered, e.g., Kaplan (2006: p. 4) gives values in Table 9.3 which are not unique but found in many other sources which are less accurate than those in Table 9.2! The reason for this may be interpreted individually. One argument could be the removal of the worst two satellites from the constellation in the SPS case which is not reported for PPS.

On the other side, Kelly (2006) speaks of a near-equivalent position accuracy of 8–60 m (95%) for SPS versus 6–20 m for PPS; note, however, this does not coincide with the Department of Defense (2001) values, which are regarded standardized. A significant improvement occurs if dual-frequency data are used since the P-code is modulated onto L1 and L2: "Real-time 3D absolute positionial accuracies of better than 10 m are attainable through use of the PPS with dual-frequency receivers" (US Army Corps of Engineers 2003: p. 2-11).

As Kelly (2006) shows from the military perspective, the individual service is a necessity regarding the existing US legislative, doctrinal, and operational directives currently in force. Some arguments are repeated here, e.g.: The National Security Council and Office of Science and Technology Policy resolves that GPS will be responsible to the US National Command Authorities as a first priority and that the maintenance of a military PPS (and use by the US military and authorized users) will be enforced. Furthermore, there exist army dictates as of 2003 not to use SPS by infantry and other military persons. Other arguments in favor of the PPS are

Table 9.3. Precise positioning and timing service based on SIS only

Accuracy standard	Probability level
22 m horizontal error	2 DRMS (i.e., 98.2%)
27.7 m vertical error	95%
200 ns time transfer error	95%
0.2 m s^{-1} velocity error	95%

9.3 GPS services

precision, vulnerability, availability, and robustness. For each of these terms some more details might be revealed to show the strength of the PPS. For example, for the precision consider the tenfold smaller chip length of the P-code compared to the C/A-code leading to a better precision in range measurements. Referring to vulnerability, the spoofing of the encrypted P(Y)-code is much more difficult than that of the faster repeating unencrypted C/A-code. Since the P-code is available on L1 and L2, while the C/A-code is only on L1, the availability is clearly dominated by the PPS. Finally, the P-code signal is more robust and resistant against jamming than the C/A-code, thus, the robustness properties of PPS are better than those of SPS. This is specifically important for military purposes as employed in a military doctrine supporting the intentional removal of SPS within the area of interest by jamming sources so that the use of SPS is selectively denied. Selective denial (SD) is the proper term also explained subsequently.

9.3.3 Denial of accuracy and access

Two techniques are known for denying civilian users full use of the system. The first is selective availability (SA) and the second is anti-spoofing (A-S).

Selective availability

During the design of GPS, the accuracy expected from C/A-code pseudorange positioning was in the range of some 400 m. Field tests achieved the surprising level of navigation accuracy of 15–40 m for positioning and a fraction of a meter per second for velocity. The goal of SA was to deny this navigation accuracy to potential adversaries by dithering the satellite clock (δ-process) and manipulating the ephemerides (ε-process). The δ-process is achieved by dithering the fundamental frequency of the satellite clock. The satellite clock bias has a direct impact on the pseudorange which is derived from a comparison of the satellite clock and the receiver clock. Since the fundamental frequency is dithered, code and carrier pseudoranges are affected in the same way. After Breuer et al. (1993) and Görres (1996: Sect. 3.2.1), Fig. 9.2 shows the different behavior of satellite clocks with and without SA. With SA activated, there are variations of the pseudoranges with amplitudes of some 50 m and with periods of some minutes. When pseudoranges are differenced between two receivers, the dithering effect is eliminated.

The ε-process is the truncation of the orbital information in the transmitted navigation message so that the coordinates of the satellites cannot accurately be computed. The error in satellite position roughly translates to a similar position error of stand-alone receivers. For baselines, the relative satellite position errors are (approximately) equal to the relative baseline errors. In Fig. 9.3, after Breuer et al. (1993) and Görres (1996: Sect. 3.2.1), the behavior of the radial orbit error with and without SA is shown. In the case of SA, there are variations with ampli-

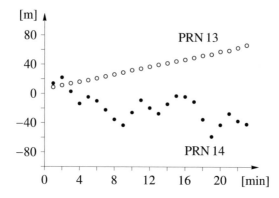

Fig. 9.2. Satellite clock behavior of PRN 13 (without SA) and of PRN 14 (with SA) on day 177 of 1991

tudes between 50 m and 150 m and with periods of some hours. The orbital errors cause pseudorange errors with similar characteristics. Thus, these errors are highly reduced when pseudoranges are differenced between two receivers.

SA was activated on March 25, 1990. According to the specifications of the DoD, the accuracy for stand-alone receivers was degraded to 100 m for horizontal position and to 156 m for height. These specifications also implied a velocity error of $0.3\,\mathrm{m\,s^{-1}}$ and an error in time of 340 ns. All numbers are given at the 95% probability level. At the 99.99% probability level, the predictable accuracy decreased to 300 m for horizontal position and to 500 m for height (Department of Defense 1995).

Due to the undermined military effectiveness of SA by applying differential

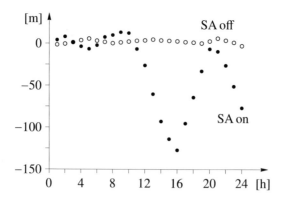

Fig. 9.3. Radial orbit error of PRN 21 on day 177 of 1992 with SA on and on day 184 of 1991 with SA off

9.3 GPS services

techniques, a joint recommendation of the US National Academy of Public Administration and a committee of the National Research Council has proposed that SA should immediately be turned to zero and fully deactivated after some years (CGSIC 1995). The official answer to this proposal was released on March 29, 1996, in form of the Presidential Decision Directive on GPS. This directive expressed the intention to discontinue the use of SA within a decade in a manner that allows adequate time and resources for the military forces to prepare fully for operations without SA. The full text of the public release statement on the Presidential Decision Directive is published, e.g., in the magazine GPS World 1996, 7(5): p. 50.

Somehow surprisingly, SA was turned off on May 2, 2000, at about 4:00 universal time (UT) after an announcement of the White House one day before. The "civilian benefits of discontinuing selective availability" are discussed in a fact sheet released by the US Department of Commerce on May 1, 2000. A prediction of the world after SA is given in Conley and Lavrakas (1999) and first experiences with SA off are discussed in Conley (2000), Jong (2000). One impressive result is presented in Fig. 9.4. Although the accuracy for stand-alone receivers is improved by a factor of ten, it must be kept in mind that despite turning off SA, military advantages are ensured by new developments. One of these developments is selective denial (SD), which will deny access to the GPS signal for unauthorized users in regions of interest by ground-based jammers.

Beyond this, the discussion to revisit the engagement of SA within modernized

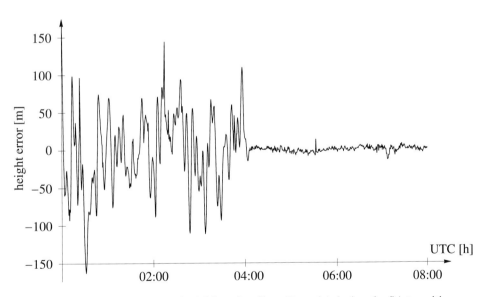

Fig. 9.4. Height variation in the IGS station Graz (Austria) during the SA transition on May 2, 2000

GPS has increased significantly. Kelly (2006): "I believe that SA does in fact have a high probability of both being retained and being reactivated due to a growing awareness of potential for misuse."

Anti-spoofing

The design of GPS includes the ability to essentially "turn off" the P-code or invoke an encrypted code as a means of denying access to the P-code to all but authorized users. The rationale for doing this is to keep adversaries from sending out false signals with the GPS signature to create confusion and cause users to misposition themselves.

Essentially, A-S is accomplished by an encrypting W-code. The resulting code is denoted as the Y-code. Thus, when A-S is active, the P-code on the L1 and the L2 carrier is replaced by the unknown Y-code. Note that A-S is either on or off. A variable influence of A-S (as was the case with SA) cannot occur.

For testing purposes, A-S was first turned on over the weekend of August 1, 1992, and later for several periods. It was expected that A-S would be switched on permanently when FOC had been attained; however, A-S was permanently implemented on January 31, 1994. In accordance with the DoD policy, no advance announcement of the implementation date was made.

The future signal structure will provide the C/A-code on the L1 and the L2C-code on the L2 carrier. Instead of the Y-code, a new military split-spectrum signal, denoted as M-code, will be introduced. Thus, anti-spoofing will need a redefinition.

9.4 GPS segments

9.4.1 Space segment

Constellation

The GPS satellites have nearly circular orbits with an altitude of about 20 200 km above the earth and a period of approximately 12 sidereal hours. The constellation and the number of satellites used have evolved from earlier plans for a 24-satellite and 3-orbital plane constellation, inclined 63° to the equator. Later, for budgetary reasons, the space segment was reduced to 18 satellites, with three satellites in each of six orbital planes. This scheme was eventually rejected, since it did not provide the desired 24-hour worldwide coverage. In about 1986, the number of satellites planned was increased to 21, again three each in six orbital planes, and three additional active spares. The spare satellites were designated to replace malfunctioning "active" satellites. The present nominal constellation consists of 24 operational satellites deployed in six evenly spaced planes (A to F) with an inclination of 55° and with four satellites per plane. Furthermore, several active spare satellites for replenishment are usually operational.

9.4 GPS segments

With the full constellation, the space segment provides global coverage with four to eight simultaneously observable satellites above 15° elevation at any time of day. If the elevation mask is reduced to 10°, occasionally up to 10 satellites will be visible; and if the elevation mask is further reduced to 5°, occasionally 12 satellites will be visible.

Satellite categories

There are various classes or types of GPS satellites. These are the Block I, Block II, Block IIA, Block IIR, Block IIR-M, Block IIF, and Block III satellites. Detailed information on launch dates, orbital position (designation letter for orbital plane plus number) and operational periods can be found on the Web site of the USNO under http://tych.usno.navy.mil/gps.html and selecting "current GPS constellation" or on the Web site www.navcen.uscg.gov/gps/status_and_outage_info.htm.

Eleven Block I satellites (weighing 845 kg) were launched in the period between 1978 to 1985 from Vandenberg air force base (AFB), California, with Atlas F launch vehicles. With the exception of one booster failure in 1981, all launches were successful. Today, none of the original Block I satellites is in operation. Considering the 4.5-year design life of these satellites, however, it is remarkable that some of the Block I satellites were operational for more than 10 years.

The Block II constellation is slightly different from the Block I constellation since the inclination of their orbital planes is 55° compared to the former 63° inclination. Apart from orbital inclination, there is an essential difference between Block I and Block II satellites related to US national security. Block I satellite signals were fully available to civilian users, while some Block II satellite signals are restricted.

The first Block II satellite, costing approximately USD 50 million and weighing more than 1 500 kg, was launched on February 14, 1989, from the Kennedy Space Center, Cape Canaveral AFB in Florida, using a Delta II rocket. The design life of the Block II satellites is 7.5 years. Individual satellites, however, remained operational more than 10 years.

The Block IIA satellites ("A" denotes advanced) are equipped with mutual communication capability. Some of them carry retroreflectors and can be tracked by laser ranging. The first Block IIA satellite was launched on November 26, 1990. Today, no distinction is made between Block II and Block IIA satellites.

The Block IIR satellites ("R" denotes replenishment or replacement) weigh more than 2 000 kg and the USD 42 million cost are about the same as for the Block II's. The first Block IIR satellite was successfully launched on July 23, 1997. These satellites have a design life of 10 years. They are equipped with improved facilities for communication and intersatellite tracking.

The first Block IIR-M ("M" denotes modernized) satellite was launched on September 25, 2005. The characteristic features of this satellite class are the new civil L2C-code on the L2 frequency, which allows for correction of the ionospheric effects, and the new military M-code on L1 and L2.

Space segment modernization

The limited design life of satellites along with the mean mission duration offers the option to modernize the space vehicles. This modernization process started with the launch of the first Block IIR-M satellite. Using the satellite's mean mission duration as an indicator, the IOC of satellites with the L2C- and the M-code might be expected in 2014 and the FOC might be achieved in 2015 (Prasad and Ruggieri 2005: p. 120).

The next generation after the Block IIR-M satellites will be the Block IIF ("F" denotes follow on) satellites. The highlights of this new generation will be the addition (compared to the previous generation) of a third civil signal designated as L5C- as well as two military M-code signals.

The Block IIF satellites will weigh more than 2 000 kg and will have a design life of 15 years. They will be equipped with improved onboard capabilities such as inertial navigation systems.

Prasad and Ruggieri (2005: p. 121) suppose the IOC of the Block IIF satellites in 2016 and FOC in 2019. These dates may likely change since the GPS Joint Program Office has announced that the launch of the first Block IIF satellite will be delayed until at least March 2008 due to technical problems encountered by the contractor (The Quarterly Newsletter of the Institute of Navigation 2006, 16(1): p. 14).

Presently, the DoD undertakes studies for the next generation of GPS satellites, called Block III satellites. A few more details are given in Sect. 9.6, subsection GPS III. These satellites are expected to carry GPS into 2030 and beyond.

9.4.2 Control segment

The operational control segment (OCS) consists of a master control station, monitor stations, and ground antennas. The main operational tasks of the OCS are the following: tracking of the satellites for the orbit and clock determination and prediction, time synchronization of the satellites, and upload of the navigation data message to the satellites. The OCS was also responsible for imposing SA on the broadcast signals. The OCS performs many nonoperational activities, such as procurement and launch activities, that will not be addressed here.

Master control station

The location of the master control station was first at Vandenberg AFB, California, but has been moved to the Consolidated Space Operations Center (CSOC) at Schriever AFB (formerly known as Falcon AFB), Colorado Springs, Colorado.

CSOC collects the tracking data from the monitor stations and calculates the satellite orbit and clock parameters using a Kalman estimator. These results are then passed to one of the ground antennas for eventual upload to the satellites. The satellite control and system operation is also the responsibility of the master control station.

Monitor stations

Before the GPS modernization, there were five monitor stations located at Hawaii, Colorado Springs, Ascension Island in the South Atlantic Ocean, Diego Garcia in the Indian Ocean, and Kwajalein in the North Pacific Ocean. Later, Cape Canaveral in Florida showed up. Each of these stations is equipped with a precise atomic time standard and receivers which continuously measure pseudoranges to all satellites in view. Pseudoranges are measured every 1.5 seconds and, using the ionospheric and meteorological data, they are smoothed to produce 15-minute interval data which are transmitted to the master control station.

Ground antennas

The four dedicated ground antennas are at the same site as the monitor stations at Ascension Island, Diego Garcia, Kwajalein, and the fourth is located at Cape Canaveral, Florida.

The ground antennas are equipped to transmit commands and data to the satellites and to receive telemetry and ranging data from the satellites. All ground antenna operations are under the control of the master control station.

The satellite ephemerides and clock information, calculated at the master control station and received at the ground antenna via communication links, are uploaded to each GPS satellite via S-band radio links. Formerly, uploading to each satellite was performed every eight hours; then the rate has been reduced to once (or twice) per day (Remondi 1991). Now the new upload strategy has returned to three uploads per day for each satellite. If a ground antenna becomes disabled, prestored navigation messages are available in each satellite to support a prediction span so that the positioning accuracy degrades quite gradually. The durations of positioning service of the satellites without contact from the OCS are given in Table 9.4.

Control segment modernization

The driving components of the control segment modernization are the reduction of operational costs and improving the system performance. Following Shaw et al. (2000) or Sandhoo et al. (2000) and Prasad and Ruggieri (2005: p. 122), the main improvements to be expected include the following:

Table 9.4. Positioning service without contact from the control segment

Block	Duration
I	3–4 days
II	14 days
IIA	180 days
IIR	>180 days

- Updating the dedicated GPS monitor stations and associated ground antennas with new equipment (receivers, computers).
- Replacement of the current master control station mainframe computers with a distributed architecture.
- Implementing the accuracy improvement initiative aiming at improvements of the broadcast navigation message and the GPS overall accuracy.
- Building a fully mission-capable alternate master control station at Vandenberg AFB in California.
- Addition of control and command capabilities and functionality for Block IIR-M and Block IIF satellites.
- Addition of direct civil code monitoring.

By the GPS modernization, six new monitor stations – located in Washington, DC (the US Naval Observatory), Argentina (Buenos Aires), Bahrain (Manama), United Kingdom (Hermitage), Ecuador (Quito), Australia (Adelaide) – were incorporated into the control segment in 2005 to provide better observability of the constellation, leading to improved data and a new data upload strategy with an increased update rate of the navigation data to the GPS satellites (three uploads per day for each satellite). Five more monitor stations located in Alaska (Fairbanks), Tahiti (Papeete), South Africa (Pretoria), South Korea (Osan), New Zealand (Wellington) were added in 2006.

Some more explanations are necessary to the item accuracy improvement initiative, also denoted as legacy accuracy improvement initiative. Three upgrades were achieved by the accuracy improvement initiative: improved accuracy of the Kalman filter state estimates (necessary to calculate accurate satellite orbits), improved accuracy of the broadcast ephemerides and clock parameters, and improved ability to observe the performance of the GPS satellites (Creel et al. 2006). The basic elements of the accuracy improvement initiative are a new software package installed on the OCS and the implementation of the National Geospatial-Intelligence Agency (NGA) monitor station network yielding improvements in monitoring the signal integrity and constellation performance and increasing the data used for satellite position and time estimation, resulting in more accurate predicted satellite orbital positions and clock data.

Note that the NGA exists since November 24, 2003. It was previously known as the National Imagery and Mapping Agency (NIMA), see Wiley et al. (2006).

9.5 Signal structure

The terminology (Table 9.5) follows the GPS JPO's approved lexicon of signal abbreviations. Table 9.5 has been adapted to account for the future civil signals on L1 and L5. Not listed are the S-band frequencies used for communication between control station and satellite, and all other frequencies. The military signals are encrypted to restrict their usage to authorized users. The civil signals, as emphasized by the US government, are provided free of charge to all users worldwide (Hudnut and Titus 2004). The GPS navigation signals are specified in the interface specification documents provided by ARINC Engineering Services (2005, 2006a, b).

When GPS became fully operational in 1995, the system started to emit naviga-

Table 9.5. GPS signal lexicon (Department of the Air Force 2001)

L1	Link 1, carrier frequency = 1 575.420 MHz
L2	Link 2, carrier frequency = 1 227.600 MHz
L3	Link 3, carrier frequency = 1 381.050 MHz
L4	Link 4, carrier frequency = 1 379.913 MHz
L5	Link 5, carrier frequency = 1 176.450 MHz
C/A	Coarse/acquisition code
P(Y)	Precision code; Y-code replaces the P-code in anti-spoofing mode
M	Military code
L1C	Civil code on L1
L2C	Civil code on L2; general reference to the code signal on L2, which consists of some combinations of C/A, L2CM, and L2CL
L2CM	Moderate-length code on L2C
L2CL	Long-length code on L2C
L5C	Civil code on L5; general reference to the civil code on L5, which consists of some combinations of L5I and L5Q
L5I	In-phase code on L5
L5Q	Quadraphase code on L5
NS	Nonstandard codes

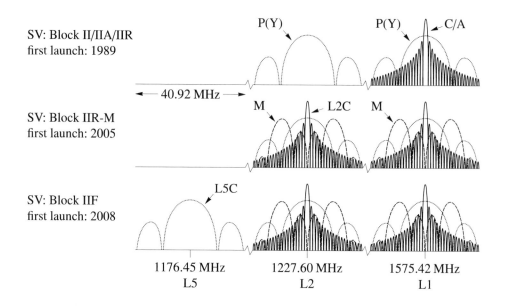

Fig. 9.5. Power spectral densities of GPS signals

nal, $c_{C/A}(t)$ is the coarse/acquisition code, and $d(t)$ corresponds to the navigation data message NAV (cf. Sect 9.5.3). The factor $a_i = \sqrt{2P_i}$ represents the power of the signal components and ω_i is the circular frequency of the respective carrier. The C/A-code is modulated only onto the L1 carrier frequency, whereas the P-code is modulated on both the L1 and the L2 frequency. Thereby, the C/A-code is placed in phase quadrature to the P-code and carries about 3 dB more power than the P-code. All three ranging codes are also modulated by the navigation message NAV. The C/A-code defines the standard positioning service (SPS) and the P-code the precise positioning service (PPS).

The modernization of GPS brought up a third carrier frequency and several new ranging codes on the different carrier links. Figure 9.5 summarizes all of them by displaying a schematic viewgraph of the power spectral density envelopes. The displayed frequency bands are different from the bandwidths allocated to GPS by ITU. Figure 9.5 additionally highlights the evolution of the signal availability due to the modernization program. Not shown is the additional civil ranging code L1C, since there exists only a baseline for the specification yet. The new PRN codes have been selected to be orthogonal to the existing ones and the modulation was chosen to spectrally separate the different signals for performance and interference reasons. All ranging codes are specified to be synchronized with the precision code.

Table 9.7 summarizes important parameters of all GPS PRN code sequences. None of the civil and military signals is optimal with respect to all performance

9.5 Signal structure

Table 9.7. GPS ranging signals

Link	PRN code	PRN code length [chip]	Code rate [Mcps]	Modulation type	Bandwidth [MHz]	Data rate [sps/bps]
L1	C/A	1 023	1.023	BPSK(1)	2.046	50/50
	P	~7 days	10.23	BPSK(10)	20.46	50/50
	M	[1]	5.115	BOCs(10,5)	30.69	[2]
	L1C$_D$	10 230	1.023	BOCs(1,1) [3]	4.092	100/50
	L1C$_P$	10 230 · 1 800	1.023	BOCs(1,1) [3]	4.092	—
L2	P	~7 days	10.23	BPSK(10)	20.46	50/50
	L2C	M: 10 230	1.023 [4]	BPSK(1)	2.046	50/25
		L: 767 250				—
	M	[1]	5.115	BOCs(10,5)	30.69	[2]
L5	L5I	10 230 · 10	10.23	BPSK(10)	20.46	100/50
	L5Q	10 230 · 20	10.23	BPSK(10)	20.46	—

[1] Encrypted [2] Not published
[3] According to ARINC Engineering Services (2006b); now changed to MBOC(6,1,1/11)
[4] Chip-by-chip time-multiplexed

parameters. Fontana et al. (2001) emphasize that the C/A-code on L1 will encounter the lowest ionospheric refraction error, the L5C-code has the highest power of the civil signals and is additionally allocated to the ARNS band, and the L2C-code has a better crosscorrelation performance than the C/A- or L5C-code. Consequently, the optimal choice for position determination will be a combined use of different signals.

Different switches in the satellite signal generators have been implemented which allow to choose different combinations of civil and military codes and data message on the different carrier frequencies. The signal definitions given in the following subsections have to be understood as baseline. Refer to, e.g., ARINC Engineering Services (2006a: p. 13) to learn more about different options.

The minimum received RF signal strengths are specified in the interface specification documents. The power levels are in the range of -153 dBW to -166 dBW, whereas the power levels for spaceborne receivers are specified to be even below -180 dBW (ARINC Engineering Services 2006b: p. 9). Langley (1998) states that the actual received signal levels may be larger than these values for a variety of reasons including satellite transmitter power output variations, satellite age, etc. The Block IIR-M and IIF satellites allow to shift the signal power from P(Y)- to M-code or vice versa. This flex-power concept has been implemented to decrease jamming susceptibility (Defense Science Board 2005). The spot beam signal option of Block III satellites will have a minimum received power of anticipated −138 dBW (Ward et al. 2006: p. 150). These high-power signals will increase antijamming protection in a local theater.

A signal with a power of about 20–30 W has to be emitted to meet the minimum received signal strength levels. The transmitted power is attenuated mainly due to the signal transmission path loss to about 10^{-16} W at the receiver antenna. The signal is said to be drown in noise, which has a noise density (4.59) of about $N_0 = -204$ dBW Hz^{-1}. The carrier-to-noise power density ratio C/N_0 therefore corresponds to about 44 dBHz. Thus the C/A signal power spectrum, considering a bandwidth of 2.046 MHz, is about 19 dB below the noise power level.

The new satellite signals L2C, L5C, and L1C come in pairs of a data and a pilot channel. The power distribution corresponds to 50% / 50% for the pairs of L2C and L5C, whereas the baseline for L1C reveals a power level of 75% for the pilot and 25% for the data channel.

To protect the user from any malfunction of the SV, the different ranging codes are replaceable by so-called nonstandard codes. The specification of these nonstandard codes is not published in order to prevent their usage.

C/A-code

The coarse/acquisition code is a nonclassified PRN code for civil use. The C/A-code is 1023 chips long with a code frequency of 1.023 ($= f_0/10$) megachips per second (Mcps). Therefore, the code duration is 1 ms and the chip length corresponds to about 297 m. The relatively short code duration allows for fast signal acquisition, but considering the maximum crosscorrelation level of −24 dB of two C/A-codes, the code is more susceptible to interference. The C/A ranging code together with the navigation message NAV is BPSK(1) modulated onto the L1 carrier frequency.

The C/A-code is a Gold code (cf. Sect. 4.2.3) generated by two 10-bit linear feedback shift registers (LFSR). The characteristic polynomials are

$$G_1 = 1 + x^3 + x^{10}, \tag{9.5}$$

$$G_2 = 1 + x^2 + x^3 + x^6 + x^8 + x^9 + x^{10}, \tag{9.6}$$

and their initial state corresponds to 1 in each register. The two code sequences are XOR-added (Fig. 9.6).

The G_2 sequence is delayed by an integer number of chips τ to produce unique code sequences. One characteristic feature of maximal length LFSR sequences is that two time offset versions of the same code XOR-added result in another version of the same code shifted by an integer number of chips (Holmes 1982: p. 311). In this way it is possible to represent a delayed G_2 register output as the output

9.5 Signal structure

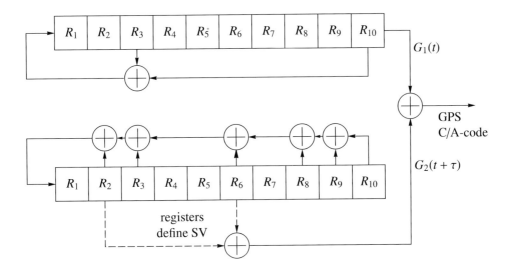

Fig. 9.6. C/A-code generation (Gold code)

of two tapped registers, as illustrated in Fig. 9.6. Hence, 45 unique C/A-codes are defined with the initialization vector having 1 in each register. An even greater family of Gold codes is defined by varying the initialization vector, as listed in ARINC Engineering Services (2006a: pp. 56c–56h).

Figure 9.6 illustrates the code generation of SV01 (= PRN01), where registers 2 and 6 of pattern G_2 have been chosen to generate the G_2-code sequence. The code delay of G_2 corresponds to 5 and the first 10-chip output of this Gold code reads in binary notation 1 100 100 000, which is equivalent to 1440 in octal notation. All C/A-codes and the respective PRN numbers are specified in detail in ARINC Engineering Services (2006a).

P(Y)-code

The P-code is generated by four 12-bit LFSR, which are regularly short cycled to their initial state. A 12-bit LFSR would generate a code sequence of length 4 095; however, the output is short cycled (i.e., truncated) at 4 092 and 4 093, respectively. In this way, two truncated sequences, $X1_A$ and $X1_B$, are used to generate an again truncated 15 345 000-chip-long $X1$ code that repeats every 1.5 seconds at a code frequency of 10.23 Mcps. The output of two other LFSR, $X2_A$ and $X2_B$, generates an again short cycled 15 345 037-chip-long $X2$ code. For details refer to the interface specification document (ARINC Engineering Services 2006a).

The P-code is generated by a XOR-addition of the codes X1 and X2. Thus, the full length corresponds to $15\,345\,000 \cdot 15\,345\,037 = 2.3547 \cdot 10^{14}$ chips. Considering the chipping rate of 10.23 Mcps, this sequence is 266.41 days or 38.058 weeks long. The P-code is partitioned into 37 unique code segments, each of 1 week ($=T$) length

$$P_k(t) = X1(t)\,X2(t + kT), \qquad 0 \leq k \leq 36. \tag{9.7}$$

These 37 code epochs are short cycled at the end of the week at midnight from Saturday to Sunday. ARINC Engineering Services (2006a: p. 6) assigns 32 of these codes for the use by SV and 5 are reserved for other purposes. 173 additional P-code PRN sequences are generated by circularly time shifting the 37 original codes by an amount of 1 to 5 days.

The P-code is a nonclassified code which is publicly available. The encrypted W-code is used to encrypt the P-code to the so-called Y-code, more commonly denoted as P(Y)-code. The clock rate of the W-code is lower than 10.23 Mcps, the clock rate of the Y-code is identical to the one of the P-code. This encryption procedure is referred to as anti-spoofing (A-S).

The P(Y)-code is very long, therefore difficult to acquire if no a priori information is available. Consequently, the C/A-code is used by military receivers for a coarse acquisition and then the receiver using the hand-over word in the data message locks onto the P(Y)-code for higher performances. Direct acquisition of the P(Y) requires accurate clock estimation, a position estimate, and ephemerides data of the satellites.

The P(Y)-code including the NAV message is BPSK(10) modulated onto L1 and L2. The modulation of one encrypted code sequence onto two carrier frequencies allows to apply codeless techniques for carrier phase and even code measurements (cf. Fig. 4.30).

L2C-code

The new civil signal on carrier L2 has been designed to meet in particular commercial needs. It is composed of the L2CM-code and the 75 times longer L2CL-code. The L2CM-code counts 10 230 chips and has a length of 20 milliseconds. 767 250 chips define the 1.5-second-long L2CL-code. The L2CL- and L2CM-code are two balanced codes with the same number of ones and zeros. Their chipping rate corresponds to 511.5 kcps. The L2CM-code is modulated by the CNAV navigation message (25 bps), whereas the L2CL-code is used as a pilot channel. The two ranging codes are finally chip-by-chip time-multiplexed to the 1.023 Mcps L2C-code which is finally BPSK(1) modulated onto the L2 carrier frequency.

According to Fontana et al. (2001) the L2C ranging code shows crosscorrelation performance equal to −45 dB.

The L2C-code is generated using a 27-bit LFSR as defined in ARINC Engineering Services (2006a: p. 37). The L2CL- and L2CM-codes are short cycled after 10 230 and 767 250 chips, respectively. Both are synchronized to the X1 epoch of the P-code.

9.5 Signal structure

The full benefits of L2C, as Dixon (2005a) states, will not be available until the majority of the current satellite constellation has been replaced with L2C capable satellites. A constellation of 24 L2C emitting satellites is expected for about 2012 (Shaw 2005).

M-code

Transmission of the military M-code started with the launch of a Block IIR-M satellite. The main characteristics of this military code are a higher resistance against jamming, increased navigation performance, higher security based on new cryptography algorithms, and the possibility of higher transmission power, denoted as flex-power. The spectral separation of the M-code and appropriate code selection guarantee orthogonality to the other civil and military code sequences. The orthogonality furthermore allows to deny civil signals without losing performance in the M-code sequence. The advantages of BOC spectral spreading and spectral separation are paid off by ambiguous discriminator functions. One major advantage beside the improved cryptography compared to the P(Y)-code is the possibility of direct acquisition. The M-code is modulated onto L1 and L2 carrier frequency using BOCs(10,5) modulation scheme. A new military navigation message MNAV has been specified to be modulated onto the M-code.

L5C-code

Starting with the launch of a Block IIF satellite presumably in 2008, GPS will for the first time emit navigation signals on a third carrier frequency. The L5C civil signal has been designed to especially meet the requirements of safety-of-life applications. The codes L5I and L5Q are in phase quadrature and QPSK(10) modulated onto the carrier frequency.

The two code sequences are generated using two 13-bit LFSR, whereas the initialization vector is unique for every SV. The codes, one of them short cycled at 8190 chips, are XOR-added and short cycled after 10 230 chips. The code rate has been specified to 10.23 Mcps. L5I is modulated with a navigation message, whereas L5Q is used as a pilot channel.

The codes L5I and L5Q are additionally modulated with synchronization sequences (ARINC Engineering Services 2005: p. 10). These low-frequency secondary codes have a length of 10 chips (i.e., 1 111 001 010) and 20 chips (i.e., 00 000 100 110 101 001 110), respectively, resulting in a composite code length of 10 ms (102 300 chips) and 20 ms (204 600 chips). The synchronization sequences are Neuman-Hoffman codes clocked at 1 kcps. The primary purpose of these codes is to reduce the narrowband interference effect. Furthermore, it reduces the cross-correlation while providing a better and more robust symbol/bit synchronization (Dierendonck et al. 2000).

The L5I signal, thus, is composed of the ranging code, the synchronization sequence, and the navigation data. The L5C ranging codes are 10 times longer than the C/A-code, consequently these codes have a better autocorrelation and cross-correlation property. In combination with a higher power the L5C ranging codes provide better interference resistance. The L5 carrier is coprimarily allocated to the ARNS band, and therefore the signals are especially useful for safety-of-life services. Additionally advantageous is the higher power level of the L5C signal compared to the others. Full operational capability with 24 Block IIF satellites will presumable not be met before 2015 (Shaw 2005).

L1C-code

The civil signals L2C- and L5C- and the military M-code represent the next steps in the GPS signal modernization program. Nevertheless, the plans to build up GPS III, the next evolution of GPS, have already started, although the first GPS III satellites are expected to be launched not before 2013 (Shaw 2005). In its baseline the US include a modernized civil signal on carrier L1. The L1C signal will be the fourth civil signal, which will not replace the C/A-code, for backward compatibility reasons, but be added to it. The L1C signal is composed of an $L1C_D$ data channel and an $L1C_P$ pilot channel. The latter is defined by a 10 230-chip-long primary code and a 1 800-chip-long secondary code.

In an agreement between the US government and the European Commission (United States of America and European Community 2004), the US and EU agreed to implement signals of a common modulation on the carrier frequency L1/E1. In this way the combined use of both systems shall be facilitated. In the agreement, furthermore, it has been decided to use the BOCs(1,1) as baseline, but at the same time to analyze further modulation options. In March 2006, a GPS-Galileo Working Group on Radio Frequency Compatibility and Interoperability recommended in an official statement the multiplexed BOC modulation MBOC(6,1,1/11), which is defined by Eq. (4.57). The multiplexing of the two orthogonal modulations BOCs(1,1) and BOCs(6,1) with sine phasing thereby adds more power to the higher frequencies (cf. Fig. 4.20) to improve tracking performance (Hein et al. 2006b). Meanwhile the MBOC has been chosen as the new baseline.

The L1C ranging codes are derived from a unique length-10223 Legendre sequence with a common 7-bit expansion Weil code sequence inserted at a PRN signal number-dependent point. Further details are given in a draft version of the interface specification document ARINC Engineering Services (2006b). The Legendre sequence and Weil codes are discussed in, e.g., Hein et al. (2006a).

9.5 Signal structure

9.5.3 Navigation messages

The navigation message essentially contains information about the satellite orbit, the satellite health status, various correction data, status messages, and other data messages. Furthermore the time offsets between GPS time and other GNSS time systems, e.g., GLONASS, Galileo, QZSS, are or will be emitted. The data rate of the navigation message is slow compared to the spreading code chipping rate. The navigation message, e.g., modulated on the L1 C/A-code has a data bit length of 20 ms, therefore every data bit contains 20 code epochs of C/A-code. Due to the low S/N ratios, low data rates are necessary to guarantee a low bit error rate (BER). The navigation message is updated and uploaded from the control stations to the SV in regular intervals.

In nominal operation, new ephemerides data will be transmitted to the user at least every two hours, whereas the data is valid for three or four hours depending on the navigation message. The broadcast ephemerides for Block I satellites with cesium clocks were accurate to about 5 m (assuming three uploads per day). For the Block II satellites, the accuracy is in the order of about 1 m. If no new data sets of ephemerides are uploaded to the satellites from the control station, the transmission interval (from the satellite to the user) and the interval of validity are steadily increasing while losing accuracy at the same time.

The almanac parameters are updated at least once every 6 days. Multiple data sets stored in the satellite memory guarantee an almanac transmission even without control station contact. The accuracy of almanac parameters, however, degrades over time.

NAV message

The original navigation message NAV is composed of a masterframe with 37 500 bits. Considering a data rate of 50 bps, the message transmission takes 12.5 minutes. The masterframe subdivides into 25 frames, each with 1500 bits with a message length of 30 seconds. One frame consists of five subframes, whereas one subframe takes 6 seconds and contains 10 words with 30 bits each. The transmission time needed for a word is therefore 0.6 seconds. One data chip, furthermore, consists of 20 460 C/A-code chips, 204 600 P-code chips, or 31 508 400 cycles of the L1 carrier.

Each subframe starts with the telemetry word (TLM) containing an 8-bit synchronization pattern (10 001 011) and some diagnostic messages. The synchronization preamble is not a unique sequence in the data message, what increases the probability of false detection. The second word in each subframe is the hand-over word (HOW). Apart from a subframe identification and some flags, this word contains the time of week (TOW) count of the leading edge start time of the next subframe. The TOW count, also called Z-count, is a multiple number of 1.5-second intervals indicating the beginning of the current GPS week which starts at midnight from Saturday to Sunday. In this way the navigation message is synchronized to the X1 interval of the P-code. The navigation message structure and content is common across all satellites.

The first subframe contains the GPS week number, a prediction of the user range accuracy, indicators of the satellite health and of the age of the data, an estimation of the signal group delay, and three coefficients for a quadratic polynomial to model the satellite clock correction. The second and third subframe transmit the broadcast ephemerides of the satellite as listed in Table 3.7.

The content of subframes 1 through 3 is repeated in every frame to provide critical satellite-specific data with high repetition rate. The content of the fourth and the fifth subframe is changed in every frame and has a repetition rate of 25 frames. The fourth subframe contains information about the ionosphere, UTC data, various flags, and the almanac data for satellites beyond the nominal 24 constellation. The rest is reserved for military use. The content of the fifth subframe is dedicated to the almanac data and the health status of the first 24 satellites in orbit. The content of the fourth and fifth subframe is broadcast by every satellite. Therefore, by tracking only one satellite, the almanac data of all the other satellites in orbit are obtained. Spilker (1996c: p. 139) states that missing almanac data slots are filled by alternating zeros and ones to improve the synchronization within the receiver.

A Hamming parity check algorithm for single-error correction and double-error detection capabilities is applied to decrease the bit failure rate. For more information concerning the data contained in the navigation message, the reader is referred to Sect. 3.3 on orbit dissemination. A detailed description of the data format is given in ARINC Engineering Services (2006a).

CNAV message

The navigation message NAV is common across all SV, fixed in structure and length. The new civil and military data messages, e.g., CNAV and MNAV, are designed to carry a new, modernized data format, which replaces the strategy of frames and subframes by a flexible data message. Consequently, the data message is composed of a header including a message type identifier, the data field, and a cyclic redundancy checkword. The CNAV message is half-rate forward error correction (FEC) encoded using a 7-bit convolutional encoder to decrease the BER.

The CNAV data set, as ARINC Engineering Services (2006a: p. 139) states, is a higher precision representation and nominally contains more accurate data than the NAV message. It is not advisable to mix the data from the different navigation messages.

MNAV message

The MNAV message structure is similar to the one of CNAV. The frame/subframe structure is changed to a data message structure. In this way it reduces inefficiencies of the NAV format while increasing the flexibility in configuration and content. Barker et al. (2000) further mention that this allows to adapt the message content with evolving military applications. MNAV, furthermore, improves the data security and integrity of the system. The military message is designed to provide the flexibility of satellite and frequency diversity as described in Sect. 4.3.4; thus, the message content is alternated depending on the satellite and the carrier frequency (i.e., L1 or L2).

Data message on L5I

The navigation data on L5I, as defined in ARINC Engineering Services (2005: Appendix II), contains the same data as NAV and CNAV but in different format. Every message consists of 300 bits composed of an 8-bit preamble, a 6-bit message type identifier (i.e., 64 different messages), the data field, and a 24-bit cyclic redundancy check. Due to the half-rate FEC encoding, one message is broadcast within 6 seconds.

CNAV-2 message

The data message modulated onto the $L1C_D$ signal is subdivided into frames, which themselves are subdivided into three subframes. The first subframe contains a time of interval identifier, whereas the second subframe carries clock and ephemerides data. The content of the third subframe is arbitrarily changed from frame to frame. Therefore, a page number is included into subframe three for unique identification. The FEC-encoded symbols will be interleaved using a block interleaver prior to being XOR-added to the L1C-code. For further details refer to ARINC Engineering Services (2006b).

9.6 Outlook

9.6.1 Modernization

Upon completion, the modernization will yield so many advantages! For civil users, the new signals provide apart from the mentioned better robustness against interference an improved compensation for ionospheric delays by the option to set up dual-frequency ionosphere-free data combinations (Sect. 5.3.2) and more precision by additional wide-lane signal combinations which may be extended to triple-frequency laning and three-carrier ambiguity resolution (Sect. 7.2.2). The interoperability between different constellations (GLONASS, Galileo, etc.) increases the capabilities of those users dealing with reduced satellite visibility which typically occurs in, e.g., urban canyons and dense forests.

For civil and military applications, all key performance elements like accuracy, availability, integrity, and reliability will be improved.

9.6.2 GPS III

The GPS modernization described above does not require a changing of the GPS architecture. Since military and civil needs and requirements will continue to grow, the next-generation program, known as GPS III, whose origin lies in an approval by the US Congress in 2000, will further enhance the performance of space-based navigation and set new standards for positioning and timing services. Another civil signal, denoted as L1C to indicate the code signal on L1, will be provided. Antijamming capabilities, improved system security and accuracy, and reliability will be the milestones of GPS III. In addition, integrity (one of the European key arguments to claim for Galileo) will be a remarkable new property compared to the existing system. This program will need a substantial redesign of the system architecture which might lead to a GLONASS- and Galileo-like three-plane constellation with some 27 to 33 satellites.

Among the military requirements, one will be a higher power of the M-code signal to improve the resistance against interference.

The schedule for IOC and FOC of GPS III is difficult to predict since it depends on so many variable factors. Anyway, the launch of GPS III satellites is expected for 2013 (The Quarterly Newsletter of The Institute of Navigation 2006, 16(1): p. 15). Prasad and Ruggieri (2005: p. 122) expect IOC and FOC in 2021 and 2023, respectively.

10 GLONASS

10.1 Introduction

10.1.1 Historical review

The abbreviation GLONASS derives from the Russian "Global'naya Navigatsionnaya Sputnikovaya Sistema", translated to its English equivalent, this means Global Navigation Satellite System. In the mid 1970s, the former Union of Soviet Socialist Republics (USSR) initiated the development of GLONASS based on the experiences with the Doppler satellite system Tsikada. Following Polischuk et al. (2002), the Academician M.F. Reshetnev's State Unitary Enterprise of Applied Mechanics has been the main contractor being responsible for the general development and implementation of the system. Moreover, the development and manufacturing of the satellites and their launch facilities and the corresponding control system belong to the tasks of this enterprise.

Subcontractors are the Russian Scientific-Research Institute of Space Industry and the Russian Institute of Radionavigation and Time. These institutes are responsible for monitoring and control, but also for a proper development of receivers and clocks.

As defined in the GLONASS interface control document released by the Co-ordination Scientific Information Center (2002), the purpose of GLONASS is to provide an "unlimited number of air, marine, and any other type of users with all-weather three-dimensional positioning, velocity measuring and timing anywhere in the world or near-earth space". On a continuous basis, meaning at any time, should be added.

Operated by the Russian military forces, GLONASS is a military system. This was the reason that almost no detailed information was publicly released. Later, this information deficit changed. In May 1988, at a meeting of the Special Committee on Future Air Navigation Systems of the International Civil Aviation Organization (ICAO), a paper with technical details of GLONASS was presented and the USSR offered the world community free use of the GLONASS navigation signals (Feairheller and Clark 2006: p. 596, Bauer 2003: p. 243). This was only the first step towards opening the system to other than Russian military users. In March 1995, the government of the Russian Federation released the Decree No. 237 entitled "On executing works in use of the GLONASS global navigation satellite system for the sake of civil users", where the Ministry of Defense of the Russian Federation, the Russian Federal Space Agency, and the Ministry of Transport of the Russian Federation "are to provide deploying of the GLONASS global navigation satellite system and the beginning of its operation with its full complement in 1995 in order to service national civil and military users and foreign civil users according to the existing commitments".

10.1.2 Project phases

Early conceptual and developmental phases

On October 12, 1982, the first GLONASS satellite together with two test satellites was launched but none of the three satellites became operational (Owen 1995). Usually, GLONASS satellites are launched three at a time. In January 1984, a test scenario of four satellites was successfully deployed. According to Polischuk et al. (2002), this test scenario belongs to the first phase extending from 1983 to the end of 1985 and covering primarily experimental tests and refinements of the system concept. During the second phase covering 1986–1993, the orbital constellation increased to 12 satellites, flight tests were completed, and the initial system operation began.

Operational capabilities

GLONASS was officially declared operational on September 24, 1993, by a decree of the President of the Russian Federation (www.spaceandtech.com/spacedata/constellations/glonass_consum.shtml) but the nominal 24-satellite constellation was completed for the first time on January 18, 1996. This date may be regarded as the GLONASS full operational capability (FOC) because the number of available satellites declined soon afterwards due to lack of funding between 1996 and 1998 (Polischuk et al. 2002). The continuously declining number of available satellites achieved its minimum in 2001 with six to eight satellites only (Feairheller and Clark 2006: p. 595).

GLONASS modernization

The GLONASS modernization program is an overall performance improvement initiative and impacts the space and the control segment (Sects. 10.4.1, 10.4.2) and, specifically, the GLONASS signals (Sect. 10.5). Referring to the satellites, the main issues are the improvement of the satellite clock stability and a better dynamical model for, e.g., the attitude determination of the satellite. Referring to the ground infrastructure, the number of monitor stations will be increased substantially. Referring to the GLONASS reference systems, the coordinate system will be refined. Also the GLONASS time keeping system will be improved by new system clocks with very high stability and the time synchronization system will be refined.

10.1 Introduction

10.1.3 Management and operation

Originally, GLONASS was developed as a (mainly) military system. However, when the FOC was achieved, GLONASS was available for civil users but restricted to one signal yet without any selective availability. Since that time, the Russian government has declared open access to and free of charge usage of the standard positioning service for civil users (United Nations 2004: p. 19).

The Russian Ministry of Transport put another important policy step in 1996 by offering the use of a GLONASS signal "for civil aviation for a period of at least 15 years without direct user fee". An enhanced offer was finally accepted by the ICAO on July 29, 1996.

The "dual-use" (military and civilian) was manifested by two basic decisions, the Presidential Directive of February 18, 1999, permitting officially the free use of the civil signal and making available the signal specification to users and industry by the interface control document and the Governmental Decision of March 29, 1999, defining the status of GLONASS as a dual-use system which is open for international cooperation.

On August 20, 2001, the Federal GLONASS Mission Oriented Program, covering the period from 2002 to 2011 with a guarantee for funding, was approved by the government. Note that in several publications this program is briefly denoted as Federal GLONASS Program.

Following Revnivykh et al. (2003), there are six State Customers of the Program: the Russian Federal Space Agency; the Ministry of Defense; the Russian Agency of the Control Systems; the Ministry of Transport; the Ministry of Industry, Science, and Technology; the Russian Mapping and Geodesy Agency.

Note that the Russian Federal Space Agency is a civil institution. Note also that there are other denotations for the same institution in use: "Roskosmos" or the formerly "Russian Aviation and Space Agency" (commonly known as "Rosaviakosmos").

To make it even more complex, the six State Customers mentioned above are not composed uniquely in different sources, e.g., in United Nations (2004: p. 20) there does not appear the Russian Agency of the Control Systems; instead, there is a splitting of the above "Ministry of Industry, Science, and Technology" into two ministries: "Ministry of Industry" and "Ministry of Science and Education".

The program covers the following directions and subprograms, where the respective responsible institutions are given within the brackets:

- the GLONASS system maintenance, modernization, deployment, operation and related research and development activities [Federal Space Agency and Ministry of Defense];
- navigation receiver and user equipment development for civil use, industry preparation for mass production of the GNSS equipment [Agency of the Con-

trol Systems according to Revnivykh et al. (2003) or Ministry of Industry according to United Nations (2004: p. 20)];

- GNSS equipment and technology implementation for transport (aviation, maritime, railroad, land transport, cars and trucks) [Ministry of Transport];

- GNSS technology applications for geodetic provision of the Russian territory and modernization of the geodetic system [Russian Mapping and Geodesy Agency];

- GNSS receiver and user equipment development for special use (military and special forces) [Ministry of Defense].

In 2002, the GLONASS Coordination Board was established including all program State Customers and chaired by a representative of the Federal Space Agency. To implement the permanent work for program coordination and to coordinate activities to implement the strategy defined by the GLONASS Coordination Board, the Executive Committee has been established including representatives of the customers, leading research institutes, and industry (United Nations 2004: p. 20). The current structure of the GLONASS Coordination Board is shown in Fig. 10.1.

In 2006, two new presidential initiatives were released also denoted as Federal GLONASS Program Update (Revnivykh 2006a). The first directive of January 18 covers three issues: (1) to ensure a minimum operational capability of GLONASS based on 18 satellites by the end of 2007; (2) to ensure the full operational capability by the end of 2009; (3) to ensure a performance of GLONASS comparable with GPS and Galileo by 2010.

The second directive of April 19 covers the mass market development to ensure the respective navigation equipment mass production.

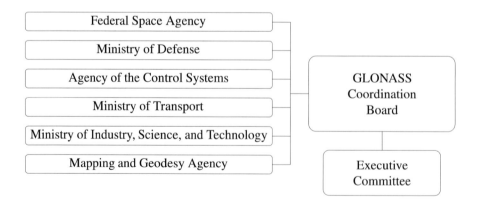

Fig. 10.1. Structure of the GLONASS Coordination Board

10.2 Reference systems

10.2.1 Coordinate system

Referring to coordinates, the GLONASS terrestrial reference system is denoted as PE-90 (sometimes also PZ-90). The first abbreviation derives from "Parameters of the Earth 1990" and the second from its respective translation into Russian "Parametry Zemli 1990". As specified in Roßbach (2001: p. 7), originally the Soviet Geodetic System 1985 (SGS-85) was employed, which was later changed to the Soviet Geodetic System 1990 (SGS-90) based on the same definition. For the sake of completeness, Roßbach (2001: p. 7) mentions that for a short period the acronym SGS was changed from Soviet Geodetic System to Special Geodetic System.

As given in Coordination Scientific Information Center (2002: Sect. 3.3.4), the reference frame as background of the PE-90 system is defined in the following way: the origin is located at the center of the earth; the Z-axis points to the conventional terrestrial pole as recommended by the International Earth Rotation Service (IERS); the X-axis results from the intersection line of the equatorial plane with the plane represented by the Greenwich meridian; the Y-axis completes a right-handed coordinate frame. Thus, the realization of the PE-90 reference frame is a geocentric system. Recall that the Cartesian coordinates related to the geocentric system are also denoted ECEF coordinates (Sect. 8.2). For the realization, 26 ground stations were established from observations to a geodetic satellite, Doppler measurements, laser ranging, satellite altimetry; also electronic and laser range measurements of GLONASS and Etalon satellites are included (Boucher and Altamimi 2001).

Associated to the PE-90 is a geocentric ellipsoid of revolution which is completely determined by four parameters. As defined in Coordination Scientific Information Center (2002: p. 16), the parameters of the PE-90 ellipsoid are given in Table 10.1.

It is interesting to compare the PE-90 with, e.g., the ITRF. This objective was one of the purposes of the International GLONASS Experiment (IGEX-98) held

Table 10.1. Parameters of the PE-90 ellipsoid

Parameter and value	Description
$a = 6\,378\,136$ m	Semimajor axis of the ellipsoid
$f = 1/298.257\,839\,303$	Flattening of the ellipsoid
$\omega_e = 7\,292\,115 \cdot 10^{-11}$ rad s^{-1}	Angular velocity of the earth
$\mu = 3\,986\,004.4 \cdot 10^8$ m^3 s^{-2}	Earth's gravitational constant

from October 1998 to April 1999. This experiment was cosponsored by the International Association of Geodesy (IAG), the International GNSS (formerly GPS) Service (IGS), the US Institute of Navigation (ION), and the International Earth Rotation Service (IERS). More than 60 stations (with ITRF coordinates) in over 25 countries participated in this first coordinated international campaign to collect and analyze GLONASS satellite data. Single- and (mostly) dual-frequency GPS/GLONASS receivers were used. In addition, laser tracking was performed in 30 satellite laser ranging (SLR) stations around the world.

The resulting parameters for a 7-parameter transformation (three components of a shift vector, three rotation angles, and one scale factor) from ITRF97 to PE-90 showed the most significant values for the Z-component of the shift vector in the amount of 0.9 m and the rotation angle about the Z-axis with −0.354 arcseconds (Altamimi and Boucher 1999). Misra et al. (1996) use a rotation about the Z-axis and a nonzero shift in the second component of the shift vector.

Comparing the WGS-84 with the PE-90, Boucher and Altamimi (2001) show that there exist many realizations of 7-parameter transformations. Some of them have almost only zero-values for the transformation parameters, e.g., Roßbach et al. (1996) reduce the transformation to a rotation about the Z-axis. Similarly, the Radio Technical Commission for Maritime (RTCM) Services recommends in its binary format RTCM-SC 104, version 2.3, which is an internationally accepted standard for the transmission of GPS and GLONASS correction data, a rotation angle about the Z-axis of −0.343 arcseconds. All of these realizations have in common a meter-level inconsistency.

Apart from the transformation parameters, the most impressive result of IGEX-98 was the determination of precise GLONASS orbits with improvements from several tens of meters to several decimeters (Weber and Fragner 1999).

10.2.2 Time system

Referring to time, the time system is maintained by the GLONASS central synchronizer using a set of hydrogen masers (Roßbach 2001: p. 31). The GLONASS time is closely related to the UTC but has a constant offset of three hours reflecting the difference between Moscow time and Greenwich time. This relation implies leap seconds for the GLONASS time. Apart from the constant offset, the difference between GLONASS time and UTC shall be within 1 millisecond (Coordination Scientific Information Center 2002: Sect. 3.3.3) arising from the keeping of the time scales by different clocks. The navigation message contains the information on this difference by the parameter τ_c. Thus, UTC can be computed from GLONASS time by

$$\text{UTC} = \text{GLONASS time} + \tau_c - 3^h. \tag{10.1}$$

The relation between UTC and TAI, the international atomic time, is given in Eq. (2.22) and reflects the variable number of leap seconds.

The correction of the GLONASS time because of leap seconds is carried out in agreement with the UTC corrections as performed by the Bureau International de l'Heure in Paris. Users may get the information on the intended correction in advance (at least three months ahead of the occurrence) via bulletins and notifications (Coordination Scientific Information Center 2002: Sect. 3.3.3).

10.3 GLONASS services

A few details on the signals are necessary here to understand the GLONASS services. More details on the the signal structure are given in Sect. 10.5.

Each GLONASS satellite continuously provides navigation signals: the standard-accuracy signal, i.e. the C/A-code (also denoted as S-code), and the high-accuracy signal, i.e., the P-code, in two subbands of the L-band, denoted as G1 and G2. Note that this denotation enables a better distinction from the GPS carriers L1 and L2. However, in the literature sometimes L1 and L2 are also used for GLONASS. The C/A-code is modulated onto G1 only, whereas the P-code is modulated onto G1 and G2. In the course of modernization of GLONASS, a standard-accuracy signal has been added on G2 in the GLONASS-M satellites.

The C/A-code of GLONASS has an effective wavelength of about 600 m and the respective value of the P-code is about 60 m.

10.3.1 Standard positioning service

Note that "standard positioning service" is not an official notation. Other terms used are "standard accuracy positioning service" (United Nations 2004: p. 19), "lower accuracy service" and "GLONASS civil accuracy" (Feairheller and Clark 2006: p. 611), or the general term "GLONASS performance" (Roßbach 2001: p. 29). Talking here now of standard positioning service implies that only the standard-accuracy signal – the C/A-code – is available.

It is important to note that there are no standardized values for the standard positioning service. On www.spacetoday.org/Satellites/GLONASS.html, a generic remark is found reading: "This GLONASS system provides accuracy that is better than GPS with SA on and worse than GPS with SA off." This translates to about 13 m $\leq$ horizontal error $\leq$ 100 m and 22 m $\leq$ vertical error $\leq$ 156 m (95% probability). Values given in the literature, are usually within these bounds.

Even if not standardized, Feairheller and Clark (2006: p. 611) give specification values but use slightly different accuracy measures (Sect. 7.3.6): 100 m (2DRMS, which corresponds to 98.2% probability) horizontal accuracy, 150 m (2σ, which corresponds to 95.4% probability) vertical accuracy, 15 cm s^{-1} velocity accuracy;

the authors also mention that in practice the GLONASS accuracy was much better than the specified values with the full 24-satellite constellation. Some tests demonstrated 26 m (2DRMS) horizontal accuracy, 45 m (2σ) vertical accuracy, 3–5 cm s^{-1} velocity accuracy.

Based on measurements from four satellites, Space Today Online gives on its Web site www.spacetoday.org/Satellites/GLONASS.html a horizontal accuracy of 180 feet (55 m) and a vertical accuracy of 230 feet (70 m).

The United Nations (2004: pp. 22, 23) give the following accuracy performances (95%): 28 m horizontal, 60 m vertical, and for the velocity 15 cm s^{-1} and for the time 1 microsecond.

Resulting from these different numbers, it is probably correct to use the generic remark at the beginning of this subsection as a guideline.

Selective availability (SA) is not foreseen for GLONASS satellites. This is confirmed in Coordination Scientific Information Center (2002: p. 7): "An intentional degradation of the standard-accuracy signal is not applied."

10.3.2 Precise positioning service

Again, this is not an official denotation. Talking here now of precise positioning service implies that the high-accuracy signal – the P-code – is available which is a military signal even if the P-code is on the one hand not encrypted, but it has on the other hand not been officially released. Furthermore, the Russian Ministry of Defense does not recommend its unauthorized use.

There is not much known about the precise positioning service. Essentially, the same military arguments as given in Sect. 9.3.2 for the United States in favor of the GPS precise positioning service will apply for Russia and the GLONASS precise positioning service but are not repeated here.

Anti-spoofing (A-S), which means to "turn off" the P-code or invoke an encrypted code as a means of denying access to the P-code to all but authorized users, might be an issue. So far no activation of A-S is reported. However, since the P-code may be changed without prior notice to unauthorized users, the option of activating A-S is available.

10.4 GLONASS segments

10.4.1 Space segment

Constellation

The GLONASS satellites have circular orbits with an altitude of about 19 100 km and a period of nominally 11 hours and 15 minutes and 44 seconds. The complete constellation consists of 24 satellites in three orbital planes, where 21 satellites

10.4 GLONASS segments

are considered active satellites and the three remaining ones are "active on-orbit spares" (Feairheller and Clark 2006: p. 597). The ascending nodes of the planes are separated by 120° with an inclination of the planes of 64.8° to the equator. In each plane, there are eight satellites equally spaced (which means that within the plane the argument-of-latitude displacement amounts to 45°). The argument of latitude between two planes is displaced by 15°. This constellation assures that at least five satellites are simultaneously visible on 99% of the sites on the earth (Habrich 1999).

Following Feairheller and Clark (2006: p. 598), a 21-satellite constellation provides a continuous and simultaneous visibility of at least four satellites over 97% of the earth's surface, whereas the 24-satellite constellation provides a continuous and simultaneous visibility of at least five satellites over more than 99% of the earth's surface. All GLONASS satellites to date have been launched by the Proton launch vehicles.

Satellite categories

There are various categories (other notations: series, classes, types) of GLONASS satellites. These are the GLONASS, the GLONASS-M, the GLONASS-K, and the GLONASS-KM satellites. The denotation within a category is refined by using blocks denoted by Roman numbers (e.g., the Blocks I and II for the GLONASS satellites) and variants denoted by small letters (e.g., GLONASS Block IIa). The main difference between the spacecraft blocks is the design lifetime. The Web site www.russianspaceweb.com/uragan.html contains detailed information on the launch dates. The current constellation is obtained under www.glonass-ianc.rsa.ru .

A detailed description of the payload is given in Feairheller and Clark (2006: pp. 600–601) which is given in abbreviated form here. The onboard navigation complex consists of an information logical complex unit, a set of three atomic clocks, a memory unit, a telemetry, tracking and control (or command) (TT&C) link receiver, and a navigation signal transmitter. The onboard navigation complex may operate in the recording and the transmission mode. The recording mode is necessary to receive the updated information of the navigation message from the uplink stations. The transmission mode generates the signals on two carrier waves. More details on these signals are given in Sect. 10.5.

As in every GNSS, the clock is the critical part of the satellite payload. The three atomic clocks of the GLONASS satellite are cesium standards produced by the Russian Institute of Navigation and Time. The operational lifetime of one clock amounts to 17 500 hours and the stability over one day is $5 \cdot 10^{-13}$. This means that one clock may operate for two years. In total, using one clock after the other, there result six years of operation.

Some data from the GLONASS satellite specification list: guaranteed lifetime 3 years (actual lifetime 4.5 years); satellite mass 1 415 kg; the mass of the navigation payload 180 kg; power supply 1 000 W; navigation payload power consumption 600 W; clock stability as previously mentioned $5 \cdot 10^{-13}$ over one day; attitude control accuracy 0.5°, solar panel pointing accuracy 5°.

Space segment modernization

Referring to the Federal GLONASS Mission Oriented Program, the space segment modernization covers three phases (Polischuk et al. 2002).

1. Phase: maintenance of the orbital constellation "at minimum required level" by GLONASS satellites.
2. Phase: development of GLONASS-M satellites and achievement of an 18-satellite constellation based on GLONASS/GLONASS-M satellites.
3. Phase: development of GLONASS-K satellites and achievement of a 24-satellite constellation based on GLONASS-M/GLONASS-K satellites followed by a global exploitation of the system.

Primarily the limited design life of satellites offers the option to modernize satellites. The next developmental step after the first-generation GLONASS satellites are the GLONASS-M satellites, where "M" indicates modified (and also implies modernization). The first GLONASS-M satellite was launched in December 2003. One year later, the next satellite of the M-generation was successfully launched and two more satellites followed in the year 2005.

Following closely Feairheller and Clark (2006: pp. 601–602), this satellite generation has several new features:

- Improved navigation performance resulting from the GLONASS-M satellite clock, whose stability has improved to $1 \cdot 10^{-13}$ (as compared to $5 \cdot 10^{-13}$ of the former generation) due to temperature stabilization. Yet, there will be only cesium clocks onboard (Polischuk et al. 2002). In addition, a better attitude control system is used and intersatellite navigation links (incorporated after the second GLONASS-M satellite) are available.
- Longer design lifetime resulting from the more stable clocks in combination with an increased loading for the propelling of the satellite, improved batteries, and modernized electronics. This increases the design lifetime to seven years (compared to the about three years of the former generation).
- Improved navigation message resulting from additional information, e.g., on the divergence of GPS and GLONASS time scales, navigation frame authenticity flags, age of data information. Also the ahead information on the correction of the GLONASS time because of leap seconds (Sect. 10.2.2) is included.
- Improved navigation signals primarily because of the second civil signal on G2 which enables the civil users to take into account ionospheric corrections.

10.4 GLONASS segments

Some data from the GLONASS-M satellite specification list: guaranteed lifetime 7 years; satellite mass 1 230 kg; mass of the navigation payload 250 kg; power supply 1 415 W; navigation payload power consumption 580 W; clock stability as previously mentioned $1 \cdot 10^{-13}$ over one day; attitude control accuracy 0.5°, solar panel pointing accuracy 2°.

Among others, a laser retroreflector, orbital correction engines, a 12-element navigation signal antenna, a cross-link antenna, and various command and control antennas belong to the additional equipment.

Three GLONASS-M satellites can be managed by one launch from the Baikonur Cosmodrome using a Proton-M launcher with a Breeze-M booster and a single satellite may also be launched from the Plesetsk Cosmodrome using a Soyuz-2 launcher with a Fregat booster (Polischuk et al. 2002).

The first satellite of the new GLONASS-K generation is scheduled for 2009. These satellites will have a design lifetime of ten years and will deliver a third civil signal on the carrier G3. Referring to the clocks, cesium and rubidium should be available (Polischuk et al. 2002). According to Gibbons (2006), the current plans for GLONASS-K include providing GNSS integrity information in the third civil signal and global differential ephemerides and time corrections. Based on this information, the mobile user may achieve submeter accuracy in real time.

Compared to the previous generations, search and rescue payload will be included.

Another feature is the reduction of the weight to about one half of the previous generation. This reduction increases the number of simultaneously launchable satellites by a factor of two.

Some data from the GLONASS-K satellite specification list: guaranteed lifetime 10 years; satellite mass 850 kg; mass of the navigation payload 260 kg; power supply 1 270 W; navigation payload power consumption 750 W; clock stability as previously mentioned $1 \cdot 10^{-13}$ over one day; attitude control accuracy 0.5°, solar panel pointing accuracy 1°.

Finally, the generation after GLONASS-K will be GLONASS-KM. There is not much known yet since this generation is still in its conceptual phase.

The constellation history back to 2001 and the forecast to 2012 is shown in Fig. 10.2.

10.4.2 Control segment

The main operational tasks of the control segment are the following: tracking of the satellites for the orbit and clock determination and prediction, upload of the navigation message to the satellites, time synchronization of the satellites, and controlling the offset between GLONASS system time and UTC. The control segment also performs many nonoperational activities, such as procurement and launch ac-

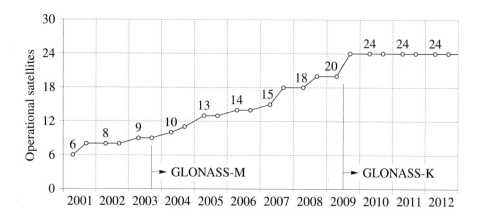

Fig. 10.2. GLONASS constellation: some history and some future

tivities, that will not be addressed here.

Note that the majority of references uses the term "ground control segment"; rarely also "ground segment" may be found (Revnivykh 2004). However, the Coordination Scientific Information Center (2002: p. 5) is taken here as the appropriate source (and because this is the same denotation as used for GPS).

The information given here is primarily taken from Feairheller and Clark (2006: Sect. 11.1.6), United Nations (2004: pp. 21–25), and Bauer (2003: Sect. 4.3). But the main problem is that the descriptions are not consistent.

Feairheller and Clark (2006: Sect. 11.1.6) primarily backing on material of 1994 distinguish between the system control center, the central synchronizer, the command and tracking stations, the laser tracking stations, and the navigation field control equipment.

United Nations (2004: pp. 21–22) do not directly refer to a specific reference but probably primarily rely on the information of an international conference held in 1999 in Vienna, Austria. They itemize into the system control center, the central system clock, four TT&C stations, and the signal monitoring system based on the direct comparison of two-way and one-way signals. This source mentions that "this architecture design ensures separate calculation of satellite orbits (ephemerides) based on two-way tracking data by TT&C stations, and clock correction determination by the direct comparison of central clock and satellite clock time scales. This procedure has been designed to simplify the orbit determination and time correction procedure."

Bauer (2003: Sect. 4.3) primarily backs on material of 1993 and partitions into two main components (1) the system control center and (2) the command tracking stations; and each of these two components has its subdivisions. Thus, the system control center subdivides into (1a) the phase control system and (1b) the central synchronizer. The command tracking stations are subdivided into (2a) four TT&C stations, (2b) four navigation field control equipment stations, and (2c) five so-called quantum optical stations (i.e., laser tracking stations). Bauer (2003: p. 248) comments that the detailed interaction of the elements of the control segment is not published.

There are many very similar presentations given from Russian authors in recent years. As one example, Klimov et al. (2005) use for the control segment the partitioning into (1) system control center, (2) central synchronizer (system clock), and (3) TT&C stations. Referring to modernization ("mission control segment development in schedule"), the deployment of one-way tracking stations is mentioned.

Sticking essentially to the Russian structuring, a few more details taken primarily from Feairheller and Clark (2006: Sect. 11.1.6) are given below.

System control center

The system control center is a military complex under the control of the Russian Space Forces and is located in Krasnoznamensk Space Center about 70 km southwest of Moscow. All system functions and operations are scheduled and coordinated at the system control center.

Central synchronizer

The central synchronizer is situated at Schelkovo in the Moscow region and is responsible for the GLONASS system time. Signals from the central synchronizer are relayed to the phase control system. Essentially, two kinds of measurements are made. First, the range to the satellites is measured by radar techniques to an accuracy of a few meters. Second, the satellite-transmitted navigation signals are compared to a reference time and reference phase by a frequency standard accurate to $1 \cdot 10^{-13}$. From these two kinds of measurements, the satellite clock offset and the phase offset are determined and also predicted. The predicted values are uploaded to the satellites at least once a day.

TT&C stations

The four TT&C stations are situated at the Russian territory in St. Petersburg region, Schelkovo in the Moscow region, Yenisseysk in Siberia, and Komsomolsk-na-Amure in Far East. These TT&C stations are supplemented by five laser stations being located in Komsomolsk-na-Amure (the only one in Russia), Balkhash (Kazakhstan), Evpatoria (Ukraine), Kitab (Uzbekistan), and Ternopol (Ukraine). The tasks of these stations are the tracking and monitoring of the satellites and providing information to the satellites via the uplink.

According to Feairheller and Clark (2006: Sect. 11.1.6), the tracking involves between three and five measurement sessions, each lasting 10 to 15 minutes. Radar is used to measure the ranges to about 2–3 m. Laser measurements, accurate to a few centimeters, are used to calibrate these radar measurements. The ephemerides are predicted 24 hours in advance and uploaded once a day. The satellite clock corrections are uploaded twice a day (Coordination Scientific Information Center 2002: Sect. 3.3.3) to the satellites. The accuracy of the uploaded clock correction is specified to amount to less than 35 ns (1σ).

The GLONASS measurements are transmitted to the system control center via radio link once per hour.

Control segment modernization

Under the title "performance modernization plan" and the item "receiving monitoring stations network extension" Revnivykh (2006b) mentions four subitems: (a) the Space Force network (3 stations), (b) the Roskosmos network (9–12 stations), (c) the Rosstandard network (3 stations at UTC sites), and (4) the cooperation with the International GNSS Service and foreign agencies. But no details are given.

Similarly, no details are given on the modernization of the GLONASS time keeping system. Only new system clocks with high stability (2 distributed clocks) and the modernization of the synchronization system are mentioned.

In Dvorkin and Karutin (2006) one transparency deals with the "GLONASS control segment modernization". The authors mention four items: (1) "one-way measurement and ephemeris computation stations network development", (2) "one-way measurement stations network creations", (3) "two-way measurement stations deployment", and (4) "communication channels modernization". But no more details are given. At the very end, the authors present a transparency on "Data collection network development program" and enumerate the eight existing stations (end of 2005) and add 11 scheduled new stations.

10.5 Signal structure

GLONASS, as a dual-use system, provides a high-accuracy signal for military use and a standard-accuracy signal for civil use free of charge. The interface control document (Coordination Scientific Information Center 2002: p. 7) lists two carrier frequencies, L1 and L2. For better differentiation from GPS, the GLONASS carrier frequencies are denoted using G instead of L. Additionally, the third carrier frequency, as announced in various publications, is denoted by G3. The abbreviations P and C/A of the accuracy signals (Table 10.2) are not standardized but are commonly used in literature.

GLONASS implements the frequency division multiple access (FDMA) technique to differentiate between the signals of different satellites. In this way the GLONASS signals are more resistant against narrowband interference, and further-

10.5 Signal structure

Table 10.2. GLONASS signal lexicon

G1	Link 1, carrier frequency = 1 602.000 MHz
G2	Link 2, carrier frequency = 1 246.000 MHz
G3	Link 3, carrier frequency = 1 204.704 MHz [1]
C/A	Standard-accuracy signal
P	High-accuracy signal

[1] subject to change

more the crosscorrelation between different GLONASS signals is low, despite short ranging codes. The crosscorrelation between two signals of adjacent frequencies has been specified to be no more than −48 dB (Coordination Scientific Information Center 2002: p. 10). FDMA, as a trade-off, requires RF front-end components with extra-wide bandwidths.

In a joint statement of the GPS/GLONASS Interoperability and Compatibility Working Group between the US and the Russian Federation, both sides noted that the user community would benefit from a common approach considering CDMA and FDMA principles. Russia declared that a decision in this regard will be made by 2007. The following sections will discuss the state of the art and the baseline as per January 2007. This could, considering the before mentioned statement, strongly change.

Since its FOC in 1996, GLONASS satellites continuously emit the standard-accuracy signal (C/A-code) and the high-accuracy signal (P-code) on two carrier frequencies, G1 and G2. Similar to GPS, the C/A-code is only modulated onto G1, whereas the P-code is modulated onto G1 and G2. Starting with the operation of the first GLONASS-M satellite in 2004, a standard-accuracy signal has also been added to G2. In parallel, additional information has been placed in formerly reserved bytes of the navigation message.

The P-code signal is not encrypted; however, the ranging code has not been officially published. Scientists were able to decipher the one-second long P-code. The Russian Ministry of Defense does not recommend its unauthorized use, because the P-code may be changed without prior notice to unauthorized users.

The modernized GLONASS-K satellites will provide a third carrier frequency G3 together with a third civil (C/A$_2$) and military (P$_2$) ranging code. The third frequency will increase the reliability and accuracy and will especially be useful for safety-of-life applications (Revnivykh 2006a). Investigations are ongoing to include integrity information on the third frequency. Furthermore, it is planned to include differential ephemerides and time corrections to achieve submeter real-time accuracy for mobile users at least for the coverage area of Russia. Several years will pass before the modernized signals and services will be available on a global scale for 24 hours, 7 days a week.

The beam pattern of the satellite antennas has been designed not only to cover the earth but also to provide navigation signals to other satellites. Especially the beam pattern of the GLONASS-M satellites is widened to serve spaceborne receivers. Additionally, the satellites vary the emitted signal power to accommodate for the different transmission path loss depending whether the satellite is, relative to the user, in zenith or at the horizon. The signal power variation shall guarantee a nearly uniform power level of different satellites.

All navigation signals are right-handed circularly polarized. The deviation of the perfect circular polarization is no more than 1.5 dB depending on the angular range ±19° from boresight (Coordination Scientific Information Center 2002: p. 11).

The synchronization of different ranging codes is a fundamental requirement to avoid biases in the run time measurements. The equipment group delay quantifies the delay between the output of the atomic frequency standard (AFS) and the transmitted signal. The delay consists of determined and undetermined components, where the undetermined component is specified not to exceed 8 ns for GLONASS satellites and 2 ns for GLONASS-M satellites (Coordination Scientific Information Center 2002: p. 11).

10.5.1 Carrier frequencies

The carrier frequencies and all timing processes are coherently derived from the AFS. The frequencies are reduced by a relative value $\Delta f / f = -4.36 \cdot 10^{-10}$ to compensate relativistic effects (cf. Sect. 5.4). The FDMA design requires unique carrier frequencies for all satellites defined by

$$\begin{aligned} f_{1k} &= f_1 + \Delta f_1 \, k = 1\,602.0000 + 0.5625\,k \quad [\text{MHz}], \\ f_{2k} &= f_2 + \Delta f_2 \, k = 1\,246.0000 + 0.4375\,k \quad [\text{MHz}], \\ f_{3k} &= f_3 + \Delta f_3 \, k = 1\,204.7040 + 0.4230\,k \quad [\text{MHz}], \end{aligned} \quad (10.2)$$

where k differentiates the frequency channels. The factors Δf_1, Δf_2, Δf_3 denote the frequency increments of satellite signals in two adjacent channels. The frequency f_3 and the frequency increment Δf_3 are not fixed yet and subject to change. A triple of frequencies f_{1k}, f_{2k}, f_{3k} is assigned to each satellite, thereby the ratio between these frequencies, $f_{1k}/f_{2k} = 9/7$ and $f_{1k}/f_{3k} = 125/94$, is constant. The factor between f_{1k} and f_{3k} is still subject to change. The carrier frequencies themselves are multiples of the frequency increments: $f_{jk} = \Delta f_j (2848 + k)$.

Originally, 24 channels ($k = 1, 2, \ldots, 24$) have been assigned to GLONASS. However, the navigation signals interfered with radio astronomy frequency bands

10.5 Signal structure

Table 10.3. GLONASS frequency bands

Link	Factor ($\cdot f_1$)	Frequency [MHz]	Increment [MHz]	Wavelength [cm]	Frequency band
G1	1	1 602.000	0.5625	18.7	ARNS/RNSS
G2	7/9	1 246.000	0.4375	24.1	RNSS
G3 [1]	94/125	1 204.704	0.4230	24.9	ARNS/RNSS

[1] subject to change

and signals of satellite communication services. The frequency 1612 MHz, for instance, is used by radio astronomy to observe the radiation emitted from certain molecules that provide clues about the evolution of the galaxy (Roßbach 2001: p. 11). Russia agreed to gradually change the channel allocation. In a first step, 1998–2005, the number of frequency channels has been reduced to 12. After 2005 the satellites have been transmitting on frequency channels $k = -7, -6, \ldots +5, +6$, where channel numbers +5 and +6 are reserved for technical purposes. Additionally, the satellites launched after 2005 implement filters to limit the out-of-band emissions to the harmful interference limit (Coordination Scientific Information Center 2002: p. 10).

The limitation to 12 channels has been possible by assigning satellites in antipodal position within the same orbital plane the same channel number. A receiver on the earth will never track both satellites at the same time. Spaceborne receivers, in contrast, have to implement discriminating functions, like Doppler checks, to distinguish the satellites in antipodal position (Branets et al. 1999).

Although a third frequency band has already been allocated to GLONASS, there is little information available about its usage. Revnivykh (2004) emphasizes that a third civil and a third military ranging code will be modulated onto G3.

10.5.2 PRN codes and modulation

GLONASS relies on the FDMA principle and therefore uses common PRN sequences for all satellites. The two ranging codes, i.e., the standard- and the high-accuracy signal denoted as C/A-code and P-code, respectively, are modulated onto the carrier frequencies in phase quadrature similar as the signals given in Eq. (4.37). The standard- and high-accuracy codes are synchronized to each other. Together with the ranging codes the navigation message is BPSK-modulated onto the carrier frequency. Note that the BPSK modulations as described in Sect. 4.2.3 have generally been referenced to 1.023 megachips per second (Mcps), thus, e.g., BPSK(1) was equivalent to BPSK(1.023 Mcps). Table 10.4 does not apply this definition.

Figure 10.3 plots the power spectral density envelopes of the GLONASS rang-

Table 10.4. GLONASS ranging signals

Link	PRN code	PRN code length [chips]	Code rate [Mcps]	Modulation type	Bandwidth [MHz]	Data rate [bps]
G1	C/A	511	0.511	BPSK(0.511 Mcps)	1.022	50
	P	5 110 000	5.11	BPSK(5.11 Mcps)	10.22	50
G2	C/A	511	0.511	BPSK(0.511 Mcps)	1.022	50
	P	5 110 000	5.11	BPSK(5.11 Mcps)	10.22	50
G3 [1]	C/A$_2$	[2]	4.095	BPSK(4.095 Mcps)	8.190	[2]
	P$_2$	[2]	4.095	BPSK(4.095 Mcps)	8.190	[2]

[1] according to Dvorkin and Karutin (2006); subject to change
[2] not published yet

ing signals. For a better visualization, only three adjacent channels are given. The frequency bandwidths displayed are different from the bandwidths allocated to GLONASS by ITU. The C/A- and P-code carry about the same power, therefore the code with the lower code rate has a higher amplitude as indicated in Fig. 10.3. The minimum received power levels are specified to be in the range of -161 dBW to -167 dBW (Coordination Scientific Information Center 2002: p. 10). Higher power levels may be caused by a variation of the nominal orbit altitude (but still within the specifications), different antenna gains as a function of the azimuth and frequency band, or as the Coordination Scientific Information Center (2002: p. 37) further emphasizes, variations in output signal power due to technological reasons, temperature, voltage, or gain variations. The maximum received power level is expected not to exceed -155.2 dBW.

Standard-accuracy signal (C/A-code)

The standard-accuracy signal (C/A-code) is specified in the interface control document (Coordination Scientific Information Center 2002). The standard-accuracy code has a chipping rate of 0.511 Mcps. The code length corresponds to 511 chips, consequently, the code period corresponds to 1 ms. One code chip has a length of about 587 m. The code is generated using a 9-bit linear feedback shift register (LFSR) with the characteristic polynomial

$$p(x) = 1 + x^5 + x^9, \tag{10.3}$$

which means that the registers R_5 and R_9 are the defining registers. The output of the 7th stage of the shift register, as indicated in Fig. 10.4, defines the ranging code

10.5 Signal structure

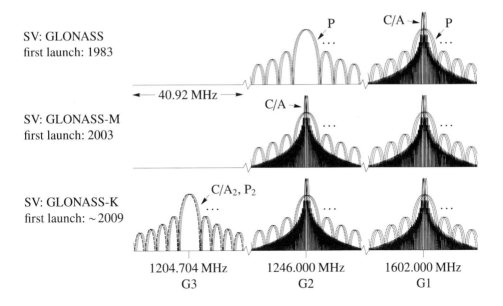

Fig. 10.3. Power spectral densities of GLONASS signals

sequence. The initialization vector corresponds to all ones. The standard-accuracy code is a maximal length code sequence with good autocorrelation property. The short code length allows fast signal acquisition; however, it produces frequency components at 1 kHz intervals, which are in particular susceptible to interference. The code rate creates spectral nulls at multiples of 511 kHz and a signal bandwidth of 1.022 MHz. Comparing it with the frequency increments of the satellite signal reveals that the mainlobes of adjacent frequency spectra overlap. The crosscorrelation between two signals in adjacent frequency bands does not exceed −48 dB.

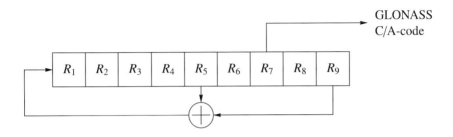

Fig. 10.4. Standard-accuracy signal LFSR

High-accuracy signal (P-code)

There is no publicly available interface control document for the high-accuracy signal (P-code). The P-code has a chipping rate of 5.11 Mcps. The code length corresponds to 5 110 000 chips. The code duration corresponds to 1 s, the chip length is equal to about 59 m. Although the ranging code is not encrypted, its use is not recommended. The high-accuracy signal is generated using a 25-bit maximal length LFSR. The characteristic polynomial, with the initialization vector all ones, corresponds to

$$p(x) = 1 + x^3 + x^{25}. \tag{10.4}$$

The LFSR generates a 33 554 431-chip-long code sequence that is truncated after 1 second. The high code rate of the high-accuracy signal positively influences the signal acquisition and tracking capability and the resistance against interference. Considering the code rate and the frequency increments of the satellite signals, the frequency bands overlap to a great extent. The short C/A-code is generally used for a coarse acquisition, whereas the longer P-code provides a higher precision. A hand-over word like the one used in GPS is not necessary due to the short period of 1-second compared to 1-week length of the GPS P(Y)-code.

10.5.3 Navigation messages

The navigation message provides information about satellite orbits, satellite health status, correction data, and various other information to the receiver. The data rate of the navigation message is slow compared to the ranging codes or compared to typical communication channels. However, the low data rates enable a low bit error rate even despite the weak signal power. The effective data rate corresponds to 50 bps, although the differential modulation generates a data rate of 100 symbols per second. Two different navigation messages are modulated onto the standard- and the high-accuracy signal.

The navigation data of the standard-accuracy signal is structured into superframes, frames, and strings. A superframe has a duration of 2.5 minutes and is subdivided into five frames of 30-second length each. A frame contains immediate data of the transmitting satellite and nonimmediate data from all other satellites.

The structural elements of a frame are 15 strings of 100-bit length (= 200 symbols) each. A string holds 85 data bits and a time mark. The time mark, or preamble, corresponds to a short cycled pseudorandom sequence generated by a 5-bit LFSR with the characteristic polynomial

$$p(x) = 1 + x^3 + x^5, \tag{10.5}$$

which is truncated after 30 symbols. The time mark is modulated with a symbol length of 10 milliseconds (symbol rate of 100 symbols per second) onto the carrier

10.5 Signal structure

frequency, consequently taking 0.3 seconds in duration. The 85 data bits, which include an 8-bit parity check, take 1.7 seconds for transmission, with a symbol length of 20 milliseconds. The data bits are XOR-added with a sequence that has a symbol duration of 10 milliseconds and looks similar to the sine-phased BOCs(1,1) modulation scheme. Due to this meander sequence, there will never be three or more symbols equal to 0 or 1 in a row except for the symbols in the time mark.

The 2-second-long strings are synchronized to the start of a day in the satellite time frame. The introduction of a UTC leap second will deteriorate the synchronization. Consequently, the generation of the string has to be recycled to account for the introduced time jump. The interface control document (Coordination Scientific Information Center 2002: Appendix 2) contains a recommendation how to proceed during UTC leap-second correction.

The first five strings contain the immediate information. The immediate data, repeated in every frame, comprises beside the ephemerides data the time tag corresponding to the beginning of the frame, a satellite health flag, the satellite clock correction, and the variation of the satellite carrier frequency from the nominal value. The ephemerides data is valid for several hours. Strings 6 through 15 contain the nonimmediate data for the 24 satellites in orbit. The almanac data of a satellite occupies two strings. The last two strings of the fifth frame of the original navigation message are empty and are used for the modernized message content. Due to the meander sequence encoding, the empty strings do not have to be filled with alternating zeros and ones (as GPS does). The nonimmediate data repeats every superframe.

The broadcast ephemerides are given by positions and velocities referred to the PE-90 reference system in contrast to GPS, which transmits Keplerian elements. Additionally, the accelerations of the satellites caused by solar and lunar gravitational perturbations are given in the same coordinate system. The values are provided in half-hour sampling intervals. The positions and velocities at intermediate times are calculated using interpolation procedures. Integration equations based on a 4th-order Runge–Kutta method are given, e.g., in Coordination Scientific Information Center (2002: Appendix 3). Zinoviev (2005) compares the Runge–Kutta method with a 5th-order Fehlberg and a 7th-order Shanks numerical integration method. According to Coordination Scientific Information Center (2002: p. 23), the accuracy of the transmitted coordinates and velocities (Table 10.5) evolves with the modernization of GLONASS satellites. Dvorkin and Karutin (2006) mention that the accuracy will further increase even down to the submeter level within the next years. In contrast to the ephemerides data set, the almanac information is similar to the one emitted by the GPS satellites (cf. Table 3.6).

Since the modernization step realized by GLONASS-M, the navigation messages also contain data to increase the interoperability of GPS and GLONASS especially with respect to the two time systems. The difference between the GPS

Table 10.5. Accuracy of transmitted coordinates and velocities

Error component	Root mean square error of:			
	Predicted coordinates [m]		Predicted velocities [cm s^{-1}]	
	GLONASS	GLONASS-M	GLONASS	GLONASS-M
Along-track	20	7	0.05	0.03
Across-track	10	7	0.1	0.03
Radial	5	1.5	0.3	0.2

and GLONASS time is emitted with a specified maximal deviation of 30 ns. The modernized navigation message will furthermore contain information about any upcoming leap-second correction, estimates of pseudorange accuracy, or the hardware delay between G1 and G2 bands (Zinoviev 2005). This information is placed in the spare bits in order to guarantee compatibility with the existing navigation structure.

The navigation message of the high-accuracy signal is longer and contains more precise information. The navigation message consists of a superframe that is subdivided into 72 frames. Each frame consists of five strings with 100 bits length each. A frame length corresponds to 10 seconds. One superframe, consequently, needs 12 minutes for transmission. Feairheller and Clark (2006: p. 611) emphasize that the repetition rate of the ephemerides information corresponds to 10 seconds for the high-accuracy signal navigation message, whereas for the standard-accuracy signal it corresponds to 30 seconds. In contrast, the time needed to transmit the almanac of all satellites corresponds to 12 minutes for the P-code and 2.5 minutes for the C/A-code.

10.6 Outlook

The Federal GLONASS Mission Oriented Program 2001–2011 is the key for the future. If the objectives can be achieved in a proper time frame, then GLONASS could play a substantial role not only locally but also in the world market. Reviewing the past few years, there occurred a remarkable development with GLONASS. The first decisive step was to declare GLONASS a dual-use system and to open it even more widely by the GLONASS Coordination Board.

As Revnivykh et al. (2003) outline, international cooperation is the next promising step ahead. Negotiations with the European Union (EU), the European Space Agency (ESA), the United States, India, China, and other countries are initiated for a future success. The main objectives of these cooperations are the following:

10.6 Outlook

- to develop GLONASS and its augmentations in such a way that compatibility and interoperability is possible with GPS and Galileo;
- to offer a reliable, accurate, and highly available navigation service for the benefit of the users;
- to establish a benefit by using satellite navigation services in the world market.

The commercial market becomes more and more attractive. This may be the reason that GLONASS might experience one of the most exciting changes or updates, the use of code division multiple access (CDMA):

"Addition of CDMA signals to the Russian system's transmissions – or possibly even conversion of GLONASS's frequency division multiple access (FDMA) signals to CDMA – appears increasingly likely. Such a move would simplify interoperability with the GPS and Galileo systems and associated user equipment." (Inside GNSS, January/February 2007, p. 18). S. Revnivykh, Deputy Director of the Mission Control Center of the Russian Federal Space Agency (Roskosmos), mentioned this as a possibility in a presentation in September 2006.

In a report of GPS Daily (www.gpsdaily.com) of January 23, 2007, the Russian cooperation with India is outlined: "GLONASS is available for India to use" and "Moscow and New Delhi had agreed to launch GLONASS-M satellites with the help of Indian carrier rockets, and to create new-generation navigation satellites".

In the same report of GPS Daily the efforts are mentioned, to speed up the process to achieve the full constellation: "President Vladimir Putin ordered in December 2005 that the system be ready by 2008 and in March this year Ivanov said GLONASS will be available to domestic consumers for military as well for civilian purposes by the end of 2007." And later: "The Agency plans to have 18 satellites in orbit by late 2007 or early 2008, and a full orbital group of 24 satellites by the end of 2009, he said."

In Averin (2006) the "projected GLONASS positioning error (95%)" is given for the years 2007–2011 by 30 m, 20 m, 10 m, 7 m, and 5 m, respectively where the reason for the performance improvement is seen in advanced orbit and time determination techniques. It may be assumed that the contribution of the scheduled increased number of satellites is also implied.

If all this comes true, GLONASS will have a much better future compared to its past.

11 Galileo

11.1 Introduction

11.1.1 Historical review

Europe early recognized the strategic, economic, social, and technological importance of satellite-based navigation. A European strategy and major actions in the field of satellite positioning and navigation have become necessary to establish trans-European networks in the fields of transport, telecommunications, and energy infrastructures in accordance with the European Community (EC) treaty (European Council 1994).

In the 80s, the European Space Agency (ESA) studied different system concepts (Wakker et al. 1987). In particular, a time division multiple access (TDMA) concept with satellites working only as bent pipes for ground station signals, thus, simplifying the satellite complexity, has been considered.

In 1994, the European Council requested from the European Commission (European Council 1994) in a resolution to respond to the challenges of information technology and take necessary initiatives to contribute to satellite navigation. Europe envisaged a two-step approach: the first step headed for the augmentation of the existing first-generation GNSS (i.e., GPS and GLONASS). This action finally resulted in the development and deployment of the European geostationary navigation overlay service (EGNOS) described in Sect. 12.4. In the second step, the European Union (EU) requested to "initiate and support the preparatory work needed for the design and organization of a global navigation satellite system for civil use". At the same time the European Council fostered the need to cooperate with the private sector for cost and risk sharing whilst maximizing the benefits. Europe also emphasized the need to closely cooperate with the International Civil Aviation Organization (ICAO) and the International Maritime Organization (IMO) in order to develop and implement a system in accordance with international standards.

The EU headed at first for a close cooperation with the US in the development of the next-generation GPS and an active participation in the control and development of it. Negotiations with the US revealed the interest of both parties in a cooperation but also brought up the reservations of the US. GPS has always been considered as safety-critical infrastructure. A participation of foreign countries in the definition and control of GPS was not acceptable for the US. Europe, in contrast, headed for a maximum of control to guarantee its own sovereignty, autonomy, and competitiveness. Europe also considered to closely cooperate with Russia but finally took the decision to develop its own GNSS.

In 1999, the European contribution to satellite navigation has been provisionally named Galileo, but meanwhile Galileo has become the synonym for the European GNSS. As a matter of fact, Galileo is not an acronym, therefore it is written in small letters in this textbook. The European satellite navigation system has been named after the Italian scientist and astronomer Galileo Galilei (1564–1642). In 1610, he discovered the first four satellites of the planet Jupiter, later named Io, Europa, Ganymede, and Callisto. Galileo Galilei furthermore described how the regular movement of the four satellites could be used for longitude determination by observing their eclipses.

In its paper concerning the involvement of Europe in a new generation of satellite navigation service, the European Commission (1999) recommended Galileo to be an open, global system, fully compatible with GPS but independent from it. The key parameters of Galileo have been identified to be the independence of any other system while maintaining interoperability, global availability, and high level of service reliability, thus implementing integrity information.

In May 1999, the Ministerial Council of ESA approved the Galileo definition phase, whose contracts have been signed in December 1999. The market studies preceding and accompanying the definition phase of Galileo promised tremendous economic and social benefits. Thereby the increased safety of transport system, the reduced congestion, or the reduced air pollution, to name but a few, contribute to the social benefit. Furthermore, a strategic benefit is expected, e.g., by the implementation of a certified system under European control for the purpose of law enforcement.

The costs for the Galileo program including an operation phase until 2020 have been estimated to be in the range of EUR 3.4 billion. The European Commission (2000) also estimated that a two-day interruption of the GPS service in 2015 would cost Europe's transport and financial sectors about EUR 1 billion, if Galileo would not be operational. The cost benefit analysis brought up a factor of +4.6 opposing benefits to costs, thereby considering only the incremental benefit of Galileo compared to the one of satellite navigation in general. The analysis also highlighted that there is a certain time window the Galileo system has to be deployed, otherwise the market prospects will be lost.

From a user perspective, the interoperability and compatibility of Galileo with the existing satellite navigation systems is essential (European Commission 2000). Also industry has committed that Galileo is not intended to replace or to compete with other GNSS, but be interoperable with them. The EC established a signal task force in March 2001 to support the signal definition and to guarantee interoperability. The work of the signal task force resulted in 2004 in the agreement signed between the US and EU to implement a common signal structure in order to facilitate the combined use of both systems (United States of America and European Community 2004). Security concerns of the US at the same time led to a separation of the military GPS M-code and the Galileo public regulated service signal in the E1 frequency band (cf. Sect. 11.5).

The Galileo definition phase was completed in 2003. Already in March 2002, the Council of European transport ministers agreed on the decision to launch the next phases of the Galileo program. At the same time the Council released the budget for the development and validation of the satellite navigation system. In the same political agreement the Council decided to create a joint undertaking to account for the joint action of the EU and ESA. The EU is responsible for the legal and political issues, whereas ESA manages the technological steps to be taken during the Galileo program. The Galileo Joint Undertaking (GJU) has been established on June 10, 2003, under the Article 171 of the European Commission treaty. The GJU, as a nonprofit entity under public law, combined the interests of the two founding members ESA and EU. The main tasks of the GJU were to start the competitive process of Galileo concession and the management of the development phase. Furthermore, GJU managed a number of research activities in the field of satellite navigation.

As planned, the GJU was disbanded at the end of 2006. Its tasks, especially the negotiations with the Galileo concessionaire, have been taken over by the European GNSS Supervisory Authority (GSA). The GSA has been established by the European Commission to represent and defend the public interests in the field of satellite positioning and navigation. The mission of the GSA is defined by the European Council (2004). The GSA is responsible for the enforcement of the concession contract and the deployment of the system according to the financial, technical, and temporal framework. The GSA, furthermore, represents the certification authority and coordinates the consideration of Galileo in various international standards.

From the very beginning, Europe followed the policy of international cooperation in order to acquire worldwide markets, define global standards, and guarantee financing. Agreements have been headed for, negotiated, and signed with a number of states (Flament 2006).

11.1.2 Project phases

The Galileo program is conducted in four phases:

- the definition phase,
- the development and in-orbit validation phase,
- the deployment phase,
- the operation phase.

The definition phase resulted in a baseline for the main characteristics and performance parameters of the Galileo mission summarized in the high-level definition (HLD) document (European Commission and European Space Agency 2002).

The baseline has been further split up into the mission requirements and system requirements, which finally end up in the signal-in-space interface control documents (ICD). Into the HLD, the European Commission and ESA consolidated the requirements, needs, and opinions of potential user groups, private investors, industry, and others. A number of projects and comprehensive studies have contributed to this phase, e.g., the GALA project dealing with the Galileo overall architecture definition.

The development and in-orbit validation (IOV) phase defines the satellites, ground components, and user test receivers in detail. In the next step the elements are manufactured and implemented. Two experimental prototype satellites are launched in order to secure the Galileo frequency filing and conduct a number of first tests. After conclusion of the first experiments, four satellites will be brought into orbit to validate the major components of the Galileo system, before the deployment phase starts. The development phase is financed by public funds. The phase is managed by the GJU/GSA. The development and IOV will last for 48 months.

During the deployment phase, the fully operational satellites will be brought into orbit and the ground infrastructure will be fully deployed. The deployment phase will last 24 months. Full operational capability (FOC) of the Galileo services, at least of the non-safety-critical ones, shall be reached in 2012/2013.

The operation phase starts with FOC and will be managed by the private sector. This phase shall be financed by win-bringing revenue streams. The operation phase is scheduled for at least 20 years.

11.1.3 Management and operation

Galileo will be managed and operated by a private concessionaire which has been selected after a competitive tendering process. Together with the GSA, the Galileo operating company (GOC) that holds the concession will establish a public-private partnership (PPP). The PPP is successfully implemented when a win-win situation results, where both the private and the public sector take benefit of the partnership while splitting the risk factors of the program. The concessionaire will be responsible for the deployment and operation phase of Galileo for a time period of 20 years.

The selection process of the long-term concessionaire started with the call for interest launched by the GJU in October 2003. Two consortia have been invited for the competitive negotiation phase in April 2004. The consortia finally joined forces in May 2005 and delivered a joint proposal which showed significant improved revenue streams while minimizing the public share of the financial contribution. The negotiations of the concession contract ran into problems in April/May 2007, and the contract has not been signed in May 2007 as expected.

11.2 Reference systems

11.2.1 Coordinate system

Galileo relies on a geocentric Cartesian reference frame as defined by the Galileo terrestrial reference frame (GTRF). GTRF will be related to the international terrestrial reference frame (ITRF), which has been established by the International Earth Rotation Service (IERS) (Hein and Pany 2002). The GTRF is specified to differ from the latest version of ITRF by no more than 3 centimeters (2σ). This will be ensured by the Galileo geodetic service provider (GGSP). Furthermore, the GGSP is responsible for the involvement of the geodetic community during the definition, implementation, and maintenance of GTRF (Swann 2006).

11.2.2 Time system

The Galileo system time (GST) is a continuous atomic time scale with a nominal constant offset (i.e., integer number of seconds) with respect to the international atomic time (TAI). With respect to the coordinated universal time (UTC), the modulo 1 second offset is variable due to the insertion of leap seconds (cf. Sect. 2.3.2).

GST will be maintained by an ensemble of atomic frequency standards (AFS), where active hydrogen maser clocks will serve as the master clock (Hahn 2005). Galileo will use the steering correction parameters provided by an external time service provider in order to steer GST towards TAI. Therefore, the external time service provider will interface to the International Bureau of Weights and Measures (BIPM), which maintains UTC time. The offset of GST from its nominal value (TAI) is specified to be less than 50 ns (2σ) modulo 1 second over 95% of any yearly interval (Bedrich 2005). The uncertainty of this offset is 28 ns (2σ).

The computation of GST is conducted by the precise timing facility of the control segment (cf. Sect. 11.4.2). The offset of GST with respect to TAI and UTC will be included in the navigation message and broadcast to the users. GST will be emitted using week numbers (modulo 4096) and the time of week that starts from Saturday to Sunday midnight. The start time of the GST week number has to be defined yet. The user segment will be able to synchronize in real time to UTC with an accuracy of 30 ns for 95% of any 24 hours of operation (Falcone et al. 2006a). Furthermore, the GPS to Galileo time offset (GGTO) will be computed and distributed to the users via the Galileo space segment. The projected accuracy of GGTO is 5 ns (2σ) modulo 1 second over any 24 hours. The user may also determine GGTO by introducing an additional unknown into the range equations (6.6). The time offset to other systems will presumably also be incorporated into the Galileo navigation message.

11.3 Galileo services

Europe has chosen a service-oriented approach for the design of Galileo. During the definition phase, the user requirements have been categorized into four different service levels. (1) The satellite-only service relies solely on the signals from the Galileo satellites. This service will be available worldwide and independent from other systems. (2) Enhancing the performance of the satellite-only service by local augmentation or assistance information is summarized by the Galileo locally assisted services. (3) The EGNOS service concentrates on the combined use of Galileo and a future evolution of EGNOS to provide a maximum of integrity. (4) Finally, the combined-service level describes the use of Galileo in combination with other GNSS or other means of navigation systems.

The user needs, operational needs, and application needs have been consolidated in the HLD document that especially defines the Galileo satellite-only service in detail. The Galileo satellite-only service is subdivided into four different navigation services and one service to support search and rescue operations. The performance requirements of the navigation services are listed in Table 11.1. The service availability represents the percentage of time averaged over the design lifetime of the system (20 years) when the service is within the specified performance with regard to accuracy, integrity, and continuity anywhere on the globe (Falcone 2006). Since always the worst-case situation is considered, the service availability has to be considered the critical parameter in the performance definition of Table 11.1.

11.3.1 Open service

The open service (OS) is accessible to all users free of charge. Since no integrity information is included and therefore no service guarantee or liability provided, the receiver may apply RAIM techniques to derive integrity information. The OS is primarily intended for the mass market providing simple positioning and timing services. Six unencrypted signals are modulated onto three different carrier frequencies to provide a competitive navigation service compared to other GNSS. The usage of several signals and frequencies increases the performance and interference resistance but at the same time increases technological requirements, if these advantages want to be exploited. The frequency bands partly overlap with the frequency bands of other GNSS to increase compatibility and interoperability. Galileo single-frequency receivers will provide a performance comparable to GPS C/A-code receivers.

11.3.2 Commercial service

The commercial service (CS) is intended to generate a revenue stream for the GOC. Therefore, the CS will rely on data included in the navigation message in all fre-

Table 11.1. Performance requirement for the Galileo services (Falcone 2006)

Satellite-only service	Open service	Commercial service	Safety-of-life service	Public regulated service
Coverage	global	global	global	global
Accuracy [1] single-frequency dual-frequency	15 m / 24 m H; 35 m V 4 m H; 8 m V		4 m H; 8 m V	15 m / 24 m H; 35 m V 6.5 m H; 12 m V
Timing accuracy (95%) [2]	30 ns	30 ns	30 ns	30 ns
Integrity: alarm limit time to alarm integrity risk	— [3]	— [3]	12 m H; 20 m V 6 s $3.5 \cdot 10^{-7}$ / 150 s	20 m H; 35 m V 10 s $3.5 \cdot 10^{-7}$ / 150 s
Continuity risk	—	—	10^{-5} / 15 s	10^{-5} / 15 s
Service availability	99.5 %	99.5 %	99.5 %	99.5 %
Access control	free open access	controlled access of ranging codes and navigation data message	authentication of integrity information in the navigation data message	controlled access of ranging codes and navigation data message
Certification and service guarantees	—	guarantee of service possible	built for certification and guarantee of service	built for accreditation and guarantee of service

[1] Positioning accuracy: horizontal (H) 95%, vertical (V) 95%;
single-frequency accuracy depends on which frequency is used;
ionospheric corrections using dual-frequency measurements or NeQuick model
[2] Offset Galileo to UTC over 24 hours
[3] Receiver autonomous integrity monitoring is possible

quency bands. The data messages will be encrypted and broadcast with a data rate of up to 500 bits per second providing an added value compared to the OS. Additionally, a service guarantee is envisaged for the CS. The access to the data message as well as to the encrypted ranging data will be controlled by the Galileo concessionaire. The Galileo ground segment will provide an interface to external service providers in order to forward any information from them to the users via the satellite signal.

11.3.3 Safety-of-life service

The safety-of-life (SoL) service relies on the same signals as the OS, but additionally adds integrity information to provide a service guarantee to the users. The frequencies out of the aeronautical radionavigation service (ARNS) bands thereby provide a maximum of signal protection. Integrity information in the navigation message will indicate any failure of the system and provide timely warnings to the user on a global scope. Beside these operations that need timely warnings (e.g., precision approach in the aviation domain), the SoL service will also provide a noncritical service level that is less time critical but still needs integrity information (e.g., open sea navigation in the maritime domain). The SoL service has been designed to be compliant to different standards in the aeronautical, maritime, and railway domains to maximize the benefit of the user community. For example, the aviation service requirements for oceanic, en route, nonprecision approach, and approach and vertical guidance operations will be met by the Galileo SoL service.

11.3.4 Public regulated service

Galileo is considered a sensitive infrastructure susceptible to misuse and endangerment of the safety and security of the public. Adequate measures have been and will be taken to protect the Galileo infrastructure, protect the Galileo signal against intentional interference, jamming, spoofing, or meaconing, and protect Galileo against misuse of the services (Galileo Joint Undertaking 2003).

The main objective of the public regulated service (PRS) is to provide a continuous, robust, and encrypted signal that will be usable even in situations of crises, while the other services will be either deactivated or intentionally jammed. Encrypted signals on two spectrally separated carrier frequencies maximize the interference resistance, while minimizing the vulnerability. The access to the PRS signals will be controlled and granted only to European public authorities responsible for civil protection, national security, and law enforcement like the European police office or the European anti-fraud office.

The envisaged service performance is comparable to the one of the OS, while the integrity provided is comparable to that of the SoL service.

11.3.5 Search and rescue service

The Galileo search and rescue (SAR) service is Europe's contribution to the international COSPAS-SARSAT system in accordance to the requirements and regulations of IMO and ICAO (European Commission and European Space Agency 2002). This system has been defined and built by Russia, US, France, and Canada to provide a means for worldwide humanitarian SAR operations (cf. Sect. 12.1.3). In future also the Galileo satellites will detect emergency signals at 406 MHz and forward the emergency message to the SAR ground segment in the SAR downlink frequency band. The emergency beacons are located by Galileo satellites using range and/or Doppler measurements (Cospas-Sarsat Secretariat 2006b), which enable the SAR control centers to determine their position. A new generation of the SAR beacons will implement a GNSS receiver in the terminal and modulate the position information onto the emergency signal. The position accuracy, thus, increases from 5 km (95%) using range and/or Doppler measurements within 10 minutes to a few meters using instantaneous measurements. The control center finally forwards the rescue information to local rescue coordination centers which alarm the rescue teams. Galileo satellites additionally provide the possibility of a return link to acknowledge the emergency signal. The acknowledge information is modulated onto the OS signals. Galileo will provide a continuous global service with nearly no latency in forwarding the position of activated beacons to the rescue centers. One Galileo satellite will have the capacity of detecting 150 signals at a time. The availability of the service is specified to be greater than 99.7%.

11.4 Galileo segments

The overall concept and the architecture of Galileo is similar to the one of GPS and GLONASS. The envisaged services, however, require an additional categorization. Apart from the user segment (cf. Sect. 13.4), Galileo defines three principal components: the global component, the regional component, and the local component (Fig. 11.1).

The global component is the core element of Galileo and is subdivided into the space segment and the ground segment. The space segment is described in more detail in Sect. 11.4.1 and the ground segment in Sect. 11.4.2.

The regional component consists of a network of integrity-monitoring stations and an integrity control center which determines a regional limited valid integrity

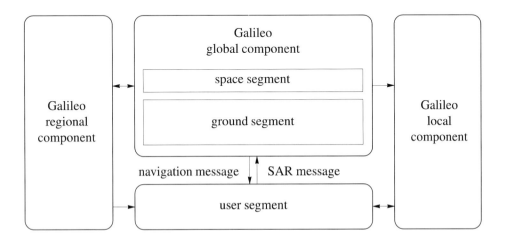

Fig. 11.1. Galileo architecture

information and directly uplinks it to the Galileo satellites via a dedicated and secured channel. The regional component complements the Galileo integrity concept and allows to meet stringent integrity requirements in fulfillment of, e.g., aviation or maritime applications of certain nations or regions. Galileo will provide the capacity to emit integrity messages for five different regions.

The local component enables the Galileo local assisted services to increase navigation performance in order to meet special application requirements. Accuracy, availability, and integrity shall be increased for locally limited areas. The notation "local" thereby covers ranges between hundreds of meters up to thousands of kilometers. Typically, local component elements are systems with spatial limited coverage for particular services (e.g., airport, maritime harbor) or for special applications (e.g., geosciences, monitoring). The HLD differentiates between four local service categories:

- local precision navigation services: differential code corrections provided through local elements will increase position accuracy to better than 1 meter;
- local high-precision navigation services: applying, e.g., TCAR techniques in conjunction with information emitted from local elements will even increase the position accuracy to below 10 centimeters;
- local assisted navigation services: transmitting the Galileo navigation message via local components will increase acquisition and tracking capability;
- local augmented availability services: pseudolites, emitting Galileo-like signals, will increase the Galileo availability and accuracy.

The Galileo segments have been tested and validated in a three-step program officially denoted as incremental development logic. (1) The objective of the Galileo system test bed (GSTB) was to test the overall architecture and to address the critical technologies. In a first stage an experimental ground segment has been evaluated on its capability to determine satellite orbits, synchronize time, and verify integrity algorithms, whereby GPS satellites have been used to simulate the Galileo constellation. (2) In a second step, experimental Galileo satellites are launched to test space technology and to verify ground elements. (3) During the IOV phase, the first four Galileo satellites will be launched and the ground infrastructure will be available in a first realization. The full deployment of the space segment and the installation of all ground elements will finally lead to the operation phase.

11.4.1 Space segment

The Galileo satellite constellation foresees, as per January 2007, 27 operational and 3 spare satellites positioned in three nearly circular medium earth orbits (MEO). The three orbital planes are 56° inclined with reference to the equatorial plane.

11.4 Galileo segments

According to Falcone et al. (2006a), the eccentricity corresponds to $e = 0.002$. The nominal value for the semimajor axis is $a = 29\,601.297$ km (European Space Agency and Galileo Joint Undertaking 2006: p. 13). This orbit parameter leads to a period of about 14 hours 4 min 45 sec and a ground track repeat cycle of about 10 days, i.e., 17 revolutions. Nine operational satellites are equally distributed by $40°$ in the orbital plane and will be complemented by one spare satellite. Repositioning the satellites within the orbital plane will take up to one week, what is still much less than launching new satellites. According to the HLD document, the spare satellites will be active. In order to increase the satellite lifetime, the spare satellites may also be inactive if the service guarantee can be met with the full satellite constellation. Falcone et al. (2006b) mention that the orbital period has been selected to be long enough to avoid gravitational resonance and to minimize the number of maintenance maneuvers and short enough to allow repeatability measurements. The offset between the actual satellite position and its nominal position will be in the range of $\pm 2°$.

The Galileo satellite constellation guarantees in nominal operation, without spare satellites and satellite failure, a minimum of 6 satellites to be in view to every user worldwide (elevation mask angle $10°$). The maximum PDOP is less than 3.3, the maximum HDOP less than 1.6.

Experimental satellites

During the second stage of the GSTB, a minimum of two experimental satellites is launched with four main objectives. First, emitting Galileo signals secured the frequencies allocated by the International Telecommunication Union (ITU). The deadline to transmit signals in the frequency bands ended in June 2006. Missing this deadline would have meant a loss of allocation and in particular a loss of competitiveness. Thus, two satellites have been developed in parallel to mitigate the risk of satellite or launch failure. The Galileo signals, furthermore, will be used to analyze the achievable performance and evaluate the conformity between simulation and real measurements. Secondly, the experimental satellites will test critical space technologies like the AFS. Third, the satellites will indicate the perturbing forces and the maneuvers to be expected. Fourth, a number of environmental parameters will be determined like the stress during satellite launch or the radiation to be expected during satellite operation.

The experimental satellites have been officially named Galileo in-orbit validation elements (GIOVE). The satellites operate in an altitude of 23 257 km and 23 222 km and have a projected lifetime of three years. The first satellite, GIOVE-A, was launched on December 28, 2005, from the Baikonur cosmodrome using a Soyuz-Fregat rocket. The satellite started emitting Galileo navigation signals on January 12, 2006. The successful operation of GIOVE-A relaxed the timeframe of the second prototype, therefore the launch has been shifted in order to allow further developments. GIOVE-A integrates two rubidium clocks, whereas GIOVE-B will also use passive hydrogen maser (PHM) AFS. A first signal analysis on Galileo GIOVE-A satellites has been published, e.g., in Montenbruck et al. (2006). By February 2007, ESA announced to build a second satellite of prototype GIOVE-A. The objective of GIOVE-A2 is to guarantee a continuous emission of Galileo signals to mitigate the risks induced by satellite failures.

Various other test beds, like the German Galileo test and development environment (GATE), have been installed to test different equipments or algorithms. In contrast to the GPS test bed in Yuma (US), where the transmitter had been placed on ground and the receivers were in airplanes, GATE is based on Galileo signal transmitters installed at mountain tops and the receivers are located on ground (Heinrichs et al. 2004). This allows acceptable DOP values while facilitating the test scenario. For further information about GATE refer to www.gate-testbed.com.

The experience gathered during the operation of the Galileo test beds will be used to settle the design and develop the Galileo space, ground, and user segment.

Operational satellites

After finalization of the first experiments using the GIOVE satellites, four fully operational space vehicles will be launched for the in-orbit validation. Simultaneously to the validation of the space segment, the ground infrastructure will undergo final testing. The four IOV satellites have a projected lifetime of 12 years, thus the IOV satellites will also be used in the full operational constellation.

The Galileo satellites carry two main payloads, namely the navigation and the SAR payload. For a block diagram refer to Fig. 11.2 as adapted from Falcone (2006). The navigation payload consists of the navigation signal generator unit (NSGU), the frequency generation and upconversion unit, amplifiers, signal filters, and the L-band antenna. The NSGU generates the navigation message using orbit and integrity information uplinked by the ground segment and modulates the message onto the ranging codes. The signals are synchronized to the GST using input from the clock monitoring unit and converted to the L-band frequencies using again reference frequencies of the AFS. Before the signals are transmitted (TX) to the users, a low-noise amplifier (LNA) increases the power level and a band-pass filter (BPF) mitigates out-of-band interferences. The signals are broadcast by the L-band antenna, which envelop the earth's surface with a power level independent of the elevation of the satellites with respect to the user position.

The NSGU receives (RX) the navigation and integrity mission data and any other data (e.g., CS data) from the C-band mission receiver that interfaces to the uplink stations of the ground segment.

The Galileo space elements will implement four redundant AFS to guarantee a high level of continuity, reliability, and security. Two rubidium atomic frequency standards (RAFS) secure short-time stability (10 ns per day), whereas the two pas-

11.4 Galileo segments 377

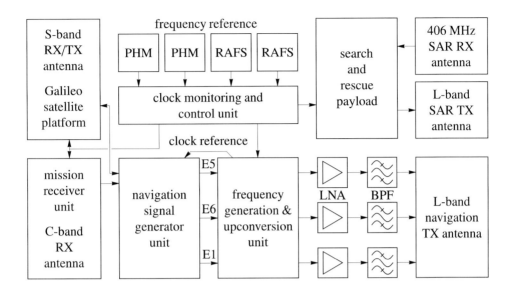

Fig. 11.2. Galileo satellite block diagram

sive hydrogen maser (PHM), Galileo's master clock, will guarantee short- and long-time stability of the frequency output (1 ns per day). The RAFS are smaller and cheaper but experience larger frequency variations, therefore clock correction updates are needed in worst-case scenarios every 100 minutes. The clock prediction validity time of the PHM clocks corresponds to about 8.3 hours.

The ephemerides information transmitted to the satellites is valid for 12 hours (Falcone 2006). The satellites will receive orbital data and clock correction parameters from the ground stations at least once per orbit revolution. A higher update rate is necessary for the integrity information to be relayed to the user.

The SAR payload will receive emergency signals and forward it to the ground facilities using L-band frequencies. The satellites include laser retroreflectors as additional payload for high-accuracy satellite orbit determination.

The telemetry, tracking, and control subsystem of the satellite interfaces via S-band antenna to the ground segment. The subsystem controls and commands the satellite and its payloads according to the data exchanged with the ground stations. The S-band antennas additionally receive, process, and transmit ranging signals, which are used to measure the altitude of the satellites with an accuracy of a few meters.

The attitude determination and control subsystem of the satellite uses the measurements of infrared sensors and visible light detectors to determine the earth-pointing and the sun-pointing direction. The first is used for the alignment of the antenna to the earth, the second to keep the solar panels pointing to the sun.

The fully operational Galileo satellites will have a size of 2.7 × 1.2 × 1.1 m and a weight of 730 kg. The spacecraft length with deployed solar arrays spans about 17.5 m. The projected lifetime of the satellites is 12 years. At the end of life the Galileo satellites will be repositioned into a graveyard orbit.

Different options have been studied to directly inject the Galileo satellites into orbit, e.g., the Ariane-5 rocket. To bring multiple satellites into orbit during a single launch will be more cost-effective but at the same time the risk in case of launch failure increases.

11.4.2 Ground segment

The ground infrastructure is composed of two ground control centers (GCC), five telemetry, tracking, and control (TT&C) stations, nine C-band mission uplink stations (ULS) and about 40 Galileo sensor stations (GSS) (Fig. 11.3). The TT&C stations are colocated with the ULS (except Kiruna), which are themselves colocated with the GSS. This global assembly is complemented by regional ground segments which determine the parameters necessary for the Galileo regional integrity service.

The ground segment is composed of two major elements: the ground control segment (GCS) will control and command the satellite constellation, while the ground mission segment (GMS) will operate navigation system control and integrity determination and dissemination services. The allocation of the different facilities to these segments, whereby the GCC incorporates facilities of the GMS and GCS, is schematically shown in Fig. 11.4.

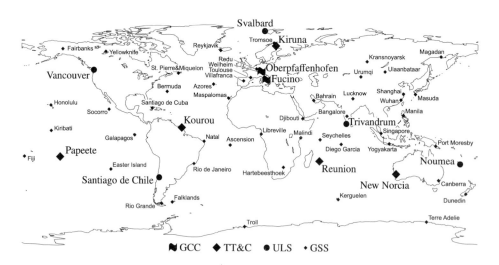

Fig. 11.3. Galileo ground infrastructure

11.4 Galileo segments

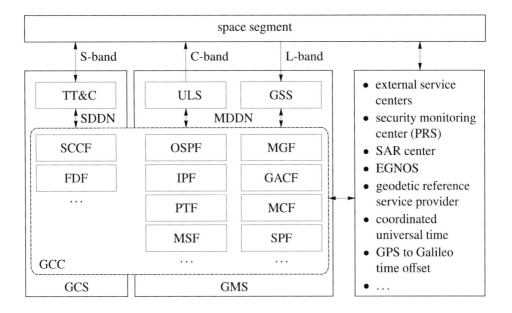

Fig. 11.4. Ground control segment and ground mission segment

The two GCC are located in Oberpfaffenhofen (Germany) and Fucino (Italy). The redundant architecture increases continuity and decreases failure rate. The facilities incorporated by the GCC are beside others (Falcone 2006):

- the orbit determination and time synchronization processing facility (OSPF) estimates the ephemerides and satellite clock correction parameters;
- the integrity processing facility (IPF) processes range measurements and compares them with the signal-in-space accuracy; integrity flags will be generated if a Galileo satellite is identified to cause failures;
- the precise timing facility (PTF) generates and keeps the GST, including two-way time transfer with UTC laboratories;
- the mission support facility (MSF) monitors, controls, and plans the mission in off-line operation;
- the message generation facility (MGF) multiplexes the mission data to be uploaded to the satellites;
- the ground asset control facility (GACF) monitors and controls the ground elements and archives all data;
- the mission control facility (MCF) monitors and controls all online and off-line operations of the mission;

- the service products facility (SPF) provides the interface to external service centers and manages the data from external users, e.g., commercial service users or time service provider;
- the spacecraft constellation control facility (SCCF) monitors and controls the satellite constellation;
- the flight dynamics facility (FDF) calculates the orbital events and maneuvers in online and off-line operation.

A global network of five TT&C stations interface the SCCF with the Galileo satellites using S-band frequencies in the 2 GHz spectrum and using 13-meter antennas, one antenna per site. The stations communicate with the facilities of the GCC via the satellite data distribution network (SDDN). The selection of the TT&C stations followed different criteria, e.g., the maximum loss of contact to a satellite in case of any station failure (cf. Table 11.2).

A worldwide distributed network of GSS permanently monitors the Galileo signal in space (SIS). The known positions of the GSS together with the measurements taken allow to derive orbit parameters, satellite clock offset, and integrity information. The measurements of the monitoring network are transmitted through the mission data dissemination network (MDDN) to the Galileo control centers. The GMS implements three separate chains: one for navigation, one for integrity, and one backup chain.

The results of the navigation and integrity chain are finally uploaded to the satellites via the network of ULS. Each station will host five to six 3-meter antennas, operated in the 5 GHz C-band.

The Galileo control centers will implement a number of interfaces to external services. The geodetic service provider supports all efforts to maintain the GTRF within the specified offset to ITRF. The GGSP uses, as Falcone et al. (2006a) further emphasize, satellite laser ranging measurements to verify and calibrate the orbit parameter estimation process. The time service providers will provide the

Table 11.2. Maximum loss of contact resulting from a failure of a TT&C station (elevation mask angle 5°)

Station failure	Maximum loss of contact [minutes]
Kiruna	193
Kourou	265
New Norcia	130
Papeete	361
Reunion	186

GST steering correction parameters. The external service centers interface between the Galileo ground segment and the commercial service providers relying on the Galileo CS signals. The regional integrity service providers will determine the integrity information using regional integrity-monitoring stations. The integrity information is provided to the users through a GCC-satellite-user link. Furthermore the Galileo ground segment will provide an interface to the COSPAS-SARSAT system, in order to disseminate the SAR returnlink message through the Galileo navigation message.

Integrity determination

One of the main characteristics of Galileo is the real-time provision of integrity information. The major element in the integrity concept is the signal-in-space error (SISE), which expresses the maximum error of the SIS in the range domain, which transforms into a position error using the satellite geometry. The Galileo integrity information splits in two components: The signal-in-space accuracy (SISA) describes low-frequency components in the SISE like gravitational, thermal, or other degradation. The integrity flag (IF) and signal-in-space monitoring accuracy (SISMA) in contrast, as further stated on the ESA Web site (www.esa.int/esaNA/galileo.html), monitor high-frequency components, which have a high variability and fatal amplitude. For the design of Galileo, the SISA has been specified to be less than 0.85 m, the SISMA 0.7 m in nominal condition, and 1.3 m in case of one GSS failure (Falcone 2006). The determination of the integrity parameters is based on the integrity and navigation chain.

The OSPF uses the range measurements of the navigation chain to determine the satellite orbit and the clock offset parameters. The satellite ephemerides sent to the satellites are valid for 12 hours. The short-time stability of the RAFS within the satellites, however, requires in worst-case scenarios regular updates in 100-minute intervals. These update intervals guarantee the contribution of the orbit and satellite clock error to the overall UERE (cf. Eq. (7.80)) to be less than 0.65 m (1σ). Furthermore, the OSPF determines the SISA, which overbounds the predicted ephemerides and satellite clock error, based on the past performance and recent measurements. The SISA neither statistically bounds nonnominal signal conditions nor does it bound user errors like multipath or receiver errors. Dixon (2003) emphasizes that the Galileo SISA is similar to the GPS user range accuracy parameter. Both parameters are used to estimate user ranging and positioning performance. The SISA parameter is uplinked in a maximum interval of 100 minutes.

The IPF uses the measurements of the ground mission segment integrity chain to determine the SISMA and the integrity flag. The IPF estimates the SISE and compares this value with the SISA emitted from the satellites. In nominal operation, the SISA will overbound the SISE using a Gaussian distribution. In a faulty operation, the SISE exceeds the SISA and the integrity flag is set to "not OK". The integrity flag is furthermore used to inform the user when the satellite signal is "not monitored" at all. The ULS transmit the IF to the real-time connected satellites, which relay the IF to the user again in real time in order to meet stringent time-to-alarm requirements. Any user receiver has to confirm at least two satellites, which are connected to the ULS, before initiating a critical operation.

Since the SISE is not known, it is rather estimated using the known position coordinates of the monitoring stations and the range measurements. The estimation process of the SISE additionally results in a statistical distribution function that is described by SISMA. The monitoring accuracy SISMA represents a standard deviation of a Gaussian distribution that shall overbound the error between estimated and real SISE. The SISMA is broadcast from the ground segment to the user nominally in 30-second intervals (Blomenhofer et al. 2005). The user receiver integrates the SISA and SISMA into a protection level which is compared to the alarm limit (Feng and Ochieng 2006).

11.5 Signal structure

The description of the signal structure is primarily based on the draft version of the ICD for open signals (European Space Agency and Galileo Joint Undertaking 2006). Although a number of parameters has been fixed, not all parameters are already settled and therefore subject to change. The reader is advised to keep a look at new releases of ICD for final specification of the parameters. Little information is available about the PRS ranging codes and governmental and commercial navigation message structure.

The Galileo signal design accounted for various influencing factors, among them signal acquisition and signal tracking characteristics, interoperability with other GNSS signals, or resistance against interference and multipath mitigation. The EU and ESA installed a signal task force whose responsibility was to analyze and define the optimal Galileo signals, while maintaining the interoperability with other GNSS.

The denomination of the carrier frequencies (frequency bands) and signal components follows the terminology given in Table 11.3. The carrier frequency E5a coincides with the carrier frequency L5 of the GPS system. Thus, E5a and L5 are used as a synonym. Galileo has been allocated to the frequency band E1, which includes the GPS frequency band L1 and the adjoining bands 1559.052–1563.144 MHz and 1587.696–1591.788 MHz. These frequency bands have been formerly named E1 and E2, ending up in the terminology E2-L1-E1 for the complete band. Meanwhile ESA changed the terminology to E1 for the complete band, and the adjoining bands are not named separately.

The selection of the L-band frequencies for navigation had been preceded by an analysis of other frequency options, especially the C-band was under investigation

11.5 Signal structure

Table 11.3. Galileo signal lexicon

E1	Carrier frequency 1 575.420 MHz; also denoted as L1 (US denomination)
E6	Carrier frequency 1 278.750 MHz
E5	Carrier frequency 1 191.795 MHz; also denoted as E5a+E5b
E5a	Carrier frequency 1 176.450 MHz; also denoted as L5 (US denomination)
E5b	Carrier frequency 1 207.140 MHz
E1A, E1B, E1C	The three signal components (A, B, C) of E1
E6A, E6B, E6C	The three signal components (A, B, C) of E6
E5a-I, E5a-Q	The in-phase and quadrature signal components of E5a
E5b-I, E5b-Q	The in-phase and quadrature signal components of E5b
SAR downlink	Frequency band 1 544.050–1 545.150 MHz
SAR uplink	Frequency band 406.0–406.1 MHz

(Irsigler et al. 2002, Hammesfahr et al. 2001). The conclusion of all these analysis has been that the C-band is a promising option for a next-generation Galileo, but the L-band is the one for the first-generation Galileo. Additionally to the carrier frequencies and signal components listed in Table 11.3, Galileo uses a number of other frequencies out of the C-band and S-band for uplink and downlink to the satellites.

The frequency bands have been allocated to Galileo during the World Radiocommunication Conferences in 2000 and 2003. In parallel, the ITU also extended the RNSS and ARNS frequency bands. According to the regulations of the ITU, a system allocated to a frequency band has to transmit signals in this band in a certain timeframe in order to avoid deallocation of the system. Galileo has avoided this threat by launching the first experimental satellite in 2005.

11.5.1 Carrier frequencies

The carrier frequencies and all timing processes are based on the fundamental frequency $f_0 = 10.23$ MHz, which is coherently derived from the onboard AFS. The fundamental frequency is intentionally reduced by $\Delta f \sim 5 \cdot 10^{-3}$ Hz in order to compensate for relativistic effects (cf. Sect. 5.4). The carrier frequencies of the navigation signals are listed in Table 11.4. Frequencies allocated to the ARNS band are especially useful for safety-critical applications, since the usage of these frequency bands is strictly regulated by national and international agreements.

Table 11.4. Galileo frequency bands

Link	Factor ($\cdot f_0$)	Frequency [MHz]	Wavelength [cm]	ITU allocated bandwidth [MHz]	Frequency band
E1	154	1 575.420	19.0	32.0	ARNS/RNSS
E6	125	1 278.750	23.4	40.9	RNSS
E5	116.5	1 191.795	25.2	51.2	ARNS/RNSS
E5a	115	1 176.450	25.5	24.0	ARNS/RNSS
E5b	118	1 207.140	24.8	24.0	ARNS/RNSS

The frequency bands listed in Table 11.4 are shared between satellite navigation systems and nonradionavigation services. Distance measuring equipment and tactical air navigation systems emit signals for aeronautical users on the frequency bands E5a and E5b. The same frequencies are also used by military systems for information transmission. Galileo signals in the E6 frequency band will overlay with primary radar, wind profilers (sensitive Doppler radars), or signals from radio amateurs (Hollreiser et al. 2005). Two frequency bands, E5a and E1, have been chosen purposely in common to GPS to increase interoperability and compatibility between Galileo and GPS. GLONASS carrier G3 will overlay with Galileo E5b.

The large frequency difference between E1 and E5 will be advantageous to compute ionospheric corrections. The frequency offset between E5a and E5b results in a carrier phase combination with a wavelength of 9.8 m, thus of particular interest for ambiguity resolution.

11.5.2 PRN codes and modulation

A number of different ranging codes and navigation messages had to be specified to meet the various application requirements of the Galileo services. Ten navigation signals have been defined in the four frequency bands E5a, E5b, E6, and E1. Three types of ranging codes are distinguished: the open-access ranging code (which is not encrypted and publicly known), the ranging codes encrypted with commercial encryption, and finally the ranging codes encrypted with governmental encryption. All satellites use the same carrier frequencies for signal transmission. The signals, thus, are differentiated by their spread spectrum using the principles of code division multiple access (CDMA).

Hollreiser et al. (2005) emphasize that the implementation of the ranging codes and signal modulation onboard the Galileo satellites is specified to be flexible, in order to allow uploading of new modulation tables and code sequences into the satellite memory.

11.5 Signal structure

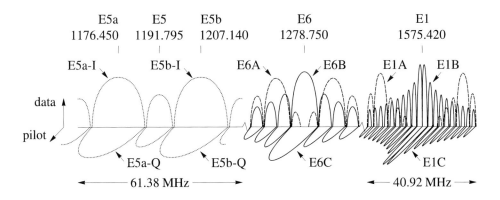

Fig. 11.5. Power spectral densities of Galileo signals

Dataless signals, also denoted as pilot channels or pilot tones, have been specified to increase tracking robustness. These signals consist of the ranging code sequence only, whereas the data channels also contain navigation messages. All Galileo signals, except the PRS components E6A and E1A, come in pairs. This is accounted for in Fig. 11.5 by plotting the data channel and pilot channel in orthogonal planes. The pilot channel allows for long coherent integration time, thereby sensing even weak signals.

Figure 11.5 shows the power spectral density envelopes of the Galileo signals. Note that the displayed frequency bandwidths, which have been chosen for better visualization, are different from the available bandwidths listed in Table 11.4.

Six signals, including three data signals and three pilot channels, are accessible to all Galileo users on the E5a, E5b, and E1 carrier frequencies for OS and SoL services. One data channel and one pilot tone are cryptographically modulated onto the carrier E6. These two signals are reserved for CS users. The CS additionally relies on commercial data bits cryptographically emitted in the navigation message of the open signals. The PRS relies on two signal components modulated with encrypted ranging codes and navigation messages on the carrier frequencies E6 and E1. The access to these signals is controlled.

The user minimum received power above $10°$ elevation is defined to be in the range of -155 to -157 dBW (European Space Agency and Galileo Joint Undertaking 2006: p. 22). The relative power between the channels of each carrier are listed in Table 11.6. Note that the power share of pilot and data channel, e.g., E1B to E1C or E5a-I to E5a-Q, is always 50%.

The ranging code sequences are either generated using a linear feedback shift register (LFSR) or they are optimized, constructed codes stored in satellite memory. LFSR code sequences are generated by a combination of two maximal length LFSR sequences that are short cycled after the code length specified in Table 11.5.

Table 11.5. Galileo ranging codes

Link	PRN code	Channel	Code length primary [chip]	Code length secondary [chip]	Code rate [Mcps]	Modulation type [1]
E1	E1A	data	(2)	(2)	2.5575	BOCc(15,2.5)
	E1B	data	4092	1	1.023	MBOC(6,1,1/11)
	E1C	pilot	4092	25	1.023	MBOC(6,1,1/11)
E6	E6A	data	(2)	(2)	5.115	BOCc(10,5)
	E6B	data	5115	1	5.115	BPSK(5)
	E6C	pilot	5115	100	5.115	BPSK(5)
E5	E5a-I	data	10230	20	10.23	BPSK(10)
	E5a-Q	pilot	10230	100	10.23	BPSK(10)
	E5b-I	data	10230	4	10.23	BPSK(10)
	E5b-Q	pilot	10230	100	10.23	BPSK(10)

[1] Multiplexing scheme (E1, E6): constant envelope;
Multiplexing scheme (E5): AltBOC(15,10)
[2] Classified information

The European Space Agency and Galileo Joint Undertaking (2006: pp. 33–45) define the characteristic polynomials for a number of ranging codes. Therefore the ICD specifies unique octal numbers, which indicate the registers to be used for the LFSR. The polynomial octal notation 40503, for instance, is first transformed into its binary equivalent 100 000 101 000 011. In the sequel, the binary number is read from right to left, whereas the rightmost bit, which is the least significant bit, is not considered. Thus the octal number 40503 indicates the register feedback cells 1, 6, 8, and 14. The ICD furthermore specifies the initial values of the LFSR given also in octal notation. After the transformation of the octal number into the binary equivalent, here, however the leftmost bit, which is the most significant bit and always 0, is not considered.

The Galileo codes are tiered code sequences, where the logic levels of a long high-frequent primary code and a short low-frequent secondary code are XOR-added. The chip length of the secondary code coincides with the code length of the primary code (Fig. 11.6). Therefore, the code length of the tiered code sequence N_t is composed of the code length of the primary code, N_p, and the code length of the secondary sequence, N_s, yielding

$$N_t = N_p N_s . \tag{11.1}$$

The long tiered code length increases the robustness of the signal, while the short repetitive period still allows for short acquisition time in favorable signal environ-

11.5 Signal structure

Table 11.6. Galileo navigation messages

Link	Services	Data rate [bps/sps]	Encryption	Relative power [%]
E1	PRS	50/100	ranging code and data	50
	OS/CS/SoL	125/250	selected data fields	25
	OS/CS/SoL	—	—	25
E6	PRS	50/100	ranging code and data	50
	CS	500/1000	ranging code and data	25
	CS	—	ranging code	25
E5	OS/CS	25/50	selected data fields	25
	OS/CS	—	selected data fields	25
	OS/CS/SoL	125/250	selected data fields	25
	OS/CS/SoL	—	—	25

ment with high signal power. This is a trade-off against losing correlation properties compared to a random noise sequence of the same tiered code length. The code lengths are defined in Table 11.5. The secondary codes are fixed sequences as defined in hexadecimal notation in the ICD of European Space Agency and Galileo Joint Undertaking (2006: pp. 42–44). Only one realization of the 4-, 20-, and 25-bit secondary codes is defined each, although for the latter two there exists a greater family of maximal length codes. The same codes are used for all associated primary codes. Unique secondary codes with a length of 100 bits, in contrast, are defined for every primary code. Assignment of primary and secondary code numbers to satellites will be published in a future update of the ICD.

Following the definitions in Sect. 4.2.3, the BOCc(10,5) modulation type (Table 11.5) specifies a code sequence with chipping rate $f_c = 5.115$ MHz and subcarrier frequency of $f_s = 10.23$ MHz. The open signals on E1 are sine-phased modulated BOC codes, whereas the PRS signals on E1 and E6 are cosine-phased modulated BOC codes.

E1-codes

Three navigation signal components are modulated onto the carrier frequency E1. The OS signals E1B and E1C are unencrypted and accessible to all users. The data channel E1B transports, beside the general navigation message, integrity information and encrypted commercial data. The data channel E1B and the pilot channel E1C support the OS, CS, and SoL services. The E1A signal component is encrypted and only accessible to authorized PRS users. The E1A ranging code is modulated by a governmental navigation message. The modulation type envisioned for E1A

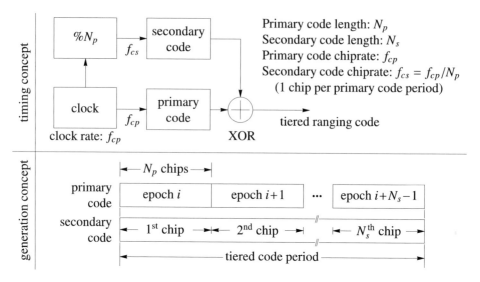

Fig. 11.6. Tiered code generation (European Space Agency and Galileo Joint Undertaking 2006)

will especially place signal power into the frequency bands covered by E1 but not by the L1 band. This spectral separation makes it resistant against narrowband interference.

The ranging sequences E1B and E1C will be modulated by MBOC(6,1,1/11). In an agreement between the US government and the European Commission (United States of America and European Community 2004), the US and EU agreed to implement signals of a common modulation on the carrier frequency L1/E1. In this way the combined use of both systems shall be facilitated. In the agreement, furthermore, it has been decided to use the BOCs(1,1) as baseline, but at the same time to analyze further modulation options. In March 2006, a GPS-Galileo Working Group on Radio Frequency Compatibility and Interoperability recommended in an official statement the multiplexed BOC modulation MBOC(6,1,1/11), which is defined by Eq. (4.57). Combining two BOC signals will add more power to the higher frequencies, thereby increasing the interference stability. Only wideband receivers taking full advantage out of the spectral spreading will also benefit from the MBOC design. Narrowband receivers only processing the BOCs(1,1) component of the MBOC modulation will have to cope with lower signal power. Meanwhile the MBOC has been chosen as the new baseline.

The Galileo ICD for open signals (European Space Agency and Galileo Joint Undertaking 2006: Annex 1) defines 50 unique pseudorandom memory code sequences for the primary codes of E1B and E1C. The code sequences for E1A will not be published.

11.5 Signal structure

All three signal components, E1A, E1B, and E1C, are modulated onto the E1 carrier frequency using a modified hexaphase modulation. Thereby the E1A signal component is modulated onto the quadrature phase signal, whereas the other two are modulated onto the in-phase signal. The composite signal reads, omitting the indication of time dependency and the indication of the frequency band E1 in the subscript of the signal components

$$s_{E1} = \sqrt{2P} \, [\alpha \, c_B \, d_B - \alpha \, c_C] \cos(2\pi f_{E1} t) - \\ \sqrt{2P} \, [\beta \, c_A \, d_A + \gamma \, c_A \, d_A \, c_B \, d_B \, c_C] \sin(2\pi f_{E1} t) \, . \quad (11.2)$$

The $c_B \, d_B$ signal component consists of the I/NAV navigation data stream, the PRN code sequence, and the MBOC subcarrier. The c_C pilot signal component consists of the PRN code sequence which is MBOC modulated. The $c_A \, d_A$ signal consists of the G/NAV navigation data stream and the encrypted ranging code sequence that are cosine-phased BOC modulated onto the carrier frequency. The second element of the quadrature phase component in (11.2) is the intermodulation product, which ensures according to Kreher (2004) the constant power spectrum envelope property of the transmitted signal. The coefficients α, β, and γ determine the power distribution among the four components, whereby γ shall be a minimum.

The components E1B and E1C are sometimes jointly denoted as E1F (freely), whereas E1P is used as a synonym for E1A.

E6-codes

Similar to E1, three ranging codes are modulated onto the carrier frequency E6. The first one is reserved for PRS, whereas the other two are designated to the CS. The navigation and commercial data is encrypted onto the data channel E6B. A data rate of 500 bps grants a high data throughput. E6C has been designed as pilot tone. E6A carries public regulated service data.

The E6B signal component results of the XOR-addition of the navigation data stream and the ranging code sequence. E6B is BPSK(5) modulated onto the carrier frequency with a chipping rate of 5.115 Mcps. The pilot channel E6C is also BPSK(5) modulated onto the carrier frequency. The E6A navigation signal component consists of the XOR-addition of the navigation data stream and the ranging code sequence that is BOCc(10,5) modulated, with a code chipping rate of 5.115 Mcps and a subcarrier frequency of 10.23 MHz. The three signal components are modulated onto the carrier frequency using similar modified hexaphase modulation as defined in (11.2).

Similar to the E1-codes, also the signal components E6A and E6C are jointly denoted as E6C (commercial), whereas E6P is used as a synonym for E6A.

E5-codes

The Galileo E5 signal carries four signal components: a pair of data channel and pilot tone in the frequency bands E5a and E5b each. All four are OS signals having a chipping rate of 10.23 Mcps. The data channel in the E5a frequency band, commonly denoted E5a-I, is modulated by an unencrypted ranging code and a freely accessible navigation message. The data message has a low data rate of 25 bps, what increases data demodulation robustness. The high chipping rate of E5a-I and E5a-Q in combination with the low data rate and the pilot tone characteristic facilitates the reception in weak signal environments, like indoors.

The data channel in the E5b frequency band, denoted as E5b-I, carries unencrypted navigation data and integrity information which is used for SoL services. Encrypted bytes within the message are additionally used by the CS. The data channel and the pilot signal of E5a and E5b have been assigned 25% of the E5 relative power each. 15% of the absolute power has been assigned to the intermodulation product (European Space Agency and Galileo Joint Undertaking 2006: p. 27).

The E5a and E5b signal components are modulated onto the E5 carrier frequency using the AltBOC(15,10) modulation method with a subcarrier frequency of 15.315 MHz and a chipping rate of 10.23 Mcps. The coherently generated composite signal can be processed jointly as a single wideband (51.15 MHz) signal with 20.46 MHz spectral mainlobes that are 30.69 MHz separated (Wilde et al. 2004). Such a wideband signal especially shows low multipath errors and high code tracking accuracy. In contrast to a true BOC modulation, the two mainlobes in the power spectrum are caused by the two different spreading codes of the E5a and E5b signal components. Alternatively the signal components can also be processed independently in single sideband methods using only 24 MHz frequency bandwidths.

The AltBOC signal can be described as an eight-phase modulation signal, thus distinguishing between eight different phase states. A generic view of the E5 signal modulation is given in Fig. 11.7 (European Space Agency and Galileo Joint Undertaking 2006: p. 24).

The E5 primary codes are either memory codes as listed in the ICD (European Space Agency and Galileo Joint Undertaking 2006: Annex 1) or generated using LFSR.

11.5.3 Navigation messages

The navigation message is generated by the ground segment and uploaded to the Galileo satellites. Five different types of data content have been envisaged for the Galileo messages: the positioning/navigation data, integrity data, the supplementary data, the public regulated data, and data for SAR operations.

The navigation data will include the parameters necessary for position determination like ephemerides information, satellite clock reading, space vehicle identi-

11.5 Signal structure

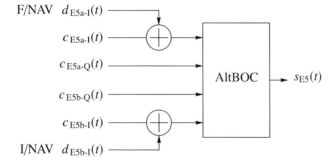

Fig. 11.7. Modulation scheme for the E5 multiplexed signal

fication, satellite status flag, and almanac information. The ephemerides for each Galileo satellite consist of 16 parameters similar as listed in Table 3.7, including beside others the Keplerian parameters and the ephemerides reference epoch. The total data size of these 16 parameters corresponds to 356 bits. The Galileo ephemerides message emitted by the satellites is valid for 4 hours. About every 3 hours, the message will be updated. The proposed Galileo almanac orbital parameters consist of the parameters similar as listed in Table 3.6. The satellite clock readings are inserted in regular intervals as time stamps in the navigation message. Demodulating the clock readings and referring to the leading edge of a particular chip of the navigation message relates the satellite signal to the GST.

The integrity data consists of the integrity information as determined by the Galileo control station (cf. Sect. 11.4.2) and constitutes the main part of the SoL service.

The five different data content types (Table 11.7) define four navigation message types (Table 11.8), whereby the terminology of the message types expresses the respective primary focus. All message types have spare fields that are reserved bytes for future use.

The navigation messages consist of a sequence of frames that in turn are composed of subframes. The basic structure to build the subframes and therefore the navigation messages are the pages. The data is spread over this three-layer data structure in dependency whether it has to be transmitted at fast, medium, or slow rates. Urgent data is transmitted by spreading it over few pages. Slow-rate data, like the ones required for cold start TTFF, is spread over greater message structures (i.e., frames).

Every message page contains a synchronization word and the data field. The synchronization word is a binary pattern that is used to gain synchronization to the data fields. This unique word of M symbols is modulated unencoded onto the

Table 11.7. Content of the Galileo message types

Message type Channels	F/NAV E5a-I	I/NAV E1B	I/NAV E5b-I	C/NAV E6B	G/NAV E1A, E6A
Positioning/Navigation	×	×	×		×
Integrity		×	×		×
Supplementary				×	
Public regulated service					×
Search and rescue		×			

satellite signal. The F/NAV message type, for instance, uses a binary pattern of 12 symbols, whereas the I/NAV message relies on a 10-symbol-long preamble.

Galileo applies a three-level error correction encoding strategy to decrease the bit error rate despite increasing data rates: a cyclic redundancy check (CRC), a half-rate convolutional forward error correction (FEC) encoding, and block interleaving. This strengthens, as Hollreiser et al. (2005) emphasize, the transmitted message while reducing the error rate.

The CRC bits added to the navigation message allow to detect corrupted data messages. The checksum does not take the synchronization pattern and the tail bits into account, since these do not add to the required information. The CRC parity block consists of 24 bits. The CRC generator polynomial is described in detail in the ICD (European Space Agency and Galileo Joint Undertaking 2006: pp. 76–77). The tail bits that consist of six zero-value bits are necessary for the FEC encoding.

The data page bit train is half-rate convolutional encoded. Therefore, the symbol rate in [sps] is twice the original data rate in [bps]. The data symbol rates are between 50 Hz and 1 kHz, whereby the chip length of one data message symbol is a multiple of the code length of the Galileo ranging codes. In this way a code ambiguity is avoided. The convolutional coding is characterized by the constraint

Table 11.8. Galileo message types

Message type	Acronym	Services	Channels
Freely accessible navigation message	F/NAV	OS	E5a-I
Integrity navigation message	I/NAV	OS/CS/SoL	E5b-I, E1B
Commercial navigation message	C/NAV	CS	E6B
Governmental navigation message	G/NAV	PRS	E1A, E6A

11.5 Signal structure

length 7 and the polynomials $G_1 = 171$ and $G_2 = 133$, both in octal notation. The convolutional encoding scheme is described in Sect. 4.3.4.

The symbols of the convolutional encoded data pages, excluding again the synchronization pattern, are reordered following the block interleaving schematic. Therefore, the symbols fill up a matrix column by column. The symbols, in the sequel, are transmitted row by row (cf. Sect. 4.3.4). The block interleaver matrix dimensions for the F/NAV message type correspond to $61 \cdot 8 = 488$ symbols, whereas the I/NAV message uses a $30 \cdot 8 = 240$ interleaver matrix.

This three-level encoding scheme requires appropriate tasks to retrieve the original message out of the received symbols. First the synchronization pattern has to be detected and the page symbols deinterleaved. The page symbols are decoded into data bits following the Viterbi decoding algorithm. Encrypted data messages have to be deciphered. Finally, the computed checksum is compared with the received one before the data bits are useable.

Subsequently, the navigation messages F/NAV and I/NAV are explained in more detail. Note that no information is yet available for the messages C/NAV and G/NAV.

F/NAV message

The F/NAV message is defined by frames with a length of 600 seconds length. A frame is subdivided into 12 subframes with a length of 50 seconds. One subframe furthermore is split up into 5 pages with a length of 10 seconds each. The general page structure as mentioned before consists of the synchronization field and the data field. The F/NAV data field consists of a page type field, the navigation data field, the CRC bits, and tail bits. The type field of six bits identifies the page content. The navigation data field with a length of 208 bits transmits the almanac, the ephemerides data, and general satellite information. The latter two are transmitted in pages 1 to 4. Page 5 contains the almanac information. The transmission of the first symbol of every page is synchronized with GST.

The almanac information of three satellites is transmitted in two consecutive subframes taking 100 seconds for transmission. The ICD (European Space Agency and Galileo Joint Undertaking 2006: p. 52) mentions that a capacity for 36 satellites has been projected resulting in a transmission time of 20 minutes for transmitting the almanac information of the entire Galileo constellation.

Every satellite, however, does not transmit the same subframe at the same instant of time, what is called satellite diversity. The subframe identification number is alternated as function of time and satellite. Different subframes contain different almanac information; in this way, when tracking more than one satellite, demodulation of almanac information from several satellites becomes possible in a short period of time. This furthermore decreases the TTFF and increases the number of tracked satellites despite the short tracking periods.

I/NAV message

The navigation message of E5b-I and E1B provide a dual-frequency service, also denoted as frequency diversity. The channels transmit the same information in the same page layout, however with different page sequencing. Demodulating the navigation message of both allows for fast reception of data. Single-frequency receiver, demodulating only one navigation message, will receive the same information; however, in about the double time period.

Two types of pages have been specified for the I/NAV message: the nominal and the alert page. Several types of nominal pages with a length of 2 seconds have been defined that are transmitted in a different sequencing on the channels E5b-I and E1B and also in a different sequencing of the satellites. The general page layout for the I/NAV channels consists of an even/odd field to indicate the part of the page, the type field that differentiates the nominal from an alert message, and the data field which consists of a total of 128 bits.

The alert page has a length of 1 second and is transmitted on both carrier frequencies in parallel. The alert content is spread over two pages. The first part is transmitted on the first channel (E5b-I), while the second part is transmitted on the second channel (E1B) at the same epoch. In the next epoch, the page content of the channels of the previous step is switched accordingly. Similar to the nominal channel, the alert message consists of an even/odd field, the type field, and a data field that is split in two pages.

11.6 Outlook

The definition phase of the Galileo program was completed in 2003. The development and in-orbit validation phase is ongoing (status April 2007). The first experimental satellite of the GSTB has been launched in December 2005 and is operating since January 12, 2006. The second experimental satellite is expected for late 2007 or early 2008. A contract has been awarded to build a third GSTB satellite (GIOVE-A2), which will be presumably launched in the second half of 2008. Four fully operational satellites will be launched in the timeframe 2008/2009. The FOC of Galileo, at least for the non-safety-critical services, has been projected for 2012/2013. The safety-critical services will be operational about two years later. Galileo will be managed and operated by a concessionaire.

In the white paper on the implementation of the European space policy considering Europe's growth, competitiveness, and employment (European Commission 2003b) the importance of satellite navigation has been highlighted. Meanwhile, the European Commission issued the first European directives, where the European GNSS will play a major role. One of the first has been the directive about the pan-European electronic toll system issued on April 29, 2004 (European

11.6 Outlook

Parliament and European Council 2004), which recommends the use of satellite positioning for electronic toll systems.

Close cooperation between the Galileo program and ICAO, IMO, and other international organizations facilitates the introduction of the Galileo services even in safety-critical applications. The agreement between Europe and other nations further highlights the importance of the European initiative and brings Galileo closer to a world standard.

Different market analysis talk about a window of opportunity for Galileo. Due to the delays in the modernization programs of the other GNSS and the slower development of the location-based service and other services, the timeframe has been extended. But the deployment of Galileo and FOC will not be shiftable arbitrarily without losing competitiveness.

The discussion for the technological evolution of Galileo has already started. The usage of the C-band (5 010 – 5 030 MHz) for ranging signals is under consideration as well as intersatellite links for autonomous operation and faster information distribution.

12 More on GNSS

Satellite navigation systems are categorized into one-way and two-way ranging systems. One-way systems either measure ranges or range rates using signals sent from earth to space (uplink) or from space to earth (downlink). Considering two-way ranging systems, the signals travel the distance between user and satellite or vice versa two times. Some system concepts rely on signals sent from ground stations via satellites to the user and back again.

A further categorization differentiates between active systems, which require the user equipment to emit signals, and passive systems, where users only receive signals. Two-way systems are always active systems. Systems have been conceived to provide navigation information for users on a global scope. Others limit their services for certain regions. Depending on the different methods applied, systems have a limited capacity or they serve an unlimited number of users. These are only some but a few features to categorize satellite navigation systems.

12.1 Global systems

12.1.1 Comparison of GPS, GLONASS, and Galileo

Chapters 9 through 11 discussed in detail the global one-way downlink satellite systems GPS, GLONASS, and Galileo. Table 12.1 summarize important system parameters. The following paragraphs highlight the issues of compatibility and interoperability and the steps to be taken for a combined navigation solution.

Compatibility and interoperability

According to the definition of the US space-based position, navigation, and time policy (www.navcen.uscg.gov/cgsic/geninfo/FactSheet.pdf), compatibility refers to the ability of two services to be used separately or together without interfering with each individual service or signal. Interoperability, in contrast, refers to the ability to use two services together to achieve better performances at user level. The different global systems have been designed to be compatible. Meanwhile, an increasing number of agreements between the operators guarantees the interoperability of systems and signals. Signals have been specified to be in common between the systems, nevertheless some signals have intentionally been separated to avoid common mode failures.

The coexistence of the three GNSS will either result in an alternative use or in a combination of the services and signals to gain a combined solution. An increasing

Table 12.1. GPS, GLONASS, and Galileo at a glance

Characteristic	GPS	GLONASS	Galileo
First launch	February 22, 1978	October 12, 1982	December 28, 2005
Full operational capability	July 17, 1995	January 18, 1996	2012/2013 [1]
Funding	public	public	public & private
Nominal number of SV	24	24	27
Orbital planes	6	3	3
Orbit inclination	55°	64.8°	56°
Semimajor axis	26 560 km	25 508 km	29 601 km
Orbit plane separation	60°	120°	120°
Phase within planes	irregular	±30°	±40°
Revolution period	11h 57.96 min	11h 15.73 min	14h 4.75 min
Ground track repeat period	~1 sidereal day	~8 sidereal days	~10 sidereal days
Ground track repeat orbits	2	17	17
Ephemerides data	Kepler elements, correction coefficients	position, velocity, acceleration vectors	Kepler elements, correction coefficients
Geodetic reference system	WGS-84	PE-90	GTRF
Time system	GPS time, UTC (USNO)	GLONASS time, UTC(SU)	Galileo system time
Leap seconds	no	yes	no
Signal separation	CDMA	FDMA	CDMA
Number of frequencies	3 – L1, L2, L5	one per two antipodal SV	3(4) – E1, E6, E5(E5a, E5b)
Frequency [MHz]	L1: 1 575.420	G1: 1 602.000	E1: 1 575.420
	L2: 1 227.600	G2: 1 246.000	E6: 1 278.750
	L5: 1 176.450	G3: 1 204.704 [1]	E5: 1 191.795
Number of ranging codes	11	6 [1]	10
Integrity transmission	No (GPS III: yes)	No (GLONASS-K: yes[1])	yes

(1) to be confirmed

12.1 Global systems

Table 12.1. (continued)

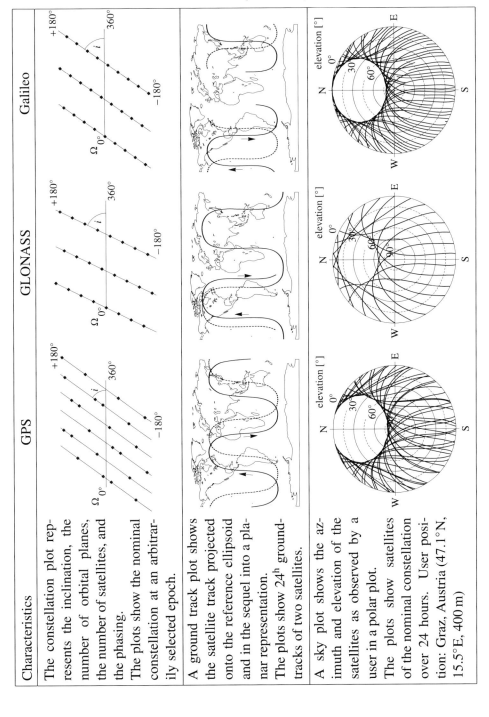

number of systems and signals will provide an increasing number of observations. Remembering Eq. (7.172) and the discussion about dilution of precision (DOP), one could conclude that in general with an increasing number of satellites the DOP values decrease, and, consequently, the position accuracy increases. Using two systems for position determination will almost double the number of navigation satellites and, thus, double the number of observations for position computation. However, this does not necessarily decrease the DOP values, although it will be the common case. Nevertheless, from the increasing redundancy in the adjustment process, the position accuracy, availability, integrity, and continuity will benefit.

Today's mass market receivers are based on the GPS C/A-code using L1 frequency only. Table 9.2 lists the achievable position accuracies. Depending on the system, services, frequencies, and ranging codes as well as depending on their combination, the positioning performance of a mass market receiver is expected to increase by a factor of 4 to 10, especially when augmentation information (Sect. 12.4) is additionally implemented. Note that in spite of all the modernization programs of GPS and GLONASS, the systems guarantee a backward compatibility of the ranging codes and navigation messages.

Combined solution

Although common reference systems would have facilitated the interoperability, the systems have been intentionally designed to use different reference frames, in order to avoid common mode failures and, thus, to increase the integrity of combined solutions (Hein et al. 2002).

Coordinate systems

The different coordinate reference frames influence the satellite coordinates. Consequently, the satellite coordinates have to be transformed into a common system before the adjustment process is applied. Any difference in coordinate frame can be considered as orbital error of the respective satellites. Transformation parameters between the different coordinate systems are given according to the Helmert transformation as described in Sect. 8.3.1. Due to the similarity of the reference systems, the transformation can be applied in differential form.

The Galileo terrestrial reference frame (GTRF) is specified to maximally differ from the latest version of international terrestrial reference frame (ITRF) by no more than 3 centimeters (2σ). Although there does not exist a similar specification for WGS-84, the difference to ITRF has been determined to be also within this range. For navigation purposes and most user requirements, the agreement between ITRF, GTRF, and WGS-84 is sufficient and no coordinate transformations have to be applied. For geosciences, surveying, and other high-accuracy applications, an appropriate transformation has to be applied.

Roßbach et al. (1996) and Zinoviev (2005) describe a number of different trans-

12.1 Global systems

formation parameters for the transformation between PE-90 and WGS-84 coordinate systems. As Roßbach et al. (1996) emphasize, any of the tabulated coordinate transformations may be used for navigation applications, where meter-level accuracy is sufficient. The Radio Technical Commission for Maritime Services (2001: Appendix H) specifies a standard transformation between PE-90 and WGS-84 using as rotation angle around the Z-axis of −0.343 arcseconds. The scale is equal to one, and all other parameters are set to zero. Averin (2006) mentions that PE-90 will in future be improved to better agree with ITRF.

Global transformation parameters will be available for all reference systems. Users may also determine their local transformation parameters if necessary.

Time systems

The time offset between the different reference time systems will be emitted in the navigation message of the systems. Various agreements, e.g., between US and EU, Russia, and Japan, already specify the time offsets and its provision to the user. For example, the agreement between the US and EU specifies the accuracy of the GPS to Galileo time offset (GGTO) to be better than 5 ns (2σ) over any 24 hours.

The time offset between two systems might either be applied as transmitted in the navigation message or be modeled as an additional unknown in the parameter estimation (Borre et al. 2007: p: 134). Considering three Galileo observations and two GPS observations, then the right side of the observation equation (7.144) reads

$$\mathbf{A}\mathbf{x} = \begin{bmatrix} a_{X_r}^{E1} & a_{Y_r}^{E1} & a_{Z_r}^{E1} & c & 0 \\ a_{X_r}^{E2} & a_{Y_r}^{E2} & a_{Z_r}^{E2} & c & 0 \\ a_{X_r}^{E3} & a_{Y_r}^{E3} & a_{Z_r}^{E3} & c & 0 \\ a_{X_r}^{G1} & a_{Y_r}^{G1} & a_{Z_r}^{G1} & 0 & c \\ a_{X_r}^{G2} & a_{Y_r}^{G2} & a_{Z_r}^{G2} & 0 & c \end{bmatrix} \begin{bmatrix} \Delta X_r \\ \Delta Y_r \\ \Delta Z_r \\ \delta_r^E(t) \\ \delta_r^G(t) \end{bmatrix}, \quad (12.1)$$

where δ_r^E denotes the receiver clock unknown with respect to Galileo system time and δ_r^G denotes the receiver clock unknown with respect to GPS time. The superscripts E, G agree with the receiver-independent exchange format (RINEX) definition (Gurtner and Estey 2006), which uses the letter 'G' to identify GPS, 'E' for Galileo, and 'R' for GLONASS. Observations of at least two satellites from one system should be available, otherwise one observation will only determine the time offset, but will not contribute to the position solution.

12.1.2 Beidou-2/Compass

The People's Republic of China carries out space activities since the 1970s, when it started its first satellite. The idea of satellite navigation has been pursued since the early 1980s, following a regional system concept similar to the one proposed and build by the US GEOSTAR corporation. The concept foresees two geostationary satellites to determine position in a two-way ranging method. In 1989, two communication satellites have been successfully used for first tests under the Twin-Star program. In 1993, China decided to implement an independent navigation system called Beidou. The name Beidou denotes the seven-star constellation also known as Ursa Major, Great Cart, or Big Dipper. This constellation has been used for centuries to identify the star Polaris, which indicates the north direction on the northern hemisphere.

Little official information is publicly available about Beidou. Interface specification documents are not provided, therefore most of the system parameters are still to be considered preliminarily.

A two-step approach has been chosen for the development of Beidou. In its first step, Beidou-1, the regional system concept of the Twin-Star program has been realized (cf. Sect. 12.2.1). In a second step, the satellites of Beidou-1 should have been complemented by a number of additional satellites to provide higher performances. Various design options have been studied for Beidou-2, incorporating geostationary as well as medium earth orbit (MEO) satellites. Meanwhile, China extended its original concepts by planning to gradually build up its own global system like GPS, GLONASS, and Galileo.

Beidou-2 will be a global passive one-way downlink ranging system. In its first development step, it will provide service to China and parts of neighboring countries by 2008. In the latest iteration of the Beidou system, referred to as Compass, the Chinese navigation satellite system will provide global coverage.

Beidou-2 will be a dual-use system. The civilian open service is designed to provide position accuracy of 10 m, velocity accuracy of $0.2\,\mathrm{m\,s^{-1}}$, and timing accuracy of 50 ns. No detailed performance parameters have been published for the authorized service. However, it has been announced that it will provide a higher integrity level.

The system architecture is similar to the other global systems. According to the Radio Regulatory Department (2006), the Beidou-2 satellite constellation will consist of 27 MEO satellites, 5 satellites in geostationary orbit (GEO), and 3 more satellites in geosynchronous orbit. Feairheller and Clark (2006: p. 624) mention that beside this latest design option, China has announced three other constellations during previous filings to the International Telecommunication Union (ITU).

The 27 MEO satellites will have an average satellite altitude of 21 500 km. A number of 24 satellites will be evenly distributed, thus, 45° separated in argument of latitude, in 3 orbital planes with an inclination angle of 55°. Three additional satellites are presumably spare ones. The geosynchronous orbit satellites have a satellite altitude of 35 785 km above the earth. The inclination angle corresponds to 55°. The satellites will be positioned in three orbital planes with right ascensions

12.1 Global systems

of ascending node $\Omega = 0°$, $120°$, $240°$ and arguments of latitude of $187.6°$, $67.6°$, and $207.6°$, respectively.

The first GEO satellite of Compass has been launched on February 3, 2007, another will follow. These two satellites, positioned at $58.75°$E and $160°$E longitude, will also complement the Beidou-1 constellation, but all five geostationary Beidou-1 satellites will be part of Compass.

The first MEO satellite of Compass has been launched on April 13, 2007. Within few days the satellite started to transmit navigation signals on three frequencies.

12.1.3 Other global systems

Some of the following systems have only position determination capability whereas others also have the capability for data dissemination.

For the sake of completeness, the early global systems Transit and Tsikada are mentioned here. For details the reader is referred to Sect. 1.2.3. Note, however, that Transit has been phased out on December 31, 1996, while Tsikada is still operational.

DORIS

The French system Doppler orbitography by radiopositioning integrated on satellite (DORIS) is a one-way uplink system mainly used for the orbit determination of satellites. DORIS became operational in 1990. According to the Centre National d'Etudes Spatiales (2007), 6 DORIS instruments are currently in service, and 60 reference stations are operational.

Reference stations emit two signals, one S-band frequency, $2\,036.25$ MHz, and a very high frequency (VHF), 401.25 MHz. The receiver onboard the satellite measures the Doppler offsets of these signals every 10 seconds. Based on the measurements, the satellites determine their position with an accuracy of 1 m and velocity with an accuracy of 2.5 mm s^{-1} in real time. If the measurements are sent back to the ground control station, the satellite position, incorporating sophisticated models, is determined with a radial position accuracy of 2.5 cm and a velocity accuracy of 0.4 mm s^{-1}.

PRARE

The German precise range and range rate equipment (PRARE) is used for orbit determination of satellites like DORIS. But in contrast to DORIS it is a two-way system measuring ranges and range rates between the ground segment and the satellites. The satellites emit two signals, one in the S-band, $2\,248$ MHz, and another one in the X-band, $8\,489$ MHz. Both signals are modulated with a pseudorandom noise (PRN) code. A number of 29 globally distributed ground stations measure the run time of the signals with an accuracy of better than 1 ns. The measurements are sent back to the satellite together with ionospheric and tropospheric data. The satellite collects the data and transmits it to the master station as soon as it comes in view. The master station determines the satellite position in postprocessing mode with an accuracy of 5 cm and the velocity with an accuracy of 1 mm s^{-1}. The system is operational since January 1, 1996.

ARGOS

The advanced research and global observation satellite (ARGOS) scientific program was started in 1978 as a joint project of the French Centre National d'Etudes Spatiales (CNES), US National Aeronautics and Space Administration (NASA), and the US National Oceanic and Atmospheric Administration (NOAA). A French and an American company, both subsidiaries of the research institutions, have been created to operate and commercialize the system. The ARGOS satellite system provides means for positioning as well as for communication.

ARGOS is a one-way uplink system based on Doppler shift measurements for position determination. The user platforms transmit a signal at 401.65 MHz modulated with a user-specific message. The satellites measure the Doppler shift and decode the user message. Measurements and messages are forwarded to ARGOS processing centers via the tracking stations. The user message has a maximum length of 256 bits. The next evolution, ARGOS-3, will provide transmission capabilities of up to 4 608 bits per message and a bidirectional communication link.

The mathematical model to determine the (static) user position relies on a cone which is defined by the satellite position as its vertex, the satellite velocity vector as its axis, and the aperture angle of the cone as a function of the measured range rate (cf. Fig. 1.3). Using two measurements and assuming the height of the user to be known, the position of the user platform can be determined. The height is either assumed to be at sea level or transmitted in the user message as, e.g., barometric height. Horizontal accuracies of 200 m and better are achievable. A GNSS receiver may be integrated into the transmitter to meet higher application requirements. The position information is included in the user message.

The ARGOS satellites operate in a polar, nearly circular orbit at an altitude of some 850 km. Accurate satellite orbit determination is based on about 50 ground reference stations which emit an ARGOS signal. The range rate measurements are then used to determine the satellite orbit.

The system has been conceived for monitoring purposes. ARGOS is used, e.g., within ocean buoys to collect oceanographic information. Furthermore, about 3 000 ARGOS transmitters are used for animal tracking, including mammals, marine animals, and birds. As per February 2007, more than 16 000 user platforms are active. For more information refer to www.argos-system.org.

12.1 Global systems

COSPAS-SARSAT

The Soviet Union (SU), USA, Canada, and France, as cited in Cospas-Sarsat Secretariat (1999), signed in 1979 a memorandum to assist search and rescue (SAR) activities on a worldwide basis by providing accurate, timely, and reliable alert and location data to the international community on a nondiscriminatory basis. The cooperative effort has led to the COSPAS-SARSAT system, which has been declared operational in 1985. COSPAS is the acronym for cosmicheskaya sistyema poiska avariynich sudov, which means space system for the search of vessels in distress. SARSAT is the acronym for search and rescue satellite-aided tracking.

In case of a distress, emergency beacons emit a signal which is detected by LEO or GEO satellites. The satellites either process the signals or directly forward them to tracking stations denoted as local user terminals (LUT), which are distributed around the globe. The LUT provide the distress alert message, attributed by position information, to the mission control centers, which again forward it to the rescue coordination centers.

The system differentiates between beacons for the aviation community, for the maritime community, and beacons for land application in particular for personal use. The beacons emit a signal either on the frequency 121.5 MHz or on 406 MHz. The latter represents the evolution step in signal design and allows to modulate a user message on it. The US SARSAT additionally detects emergency signals emitted on 243 MHz. For various reasons, the operators of COSPAS-SARSAT decided to phase-out the 121.5/243 MHz services on February 1, 2009 (Cospas-Sarsat Secretariat 2006a).

The LEO satellites are in nearly circular and nearly polar orbits with an altitude of 850 km and 1000 km, respectively. A minimum of four LEO satellites is required to meet the COSPAS-SARSAT specifications. The LEO satellites, providing a global coverage, determine the position of the distress beacon using Doppler measurements. The GEO satellites cannot rely on position determination using Doppler measurements. Thus, they either require the position information determined by the LEO subsystem, or the distress beacons integrate a GNSS receiver whose position information is modulated onto the distress signal. The latter is provided by second- and third-generation emergency beacons. The coverage of the GEO subsystem is determined by the footprint of the geostationary satellites, thus there is poor coverage in polar regions.

Using the 121.5 MHz signal for position determination enables a position accuracy of about 20 km. With the 406 MHz signal the position accuracy increases to about 5 km. Tests conducted in 1990 with three operational satellites showed an average notification time of about 90 minutes. This time delay comprises the waiting time until the satellites are in view plus the time needed for position determination and position transmission. There are nearly no delays in case of the GNSS position forwarding with GEO satellites.

In future, the COSPAS-SARSAT system will be supported by the constellation of all three major GNSS (Cospas-Sarsat Secretariat 2006b). The MEO satellites will detect the 406 MHz signals of a distress beacon and downlink the information to the LUT using the 1 544 MHz SAR frequency band to guarantee COSPAS-SARSAT compatibility. The LUT extract the position information out of the distress signal or derive it by Doppler and time-of-arrival (TOA) measurements made by the GNSS satellites. The COSPAS-SARSAT support is manifested in Galileo's search and rescue service. GPS will implement the distress alerting satellite system (DASS). Russia intends to implement a similar system payload into the GLONASS satellites, denoted as GLONASS SAR.

Galileo has the capability to provide a returnlink to the emergency beacon in order to provide an acknowledgement to the persons in distress. The returnlink option is still to be decided for DASS and GLONASS SAR (Cospas-Sarsat Secretariat 2006b). Only third-generation emergency beacons will be able to receive this message, what differentiates them from second-generation beacons.

As per December 2006, 7 LEO satellites and 5 GEO satellites are operational. About 1 000 000 distress beacons are in use. Cospas-Sarsat Secretariat (2006c) further lists that from September 1982 to December 2005, the system provided assistance in rescuing 20 531 persons in 5 752 emergency events. More information is available at www.cospas-sarsat.org.

12.2 Regional systems

12.2.1 Beidou-1

Beidou-1 is the predecessor of the satellite navigation system Beidou-2/Compass, described in Sect. 12.1.2. Beidou-1 is a two-way ranging system, thus signals are emitted from a control center to the satellites and forwarded to the user equipment. The signal is then sent back to the control center again via the satellites. The control center transforms the run-time measurement into position information which is then provided to the user.

Reference systems

The Beijing 1954 coordinate system has been defined as coordinate reference frame for Beidou-1 (Bian et al. 2005). The corresponding ellipsoid is the Krassovsky ellipsoid with a semimajor axis of $a = 6\,378\,245$ m and a flattening of $f = 1/298.3$.

Bian et al. (2005) emphasize that the Chinese coordinated universal time, as maintained by the atomic clocks in the control center at Beijing, is used as time reference system.

12.2 Regional systems

Beidou-1 services

Beidou-1 is a dual-use system to which access is controlled by the Chinese government. Two private companies, as Feairheller and Clark (2006: p. 621) mention, have been contracted to provide the positioning and fleet management services to the user community. Beidou-1 is available for civilian users since April 2004.

Position determination, time synchronization, and communication are the three major services. Roughly, the geostationary satellites provide a position service for the Asian subcontinent on the northern hemisphere. The position information determined by the Beidou-1 system has an accuracy of about 20 to 100 m depending whether the user is in the coverage area of the calibration stations or not and whether ground correction information is applied. According to Shi and Liu (2006), the accuracy decreases with decreasing latitude as a consequence of the geometry between user and satellite constellation. The capacity of the positioning service is 540 000 users per hour. According to Bian et al. (2005), to one group of users position updates are provided every 5 to 10 minutes and a second smaller group receives position updates every 10 to 60 seconds.

The time synchronization capability of Beidou-1 allows to synchronize to the reference time with a precision of 100 ns in case of the one-way synchronization concept and 20 ns applying two-way methods (Wei et al. 2004).

Beidou-1 has been specified to transmit encrypted short text messages of less than 120 Chinese characters (Bian et al. 2005). The communication service is furthermore specified to be used for fleet management services. Here either the position of Beidou-1 or position determined by other means is attributed by fleet management relevant information and transmitted to the control center.

Another extension to the communication service is to use the Beidou-1 channel to transmit augmentation information. This information can be used to increase the accuracy and integrity of GNSS position solutions. The Beidou wide-area differential GPS system is estimated to provide an accuracy of 5m (Shi and Liu 2006).

Method of positioning

The central control station transmits a message via the geostationary satellites to a subset of users. The user segment responds to the signal via all geostationary satellites in view. The control station determines the transmit and receive time and transforms this run-time measure into a position estimation. Therefore, the system requires accurate position information of the control center and of the geostationary satellites.

The Beidou-1 concept foresees measurements to two satellites which are complemented by a digital elevation model (DEM), reducing the three-dimensional position determination problem to a two-dimensional one. Although three geostationary satellites are in orbit and could be used for three-dimensional position determination (Forden 2004), the third satellite is used as an active backup. According

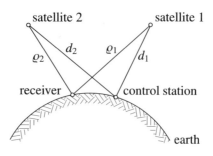

Fig. 12.1. Method of positioning

to Bian et al. (2005), the DEM is stored in a grid of longitude and latitude with a resolution of 1 or 2 arcseconds. Nongrid heights are computed using a quadratic interpolation. The user may also provide height information using, e.g., a barometer.

The geometrical concept of the positioning method is schematically shown in Fig. 12.1. In a simplified mathematical model, the observation equations read

$$\ell_1 = 2\varrho_1 + 2d_1,$$
$$\ell_2 = \varrho_2 + d_2 + \varrho_1 + d_1, \quad (12.2)$$

where ℓ_1 and ℓ_2 are the run-time measurements, ϱ_1 and ϱ_2 correspond to the respective distance from the user receiver to the satellites, and d_1 and d_2 correspond to the distance from the control center to the satellites. Bian et al. (2005) further introduce the terrain height, which is a function of the position. The third (fictitious) measurement, thus, reads

$$\ell_3 = N + h = \sqrt{X_r^2 + Y_r^2 + (Z_r + Ne^2 \sin \varphi_r)^2}, \quad (12.3)$$

where N is the radius of curvature in prime vertical as defined in (8.2), h is the ellipsoidal height, X_r, Y_r, Z_r are the Cartesian coordinates of the user position (cf. Eq. (8.1)), and φ_r is the corresponding latitude of the user position. The latitude φ_r has to be known in advance for the computation of N. A problem which is solved in an iterative procedure, where ℓ_3 is approximated in a first step, the user position is then solved, and ℓ_3 corrected in the next iteration step.

Forden (2004) mentions that the positioning method of Beidou-1 may not be usable for fast-moving objects with a velocity greater than 280 m s^{-1}.

Beidou-1 segments

China launched the first two Beidou-1 navigation satellites in 2000. A third satellite, an active backup, was launched in 2003. The satellites are denoted as Beidou-1A, 1B, and 1C, respectively, in the literature also named Beidou navigation test satellites (BNTS). All three have been placed in (near) geostationary orbits.

12.2 Regional systems

Table 12.2. Beidou-1 satellite parameters

Name	Launch date	Longitude	Perigee / Apogee	Inclination
Beidou-1A	Oct 30, 2000	139.9°E	35 770 / 35 804 km	0.05°
Beidou-1B	Dec 20, 2000	80.2°E	35 773 / 35 801 km	0.07°
Beidou-1C	May 24, 2003	110.4°E	35 747 / 35 829 km	0.15°

The coverage area is specified to be between latitude 5°N to 55°N and longitude 70°E to 140°E. The satellites Beidou-1A and 1B provide the best geometry for position determination; however, any pair of the satellites can be used. Table 12.2 lists the constellation parameters as described in Forden (2004).

Since Beidou-1 applies the two-way ranging concept with the signals originating in the control center, the satellites do not require a high-accuracy clock onboard. Although the satellites are in geostationary orbits, they underlie small variations in position over time. For accurate position determination of the user, the positions of the satellites have to be continuously monitored and determined.

Beidou-1C additionally includes a transponder payload to provide space-based augmentation information for GPS similar to the systems described in Sect. 12.4 (Feairheller and Clark 2006: p. 623).

The ground segment is composed of the control station located at Beijing, the ground tracking stations for orbit determination, and calibration stations distributed throughout China. The measurements of the calibration stations are used to calibrate the errors in the Beidou-1 transmission time determination.

The ground segment communicates with the geostationary satellites using the S-frequency band, 2491.75 ± 4.08 MHz. The Beidou-1 receivers send signals in the 1 615.68 MHz L-band to the satellites (Bian et al. 2005).

12.2.2 QZSS

The Quasi-Zenith Satellite System (QZSS), developed by Japan, provides a regional satellite navigation service in East Asia and Oceania. QZSS was designed to provide position service in urban canyons and mountainous environments. Although QZSS is primarily an augmentation and complementary system to GPS, it also has the potential to operate in stand-alone mode providing a regional service, however with diminished positioning performance. Nevertheless, the system has the capability to be extended to a fully operational, high-performance Japanese regional system in the future.

The technical details in this section are based on the interface specification document (draft version 0.1) released in January 2007 (Japan Aerospace Exploration

Agency 2007). An update of this document has been available in mid 2007, incorporating user feedback.

Introduction and history

Japan requested an autonomous satellite positioning system in case that the performance of the other GNSS degrades due to any unforeseen anomaly or failure. Such a system has been considered essential for national security and crisis management.

QZSS is a joint initiative of the government, represented by four ministries, and the private sector. Various companies have founded the Advanced Space Business Corporation, which is mainly involved in the communication and broadcasting service of QZSS. The responsibility of the public entity lies within the positioning component. A GPS/QZSS Technical Working Group was established in October 2002 for a joint specification of the QZSS signals to guarantee interoperability with GPS.

In 2003, the conceptual study phase has been started. The definition and design phase followed in the years 2004 and 2005. The system development phase began in 2006 and shall be concluded in 2008. The first satellite shall be launched in 2008, the second and third satellites in 2009. The verification phase in 2009 precedes the operation phase and the commercialization of QZSS starting in 2010 (Gomi 2004).

Reference systems

The QZSS coordinate system is known as the Japanese geodetic system (JGS). The JGS is expected to approach to the latest realization of the ITRF. The JGS uses the Geodetic Reference System 1980 (GRS-80) ellipsoid, including its definition of the direction of axes and the earth's center of gravity. The semimajor axis corresponds to $a = 6\,378\,137$ m, the flattening is defined by $f = 1/298.257222101$. According to Japan Aerospace Exploration Agency (2007: p. 34), the offset between the QZSS coordinate system and WGS-84 and GTRF, respectively, will in future be less than 0.02 m.

The QZSS time scale is aligned to the international atomic time (TAI), having the same integer second offset to TAI as GPS, thus, TAI is 19 seconds ahead of the QZSS time scale. The offset between the QZSS time scale and GPS time will be emitted in the navigation message of GPS and QZSS. A similar interface is envisioned for the offset to Galileo system time.

QZSS services

QZSS will provide three major services. The first is to complement GPS by broadcasting navigation signals compatible and interoperable with GPS. This shall increase the availability, continuity, and accuracy of the navigation service. Therefore, the satellites will transmit L1C/A, L2C, L5C, and already L1C signals as specified by GPS. The achievable positioning performance is comparable to standalone GPS. Additionally, the system will emit augmentation information to correct the GNSS signals for atmospheric effects, orbital and clock errors. In this way the position accuracy shall be in the submeter range (1σ). The augmentation information will also contain integrity information. The third service to be provided is a broadcasting and communication service, in order to enable, similar to the navigation objectives, communication in dense urban and mountainous environments. The communication channel may also be used to transmit assistance information, thus facilitating navigation signal acquisition and tracking.

QZSS segments

The space segment comprises the quasi-zenith satellites (QZS) which operate in highly inclined elliptical orbits (HEO). The ground segment consist of the monitoring stations (MS), a master control station (MCS), and tracking control stations (TCS). The MS monitor the signals of the QZS and of all other GNSS satellites. Approximately ten MS will be distributed in Japan, East Asia, and Oceania. The observations are processed and forwarded to the MCS. Here the satellite orbits and time synchronization parameters are determined. In addition, the MCS determines any anomaly in the satellite operation or received signals. The MCS generates the navigation message sent to the TCS, which uplink the information to the QZS. Finally, the navigation message is modulated in the satellites onto the ranging signals in the L-band.

At least three satellites in three orbital planes with geosynchronous period guarantee signal availability even in urban canyons and mountainous areas. The orbit parameters are given in Table 12.3. Because of the eccentricity, the HEO constellation generates a characteristic asymmetric figure-of-eight ground track orbit with an average central longitude of 135°E. One of the three satellites will always be positioned over Japan, thus providing service from elevation angles greater than 70°. Figure 12.2, furthermore, indicates that for the higher latitudes the satellites have a slower ground track speed.

The satellites transmit navigation signals continuously covering the earth in a signal envelope of equal power. Rubidium atomic clocks are used as frequency standard onboard the satellites. The lifetime of the QZS has been specified to be ten years. The maximum relative velocity between the satellites and a stationary user is $600 \, \mathrm{m \, s^{-1}}$ resulting in a maximal Doppler shift of 3.2 kHz for a carrier frequency of 1 575.42 MHz

Signal structure

The QZS will emit ranging signals in the L1, L2, and L5 frequency band compatible and interoperable with the GPS signals. The QZS will furthermore emit signals

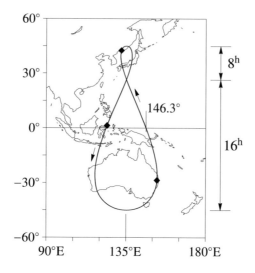

Fig. 12.2. QZS ground track plot

in a fourth frequency band in overlay to the E6 band of Galileo. This signal, called LEX, is intended for experimental purposes. The QZS will emit eight different signals in these four frequency bands. All QZS signals are right-handed circularly polarized signals. As GPS, QZSS will apply the CDMA concept. The power level of the received signals has been specified to be in the range of -152 to -160 dBW depending on the signal and the satellite position.

Carrier frequencies
The carrier frequencies are listed in Table 12.4. The QZSS will furthermore use signals in other bands for other means, e.g., high-speed communication or two-way satellite time and frequency transfer. In order to compensate for the relativistic effects, the fundamental frequency $f_0 = 10.23$ MHz will be intentionally offset by

Table 12.3. QZSS orbit parameters

Parameter	Value
Semimajor axis (average)	$a = 42\,164$ km
Eccentricity (maximum)	$e = 0.099$
Orbit inclination	$i = 45°$
Right ascension of ascending node	$\Omega = 88.09°, 208.09°, 328.09°$
Argument of perigee	$\omega = 270°$
Longitude of ascending node	$\ell = 146.3°$E

12.2 Regional systems

$\Delta f \sim 5.5232 \cdot 10^{-3}$ Hz. The elliptical orbit of the satellites causes a variation of the relativistic effect which is accounted for in the satellite clock parameters, which are emitted in the navigation message.

PRN codes and modulation

The QZSS satellites transmit eight ranging signals. Six of them are referred to as positioning availability enhancement signals since they complement the GPS signals. The two others (L1-SAIF, LEX) provide augmentation information, thus, they are commonly referred to as positioning performance enhancement signals. The QZSS applies for the L1C/A signal the same ranging code modulation as GPS. The PRN codes are assigned in coordination with GPS. Furthermore the same navigation message, as specified by GPS, is modulated onto the L1C/A signal (ARINC Engineering Services 2006a). The QZSS L1C ranging codes and navigation messages are in accordance to the codes and messages envisioned for the GPS L1C signals (ARINC Engineering Services 2006b). The QZSS L2C will be a replica of GPS L2C as specified in ARINC Engineering Services (2006a), and similarly the QZSS L5I and L5Q signals will be a replica of the GPS L5C components as specified in ARINC Engineering Services (2005).

The QZSS L1-SAIF (submeter accuracy with integrity function) signal will be used to transmit augmentation information for the navigation satellite signals. Thus, the ranging code will be similar to GPS C/A, whereas the navigation message encoding underlies the specification of Radio Technical Commission for Aeronautical Services (2006). The QZSS LEX signal will be generated using two different ranging sequences. A long ranging code and a short one will be time-multiplexed BPSK modulated onto the carrier frequency. The short code will be used to transmit data with a data rate of 2 000 bits per second.

Table 12.4. QZSS ranging signals

Link	Frequency [MHz]	factor ($\cdot f_0$)	PRN code	Code rate [Mcps]	Modulation type	Data rate [sps/bps]
L1	1575.42	154	L1C/A	1.023	BPSK(1)	50/50
			L1C$_D$	1.023	BOCs(1,1)	100/50
			L1C$_P$	1.023	BOCs(1,1)	—
			L1-SAIF	1.023	BPSK(1)	500/250
L2	1227.60	120	L2C	1.023	BPSK(1)	50/25
L5	1176.45	115	L5I	10.23	BPSK(10)	100/50
			L5Q	10.23	BPSK(10)	—
E6	1278.75	125	LEX	5.115	BPSK(5)	2000 bps

Navigation messages

The TCS uplink the navigation message in intervals in dependence of the message content. According to Japan Aerospace Exploration Agency (2007), ephemerides and almanac data, excluding satellite clock parameters and differential data, are updated every 3 600 s. Clock parameters are updated every 750 s, differential data every 300 s. Integrity data is updated with a higher frequency to guarantee a low time to alarm. For example, for the L1C/A signal, the maximum notification time will last 24 s in a worst-case scenario (Japan Aerospace Exploration Agency 2007).

The SAIF navigation message is similar to the one of satellite augmentation systems as described in Sect. 12.4. The message types 52–60 are used for QZSS-specific parameters as described in Japan Aerospace Exploration Agency (2007). No information is available yet about the LEX message structure.

12.2.3 Other regional systems

IRNSS

In May 2006, India approved to implement the Indian Regional Navigation Satellite System (IRNSS) to provide an autonomous navigation system for the Indian subcontinent. The space segment of the system will consist of seven satellites. Three will be in geostationary orbits located at longitude 34°E, 83°E, and 132°E. Four satellites will operate in geosynchronous orbits with an inclination angle of 29° to the equatorial plane and a central longitude of ground track at 55°E and 111°E, respectively. The phasing of orbital planes will be 180°, the in-plane phasing 180°, and relative phasing between the satellites in the orbital planes 56° (Kibe 2006). All satellites are planned to be in orbit by 2013. Singh and Saraswati (2006) note that the service area has been specified between 40°E to 140°E in longitude and between ±40° in latitude, thus covering India and 1 500 km beyond it.

IRNSS will provide dual-frequency service using the L-band in coallocation with GPS L5 and Galileo E5a, and the S-frequency band 2 483.5 to 2 500.0 MHz. This shall enable a position accuracy of 10 m over India and adjacent countries, and 20 m over the Indian Ocean. Higher accuracies shall be achievable implementing augmentation information. According to Singh and Saraswati (2006), one signal in S-band and three more in the L-band are planned for downlink. The ground segment consists of two master control stations and about 20 IRNSS ranging and integrity monitoring stations. The positioning concept is comparable to Beidou-1.

GEOSTAR and LOCSTAR

The private-sector system GEOSTAR provided space-based positioning and communication services from 1983 until 1991 (Pace et al. 1995: p. 230). GEOSTAR relied on a two-way ranging system including a bidirectional communication link. The ground stations emitted a signal consisting of a PRN ranging code and a

12.3 Differential systems

data message. The signal was relayed from geosynchronous satellites to the user equipment. The user equipment responded to the signal via the geostationary satellites. Finally, the master ground station computed the user position using two-way run-time measurement, implementing either measurements from three satellites or additionally implementing height information from a DEM or a barometer. The GEOSTAR system worked as a model for the Twin-Star program of China, which later became Beidou-1 (Feairheller and Clark 2006: p. 616). For various reasons, the GEOSTAR company run into financial problems, and finally the system was phased out. The European equivalent, LOCSTAR, never became operational since not enough funding could be raised.

OmniTRACS and EutelTRACS

The US company Qualcomm launched the OmniTRACS system in 1988 to provide a space-based service for position determination and data dissemination for America. EutelTRACS is the European equivalent which is operated in cooperation with the European Telecommunication Satellite Organization. The systems have been developed mainly for fleet management purposes. Both systems rely on a concept similar to GEOSTAR. A ranging signal is transmitted from a ground station via geostationary satellites to the user and back again. The position is derived from run-time measurements, incorporating height information of an elevation model (Colcy et al. 1995). The subsystem for position determination is denoted as Qualcomm automatic satellite position reporting (QASPR). Meanwhile OmniTRACS and EutelTRACS also integrate GNSS receivers in the user equipment to increase the positioning performance. The bidirectional communication link transmits the GNSS position to the service center. More information is available under www.omnitracs.com or www.euteltracs.org.

12.3 Differential systems

The degradation of the GPS point positioning accuracy by selective availability (SA) has led to the development of differential techniques. Meanwhile the degradation has been deactivated but, nevertheless, the methods still evolve to increase the performance parameters of GNSS.

12.3.1 Principles

The basic concept of differential GNSS (DGNSS) relies on two or more receivers. From the known reference receiver location and the satellite position there results in a computed range, which is opposed to the measured range. A reference station or a reference network broadcasts differential corrections which are used by the roving receiver to eliminate GNSS inherent errors in the measurement (Fig. 6.1).

Three groups of GNSS range biases, as described in detail in Sect. 5.1.4, are distinguished: satellite-specific biases, signal propagation errors, and receiver-specific biases. DGNSS eliminates or reduces the satellite-specific biases and the signal propagation errors.

The differential corrections are determined by a single or by multiple reference stations. The single-reference station concept is simple but the position accuracy decreases with increasing distance from the reference station. The rover receives the differential corrections from one reference station using a dedicated data link. The multiple-reference station concept relies on several spatially distributed reference stations which determine differential corrections and emit them to the user receiver. The network of reference stations guarantees a homogeneous service performance over a larger coverage area. Disadvantages of the multiple-reference station concept are the increased complexity and costs for the service provider, and an increasing latency of the correction data due to the network communication.

12.3.2 Differential correction domains

Three different approaches to compute differential corrections exist.

Solution-domain approach
In the solution-domain approach, the coordinate difference between the measured position and the known position of the reference station is transmitted to the rover. These differences are used to correct the rover position. As long as both receivers use measurements to the same satellites and the same atmospheric models, this simple method will provide good results. Local obstructions as well as an increasing distance between reference stations and rover will result in a different satellite constellation. This concept is mainly used in conjunction with the single-reference station concept.

Measurement-domain approach
The measurement-domain approach relies on a comparison of the measurements and the computed ranges. According to the equations of Sect. 6.2, the difference in pseudoranges determines the pseudorange corrections (PRC) and the range rate corrections (RRC). The rover uses PRC and RRC to correct its own measurements. PRC and RRC are either a weighted average of the reference station network, as mentioned by Szabo and Tubman (1994), or direct observations of one reference station. In case of DGNSS with a single reference station, the accuracy decreases as a function of distance from the reference station at a rate of approximately 1 cm per 1 km.

Conventional DGNSS system concepts rely on raw or smoothed code pseudoranges. Position accuracies in the meter level are achievable using code measure-

12.3 Differential systems

ments. Phase smoothed code ranges are necessary to obtain the submeter level. Precise DGNSS services, in contrast, use phase pseudoranges. In this way, accuracies at the subdecimeter level can be obtained in real time for ranges up to some 20 km (DeLoach and Remondi 1991). The compromise of higher positioning accuracy of precise DGNSS is the necessity of ambiguity resolution. Commonly the ambiguities must be resolved on-the-fly and therefore (generally) dual-frequency receivers are required.

DGNSS systems implementing the measurement-domain approach are commonly denoted as extended or local-area differential (LAD) GNSS.

State-space approach

In the state-space approach, the reference network models the various error sources and transmits the model parameters to the rover. This concept is particularly applied in augmentation systems but also in precise point positioning, where accurate ephemerides and clock parameters are estimated. The use of dual-frequency receivers simplifies the derivation of the ionospheric influence. Commonly a grid of ionospheric vertical delay corrections is modeled. The remaining calibrated residual between known range and measured range is projected onto the orbit and satellite clock error. The high number of model parameters requires a network of reference stations. The density of the network is lower compared to the measurement-domain approach. Systems following the state-space approach are designated as wide-area differential (WAD) GNSS. Algorithms for the state-space approach can be found in Mueller et al. (1994) or Kee (1996).

12.3.3 Examples of differential systems

Space-based communication

OmniSTAR, SkyFix, and StarFix

The Fugro company introduced the commercial services OmniSTAR, SkyFix, and StarFix for differential positioning. The use of OmniSTAR is exclusively allowed to be used for onshore applications, whereas the other two have been introduced for marine applications.

OmniSTAR provides three levels of services: a VBS (virtual base station) submeter positioning service, a better than 20 cm service, and a high-performance better than 10 cm service level. For all services, about 100 terrestrial reference stations are located around the world. In VBS mode, the reference stations track the GPS satellites and compute differential code corrections, every station individually. The differential corrections are encoded in a Radio Technical Commission for Maritime Services (RTCM) format and transmitted to the user receiver. The receiver then applies an inverse distance-weighted least-squares solution to generate corrections of a virtual reference station (cf. Sect. 6.3.7). The medium service relies on phase measurements and precise point positioning methods, thus integrating precise ephemerides and clock information which are estimated based on the reference station network. The ionospheric influence is eliminated by dual-frequency receivers. In the high-performance mode, phase corrections and dual-frequency methods are applied.

SkyFix and StarFix are similar in concept but rely on different technologies, thus, providing two redundant systems for increased integrity (Łapucha et al. 2004). The SkyFix and StarFix technique apply the precise point positioning concept, which Fugro denotes as satellite differential GPS (SDGPS).

The differential information is provided to the user via several commercial geostationary satellites which provide worldwide coverage of about 90% excluding the north and south poles. The L-band satellite signals in the frequency band 1 531–1 559 MHz are decoded by dedicated receiver modules and the correction information is provided to the GNSS receivers. Further information is available at www.omnistar.com or www.skyfix.com.

StarFire
The StarFire global space-based differential system is a commercial precise point positioning service developed by John Deere and Company. StarFire relies on 60 globally distributed monitoring stations and two independent control centers both located in the US. The space segment consists of three Inmarsat satellites which provide global coverage except for the north and south poles. The satellites are positioned at 98°W, 25°E, and 109°E longitude and transmit correction signals in the L-band. The differential corrections consist of precise orbit and clock values which are used in a dual-frequency receiver to achieve position accuracy in the submeter range. Dixon (2005b) states that even subdecimeter performance is achievable. Further information is available at www.navcomtech.com/StarFire.

GDGPS
The Jet Propulsion Laboratory (JPL) of NASA developed a global differential GPS (DGPS) named GDGPS to provide high-accuracy GPS differential information in real time to support NASA science missions. Dual-frequency receivers which eliminate the ionospheric error integrate the ephemerides and clock differential corrections to achieve higher position accuracies. Correction data transmission rates of up to 1 Hz are possible (Bar-Sever et al. 2004). A global network of 100 stations operated by JPL and international partner organizations allows to model the ephemerides and clock corrections. At the same time the performance of the GPS satellites is monitored. Thus, any GPS satellite failure is reported to the user receiver in nearly real time. The correction information is accessible, e.g., via Internet using the Internet-based global DGPS service. The real-time differential correction message is also planned to be transmitted using S-band frequencies of the tracking

12.3 Differential systems

and data relay satellite system (TDRSS). This service is especially designed for spacecraft which determine their position using GPS signals.

Ground-based communication

Maritime DGNSS

The maritime DGNSS service broadcasts differential correction information on marine radio beacon frequencies. Actually the beacons support GPS corrections, but any GNSS differential correction could be implemented. The maritime beacons transmit around the world on frequencies in the 283.5 through 325.0 kHz frequency band. The exact transmission band differs in various regions of the world. The real-time differential corrections are provided in an unencrypted RTCM format. Depending on the distance between the rover and the reference station, which might range up to several hundred kilometers as a function of transmitter power, the position accuracy is between 1 to 10 m (95%). The data transmission rate is typically 100 bps but may vary between 50 through 200 bps.

In the US, the concept of radio beacons has been extended to the nationwide DGPS (NDGPS). Beacons have been installed all over the country. NDGPS is operated by the US Coast Guard, the Federal Railroad Administration, and the Federal Highway Administration. The system provides increased accuracy and integrity of GPS information to users on land and in water. The original concept of NDGPS has been extended to the high-accuracy NDGPS (HANDGPS), which is specified to provide positioning accuracy of 10 to 15 cm throughout the coverage area. Department of Defense, Department of Homeland Security, and Department of Transportation (2005) emphasize that HANDGPS anticipates also to provide an integrity information with a time to alarm of 1 to 2 seconds. For various reasons the state financial support for NDGPS has been stopped for 2007. The future of the system is pending.

National DGNSS

Numerous national, regional, and urban DGNSS services have been created which provide differential information at a service fee. Various companies, e.g., power suppliers, create their own DGNSS network for their special fields of application and requirements. National DGNSS services are exemplarily named and shortly described in the following paragraphs.

The Austrian positioning service (APOS) integrates GNSS measurements of about 40 reference stations. In addition, measurements of reference stations of neighboring countries are used to increase the accuracy and homogeneity of the data. APOS provides different service levels, beside others RTK and DGPS services. The data streams are accessible via Internet or cellular networks.

The satellite positioning service (SAPOS) provides DGNSS services for the German territory. Three service levels are offered providing real-time positioning service, high-precision real-time positioning, and geodetic high-precision positioning service. The data is transmitted in the VHF band or using cellular networks. The high-precision service provides position accuracies in the 1 to 5 cm range. More information is available at www.sapos.de.

The objective of the European position determination system (EUPOS), an initiative started in March 2002, is to establish a uniform DGNSS basis infrastructure in central and eastern Europe. EUPOS seeks to enhance the crossborder compatibility of the DGNSS systems. The goal is to integrate 870 reference stations in 14 countries in a common network. The correction information shall be useable for land, marine, and air applications but also for geodetic point positioning, i.e., providing submeter real-time services as well as subcentimeter in postprocessing. The EUPOS data will be accessible via Internet, cellular networks, or VHF. More information is available at www.eupos.org.

For a description of the Russian system for differential correction and monitoring (SDCM) refer to the next section.

12.4 Augmentation systems

Current GPS and GLONASS positioning services are not suitable for, e.g., civil aviation, in particular for critical flight operations. Also critical maritime applications like harbor entrance, harbor approach, or inland waterway navigation may not rely on the GPS/GLONASS-only performance. Neither the position accuracy nor the integrity meet user requirements. Thus, augmentation systems have been envisioned to provide an increased performance.

Following the definition as stated in the US space-based position, navigation, and time policy (www.navcen.uscg.gov/cgsic/geninfo/FactSheet.pdf), augmentation refers to the provision of additional information to enhance the performance of space-based positioning, navigation, and timing signals. The respective performance parameters are accuracy, availability, integrity, and reliability, with independent integrity-monitoring and alerting capabilities for critical applications. The increased performance and especially the integrity information allow GNSS signals to be used for safety-critical operations. For example, precision approaches will become possible, or GNSS landing systems will enable curved landing maneuvers of airplanes.

An augmentation system might be considered as differential system with the additional feature of providing integrity information. Although the differential systems presented in Sect. 12.3.3 have the potential or transmit even by definition integrity information, they are not certified for safety-critical applications, a characteristic which is pursued by every augmentation system.

12.4 Augmentation systems

12.4.1 Space-based augmentation systems

Principles

The space-based augmentation systems (SBAS) use a network of terrestrial monitoring stations to perform GNSS ranging measurements. The observations are forwarded via a wide-area network to the processing facilities. The master stations use the measurements to generate correction parameters for the satellite orbits, the satellite clocks, and the ionospheric influence. The latter is accounted for in the wide-area differential (WAD) corrections. There is no possibility for SBAS to reduce the receiver-specific errors, like multipath, but also the local characteristic of tropospheric effects cannot be accounted for in the SBAS information. Additionally, the master stations perform several integrity checks to validate the satellite signals. The GNSS signals as well as the SBAS operation in general have to be continuously monitored to provide a high level of integrity. Any anomalous behavior of any element results in an integrity information for the user. The corrections together with integrity information are then transmitted using C-band signals to the satellites, which relay the information using L-band signals to the user. The space segment modulates the augmentation information onto a GPS-like PRN ranging code which can be used as an additional observation in the position algorithm.

Analyzing the system concept, SBAS provides three major components of information for performance enhancement. First, the corrections increase the accuracy of the position solution. Second, the GPS-like signals from the SBAS geostationary satellites increase the availability and continuity but also the accuracy of position solution (GEO-ranging). Third, the integrity information of the SBAS signals enhance the safety by alerting users within 6 seconds of any malfunction in the GNSS but also in the SBAS functionality.

Space segment

The space segment of SBAS consists, for reasons of redundancy, of at least two geostationary satellites. The satellites basically work as bent-pipe transponders. They modulate the information transmitted by the ground segment onto 1 575.42 MHz L-band ranging signals, i.e., GPS L1 together with the C/A ranging code. Additionally, signals in the C-band are emitted which are used in combination with the L-band signals by the ground segment for enhanced WAD parameter estimation. The received SBAS signals have power levels comparable to those of GPS signals in order to avoid interference. In future, augmentation information will also be emitted on and for, respectively, the L5C navigation signals.

The position of the SBAS satellites, although in geostationary orbit, vary with time and are estimated using the measurements of the monitoring stations (Meindl and Hugentobler 2004). The exact satellite positions are needed for the additional ranging capability of the SBAS satellites. The ephemerides and almanac information is part of the SBAS data message.

The geostationary position of the SBAS satellites is disadvantageous for users in urban environment or in mountainous regions. The low-power signals of the low-elevation satellites will be shadowed. Thus, concepts have been developed to transmit the SBAS information using also other means of communication, e.g., Internet.

User segment

The user segment incorporates the augmentation information together with the GNSS measurements into a navigation solution of increased performance. The positioning accuracy improves to about 1 to 3 m (95%) horizontally and 2 to 4 m (95%) vertically. Time accuracy is enhanced to better than 10 ns. The position accuracy is related to the high level of integrity of $2 \cdot 10^{-7}$ per any 150 seconds or 10^{-7} per hour, respectively, depending which service level applies. For non-safety-critical applications, position accuracy may be more interesting than integrity. Kim et al. (2006) show how the SBAS information can be used to further increase position accuracy, however not with the same level of integrity. Mathur et al. (2006) emphasize that in situations with limited GPS visibility, i.e., 4 satellites in view, the SBAS information provides a higher performance gain than in situations where 8 or 9 satellites are visible.

SBAS data message

SBAS use a GPS-like ranging code; however, they rely on a higher data rate compared to the GPS C/A-code. The SBAS message consists of an 8-bit preamble for frame synchronization, a 6-bit message type identifier, a 212-bit data field, and finally 24-bit parity information. The 250 bits are half-rate convolutional encoded (cf. Sect. 4.3.4). The 500 sps are BPSK modulated onto the ranging code.

SBAS differentiate between 64 message types, whereas not all are specified but reserved for future use. Message type number 0 is used to distinguish between three modes of operation. In test mode the message type 0 is emitted in the SBAS signal filled with all zeros. This indicates that the SBAS signal is not usable. In non-safety-of-life mode, the message type 0 is emitted, however filled with the information of message type 2. The SBAS information might be used for non-safety-of-life applications, although the system is still under test. In safety-of-life mode, message type 0 is not broadcast at all, whereas message type 2 is broadcast nominally.

The data message contains SBAS satellite ephemerides and almanac data, fast and long-term correction data, integrity information, ionospheric correction data, timing information, and different service level data. The sequence of SBAS message type transmission is variable. However, the different messages have a specified time-out and, thus, have to be refreshed within a predefined time interval. The most

12.4 Augmentation systems

stringent requirement applies to the integrity information and the fast corrections. The SBAS signal format is described in detail in the minimum operational performance standards (MOPS) as given by Radio Technical Commission for Aeronautical Services (2006: Appendix A).

SBAS augmentation information

The fast corrections model errors with high temporal decorrelation like GNSS clock errors. The long-term corrections model the low-frequency components of the different errors. Integrity information is provided at two levels (European Space Agency 2005a). A use/don't use parameter indicates which satellite signal should not be used. Furthermore, two statistical parameters estimate the remaining errors after applying the SBAS corrections. The user differential range error (UDRE) overbounds the satellite clock error σ_{sc} and the ephemerides errors σ_{eph} and the grid ionospheric vertical error (GIVE) overbounds the ionospheric error σ_{iono} (refer to Eq. (7.80)).

The SBAS models the vertical delays of the ionosphere at the ionospheric grid points (IGP), which commonly span a regular raster of $5° \times 5°$. The raster defines the coverage area of the WAD corrections. The ionospheric corrections applying to these grid points, denoted as grid ionospheric vertical delays (GIVD), are broadcast to the users. The GIVD apply to the L1 frequency. The user receiver estimates the ionospheric delays for every satellite in a four-step process. In the first step, the user receiver determines the ionospheric point (*IP*), as indicated in Fig. 5.3, of the satellite–user vector with a sphere at $h_m = 350$ km altitude. In the second step, the receiver determines the four adjacent IGP of the ionospheric point. In the third step, the vertical delay at the pierce point is estimated applying a bilinear interpolation using the GIVD of the adjacent IGP. Finally, an obliquity factor, which is a function of satellite elevation, is applied to project the ionospheric vertical delay onto the satellite-user line-of-sight slant delay. This ionospheric correction only applies for single-frequency receivers. Dual-frequency receivers apply Eqs. (5.80) and (5.83) to eliminate the ionospheric error.

In a similar four-step algorithm, the receiver uses the GIVE, which are also given at the IGP, to estimate the variance of the remaining ionospheric error. The variance of the satellite clock and ephemerides corrections are computed using the UDRE. The variances of the residual errors are finally estimated using Eq. (7.80). The range error is further used as a weighting factor in the parameter estimation. The covariance matrix of the parameters (7.76) finally determines the protection level (Eqs. (7.178) and (7.179)) which is used for integrity determination (Fig. 7.11). SBAS corrections guarantee a Gaussian normal distribution with zero mean of the remaining ranging errors, except for local effects like multipath.

Table 12.5. SBAS geostationary satellites

SBAS	Satellite	Longitude	PRN
EGNOS	Inmarsat-3-F2 / AOR-E	15.5°W	120
	ESA Artemis	21.5°E	124
	Inmarsat-3-F5 / IOR-W	25°E	126
GAGAN	INSATNAV [1]	55°E	128
	GSAT-4 [2]	82°E	127
MSAS	MTSAT-1R	140°E	129
	MTSAT-2	145°E	137
WAAS	Inmarsat-3-F3 / POR	178°E	134
	Inmarsat-3-F4 / AOR-W	142°W	122
	Intelsat Galaxy XV	133°W	135
	TeleSat Anik F1R	107.3°W	138

[1] To be launched 2008   [2] To be launched 2007

SBAS compatibility

The US wide-area augmentation system (WAAS), the Japanese multifunctional transport satellite space-based augmentation system (MSAS), the European geostationary navigation overlay service (EGNOS), and India's GPS and geoaugmented navigation (GAGAN) will provide a nearly worldwide SBAS service. Others will follow in the near future and complement the scenario. Selected satellite parameters used by these systems are listed in Table 12.5.

International standards guarantee compatibility and interoperability of SBAS. The standards and recommended practices (SARPS) have to be taken into account by the system developers, whereas the receiver manufacturers have to guarantee conformance to the MOPS. The various SBAS show small variations to these standards; however, interoperability is guaranteed.

The SBAS have optimized the augmentation information for their coverage areas. However, note that the SBAS ranging signal can be used by any user, independent whether the user position is within the coverage area or not.

WAAS

The US WAAS has been developed by the Federal Aviation Administration (FAA). The WAAS signal was made available for non-safety-of-life applications in 2000. The IOC started in July 2003. The full operational capability is expected for 2007. The WAAS architecture comprises 38 wide-area reference stations (WRS). Twenty are in the conterminous US, seven in Alaska, four in Canada, one in Hawaii, one in Puerto Rico, and five in Mexico. The WRS forward GPS signal measurements to the three wide-area master stations (WMS). The WMS determine the WAD cor-

12.4 Augmentation systems   425

rections and the integrity information which is uplinked to the four geostationary satellites via the three ground earth stations (GES).

EGNOS

On June 18, 1996, a formal agreement between the members of the European Tripartite Group, i.e., the European Commission, the European Space Agency, and Eurocontrol, was the kickoff of EGNOS. First experimental signals have been emitted starting in 2000 as part of the EGNOS system test bed (ESTB). The initial operational capability (IOC) has been declared in July 2005. The full operational capability is expected for 2007. EGNOS will provide full service for at least 20 years.

The space segment consists of three geostationary satellites. The GNSS signals are processed at 34 receiver integrity monitoring stations (RIMS). The observations are sent via the EGNOS wide-area network (EWAN) to the four master control centers (MCC). One of them is active, whereas another one serves as hot backup. This avoids single-point failures. The other two are cold backups and activated if problems occur. The augmentation information is uplinked to the GEO via the navigation land earth stations (NLES). For each GEO two NLES will be installed, one active and one as a backup.

In a first evolution, EGNOS provides coverage for the area defined by the European Civil Aviation Conference (ECAC). In the next evolution step, the coverage area will be extended to Africa. For the third major evolution step, EGNOS will implement full GPS L5 augmentation service and may also include Galileo and modernized GLONASS augmentation services (European Space Agency 2005b).

The European Space Agency developed the signal in space over Internet (SISNeT) concept to provide EGNOS information also over Internet. The SISNeT concept has meanwhile been extended to a commercial data distribution service denoted as EGNOS data access system (EDAS). EDAS additionally provides, e.g., raw RIMS measurements, via various means of communication.

MSAS

Japan's space-based augmentation system (MSAS) is a payload of the multifunctional transport satellites (MTSAT). MTSAT are owned and operated, respectively, by the Japanese Meteorological Agency and the Japanese Ministry of Land, Infrastructure, and Transport. The first geostationary satellite was launched on February 26, 2005. The second followed on February 18, 2006. MSAS is planned to reach FOC in 2008. The coverage area is limited to Japan, since there are no reference stations in other parts of East Asia and Oceania.

GAGAN

The Indian Space Research Organization in collaboration with the Airports Authority of India implements India's SBAS GAGAN, where GAGAN is one of the Hindi expressions for sky. The first geostationary satellite is planned for 2007, the second one in 2008. In 2009 the system will meet FOC. The system will provide coverage for the Indian airspace. The ground segment consists of 8 reference stations, one master control center, and one navigation land uplink station.

SNAS

A space-based augmentation system will presumably be implemented by China. The Chinese SBAS system is denoted as satellite navigation augmentation system (SNAS), although an official name has still to be published. The augmentation system has been studied and proposed to be implemented in the Beidou-1 satellite navigation system, using the communication channel for augmentation information transmission (Liu et al. 2006). Another concept uses dedicated SBAS payload onboard of Beidou satellites to transmit SBAS information similar to WAAS.

SDCM

The Russian Federal Space Agency has launched a project for the development of the system for differential correction and monitoring (SDCM). According to Averin (2006), the system will be operational by 2011. The system will finally consist of a network of 19 monitoring stations on the Russian territory. The monitoring stations will perform raw measurements of all GNSS with a measurement rate of 1 Hz. The measurements are delivered in a secure network to the central processing center. The provision of the augmentation information to the user is based on an Internet link, via TV channel, or cellular networks. Another option foresees to transmit integrity information together with satellite ephemerides and clock errors via the third frequency band G3 of GLONASS. Uplink stations forward in this case the data to the GLONASS-K satellites. The augmentation information will furthermore be provided via geostationary satellites. The anticipated position accuracy is better than 0.5 m. A high-accuracy service in conjunction with local ground station support may even provide a position accuracy between 0.02 and 0.5 m (Dvorkin and Karutin 2006). A cooperation between the EGNOS and SDCM programs exists. More information is available in Russian at http://sdcm.rniikp.ru.

12.4.2 Ground-based augmentation systems

The concept of ground-based augmentation systems (GBAS) has been conceived to meet in particular the stringent requirements of the aviation community. Augmentation information is commonly provided for a local limited area, e.g., the vicinity of airports. Extending the services to larger areas may result in a system referred to as

12.4 Augmentation systems

ground-based regional augmentation system (GRAS). Thus, GRAS is a blending of SBAS ground network with ground-based communication via, e.g., VHF data broadcast (VDB) channels.

LAAS

The local-area augmentation system (LAAS) is one realization of the GBAS, defined by the International Civil Aviation Organization (ICAO). The LAAS requirements are driven by the Category I, II, and III precision approach specifications (Table 13.11).

A reference network on ground, which uses four or more redundant receivers, computes differential corrections. The reference station measures range data from all available range sources, i.e., navigation satellite signals, SBAS signals, or pseudolite signals as described in the subsequent paragraph. Local integrity monitoring facilities evaluate the integrity of the signals and of the computed differential corrections. Separate monitoring facilities monitor the functionality of the system. The augmentation information is transmitted in standard format, e.g., RTCM SC-104, to the airborne receiver using a dedicated secure data link, e.g., VDB at the frequency band 108–117.975 MHz. The LAAS are commonly designed to provide augmentation information to users up to a distance of 45 km. The achievable position accuracy is less than 1 m (95%) combined with a high level of integrity of 10^{-7} per any 150 seconds or even higher (Federal Aviation Administration 1999). Integrating the LAAS augmentation information into the position solution will enable, e.g., curved approach paths, precision approaches, or multiple approach capabilities.

Pseudolites

The concept of pseudosatellites, or pseudolites in short, has already been applied in the 1970s. During first GPS design tests, ground-based transmitters have been installed to emit a GPS-like satellite signal. Today, the concept of pseudolites has been studied for various applications, but especially the airport pseudolites are making their way into operational application. Note that pseudolites could also be used to build up a stand-alone navigation system, used, e.g., for navigation in cities or industrial facilities.

The pseudolites, which are commonly static ground-based transmitters, emit GNSS-like signals, using similar GNSS ranging codes and carrier frequencies. The data message is either similar to the one of the respective GNSS signal or changed in order to transmit augmentation information of all other GNSS signals. The data rate is increased accordingly up to 1000 bps and more.

The additional GNSS-like signal improves the availability and continuity of position information at the user receiver. At the same time, pseudolites enhance the constellation geometry, especially in vertical direction when considering aeronautical applications. The accuracy is increased by the additional signal and in particular enhanced by the differential corrections.

The pseudolite signals are not affected by ionospheric delay errors, and also tropospheric influences are reduced. The static position of the pseudolites allows to minimize the orbital error by calibration. The small distances between pseudolites and user receivers will result in fast geometry changes in kinematic applications. This favors carrier phase ambiguity resolution techniques.

Apart of these advantages, the pseudolite concept has to deal with a number of problems. The near/far interference problem describes the effects of different power levels between pseudolite signals and GNSS signals. As described in Sect. 4.1.2, the power decreases with the square of the distance. Thus, a received pseudolite signal which has a GNSS comparable power level of -160 dBW at a distance of 30 km will increase to -130 dBW in 1 km distance, therefore drowning all other GNSS signals. Three different methods, the code, time, and frequency division multiple access approaches, have been designed to deal with this problem. In the time division multiple approach, which has been chosen for airport operation, the pseudolite signals are transmitted in low duty-cycle pulses, where signals are transmitted for a duration of, e.g., about 0.1 ms, while during the following 0.9 ms no signals are broadcast at all. This procedure is continuously applied. During the 0.1 ms the GNSS signals are interfered. The GNSS receivers however are still able to acquire and track the signals due to the 0.9 ms long interference-free period. The same is true for the pseudolite signals, where the short 0.1 ms period is compensated by a high power level and long integration times.

Another problem is caused by multipath effects. Pseudolite antennas have to be designed and arranged in a way that no multipath at the transmitter location occurs. In contrast to GNSS signals, multipath is constant in time and only changes with the user position. Finally, time synchronization to GNSS time has to be accomplished, otherwise the position solution will lose accuracy.

Low-frequency augmentation systems

A number of low-frequency augmentation systems have been studied. One of them is Eurofix, which provides GNSS augmentation information using the Loran-C infrastructure. The system has been developed at Delft University of Technology in the Netherlands. The pulsed Loran-C signals are modulated by the augmentation information. In this way, 30 bps of augmentation information can be transmitted without affecting the Loran-C terrestrial navigation performance. The achievable horizontal accuracy is in the range of 1 to 3 m (95%) dependent on the number of receivable sites. The coverage area of Eurofix corresponds to about 1 000 km around one Loran-C transmitter. Hofmann-Wellenhof et al. (2003: p. 210) emphasize that a similar system, the Loran GNSS interoperability channel (LOGIC), has been investigated in the US.

12.5 Assistance systems

The concept of assisted GNSS (AGNSS) has been conceived already in the early 1980s. The architecture of AGNSS foresees a navigation module interacting with a communication module. Today, AGNSS is mentioned in particular in conjunction with cellular networks. The requirements to determine the location of a cellular phone in case of an emergency call enforced the development of the AGNSS technologies.

The following assistance data, without being exhaustive, could be provided to the GNSS receiver via the separate communication link:

- approximate user position,
- time information,
- almanac data,
- ephemerides data,
- all other information as emitted in the GNSS navigation message.

Depending on the availability of these different data sets, the receiver is assisted in its operation in the one or the other way. Relying on these data sets, the receiver does not have to demodulate the navigation message. Consequently, the signal power level can be lower since, as mentioned in Sect. 4.3.4, the power level for data demodulation has to be higher than for signal tracking. This favors positioning in weak signal environments.

Another advantage of the assistance concept is that pilot signals (Sect. 4.2.3) could be used independently of the navigation message for position determination. Thus, long coherent integration times allow to track weak satellite signals.

Since the communication channels generally have a much higher transmission rate than GNSS signals, the transfer of the navigation message is conducted in a short period of time. Including the other assistance information, i.e., approximate position and time information, this will result in shorter time to first fix (TTFF) intervals. The TTFF parameter defines the time needed for a GNSS receiver between power on and providing the first position information.

The AGNSS concept does not remarkably increase the position accuracy; however, the communication link could also be used to transmit differential corrections or integrity information.

The close integration of navigation module and communication system opens the doors for numerous assistance concepts and for performance enhancements, while enabling at the same time a vast variety of applications.

The approximate user position can be used in the adjustment algorithm to get a position solution using less iterative loops. The almanac data together with the reference time and the approximate position are used to compute which satellites are in view. Additionally, the satellite position together with the approximate user position and time allows to estimate the Doppler of the satellite signals. This allows to remove large parts of the frequency-offset uncertainty as it is described in Sect. 4.3.3. Thus, the two-dimensional search space is limited. This is particularly interesting for long ranging codes or for weak-signal acquisition.

If the time information is highly accurately aligned to GNSS time, the receiver may also reduce the code shift uncertainty. Therefore, the receiver computes the approximate user–satellite range, implements the accurate time information, and gets approximate code shift information. The determination of the approximate user position can be based on different technologies. In cellular networks, the cell identification number is sufficient to determine the user position with an accuracy of better than 20 km. Hofmann-Wellenhof et al. (2003: Sect. 8.3.5) describe a number of different geometrical concepts of cellular network positioning which could be used for position determination.

The almanac and ephemerides data is either a copy of what is provided by the GNSS or the past and present ephemerides information is used to predict long-term orbits. The predicted orbits might be valid for several days; however, the accuracy decreases with increasing time. Predicted orbits avoid frequent data download and enable orbit usage when the receiver has been deactivated for several hours or days.

Sage and Pande (2005) emphasize that the "A" in AGNSS can also refer to aided GNSS. In this mode the receiver relies on aiding information provided by external sensors to facilitate acquisition and tracking of the signal in space. Velocity information, e.g., is provided to the tracking loops.

12.6 Outlook

Glancing at all the different systems and services available today and in the near future, one might either be filled with enthusiasm or others might be puzzled by the crowd. Indeed, a great potential is offered by the numerous signals available. Mass market applications up to the geosciences will take full advantage of the numerous navigation systems. One question however will arise: what is the best combination of signals and systems to meet the requirements of your application?

Dixon (2007) emphasizes that all systems and services presented rely on similar methods and frequencies. Expressing it more dramatically with a metaphor – several birds could be killed with a single stone.

13 Applications

No textbook can address all the GNSS applications, especially when users create new ones almost every day. Consequently, first the different products derived from satellite measurements are discussed. Then the data exchange between GNSS receivers and other system components is highlighted followed by performance enhancements to meet the stringent user requirements which can be achieved by integrating GNSS systems with other systems and technologies. Finally, a short introduction into the user segment is supplemented by selected applications.

13.1 Products of GNSS measurements

13.1.1 Satellite coordinates

Based on broadcast ephemerides, the computation of the satellite coordinates follows the formalism introduced in Chap. 3 and summarized in Table 13.1. These equations apply to the computation of the coordinates of GPS and Galileo satellites, where the satellite coordinates are first computed in the orbital plane and then transformed into the earth-centered, earth-fixed (ECEF) system. The equations in Table 13.1 refer to the observation epoch t. Referring to GLONASS, the ECEF coordinates are transmitted for equidistant epochs. Thus, an interpolation algorithm has to be applied. The satellite coordinates and the satellite trajectories are commonly visualized using sky plots as shown in Fig. 8.4, the ground track plot as described in Table 12.1, or an elevation plot, where the elevation is plotted versus time.

Remondi (2004), furthermore, shows how the broadcast ephemerides information can be used to compute the velocity of the satellite. The position and the velocity of the satellite (PVS) are the first step to compute the user position.

13.1.2 Position determination

Position determination is above all based on pseudorange measurements. The biases influencing the range measurements and thus the position solutions are described in Chap. 5. The different methods for position determination have been described in Chap. 6, the data processing steps in Chap. 7.

Point positioning
Using GPS C/A-code pseudoranges on epoch-by-epoch basis will result in a horizontal position accuracy of 13 m at the 95% probability level. GLONASS provides

Table 13.1. Satellite coordinate computation

Parameter	Description
Numerical constants	
$\mu = 3.986\,005 \cdot 10^{14}$ [m^3 s^{-2}]	geocentric gravitational constant (WGS-84)
$\omega_e = 7.292\,115\,146\,7 \cdot 10^{-5}$ [rad s^{-1}]	earth rotation rate (WGS-84)
$\pi = 3.141\,592\,653\,589\,8$	("exactly"; WGS-84)
Broadcast ephemerides	
$\sqrt{a},\, e,\, M_0,\, \omega_0,\, i_0,\, \ell_0,\, \Delta n,\, \dot{i},\, \dot{\Omega},$ $C_{uc},\, C_{us},\, C_{rc},\, C_{rs},\, C_{ic},\, C_{is},\, t_e$	cf. Table 3.7
Computation formulas	
$t_k = t - t_e$	time from ephemerides reference epoch t_e
$a = (\sqrt{a})^2$	semimajor axis
$n_0 = \sqrt{\mu/a^3}$	computed mean motion
$n = n_0 + \Delta n$	corrected mean motion
$M_k = M_0 + n\, t_k$	mean anomaly
$E_k = M_k + e \sin E_k$	eccentric anomaly (solved by iteration)
$v_k = \arctan \frac{\sqrt{1-e^2}\,\sin E_k}{\cos E_k - e}$	true anomaly
$u_k = \omega_0 + v_k$	argument of latitude
$\delta u_k = C_{uc} \cos 2u_k + C_{us} \sin 2u_k$	argument-of-latitude correction
$\delta r_k = C_{rc} \cos 2u_k + C_{rs} \sin 2u_k$	radius correction
$\delta i_k = C_{ic} \cos 2u_k + C_{is} \sin 2u_k$	inclination correction
$\omega_k = \omega_0 + \delta u_k$	corrected argument of perigee
$r_k = a(1 - e \cos E_k) + \delta r_k$	corrected radius
$i_k = i_0 + \dot{i}\, t_k + \delta i_k$	corrected inclination
$x_k = r_k \cos(\omega_k + v_k)$	x-coordinate in the orbital plane
$y_k = r_k \sin(\omega_k + v_k)$	y-coordinate in the orbital plane
$\ell_k = \ell_0 + \dot{\Omega}\, t_k - \omega_e (t - t_0)$	corrected longitude of ascending node [(1)]
$X_k = x_k \cos \ell_k - y_k \sin \ell_k \cos i_k$	ECEF satellite X coordinate
$Y_k = x_k \sin \ell_k + y_k \cos \ell_k \cos i_k$	ECEF satellite Y coordinate
$Z_k = y_k \sin i_k$	ECEF satellite Z coordinate

[(1)] t_0 denotes the beginning of the current week

a slightly worse position accuracy. Galileo and the modernization programs of GPS and GLONASS, but especially the availability of more than one civil signal, will result in position accuracies of better than 5 m (95%). These accuracies refer to instantaneous position solutions in favorable signal environments. Multipath, signal blockage, or high atmospheric turbulences can deteriorate these results up to a factor of 10 or even more.

An increasing redundancy of pseudorange observations commonly increases the position accuracy even in unfavorable conditions. An appropriate weighting

13.1 Products of GNSS measurements

strategy in the adjustment process additionally mitigates measurement biases and errors (cf. Sect. 6.1.4). The weights are selected according to, e.g., S/N ratios (Wieser 2007a).

A higher accuracy is achievable by applying various filter techniques. Smoothing the code ranges by means of phase ranges (Sect. 5.2.2) will eliminate most of the noise of the code ranges. Kalman filter techniques smooth the position solution even in dynamic applications.

Precise point positioning
Another method to increase the position accuracy in single-point positioning mode is to use accurate satellite clock information and accurate ephemerides data. This method is denoted as precise point positioning (PPP) and is described in Sect. 6.1.4. This technique has been originally introduced for efficient analysis of GPS data from large networks (Zumberge et al. 1997). The accurate data sets will eliminate the satellite-specific biases as listed in Table 5.3, which are significant error sources in GNSS positioning. In conjunction with a dual-frequency receiver that allows to eliminate the ionospheric influence, the only remaining biases in the position solution are receiver-specific biases and the tropospheric component. Various control networks, as described below, provide high-accuracy satellite clock and ephemerides information, however with a delay. Real-time data are also available but with a diminished accuracy level.

Gao (2006) describes that using the final products of control networks, computing the position solutions in postprocessing, and implementing filter techniques like Kalman filtering, accuracies in the decimeter down to the subcentimeter range are achievable. These accuracy levels are attainable for static and dynamic applications (single-point positioning) relying on dual-frequency carrier phase measurements. Note that the filters applied in PPP need an initialization time of several minutes to converge to the centimeter accuracy. The ambiguities, however, cannot be fixed, as Gao (2006) further emphasizes, because the measurements are corrupted by satellite and receiver initial phase biases.

Witchayangkoon (2000: p. 18) mentions that PPP does not eliminate or reduce biases unlike in relative positioning. Therefore, a number of corrections have to be applied to eliminate variations in the undifferenced code and phase observations induced by systematic effects, satellite antenna offset and variation, or site displacement due to solid earth tides and ocean loading.

Example of single-point positioning
For a numerical example, Table 13.2 lists ephemerides data of four satellites together with pseudorange data R. The ephemerides are used to compute the satellite coordinates at the observation epoch $t = 129\,600$ s applying the equations as listed in Table 13.1. The result is given in Table 13.3.

Table 13.2. Example of satellite broadcast ephemerides data and pseudorange measurements R

		SV 06	SV 10	SV 16	SV 21
t_e	[s]	1.295 840E+05	1.296 000E+05	1.296 000E+05	1.295 840E+05
$\sqrt{a}$	[m$^{1/2}$]	5.153 618E+03	5.153 730E+03	5.153 541E+03	5.153 681E+03
e	[ ]	5.747 278E−03	7.258 582E−03	3.506 405E−03	1.179 106E−02
M_0	[rad]	−2.941 505E+00	4.044 839E−01	1.808 249E+00	3.122 437E+00
ω_0	[rad]	−1.770 838E+00	4.344 642E−01	−7.600 810E−01	−2.904 128E+00
i_0	[rad]	9.332 837E−01	9.713 110E−01	9.624 682E−01	9.416 507E−01
ℓ_0	[rad]	2.123 898E+00	−2.006 987E+00	1.122 991E+00	−3.042 819E+00
Δn	[rad s^{-1}]	5.243 075E−09	4.442 685E−09	4.937 348E−09	4.445 542E−09
$\dot{i}$	[rad s^{-1}]	−6.853 856E−10	2.521 533E−10	2.367 955E−10	−4.035 882E−11
$\dot{\Omega}$	[rad s^{-1}]	−8.116 052E−09	−8.495 353E−09	−8.054 621E−09	−7.757 823E−09
C_{uc}	[rad]	−1.184 642E−06	4.714 354E−06	9.499 490E−07	6.897 374E−06
C_{us}	[rad]	7.672 235E−06	−1.825 392E−07	5.437 061E−06	1.069 344E−05
C_{rc}	[m]	2.146 562E+02	3.868 750E+02	2.709 062E+02	1.630 625E+02
C_{rs}	[m]	−2.140 625E+01	8.978 125E+01	1.515 625E+01	1.329 375E+02
C_{ic}	[rad]	2.980 232E−08	3.725 290E−09	6.332 993E−08	−1.080 334E−07
C_{is}	[rad]	−1.117 587E−08	8.940 696E−08	−2.421 438E−08	−8.009 374E−08
R	[m]	20 509 078.908	23 568 574.070	23 733 776.587	22 106 790.995

The approximate user position is

$$\varphi_{r0} = 47.1°,$$
$$\lambda_{r0} = 15.5°,$$
$$h_{r0} = 400 \text{ m},$$

which is transformed into geocentric WGS-84 coordinates using Eq. (8.1) leading to the approximate coordinates

$$X_{r0} = 4\,191\,621.710 \text{ m},$$
$$Y_{r0} = 1\,162\,439.580 \text{ m},$$
$$Z_{r0} = 4\,649\,632.607 \text{ m}.$$

Table 13.3. Satellite WGS-84 coordinates ($t = 129\,600$ s)

	SV 06	SV 10	SV 16	SV 21
X^s [m]	13 736 749.018	−2 156 464.014	5 780 040.699	25 897 345.749
Y^s [m]	8 001 485.736	20 642 907.598	−17 694 953.977	5 369 544.851
Z^s [m]	21 462 886.878	16 289 053.551	18 974 539.869	4 763 893.950

13.1 Products of GNSS measurements

The satellite coordinates together with the approximate user position are used to compute the design matrix $\mathbf{A}$ by Eq. (7.143), which yields

$$\mathbf{A} = \begin{bmatrix} -\dfrac{X^6(t) - X_{r0}}{\varrho_{r0}^6(t)} & -\dfrac{Y^6(t) - Y_{r0}}{\varrho_{r0}^6(t)} & -\dfrac{Z^6(t) - Z_{r0}}{\varrho_{r0}^6(t)} & c \\ -\dfrac{X^{10}(t) - X_{r0}}{\varrho_{r0}^{10}(t)} & -\dfrac{Y^{10}(t) - Y_{r0}}{\varrho_{r0}^{10}(t)} & -\dfrac{Z^{10}(t) - Z_{r0}}{\varrho_{r0}^{10}(t)} & c \\ -\dfrac{X^{16}(t) - X_{r0}}{\varrho_{r0}^{16}(t)} & -\dfrac{Y^{16}(t) - Y_{r0}}{\varrho_{r0}^{16}(t)} & -\dfrac{Z^{16}(t) - Z_{r0}}{\varrho_{r0}^{16}(t)} & c \\ -\dfrac{X^{21}(t) - X_{r0}}{\varrho_{r0}^{21}(t)} & -\dfrac{Y^{21}(t) - Y_{r0}}{\varrho_{r0}^{21}(t)} & -\dfrac{Z^{21}(t) - Z_{r0}}{\varrho_{r0}^{21}(t)} & c \end{bmatrix}.$$

The computation of the position dilution of precision (PDOP) applying Eqs. (7.172) through (7.175) results in

$$\text{PDOP} = 2.6$$

or applying the rotation matrix $\mathbf{R}$ of the local-level frame as given in Eq. (7.176), will result in

$$\text{HDOP} = 1.4, \qquad \text{VDOP} = 2.1.$$

The code pseudorange measurements R to the four satellites (Table 13.2) result in the following position solution by applying Eq. (7.66):

$$X_r = 4\,195\,408.251 \text{ m},$$
$$Y_r = 1\,159\,775.764 \text{ m},$$
$$Z_r = 4\,646\,945.784 \text{ m}$$

or in ellipsoidal coordinates

$$\varphi_r = 47.064\,188\,72°, \qquad \lambda_r = 15.452\,891\,37°, \qquad h_r = 433.278 \text{ m}.$$

The solution of the equations gives also an estimate of the receiver clock bias $\delta_r(t) = 21.45$ ns. An iteration might be necessary to gain the final results. The simplified mathematical steps do not show how to account for the various effects and biases like (without being exhaustive) the signal run time, satellite clock correction parameters, atmospheric corrections, relativity, or earth rotation as described in detail in Chap. 5. These have already been implicitly included into the code pseudoranges. The implementation of the correction terms is necessary to achieve the accuracies discussed before.

Differential positioning

The principle of differential GNSS has been introduced in Sect. 6.2 and DGNSS systems have already been discussed in Sect. 12.3. DGNSS relies on two or more receivers: a reference receiver and a roving receiver. The correlation between the errors in the reference station and in the rover receiver enables to eliminate or at least to reduce common errors. The performance of the differential system decreases with increasing temporal and spatial decorrelation of the errors. Any failure in the determination of the differential corrections will deteriorate the position accuracy of the rover, since these errors are uncorrelated to the receiver biases. Thus, especially reference receiver-specific biases, e.g., multipath, have to be calibrated and eliminated before differential corrections are determined. Following the discussion in Sect. 5.1.4 and extending Table 5.3 according to Zogg (2006), then applying DGNSS techniques for code pseudoranges will result in reduced user equivalent range errors (UERE) as given in Table 13.4. As mentioned in Sect. 5.1.4, too many factors influence the errors, thus the sizes of the errors given in Table 13.4 have to be understood as an example.

Instead of transmitting the differential corrections from the reference to the rover, which is called navigation mode, the surveillance mode relies on a transmission of the coordinates and/or measurements from the rover to the reference network. The position is then computed at the reference stations and broadcast back to the user.

Three different approaches to compute differential corrections have been introduced in Sect. 12.3. An overview of the present DGPS performance considering the measurement-domain approach is given in Table 13.5, where the individual values are matched for mnemonic reasons. The accuracy of the heights is worse by a factor of 1.5 to 2.

Table 13.4. UERE computation

Error source	GNSS [m]	DGNSS [m]
Ephemerides data	2.1	0.1
Satellite clock	2.1	0.1
Ionosphere	4.0	0.2
Troposphere	0.7	0.2
Multipath	1.4	1.4
Receiver measurement	0.5	0.5
UERE [m]	5.3	1.5

13.1 Products of GNSS measurements

Table 13.5. Accuracies for DGPS

Observable	Station separation	Horizontal accuracy
Code range	1 000 km	<10 m
Smoothed code ranges	100 km	<1 m
Carrier phases	some 10 km	<0.1 m

Relative positioning

The objective of relative positioning is to reduce or even eliminate error sources by differencing GNSS measurements taken at different stations at the same epoch. Best accuracies are achieved in the relative positioning mode with observed carrier phases (cf. Sect. 6.3). Originally, relative positioning was conceived for postprocessing. Thus, this method was and still is particularly used for surveying and geosciences. Today, real-time data transfer is routinely applied, which enables real-time computation of baseline vectors and has led to the real-time kinematic (RTK) technique.

Static relative positioning
The static relative positioning method is most commonly used for geodetic surveys. The required observation periods depend on the baseline length, the number of visible satellites, the number of carrier frequencies, and the geometric configuration. The accuracy is correlated with the baseline length and amounts to 1 to 0.1 ppm for baselines up to some 100 km and even better for longer baselines.

Static relative positioning also includes the rapid static technique based on fast ambiguity resolution techniques. These techniques generally use code and carrier phase combinations on all frequencies. Restricting the method to 20 km baselines, accuracies at the subcentimeter level can be achieved.

Standard values for the session lengths of static observations (particularly for baselines up to some 20 km) are listed in Table 13.6. These values are based on the visibility of four satellites, good geometry, and normal atmospheric conditions. Note that an additional satellite may reduce the session lengths by 20%. The numbers may be regarded as too conservative; however, they assure correct ambiguity resolution and, thus, high accuracies.

Typical applications of static surveying include state, county, and local control surveys, photo-control surveys, boundary surveys, and deformation surveys.

Pseudokinematic relative positioning
The pseudokinematic survey, developed by B.W. Remondi, is also named intermittent static or reoccupation method. These surveys require less occupation time

Table 13.6. Session lengths for static surveys

Receiver	Conventional static	Rapid static
Single-frequency	30 min + 3 min/km	20 min + 2 min/km
Dual-frequency	20 min + 2 min/km	10 min + 1 min/km

but one must occupy the "point-pair" two times. A typical scenario would involve occupying a pair of points for five minutes, moving to other points, and finally returning to the first point pair about one hour after the initial occupation for a second 5-minute occupation. Subcentimeter accuracy can be obtained with the pseudokinematic method. This high accuracy is achievable because the integer biases can be fixed due to the fact that the satellite geometry changes in the hour between occupations. There is no requirement for maintaining signal lock between the reoccupation of points. The main disadvantage is the necessity of revisiting the site.

Kinematic relative positioning
Kinematic surveys are the most productive in that the greatest number of points can be determined in the least time. The drawback is that after initialization of the survey a continuous lock on at least four satellites must be maintained.

The semikinematic or stop-and-go technique is characterized by alternatively stopping and moving one receiver to determine the positions of fixed points along the trajectory. The most important feature of this method is the increase in accuracy when several measurement epochs at the stop locations are accumulated and averaged. Relative position accuracies at the centimeter level can be achieved for baselines up to some 20 km.

The kinematic technique requires the resolution of the phase ambiguities before starting the survey. The initialization can be performed by static or kinematic techniques. Although the on-the-fly techniques (cf. Sect. 7.2.3) allow to initialize while the receiver is in motion, the initial integer resolution can be accomplished more rapidly with the receiver being static. Dual-frequency receivers require up to 1–2 minutes of observations for baselines up to 20 km to resolve the ambiguities kinematically. After initialization no loss of lock should occur, otherwise initialization has to be restarted. Triple-frequency receivers will allow to resolve the ambiguities instantaneously, i.e., epoch by epoch. Solutions of several epochs, however, are needed to mitigate the risk of false ambiguity fix.

Real-time kinematic relative positioning
Transmitting the phase measurement in real time from the base station to the rover, the ambiguities can be also solved in real time, leading to the RTK method.

13.1 Products of GNSS measurements

The decorrelation of the error sources limits the relative methods to about 20 km. The wide-area RTK (WARTK) method avoids a fast decorrelation of the error sources by implementing ionospheric corrections. In this way the ionospheric influence is significantly reduced or even eliminated and the integer nature of the ambiguities is preserved. The later characteristic allows to fix wide-lane ambiguities and to achieve subcentimeter accuracies despite the baseline lengths of up to 400 km. Similar to RTK, also WARTK takes several minutes before the solution converges using dual-frequency receivers. Triple-frequency receivers allow to instantaneously solve the ambiguities for a baseline length of about 20 km in traditional RTK mode and up to 400 km in WARTK operation (Hernández-Pajares et al. 2004).

Typical applications for RTK techniques include construction, reference benchmarks for land survey, or robotic guidance.

Achievable accuracies
It is best practice to use a mixture of the methods when using single-frequency receivers. For example, static and pseudokinematic methods can be used to establish a broad framework of control and to set points on either side of obstructions such as bridges. Kinematic surveys can then be employed to determine the coordinates of the major portion of points, using the static points as control and check points.

Relative (static) positioning can be performed on baselines of arbitrary length up to some thousands of kilometers. In many cases, however, baselines do not exceed 20 km. For such baselines, the achievable accuracies in horizontal position (1σ level) are listed in Table 13.7. The accuracy in height is worse by a factor of 1.5 to 2. The accuracies are based on single-frequency receivers tracking 5 satellites with reasonable geometry and under normal ionospheric conditions. Furthermore, it is assumed that the ambiguities have been resolved. The accuracies in Table 13.7 are rather conservative and the values were again matched for mnemonic reasons. Take as an example a 10 km baseline, which will result in a position accuracy of 1 cm in static mode.

In static mode, the session lengths can be substantially reduced with dual- and triple-frequency receivers; however, the position accuracy does not improve for distances up to 20 km. For longer baselines, dual-frequency receivers are required to mitigate of ionospheric biases. The relative accuracy for baselines of 100 km is

Table 13.7. Accuracies for relative positioning

Mode	Horizontal accuracy
Static	5 mm + 0.5 ppm
Kinematic	5 cm + 5 ppm

in the 0.1 ppm range. For 1 000 km baselines, 0.01 ppm accuracies are achievable.

In kinematic mode, the position accuracy can be improved by the factor $\sqrt{2n}$ when dual-frequency receivers are used and data are collected during n epochs in the stop-and-go mode. Take as an example a 10 km baseline and dual-frequency observations at dozen epochs, this will result in a position accuracy of 2 cm. Compared to the static mode, the accuracy in the kinematic mode is worse mainly due to multipath and DOP variations. These effects are more or less averaged in the static mode.

13.1.3 Velocity determination

The determination of velocity is either based on a simple differentiation of the position solutions over time, or the velocity is derived from the Doppler observations of the receiver. Following Eqs. (1.4) and (4.13), the Doppler shift Δf reads

$$\Delta f = f_r - f^s = -\frac{1}{c} f^s v_\varrho = -\frac{1}{c} f^s \frac{(\varrho^s - \varrho_r)}{\varrho} \cdot (\dot{\varrho}^s - \dot{\varrho}_r) = -\frac{1}{c} f^s \varrho_0 \cdot \Delta\dot{\varrho}, \tag{13.1}$$

where f_r denotes the received frequency, f^s the emitted frequency, c the speed of light, v_ϱ the relative radial velocity, ϱ^s the satellite position, ϱ_r the receiver position, and ϱ the distance between satellite and receiver. Finally, $\dot{\varrho}^s$ quantifies the satellite speed and $\dot{\varrho}_r$ the speed of the user receiver. The measured Doppler shift $\Delta \bar{f}$ additionally integrates the receiver frequency drift δf_r. If the receiver position is known, then the observation equation reads

$$\Delta \bar{f} + \frac{1}{c} f^s \varrho_0 \cdot \dot{\varrho}^s = \frac{1}{c} f^s \varrho_0 \cdot \dot{\varrho}_r + \delta f_r, \tag{13.2}$$

where δf_r is treated as an additional unknown (Mansfeld 2004: pp. 165–167). The receiver position has to be known with an accuracy of less than 10 km to get a velocity of better than 1 m s^{-1}. The combination of the Doppler information with respect to several satellites allows to determine a three-dimensional velocity vector $\dot{\varrho}_r$ together with a frequency drift.

The accuracy of velocities determined by the differentiation of the position solutions, i.e.,

$$\dot{\varrho}_r = \frac{\|\varrho_r(t_2) - \varrho_r(t_1)\|}{t_2 - t_1}, \tag{13.3}$$

depends on the accuracy of the position. Thus, taking into account the law of error propagation and assuming position accuracies of σ_ϱ, the velocity accuracy derives from

$$\sigma_{\dot{\varrho}} = \frac{\sqrt{2}}{t_2 - t_1} \sigma_\varrho. \tag{13.4}$$

13.1 Products of GNSS measurements

For a numerical example assume GPS C/A-code single-point solutions with an accuracy of 13 m measured in 20-second intervals. This leads to a velocity accuracy of $0.9\,\mathrm{m\,s^{-1}}$. The correlation of the accuracies in the three coordinate axes and the fact that the vertical component is less accurate than the horizontal one have not been taken into account in the simplified derivation and the numerical example.

13.1.4 Attitude determination

The principle of attitude has been described in Sect. 1.3.3. Attitude is defined by the three angles: roll r, pitch p, and yaw (or heading) y.

Theoretical considerations

Three antennas A_i in a proper configuration define the plane of a (rigid) platform in space. Two independent baselines provide six equations to determine the three attitude parameters. Thus, three conditions exist: the length of the baselines and the spatial angle between them are invariant against attitude variations.

For the mathematical formulation of three-dimensional attitude determination, the position vector of an antenna A_i in the local-level frame is denoted by $\mathbf{x}_i$ and the corresponding position vector in the body frame by $\mathbf{x}_i^b$. The vector $\mathbf{x}_i$ results from relative positioning with GNSS carrier phase measurements, where the accurate geocentric coordinates of the antennas with respect to each other are transformed to the local-level system. The body frame may be realized by the position vectors $\mathbf{x}_i^b$ of three antennas A_i as shown in Fig. 13.1. The three attitude parameters correspond to the three Euler angles rotating the local-level frame into the body frame. Thus, attitude determination is defined by

$$\mathbf{x}_{ij}^b = \mathbf{R}_2\{r\}\,\mathbf{R}_1\{p\}\,\mathbf{R}_3\{y\}\,\mathbf{x}_{ij}\,, \tag{13.5}$$

where relative vectors (i.e., baseline vectors) $\mathbf{x}_{ij}$ and $\mathbf{x}_{ij}^b$ have been introduced in-

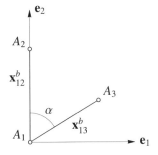

Fig. 13.1. Definition of the body frame

stead of absolute vectors $\mathbf{x}_i$ and $\mathbf{x}_i^b$ to eliminate the shift vector. The three consecutive rotations can be expressed by a single rotation matrix $\mathbf{R}\{r, p, y\}$. Explicitly,

$$\mathbf{R}\{r, p, y\} = \begin{bmatrix} \cos r \cos y & \cos r \sin y & -\sin r \cos p \\ -\sin r \sin p \sin y & +\sin r \sin p \cos y & \\ -\cos p \sin y & \cos p \cos y & \sin p \\ \sin r \cos y & \sin r \sin y & \cos r \cos p \\ +\cos r \sin p \sin y & -\cos r \sin p \cos y & \end{bmatrix} \quad (13.6)$$

is obtained (Lachapelle et al. 1994).

Direct computation of attitude
The attitude parameters can directly be computed from the local-level coordinates without knowledge of the body frame coordinates (Lu et al. 1993). The yaw y and pitch p (corresponding to azimuth and elevation angle) are determined from one baseline only. Selecting, for example, the baseline between A_1 and A_2 (Fig. 13.1), the relations

$$\tan y = \frac{e_{12}}{n_{12}},$$

$$\tan p = \frac{u_{12}}{\sqrt{e_{12}^2 + n_{12}^2}} \quad (13.7)$$

are obtained, where the $\mathbf{x}_{ij}$ baseline components are introduced in the north, east, and up local-level frame (cf. Sect. 8.2.2).

In order to obtain roll r, the baseline vector $\mathbf{x}_{13}$ is first rotated with respect to yaw and pitch resulting in the vector $\mathbf{x}'_{13} = \mathbf{R}_1\{p\} \mathbf{R}_3\{y\} \mathbf{x}_{13}$. A third rotation (i.e., roll) rotates this vector to the body frame yielding

$$\begin{bmatrix} a_{13} \sin \alpha \\ a_{13} \cos \alpha \\ 0 \end{bmatrix} = \begin{bmatrix} \cos r & 0 & -\sin r \\ 0 & 1 & 0 \\ \sin r & 0 & \cos r \end{bmatrix} \begin{bmatrix} e'_{13} \\ n'_{13} \\ u'_{13} \end{bmatrix}, \quad (13.8)$$

where a_{13} corresponds to the length of vector $\mathbf{x}_{13}$, which is identical to the length of $\mathbf{x}_{13}^b$. The roll angle can be computed from the third line of the above equation by

$$\tan r = -\frac{u'_{13}}{e'_{13}}. \quad (13.9)$$

The direct computation method is a function of three antennas only and (as mentioned) does not require a priori knowledge of the body frame. The disadvantage of the method is that redundant antennas are not used.

13.1 Products of GNSS measurements

Least-squares estimation of attitude
Least-squares estimation of attitude is based on the knowledge of the body frame coordinates of the antennas. These coordinates may be obtained through a survey or initialization process. The GNSS-derived local-level coordinates can then be treated as observations.

In a slightly different form, Eq. (13.5) may be written as

$$\mathbf{x}_{ij}^b = \mathbf{R}\{r, p, y\} \mathbf{x}_{ij}, \tag{13.10}$$

where the three rotation matrices are combined into one matrix. Each baseline with known body frame coordinates and GNSS-derived local-level coordinates gives rise to three equations.

In the most general case, the matrix $\mathbf{R}$ could be replaced by a matrix where all nine elements are unknown. This corresponds to an affine transformation. At least three baselines (or four antennas) are needed to solve for the nine unknowns.

In the case of rigid platforms, similarity transformations are sufficient. The minimum of three antennas forming two (noncollinear) baseline vectors provide six equations for the three unknown attitude parameters. Thus, the problem is redundant and can be solved by least-squares estimation. Common least-squares estimation requires linearization of Eq. (13.10) with respect to the attitude parameters. Another approach is the generalized inversion of Eq. (13.10). Denoting n as the number of independent baselines, Eq. (13.10) may be written in the form

$$\mathbf{A}^b = \mathbf{R}\{r, p, y\} \mathbf{A}, \tag{13.11}$$

where $\mathbf{A}^b$ is a $3 \times n$ matrix with n baseline vectors in the body frame as column vectors. Analogously, matrix $\mathbf{A}$ contains the baseline vectors in the local-level frame. Solving for the rotation matrix involves multiplying Eq. (13.11) by the (generalized) inverse of the matrix $\mathbf{A}$. The result given in Graas and Braasch (1991) reads

$$\mathbf{R} = \mathbf{A}^b \mathbf{A}^T (\mathbf{A} \mathbf{A}^T)^{-1}. \tag{13.12}$$

Denoting the elements of matrix $\mathbf{R}$ by R_{ij}, the attitude parameters are obtained by

$$\begin{aligned} \tan r &= -\frac{R_{13}}{R_{33}}, \\ \tan p &= \frac{R_{23}}{\sqrt{R_{21}^2 + R_{22}^2}}, \\ \tan y &= -\frac{R_{21}}{R_{22}}, \end{aligned} \tag{13.13}$$

which may be verified by inspecting the matrix $\mathbf{R}$, cf. Eq. (13.6).

The advantage of attitude estimation by the least-squares technique is that the computation of the attitude is more rigorous since it computes the best estimate for the attitude parameters.

Modeling of wing flexure

If four antennas are mounted, for instance, one on each wing tip of an aircraft and the remaining two atop the fuselage, then the roll angle could be determined independently from heading and the pitch angle. In practice, however, the movement of the aircraft's wings during flight may completely prevent the precise determination of the roll angle in this way.

Another technique is reported by Cannon et al. (1994). The nonrigidity of the body frame is taken into account by modeling the wing flexure. Constraining the wing flexure in the vertical component of the body frame leads to the relation

$$\mathbf{x}_i^b = \mathbf{x}_{0i}^b - f\,\mathbf{e}_3 \,, \tag{13.14}$$

where f is a scalar which is estimated in the least-squares adjustment. The position vectors $\mathbf{x}_{0i}^b$ refer to the situation without wing flexure (aircraft at rest) and can be measured directly using a theodolite or can be determined by static GNSS observations prior to takeoff.

Practical considerations

An accuracy of 1 mm in the relative position of the antennas corresponds to one milliradian or $0.057°$ in attitude accuracy for an antenna separation of 1 m. For larger antenna separations, the angular accuracy increases. However, for longer baselines, the search for the phase ambiguities becomes more difficult.

Multipath is a significant limiting error source in attitude determination. This effect can be reduced by installing a common ground plane for all antennas or by the use of, e.g., choke ring antennas. Other error sources are alignment errors, which transform into an attitude error. Furthermore, uncalibrated phase center offsets additionally deteriorate the attitude result.

The key requirement for attitude determination is ambiguity resolution on-the-fly. This is particularly valid for airborne or marine applications. Ambiguity resolution can be speeded up by special antenna configurations as proposed, for example, by El-Mowafy (1994). Another approach is to incorporate geometric constraints due to the known antenna array geometry (Landau and Ordóñez 1992, Lu 1995).

Attaching four antennas to one receiver builds up a dedicated system. In this configuration, single-differencing is sufficient since the receiver clock offset is common to all antenna measurements. Thus, only three satellites are required to solve the interantenna vector. The key point for the nondedicated system, i.e., four independent receivers, is that off-the-shelf receivers can be used, which provides flexibility since these receivers can then be used for other applications as well. In this case, double-differencing must be applied to account for the receiver clock offsets.

13.1.5 Time transfer

Another use of GNSS is the determination of accurate time. Inexpensive single-frequency GNSS receivers operating on known stations provide a timing accuracy of about 30 nanoseconds (95% probability level) with only one satellite in view. With more sophisticated techniques, one can globally synchronize clocks even more precisely and an accuracy of less than 1 nanosecond is achievable.

Highly accurate time synchronization and time stamps are needed for diverse applications, e.g., coordinating seismic monitoring or other global geodynamical measurements. Communication systems and power plants need exact time synchronization of their processes. The European Space Agency (2005c) further emphasizes that certified time stamps are indispensable for applications like electronic banking, e-commerce, or stock exchange.

13.1.6 Other products

The scientific community took advantage of the position determination and timing capability of GNSS. Various short-term and long-term measurements on a regional or global scope allow to monitor, e.g., geophysical effects.

Other geosciences use the GNSS signals as a tool for atmospheric remote sensing, following the paradigm that one man's noise is the other man's signal. GNSS signals passing the earth atmosphere are refracted as described in Sects. 4.1.2 and 5.3. Additionally, the signals are delayed while propagation through the atmosphere. Measuring GNSS signals at a known reference location and eliminating all other effects, like the geometric range, relativistic effects, or clock errors, allows to model the atmospheric effects. The degree of refraction and delay is a function of the temperature, pressure, water vapor, humidity, and electron content; thus, measuring the influence on GNSS signals allows to derive these parameters. The changing satellite constellation enables to derive the parameters for different layers resulting in a tomography of the atmosphere (Sect. 5.3.4). This knowledge then is used for, e.g., weather forecasting or climate monitoring (European Space Agency 2005c)

The longer the distance traveled through the atmosphere the higher the atmospheric influence. Therefore, systems have been developed where GNSS receivers are installed in low earth orbit (LEO) satellites (Fig. 13.2).

In a simplified consideration, v_L and v_G denote the velocities of the LEO (L)

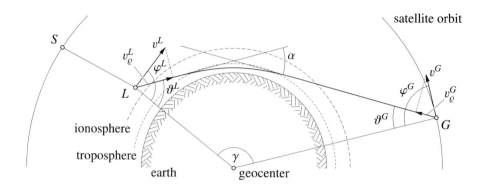

Fig. 13.2. Principle of atmospheric remote sensing

and GNSS (G) satellite. The radial velocity components are

$$v_\varrho^G = v^G \cos(\varphi^G - \vartheta^G),$$
$$v_\varrho^L = v^L \cos(\varphi^L - \vartheta^L), \quad (13.15)$$

where the angles φ^L and φ^G are known from geometry. Assuming spherical symmetry (García-Fernández 2004), ϑ^L may be expressed as a function of ϑ^G (or vice versa). Thus, the relative velocity between the LEO satellite and the GNSS satellite,

$$v_\varrho = v_\varrho^G + v_\varrho^L, \quad (13.16)$$

contains only one unknown, which can be computed from the Doppler frequency shift $\Delta f = -v_\varrho f^s/c$. Here v_ϱ denotes the relative (radial) velocity in the ray path direction rather than in the line-of-sight direction. The bending angle α, caused especially by the troposphere (Fig. 13.2), follows from

$$\alpha = \vartheta^G + \vartheta^L + \gamma - \pi \quad (13.17)$$

and, thus, the measured Doppler frequency can directly be related to the bending angle. Using the Abel inversion technique, finally, results in the neutral gas parameters as a function of the bending angle α (Jakowski 2001).

García-Fernández (2004) additionally details how the measurements taken at the LEO satellites can be used to model the ionosphere.

The orbit of the LEO satellite is determined using GNSS measurements, which are not influenced by the atmosphere (e.g., measurements to satellite S) or using other positioning systems, like DORIS or PRARE.

Various satellite missions have a dedicated payload for atmospheric tomography measurements onboard, e.g., CHAMP or GRACE.

This short introduction into atmospheric sounding describes one method to use GNSS for scientific applications. A great variety of others exist how GNSS signals could be analyzed to derive other key parameters apart from position, velocity, attitude, or time.

13.2 Data transfer and formats

The transmission of differential corrections, raw measurements, or position solutions from or to the user receiver can be based on various means of communication. The objective, however, is to minimize the extra system cost and reduce the receiver complexity of a combined communication and navigation module. One-way communication is sufficient, some systems may require bidirectional implementation.

Space-based communication links are above all realized using GEO satellites which orbit the earth at an altitude of about 36 000 km. The major drawback of the satellites is the fast-decreasing elevation angle with increasing latitude of the user receiver. Consequently, topography, vegetation, or man-made objects shadow the weak satellite signals.

Terrestrial communication links rely on amateur data radio, cellular communication networks, digital trunked radio, or even Internet techniques to name but a few. The transmission rates of ground-based communication links vary between 30 bps and up to 2 Mbps and more for wireless communication techniques. Some existing terrestrial radionavigation infrastructures have been modified in order to provide DGNSS services. For instance, maritime radio beacons have been successfully adapted, or the Loran-C network has been explored as another option of DGNSS transmission.

Various data formats have been specified for the exchange of satellite navigation data. Three formats, as described in detail below, became internationally accepted and generally supported by all receiver manufacturers. The data format defined by the Radio Technical Commission for Maritime Services (RTCM) is used for real-time transmission of measurements and differential corrections. The receiver-independent exchange (RINEX) format is used for the exchange of raw data, especially for postprocessing applications. The National Marine Electronics Association (NMEA) defined an ASCII data format which is particularly used for the transmission of position solutions.

13.2.1 RTCM format

Although some receiver manufacturers have devised their own proprietary formats, the transmission of correction data between a reference receiver and a remote receiver has been standardized in 1985 according to the proposals of the US Radio

Technical Commission for Maritime Services, Special Committee 104. The current version 3.1 is denoted as RTCM standard 10403.1. Since messages of version 3.1 are not compatible with version 2.x, the Special Committee 104 explicitly states that the version 3.1 and version 2.3 are the current ones for reasons of compatibility. Although the RTCM format has originally been introduced for differential operations in maritime applications, meanwhile it is used in all fields of applications for the transmission of all kinds of GNSS data.

The RTCM standard version 2.3 (i.e., RTCM standard 10402.3) defines 64 different message types, some of them are listed in Table 13.8 (Radio Technical Commission for Maritime Services 2001: Table 4-3). The messages consist of a sequence of words with 30-bits each. The last six bits in each word are parity bits. Every message starts with a header which is two or three words long. The first word contains a fixed preamble, the message type identifier, and the reference station identifier. The second word contains the frame time tag, the sequence number, the message length, and a reference station health indicator. In some messages a third word is added to the header. The total message has a maximum length of 33 words.

Conventional DGPS requires message types 1, 2, and 9 to provide meter accuracy. RTK operation relies on message types 18 through 21 to provide centimeter accuracy. Various systems use the RTCM message format to transmit proprietary

Table 13.8. Selected message types of RTCM version 2.3

Message type	Title
1	Differential GPS corrections
2	Delta differential GPS corrections
3	GPS reference station parameters
9	GPS partial satellite set
10	P-code differential corrections
11	GPS C/A-code L1, L2 delta corrections
15	Ionospheric delay message
17	GPS ephemerides
18	RTK uncorrected carrier phases
19	RTK uncorrected code pseudoranges
20	RTK carrier phase corrections
21	RTK code pseudorange corrections
31	Differential GLONASS corrections
32	Differential GLONASS reference station parameters
59	Proprietary message

13.2 Data transfer and formats

information. Message type 59, in particular, can be used as a communication channel to transmit, e.g., short messages.

The message types 1 through 17 were available already in version 2.0, while the messages 18 through 21, each with a three-word header, were added starting with version 2.1. Messages related to GLONASS are available since version 2.2. Version 2.3 added further messages to improve RTK operation.

RTCM version 3 has been defined to increase the efficiency of information transmission and to increase the integrity of the parity operation. Version 3 has been in particular designed for RTK and network RTK operations, where a large data volume has to be transmitted. The message consists of an 8-bit preamble, 10-bit message length identifier, and 6 additional bits in the header reserved for future use. The data field has a maximum length of 1024 bytes followed by a 24-bit cyclic redundancy check (CRC).

For the transmission of RTCM data over Internet the networked transport of RTCM via Internet protocol (NTRIP) format has been defined by the German Federal Agency for Cartography and Geodesy. NTRIP is based on the hypertext transfer protocol (HTTP). Meanwhile, the NTRIP format has been officially taken over by RTCM.

The RTCM standards 10402.3 and 10403.1 can be ordered at www.rtcm.org.

13.2.2 RINEX format

The RINEX format has already been mentioned in Sect. 7.1.1. In 1989, W. Gurtner of the Astronomical Institute of the University of Berne, Switzerland, defined the RINEX format to facilitate the exchange of the GPS data of different receiver types. Meanwhile RINEX has been updated several times to account for the different developments in GNSS signals and also applications. The current standard is version 3.0, published in January 2006 (Gurtner and Estey 2006).

The most commonly used RINEX (ASCII) file types are the navigation file, which contains the ephemerides data of the satellites, and the observation file. The latter basically contains the carrier phases, code ranges, Doppler measurements, and signal-to-noise ratios. Observations of the different GNSS are designated using the character 'G' for GPS, 'R' for GLONASS, 'E' for Galileo, and 'S' for SBAS satellites. RINEX version 3.0 specifies 228 different observation codes for the various GPS, GLONASS, Galileo, and SBAS observables and signal-to-noise ratios.

Y. Hatanaka of the Geographical Survey Institute, Tsukuba, Japan, defined the compressed version of the RINEX format. Here only the changes from epoch to epoch are stored.

13.2.3 NMEA format

The US National Marine Electronics Association (NMEA) proposed in 1983 the NMEA-0183 interface specification. Although originally defined for the interface of marine electronic devices, meanwhile it became a voluntary industry standard interface for all GNSS receivers. The NMEA data format specifies the exchange format of position information including quality indicators, course over ground, or speed over ground data. The transmission of differential correction, although specified today, was not part of the original objectives of NMEA-0183.

The NMEA-0183 specifies a serial data transmission with a data rate of 4 800 bps and an 8-bit ASCII format of the data. GNSS receivers commonly offer also higher transmission rates.

An NMEA data set, denoted as string or sentence, consists of 82 characters maximally. Every sentence starts with the dollar sign "$" followed by the address field. The latter is commonly subdivided into a two-character-long talker field and a three-character-long sentence type field. Meanwhile, about 60 different sentence types have been approved. The talker field uses a "GP" to indicate GPS data, and "GL" for GLONASS information. The address field is followed by a variable number of data fields which are separated by comma deliminators. The last data field is followed by an asterisk "*" and an optional checksum. Only some message types mandatorily require the checksum, which is computed using an XOR operation of the characters starting after the "$" until but excluding the "*". The checksum, given in hexadecimal format, finally is followed by a carriage return and line-feed character. Various messages and their content are defined in Table 13.9. The current version 3.01 of NMEA-0183 has been published in 2002.

The NMEA-2000 specification extends the single-talker–multiple-listener interface specification of NMEA-0183 to a serial data networking specification for marine electronic devices. Both standards can be ordered at www.nmea.org.

13.3 System integration

All three major GNSS systems together, i.e., GPS, GLONASS, and Galileo, will count a total of 75 satellites in nominal constellation. Taking into account that all three systems will additionally provide spare satellites, and also considering SBAS satellites, then up to 90 satellites will simultaneously emit navigation signals within the next years. In Asia and Oceania, the number will even be higher taking into account IRNSS, Beidou-1, and QZSS. Beidou-2 / Compass will add about another 30 satellites to this navigation constellation.

Despite this great number of systems, satellites, and signals, satellite navigation cannot meet the requirements of all fields of application. Especially considering the stringent requirements of safety-of-life applications, the capabilities of

13.3 System integration

Table 13.9. Some GNSS-related NMEA sentences

Sentence	Description of content
ALM	Almanac data
GGA	Position fix data (time, ellipsoidal coordinates, number of satellites, dilution of precision, quality indicator, geoidal height)
GLL	Reduced fix data (time, latitude, longitude, status)
GSA	Active satellites (used for the position computation) including DOP factors
GSV	Satellites in view, azimuth, elevation, and signal-to-noise ratios for each satellite
VTG	Navigation data (course over ground, speed over ground)

satellite navigation are limited. All systems are too similar, thus suffering from the same conceptual deficiencies as highlighted in a report by the Volpe National Transportation Systems Center (2001). System integration aims to provide parallel, complementary, dissimilar, or analytical redundancy to combine the strengths of several systems (Hofmann-Wellenhof et al. 2003: Sect. 13.3). This will increase the performance and functionality of the navigation system. Thus, the strategy is changed from using satellite navigation as sole means of navigation to use it as prime means of navigation. System integration especially becomes necessary when going from the clear sky to the deep indoor area, since GNSS, including augmented, assisted, differential, or relative system concepts, poses one stringent prerequisite in its measurement principle to get high position accuracy: direct line of sight between satellite and receiver.

Five levels of system integration are commonly distinguished:

- Separately: two systems are used completely separately. As soon as the prime means of navigation does not output position, the other system is used instead as a fallback solution.
- Loosely: two systems output position data, which is combined to one solution in, e.g., an average process.
- Closely: the systems output measurements which are integrated in a common adjustment/filter process to one position solution. In this way it is possible to benefit from the strengths of satellite navigation, even when only two satellites are visible.
- Tightly: the combined position solution is used to aid the measurement process (e.g., the combined position solution is used to predict Doppler and time shift in the GNSS module).

- Deeply: the observations of the navigation systems are not only used to compute a combined position solution, but the information is also used in the other system to aid the measurement process (e.g., inertial measurements are used to aid the tracking loops of the GNSS).

An almost infinite variety of system integrations exists. Commonly, GNSS is combined with inertial navigation systems (INS), barometer, magnetic compass, odometers or other range measuring systems, terrestrial radionavigation systems, cellular network positioning, image-based navigation, or means of map aiding or map matching – a list being by far not complete (Hofmann-Wellenhof et al. 2003: Chap. 13).

The pure position output, even when attributed with time, velocity, and attitude, will be of little use for most mass market applications. The navigation system, thus, is only one component in a system compound, integrating means of navigation, communication, visualization, and geoinformation. In this course the interoperability of the various systems is of particular importance.

13.3.1 GNSS and inertial navigation systems

Discussing the varieties of integrating satellite navigation with other systems would go far beyond the scope of this textbook. One system, however, shall briefly be mentioned: the integration of INS with GNSS which has a long history.

Basically, a (strapdown) INS consists of two components: (1) gyros to monitor the angular motion of the body frame axes with respect to the local-level frame and (2) accelerometers placed on the body frame axes to measure accelerations (i.e., velocity rates). Starting from a known position and integrating (corrected) accelerations twice over time yields differences in position which determine the trajectory of the vehicle. This simple principle becomes fairly complicated in practice, and the reader is referred to the voluminous literature.

Inertial systems are autonomous and independent of external sources. Also, there is no visibility problem as for GNSS. Furthermore, INS provides accuracy similar to GNSS when used over short time intervals. The error of an (unaided) INS generally increases with the square of time due to the double integration over time. GNSS, in contrast, shows long-term stability. Consequently, the strengths of both are exploited to get a high-performance navigation system.

13.3.2 Radionavigation plans

In 1980, the US Department of Transportation (DoT) and Department of Defense (DoD) jointly produced the federal radionavigation plan (FRP) in order to select a mix of common-use navigation systems. These efforts should increase accuracy, coverage, and reliability of available navigation systems, while costs are reduced.

Since 1980, DoT is conducting open meetings for all users of radionavigation systems provided by the US government to revise and update the FRP in regular intervals. It is also the responsibility of the FRP to study the feasibility of replacing some of the existing systems by satellite navigation. The current version is the 2005 FRP.

Similar radionavigation plans have been established for, e.g., Austria, Germany, or Switzerland. Recommendations towards the development of a European Union radionavigation plan (ERNP) have been consolidated in 2004 (refer to www.helios-tech.co.uk/ERNP). The ERNP shall harmonize the national patchworks to a common radionavigation service of the European Community.

All radionavigation plans shall help for long-term application planning, especially in the aviation, railway, and maritime domain, where the product cycles are extremely long. These application domains carefully assess the phase-in and phase-out of systems.

13.4 User segment

The main task of the user segment is to transform the products delivered by the GNSS infrastructure, i.e., signals, into services that users are mainly interested in.

13.4.1 Receiver features

The GNSS space segment provides a number of different frequencies, ranging codes, and navigation messages. Most of them have been assigned to certain services; however, it is the decision of the receiver manufacturer to process the signals for the best performance to be provided to the user. Beside the positioning performance, the receiver manufacturers will also have to take into account other design criteria, like power consumption, size, or prize.

Berg and Dieleman (2002) mention that the responsibility of the GNSS service providers is restricted to signal in space (SIS) and constellation geometry. Not included are receiver certification or even user operations approval. The wide variety of receiver implementations and configurations using GNSS satellite-only services and the even endless possibilities of integrating different sensors and systems, Berg and Dieleman (2002) further emphasize, makes it unfeasible to address all certification procedures by the GNSS service provider. A certified SIS with guaranteed service levels, however, will be of minor importance if there are no certified user receivers.

The magazine GPS World provides in its "receiver survey" a yearly overview on the market situation of GNSS receivers. In the January 2007 issue, information is provided by 73 manufacturers of 542 receivers. The most important equipment features given there are the following:

- Manufacturer and model: indicate the company name and the receiver model.
- Channel: gives the number of channels to track satellites; usually, one channel per satellite and frequency is used. A typical number for a GPS C/A-code pseudorange receiver is 12 channels. There are models available with 72 channels, which usually include the ability to track GLONASS too.
- Signal tracked: specifies the code and the frequency. Typical examples are "L1 only, C/A-code" which means the receiver is a GPS C/A-code pseudorange receiver. Also more complex descriptions may be found, e.g., "WAAS, EGNOS, MSAS" indicating the option to use the respective augmentation signals.
- Maximum number of satellites tracked: this number is correlated with the number of channels and the signals tracked. Thus, if 12 satellites can be tracked for a dual-frequency receiver, usually 24 channels are required. The number of maximum satellites tracked ranges from 6 to "all in view".
- User environment and application: the user is informed on the dedicated applications for the respective model like aviation, marine, land, navigation, survey/geoinformation, meteorology, recreational, defense, and some more. In addition, this property also contains information whether the receiver is an end-user product or only a board/chipset/module for original equipment manufacturer (OEM) applications. This information influences the next properties, size and weight.
- Size and weight: as just mentioned, the given quantities strongly vary depending on the product.
- Position accuracy: this is a rough indicator and, dependent on the model, it may refer to autonomous code, real-time differential (code), postprocessed differential, real-time kinematic.
- Time accuracy: typical values are in the range of a few nanoseconds up to 1 000 nanoseconds.
- Position fix update rate: this quantity is given in seconds with typical examples in the range from 0.01 s to 1 s.
- Cold start: indicates how long it takes to determine the position if almanac, initial position, and time are not known. Typical values are several tens of seconds up to a few minutes.
- Warm start: indicates how long it takes to determine the position if the receiver has a recent almanac, an initial position, and current time but no up-to-date ephemerides. Usually, the data for the warm start are slightly better than the data of the cold start.
- Reacquisition: this quantity is given in seconds and is defined as reacquisition time based on the loss of the signal for at least one minute. Very good values are 0.1 s and better, 1 s and a few seconds are typical values.

13.4 User segment

- Number of ports, port type, baud rate: these parameters are important for the data transfer. Various port types like serial, Bluetooth, etc. are used; and the transmission rate in bits per second is usually in the range from 4 800 to 115 200 but may be much higher if Ethernet is used.
- Operating temperature: should be somewhere in the range from $-30°C$ to $80°C$.
- Power source and power consumption: for the power source it is primarily distinguished between internal or external source, also solar battery is available in rare cases.
- Antenna type: passive or active are some typical representatives.

Sometimes additional comments are given for specific features. The receiver overview does not mention the prize. The user needs to get into direct contact with the respective manufacturer.

A categorization not yet reflected in this list is the differentiation between software and classical hardware receivers. Although most of the software receivers are still subject to research and development, some are already on the market.

Receiver calibration

In general, GNSS receivers are considered to be self-calibrating and users do not normally perform equipment calibration. One simple test that can be performed, however, is a zero baseline measurement. This measurement is made by connecting two or more receivers to one antenna. Care should be taken in doing this to use a special device that blocks the voltage being fed to the antenna from all but one receiver. Also, a signal splitter must be used to divide the incoming signal to the multiple receivers.

The baseline is computed in the normal manner. Since a single antenna is used, the baseline components should all be zero. This measurement essentially checks the functioning of the receiver circuits and electronics and is a convenient method of trouble shooting receiver problems independent from antenna biases. The zero baseline test is also one way to satisfy specifications which call for equipment calibration.

GPS-specific receivers

Based on the type of observables and on the availability of codes currently used, i.e., C/A-code, P-code, or Y-code, one can classify GPS receivers into three major groups. Note that the ever evolving receiver technology and the increasing number of carrier frequencies and ranging codes will on the one hand result in an increasing diversity and on the other hand an increasing integration of different techniques, e.g., carrier smoothing.

The civil code pseudorange receivers use the freely available C/A-code to perform pseudorange measurements. The specified horizontal accuracy of these receivers, as listed in Table 9.2, is 13 m (95%). The future civil codes L2C, L5C, and L1C will allow to apply multifrequency techniques, i.e., eliminating the ionospheric influence, and result in higher position accuracies.

The civil carrier phase receivers use carrier phase measurements for position determination. These receivers are commonly used for all types of precise surveys. Therefore, the receivers generally store the measurements in memory for postprocessing analysis. The encryption of the P-code using the Y-code required to develop codeless and quasi-codeless techniques in order to take benefit of the second frequency L2 (cf. Sect. 4.3.3). The drawback is that the signal-to-noise ratio (S/N) of the L2 measurements is considerably lower than for the measurements on L1. Normally, the L2 phase is used in combination with the L1 measurement to reduce the ionospheric effect on the signal and, thus, provide a more accurate vector determination (especially for long baselines). The already started deployment of a civil ranging code on L2 will make the codeless and quasi-codeless techniques obsolete, which increases the S/N of L2 measurements. The implementation of the third carrier L5 including civil ranging codes will further enhance the carrier phase receiver performance in future.

P(Y)-code receivers provide access to the P-code with anti-spoofing (A-S) invoked. Thus, the code ranges and phases can be derived from L1 and L2 signals by the correlation technique. Military receivers, furthermore, implement hardware and software modules to remove a possible selective availability (SA). Although SA has been turned off, military receivers are still designed to get rid of it, if it would ever be activated again. A-S and SA have originally been handled by the auxiliary output chip (AOC) and the precise positioning service–security module (PPS-SM). Meanwhile, the US Department of Defense developed for higher security reasons the selective ability anti-spoofing module (SAASM) for implementation in military receivers.

GLONASS-specific receivers

As announced by presidential decision, GLONASS is like GPS a "dual-use system" which means it is designated for military and civil users. The same categorization as described for the GPS user segment can also be applied to the GLONASS receivers.

In contrast to GPS, the GLONASS P-code is in principle freely accessible. However, the P-code is, according to the Coordination Scientific Information Center (2002: Sect. 3.1), "modulated by special code, and its unauthorized use (without permission of Ministry of Defense) is not recommended". The free accessibility simplifies the implementation of dual-frequency techniques, although following the recommendation, these receivers are not applicable for safety-critical applications or for the mass market in general.

Galileo-specific receivers

The Galileo program includes initiatives for test user segment specification and implementation. These test user elements will be used to provide a proof of the system performance and to develop a principle receiver system concept to meet minimum operational requirements of Galileo. In future, the features of the Galileo receivers will closely follow the categories of the Galileo services (cf. Sect. 11.3).

13.4.2 Control networks

GNSS is, in fact, a geodetic tool in that it provides precise vector measurements over long and short distances. Virtually all GNSS processing software employ a three-dimensional model, and results are given in geodetic latitude, longitude, and ellipsoidal height. If the GNSS datum does not correspond to the national datum, a three-dimensional similarity transformation must be performed. Afterwards, the ellipsoidal values are mapped onto the plane using a projection.

There is a problem in using GNSS to establish heights, since the heights obtained with GNSS are referenced to an ellipsoid and not to the geoid to which conventionally leveled heights refer. As explained in more detail in Sect. 8.2.4, the separation between the ellipsoid and geoid differs for every point on the earth and can be extracted from existing global or local geoid models. One way to model the local or regional geoid is to run levels to a point of known ellipsoidal height. This is normally done for points in a supernet so that each point has an accurate ellipsoidal height and elevation. Based on these points, geodesists are able to refine their geoidal height prediction model to provide a more accurate determination.

GNSS provides a reference datum by two methods: (1) the passive control networks and (2) the active control networks.

Passive control networks

Virtually all civilized areas have some type of geodetic control networks generally surveyed by triangulation, traverse (or a combination of the two methods), and by spirit leveling. These control networks were used to scale and orient maps to provide a unified reference frame for large scale projects and to provide a common datum for property surveys. Geodetic control networks basically serve two functions: (1) they provide a "seamless" datum spanning large areas, and (2) they provide a reference framework which appears to be errorless to its users (surveyors). With the advent of GNSS, the concept of supernets or sparse arrays of high-accuracy control networks began to evolve. For supernets an array of new control points spaced between 25 and 100 km at locations suitable for GNSS occupation has to be established. The internal accuracy of the supernets is normally 0.1 ppm so that in effect they appear to be errorless to users.

The accuracy of supernets is controlled by referencing them to ITRF points. This international frame provides the ultimate geodetic framework, in as much as the interrelationship of the network sites is known to a higher accuracy than any other global array of points. The global supernet observations using dual-frequency receivers are routinely made by the International GNSS Service for Geodynamics (IGS) stations (Sect. 3.4.1). The positions of the points in a state supernet can be determined to a few centimeters accuracy with respect to the IGS stations.

Active control networks

For postprocessing or real-time DGNSS applications, the monitor stations can be arranged in permanently operating networks which are called active control networks.

Since the reference stations may be very distant from the user location, the virtual reference station (VRS) concept has been developed (cf. Sect. 6.3.7). This concept is a prerequisite mainly for RTK applications which require short distances to reference stations to facilitate ambiguity resolution.

Today, most of the active control networks offer DGNSS or (for shorter distances from the reference station) RTK services where pseudorange corrections or observed carrier phases are transmitted in real time to the user. These networks are commonly not usable for navigation applications, since they do not provide the level of integrity required for safety-of-life applications.

Active control networks have already been introduced in the course of the discussion about national DGNSS systems (cf. Sect.12.3.3). Another example for an active control network is the continuously operating reference station (CORS) network managed by the US National Oceanic and Atmospheric Administration (NOAA). The CORS network provides GPS code range and carrier phase data from more than 650 stations, which are used for precise positioning and atmospheric modeling applications (Department of Defense et al. 2005). The data is provided either in real time or in postprocessing.

The IGS network serves as both a passive and an active control network on a global scale. The IGS network provides among others coordinates and velocities of the IGS tracking stations (Table 13.10). Furthermore, the measured code and phase range data at each tracking site are made available to all users to allow for relative positioning in postprocessing mode.

13.4.3 Information services

Several governmental and private information services have been established to provide GNSS status information and data to the civilian users. Generally, the information contains constellation status reports and scheduled outages. Orbital data are provided in the form of an almanac suitable for making GNSS coverage and

13.4 User segment

Table 13.10. Selected IGS products

Product	Accuracy[1]		Latency
Satellite ephemerides and satellite clocks (specified for GPS)			
	orbits	satellite clocks	
Broadcast orbits[2]	~160 cm	~7 ns	real time
Ultra-rapid (predicted half)	~10 cm	~5 ns	real time
Ultra-rapid (observed half)	<5 cm	~0.2 ns	3 hours
Rapid	<5 cm	0.1 ns	17 hours
Final	<5 cm	<.1 ns	~13 days
Geocentric coordinates of IGS tracking stations			
	horizontal	vertical	
Final positions	3 mm	6 mm	12 days
Final velocities	2 mm/year	3 mm/year	12 days

[1] based on comparisons with laser ranging results; precision is better
[2] included for comparison.

satellite visibility predictions, and as precise ephemerides suitable for making the most precise vector computations. General information is also provided by listing the various GNSS papers and meetings.

The official source for GPS civilian information is the navigation information service (NIS), formerly the GPS Information Center, operated by the Navigation Center (NAVCEN). The NAVCEN civilian information is disseminated on the Internet via www.navcen.uscg.gov. GPS users are informed about changes in operation of GPS by the notice advisories to NAVSTAR users (NANU) generally within 60 minutes.

The official source for GLONASS civilian information is accessible via the Web site www.glonass-ianc.rsa.ru, which is operated by the Information Analytical Center (IAC) of the Russian Space Agency. GLONASS users are informed about changes in operation by the notice advisories to GLONASS users (NAGU).

An overall source is provided by the IGS information system. Apart from the coordinates and the velocities of the tracking stations, the IGS network provides high-accuracy GNSS ephemerides, satellite clock and tracking station clock information, earth rotation parameters, and ionospheric and tropospheric information. The IGS products are listed at http://igscb.jpl.nasa.gov/components/prods.html, and some are given in Table 13.10.

13.5 Selected applications

GNSS provides a global continuous service. An accuracy at the 10 m level suffices for many applications, particularly in navigation. The modernized GNSS will even provide meter level in real time in stand-alone mode. For higher accuracy requirements, augmented, differential, and relative techniques are adequate for local, regional, and even global applications. Accuracy, however, is not the only performance parameter to be taken into account. Safety-critical applications especially consider the integrity parameter. Furthermore, continuity, availability, or reliability are of concern.

As mentioned in the introductory part of this chapter, it is not possible to discuss all applications of GNSS. Here the applications have been categorized in a first level into navigation, surveying, and scientific applications. The navigation domain is further differentiated, without being exhaustive, into the aviation domain, maritime applications, road and rail domains, and location-based services (LBS).

13.5.1 Navigation

The classical application of GNSS is navigation in the field of transportation. GNSS can reduce the travel time and thus the transportation cost-increasing at the same time economic efficiency and safety. The reduced travel time will also reduce congestion. In the future, various regulations will mandate the use of GNSS. One example is a recommendation of the European Commission that future electronic toll collection systems shall be based on GNSS (European Parliament and European Council 2004).

Aviation applications

Satellite navigation provides the means for aircraft navigation for all phases of flight. Integrating local augmentation systems, GNSS will even provide the required level of performance for high-accuracy landing operations. Beside accuracy, the high level of integrity, continuity, and availability are critical factors to be met by the navigation systems. The performance requirements are specified by the International Civil Aviation Organization (ICAO) and national organizations like the US Federal Aviation Administration (FAA). Selected requirements are listed in Table 13.11 as extracted from Volpe National Transportation Systems Center (2001: Table 2-1).

The advantages of GNSS in the aviation domain are manifold. Satellite navigation allows for more efficient and flexible route selection. The high position accuracy additionally favors increased system capacity while providing enhanced safety levels throughout a region. Thus, the flight times can be reduced, saving time and fuel consumption. The global systems, furthermore, make local or regional in-

Table 13.11. GNSS aviation position
accuracy requirements in meter (95%)

Phase	Category	Position	Height
En route / terminal	–	≥ 100	≥ 100
Approach	I	16.0	4.0–6.0
and	II	6.9	2.0
landing	III	6.2	2.0

stallations obsolete, saving not only installation but also maintenance costs. The transmission of the position between airplane and ground station and airplane to airplane increases the situational awareness and enhances the surveillance methods. Furthermore, satellite navigation will increase the landing capacity at airports while enabling curved airport approaches. Despite its potential, satellite navigation will also in the future not serve as a sole means of navigation at least for reasons of redundancy.

Maritime applications

Marine navigation distinguishes between five major phases: the ocean navigation, coastal navigation, the port approach and operation in restricted waters, the marine navigation in a port, and navigation in inland waterways. Satellite navigation provides a wide range of applications in the maritime domain. The position, velocity, and attitude determination capability of GNSS is used, e.g., in conjunction with river information service (RIS) to increase the situational awareness at inland waterways.

The required navigation performance (RNP) parameters, i.e., accuracy, availability, continuity, and integrity (expressed by the time to alarm/alert (TTA)), are partly listed in Table 13.12 following the European Commission (2003a). Slightly different requirements may apply for, e.g., high-speed boats.

Rail applications

The railway systems have been traditionally developed and built under national sovereignty. The result was, especially in Europe, a patchwork of different and also incompatible signaling and train positioning systems. Europe has started initiatives to increase the interoperability between the systems and to increase the efficiency and the railway competitiveness. The European rail traffic management system (ERTMS) distinguishes between two components: the European train control system and the European traffic management layer. Both layers will benefit

Table 13.12. Minimum maritime user requirements for general navigation

Phase	Horizontal accuracy [m]	Integrity TTA [s]	Availability % per 30 days
Ocean	10	10	99.8
Coastal	10	10	99.8
Port approach & restricted waters	10	10	99.8
Port	1	10	99.8
Inland waterways	10	10	99.8

from an introduction of GNSS in ERTMS. High-density lines, but especially low-density lines will increase the performance, e.g., capacity, while reducing the costs.

In a first step the ERTMS standardizes the different train control systems. In a second step the trackside-based train control shall be transferred into a radio-based train control system, to decrease the costs for operation and maintenance. In this mode the trains still rely on balises (i.e., transmitters) that are installed in regular intervals along the track. The train determines its position, e.g., based on the balise information and on odometers. The position is then reported to the train control systems using wireless secure communication. Implementing GNSS will even allow to completely eliminate trackside installations and solely rely on trainborne equipment. The expensive maintenance of the trackside installation, thus, becomes obsolete.

Railway operators distinguish four different levels of integrity. A high safety integrity level (SIL) shall bound disastrous impacts by a low acceptable failure rate (International Electrotechnical Commission 2005). The different levels of integrity as summarized in Table 13.13 express the tolerable hazard rate (THR) of the complete system (Wigger and Hövel 2002). Thus, the requirements for the navigation system are even higher. The THR describes the tolerable rate of dangerous failures and is, thus, a description of acceptance of risk. To meet the high level of integrity,

Table 13.13. Tolerable hazard rates

SIL	THR per hour and per function	Impact on the community
1	$10^{-6} - 10^{-5}$	Negligible: minor impact
2	$10^{-7} - 10^{-6}$	Marginal: potential individual injuries
3	$10^{-8} - 10^{-7}$	Critical: multiple injuries
4	$10^{-9} - 10^{-8}$	Catastrophic: disastrous impact

13.5 Selected applications

GNSS cannot operate as sole means of navigation. Especially when additionally considering the environmental conditions, i.e., shadowing of the satellite signals in urban canyons, mountainous areas, or tunnels, a hybridization of systems becomes necessary.

Road applications

The road applications encompass traditional areas like navigation, guidance, or fleet management, but in future the GNSS will also be part of more sophisticated and intelligent systems, like automatic driver assistance systems or speed limit enforcement.

How such intelligent driver systems could operate has been shown during the grand challenge of the Defense Advanced Research Projects Agency (DARPA), sponsored by the US Department of Defense. In a competition, driverless vehicles had to travel a distance of about 200 km. Along the route, which was not known to the competitors in advance, a number of natural and man-made obstacles had to be passed. The longest traveled distance during the first grand challenge held in 2004 was less than 10% of the required route. One year later, five of the 23 qualified vehicles completed the course. All participants of this challenge integrated various systems, whereas GNSS has been used as prime means of navigation providing position information in a global reference frame.

In future, intelligent navigation systems shall make the roads safer, minimizing the travel time while reducing congestion.

Location-based services

Location-based services (LBS) are said to be the major market for GNSS in terms of sales numbers. In its most general definition, LBS encompass all services where information about the location of the consumer is needed to provide the appropriate level of service. Take as an example someone asking the way to the nearest hospital. The LBS system determines the position of the user, compares, in a simple realization, the position with a database of hospital coordinates and guides the user to the nearest hospital. More complex systems would additionally take into account the accessibility, traffic information, or even real-time hospital information like presence of medicals. LBS are especially pushed by the close combination of positioning technologies, in particular GNSS, and means of communication. The third major element in LBS is geoinformation.

Integrating navigation, communication, and geoinformation opens the door for a wide range of applications. The integration of these technologies even launched applications which could not be imagined before: geocaching, GNSS gaming, or GNSS drawing. There is no pragmatic thematic clustering of the wide range of LBS applications. Swann et al. (2003), furthermore, mention that the broad range of LBS applications makes it impractical to define one set of requirements or even one technological solution to meet them. From a business perspective, applications could be grouped under three main categories: consumer, enterprise, and emergency applications.

The industry further differentiates between a client pull, where the user requests a reply, and a server push scenario, where the server sends information without a particular consent of the user for this particular message. Additionally, the category of tracing service may be added, where the position of the mobile customer is automatically transmitted to the server for, e.g., management purposes. In all three models the aspect of privacy and legal issues arises.

The combination of satellite navigation, wireless communication, and geoinformation seems to be a perfect mix of complementary technologies. However, the potential of satellite navigation is not sufficient to meet the requirements in particular considering personal mobility in urban environments. The satellite signals are attenuated or shadowed and position determination becomes difficult or impractical. High-sensitivity GNSS technologies are able to track very weak satellite signals, and in this way position determination is possible even indoors. However, these position solutions are influenced by multipath effects. Filter techniques, e.g., Kalman filter, are necessary to get acceptable position solutions. The integration of various technologies, especially of cellular network positioning in combination with assisted GNSS (AGNSS) further enhances the positioning performance. Even higher positioning performances can be met by pedestrian navigation systems which integrate autonomous sensors. Lachapelle (2004) emphasizes that the need for indoor location determination was initially driven by a decision of the US Federal Communications Commission (FCC) requesting from the mobile phone service providers to locate emergency (E911) callers with a defined level of accuracy. Meanwhile, similar mandates have also been recommended and agreed in other countries, e.g., Europe (E112).

13.5.2 Surveying and mapping

The different methods of space-based navigation allow to achieve position accuracies down to the millimeter level. These are adequate means for the purposes of surveying and mapping, in particular of cadastral surveying, geodetic control networks, and deformation monitoring in both a local and a global scope.

Although for surveying and mapping purposes the GNSS measurements are routinely processed in postprocessing, the RTK methods also allow for real-time applications. Highest accuracies are achieved using carrier phase observations over long time intervals in relative positioning mode. Depending on the performance requirements also kinematic or pseudokinematic strategies may be applied.

For moderate-length baselines up to about 20 km, single-frequency receivers

13.5 Selected applications   465

provide equivalent results to the dual-frequency receivers because the ionospheric refraction (mostly) cancels by differencing the phase measurements between the baseline sites. During periods of moderate solar activity, lines of up to 100 km have been accurately measured with single-frequency receivers by observing for several hours. The ionospheric aberrations mostly have a highly central tendency and tend to average out. Baseline lengths must be reduced in periods of high sunspot activity. This activity cycle has a repetition rate of about 11 years with the next maximum in 2012 (cf. Fig. 4.8). Dual-frequency receivers eliminate ionospheric refraction by the ionosphere-free combination of the two carrier phases. Triple-frequency receivers will in particular reduce the observation times.

The use of GNSS for surveying purposes requires adequate presurvey planning operations. The point selection, the determination of observation windows, and the choice of the number and length of sessions depend on various design criteria. Furthermore, the design of the survey network, in conjunction with the observation sessions, influences the accuracy of the surveyed points. Hofmann-Wellenhof et al. (2001) discuss in detail the application of GNSS in the fields of surveying and mapping.

The advent of GNSS also became especially useful for geographic information systems (GIS). A GIS captures, stores, manages, analyzes, and displays all forms of geographically referenced information. The interested reader is referred to the voluminous literature on GIS.

13.5.3 Scientific applications

GNSS is also a gorgeous tool for scientific applications. How GNSS signals can be used to extract information other than position, velocity, attitude, and time has been discussed in Sect. 13.1.6. Continuous GNSS measurements and position determination allow to monitor long-term geodynamical phenomena. Applications include measuring earth rotation, crustal deformations, plate tectonics, postglacial rebound, or volcanic uplift. GNSS position determination is also a cost-effective solution for seismology, glaciology, geology, meteorology, environmental monitoring, to name but a few. The highly accurate position determination of spaceborne receivers is useful for satellite missions dedicated to altimetry, gradiometry, or magnetic field campaigns. Furthermore, GNSS allows to provide a high accuracy time and frequency standard for diverse scientific applications.

Another application, for example, is GNSS altimetry, which is based on signals reflected by the oceans or by the ground. Airplanes or LEO satellites equipped with two receivers, one on the top pointing to the zenith, the other on the bottom pointing to the nadir, can measure the time of arrival of the direct line-of-sight signal coming from the GNSS satellite and the reflected signal coming from the ground. The difference of both signals allows to determine sea level height, wave height, or other parameters as mentioned by Rosmorduc et al. (2006), Martín-Neira et al. (2005), or Martín-Neira and Buck (2005).

GNSS, to give another example of the wide range of applicability of satellite navigation in science, is the first measuring device that can be used to accurately measure height differences in real time. Repeated measurements between stable and subsiding points provide accurate measures of the subsidence. Industry has used GNSS to measure, e.g., the subsidence among groups of offshore oil platforms.

14 Conclusion and outlook

The future applications of GNSS are limited only by one's imagination. For readers being not satisfied with this pretty general statement, a few more specific answers are given of five leading companies in an article on "The future of GNSS applications" published in the January/February 2007 issue of the GeoInformatics Magazine for Surveying, Mapping & GIS Professionals.

The modernization of the existing systems GPS and GLONASS as well as the introduction of Galileo will provide "exciting new developments for the survey market over the next five years", where "the benefits for the high-precision user will be improved position reliability, precision, and, ultimately, productivity". All answers have in common that the users need access to as many as possible satellites and signals for precise positioning.

Another developmental step will be "smaller, lighter, and more capable GNSS receivers" as a consequence of the increased number of GNSS signals and codes and, even more challenging, sensor integration to overcome the problem of signal obstruction in urban canyons and other problematic sites.

This sensor integration (also denoted as sensor fusion or, sometimes, integration of technologies) is one of the key words for the future.

Some examples for applications on land are vehicle navigation and information systems including intelligent vehicle highway systems (IVHS) and intelligent transportation systems (ITS), e.g., anticollision systems in railway and land transport, fleet management, journey data logging, car theft reconnaissance, road and railroad surveys.

Another land application is the use of GNSS to automate various types of machinery. For example, it should be possible to automate the grading and paving equipment used for road building. Equipment could be run around the clock without operators with GNSS performing all motion operations based upon a digital terrain model stored in the computer of the equipment.

A similar application is precision farming, where the positioning is regarded as an evident necessity but the additional data on the vehicle status, soil nutrition, crop health, and fertilizer requirements provide the real value of the application. As outlined in the mentioned article, "ultimately, these technological developments are enabling businesses across our markets to optimize efficiencies in both material and operating costs and offer a host of additional benefits ranging from environmental concerns to workforce safety".

Less clear – as to be expected – are the answers of the companies on possible "killer applications". The sensor fusion is mentioned again, e.g., an integrated compass and a camera in a handheld GNSS unit or the combination of laser technology with GNSS. Another general statement expects applications for which positioning, wireless communications, mobile computing, and application-specific software "for a very broad range of industries" are combined.

The real-time aspect for highly accurate results is also an issue. In former times, postprocessing techniques were accepted for many high-precision applications. Now the appetite of the users for real-time results increases. "They expect the industry to offer new techniques at affordable prices to give them in the field what they formerly had to achieve through a cumbersome process when back in the office."

There will also be numerous applications in survey and geodesy as well as precise time determination and – one of the fastest growing GNSS applications – time transfer. Available resources of telephone companies, power companies, and many others are enhanced by precise timing.

Other future GNSS applications concern atmospheric sounding. The data will contribute to a better understanding of the structure of the atmosphere leading, for example, to improved models for weather analysis.

Similarly, environmental monitoring and air pollution information systems utilizing mobile air quality measurement sites will increasingly gain attraction.

Marine applications will include vessel navigation and information systems, precise harbor entrance systems, and oceanography in general. DGNSS will mostly be used for these purposes, and dense networks of monitor stations are to be established along the coasts.

For aviation, GNSS will be integrated into other navigation systems like INS to fulfill the high reliability and integrity requirements. Applications will include en route navigation and surveillance, approach and landing, collision avoidance, and proximity warning. Aircraft could be operated in an automated mode with takeoffs and landings being performed by integrating GNSS and board computers.

The use of GNSS will increasingly be extended to space for precise positioning of (e.g., earth remote sensing) satellites, for attitude determination of spacecraft, and for missile navigation.

A huge market developing very fast is the use of GNSS chipsets in mobile phones which may be used, e.g., for emergency services. If the user is in a situation of distress, transmitting his position to an alarm unit will speed up the search process by the rescue team.

Despite this enormous potential of applications, there are also lurking some possible troubles due to the development of so many different systems, especially the local ones. There might evolve a kind of local isolation if not clearly defined standardization rules take place. The compatibility of all global and local systems is the challenge for the future which may only be solved on an international level. This calls for an international organization tackling these questions.

The International Committee on Global Navigation Satellite Systems might be the proper candidate. As published in A/AC.105/879 of the General Assembly of the United Nations, this committee has outlined these issues during the meeting on November 1–2, 2006, in Vienna:

"Global navigation satellite systems (GNSS) have evolved from an early period of limited programmes to a point where a number of systems and their augmentations are operating or planned. In the future, a number of international and national programmes will operate simultaneously and support a broad range of interdisciplinary and international activities. Discussions taking place at national, regional and international levels have underscored the value of GNSS for a variety of applications. The emergence of new GNSS and regional augmentations has focused attention on the need for the coordination of programme plans among current and future operators in order to enhance the utility of GNSS services.

The representatives of GNSS core system providers, GNSS augmentation providers, and the international organizations primarily associated with the use of GNSS and representatives of international projects in developing countries,

- aware of the overlap of GNSS mission objectives and of the interdisciplinary applications of GNSS services,
- recognizing the advantages of ongoing communication and cooperation between operators and users of GNSS and their augmentations,
- recognizing the need to protect the investment of the current user base of GNSS services through the continuation of existing services,
- aware that the complexity and cost of user equipment should be reduced whenever possible,
- convinced that GNSS providers should pursue greater compatibility and interoperability among all current and future systems in terms of spectrum, signal structures, time and geodetic reference standards to the maximum extent possible,
- desiring to promote the international growth and potential benefits of GNSS,
- noting that General Assembly resolution 59/2 (paragraph 11) invites GNSS and augmentation providers to consider establishing an international committee on GNSS in order to maximize the benefits of the use and applications of GNSS to support sustainable development,
- have agreed to establish on the basis of these nonbinding terms of reference, the International Committee on GNSS for the purpose of promoting the use and application of GNSS on a global basis."

If at all, only an internationally and globally acknowledged committee will have the potential to bundle and focus all these very fast ongoing developments; if this succeeds, everybody individually will profit tremendously from the rich world of GNSS.

So far this outlook has primarily covered the user's point of view. How should an idealized GNSS look like? Hein et al. (2007) envision the "system of systems" issuing several aspects. The separation of civil and military users may lead to benefits for both sides. Apart from a separation of the respective satellite payload components, also separating military from civil signals in the frequency and signal plan. As a consequence, also a separation of the control centers should be discussed. However, there remains a warning. If this separation is carried out consequently, finally completely separate systems will result; and as the past teaches, this can never be advantageous for the civilian users compared to the military users.

The call for an ideal GNSS space segment constellation is easier to accomplish. In the sense of the previous discussion, this would be a true task of the International Committee on GNSS. Currently, GPS, GLONASS, and Galileo do not take into account one of the other systems. The committee should get the responsibility to coordinate a harmonization process ending up with an ideal constellation with respect to the number of orbital planes, inclination, and altitude of the satellites. Also the replacement of the satellites should be coordinated accordingly.

Omitting all territorial and national concerns, the ideal control segment could only be realized if the monitoring stations are regularly spread over the globe and not restricted to local (national) areas. This sounds very logic but is probably – at least today – a very futuristic view of the world because it is difficult to imagine that, e.g., GLONASS satellites are controlled by a US facility and vice versa.

The user segment will profit from the continuous receiver development and miniaturization. Following the current trend of software-based receiver technology, then, after 20 years, the receiver will – figuratively speaking – be replaced by software and require no physical space at all. Certainly, all signals of all the satellites may be received and processed in a single device by a common approach.

More progressively, Hein et al. (2007) ask how the user segment might look like in 20 years: "Will it be a piece of software running on a generic computer implant under our skins powered by bio-energy, broadcasting people's positions permanently to the government or somebody else?"

Today, this may sound provocative, but after 20 years even this might be an old-fashioned piece of development stored somewhere in a virtual museum – where it does not require much space!

References

Abdel-salam MA (2005): Precise point positioning using un-differenced code and carrier phase observations. Department of Geomatics Engineering, University of Calgary, Canada, UCGE Reports No. 20229.

Abidin HZ (1993): On the construction of the ambiguity searching space for on-the-fly ambiguity resolution. Navigation, 40(3): 321–338.

Abidin HZ, Wells DE, Kleusberg A (1992): Some aspects of "on the fly" ambiguity resolution. In: Proceedings of the Sixth International Geodetic Symposium on Satellite Positioning, Columbus, Ohio, March 17–20, vol 2: 660–669.

Alban S (2004): Design and performance of a robust GPS/INS attitude system for automobile applications. PhD dissertation, Stanford University, California.

Altamimi Z, Boucher C (1999): GLONASS and the international terrestrial reference frame. In: Workshop Proceedings, Slater JA, Noll CE, Gowey KT (eds): International GLONASS experiment (IGEX-98), IGS Central Bureau, Pasadena, California: 37–46.

Altmayer C (2000): Cycle slip detection and correction by means of integrated systems. In: Proceedings of the 2000 National Technical Meeting of the Institute of Navigation, Anaheim, California, January 26–28: 134–144.

Arbesser-Rastburg B (2001): Signal propagation for SatNav systems. ESA-ESTEC/TOS-EEP, The Netherlands. In: Course books for the summer school in Alpbach, Austria, July 17–26.

Arbesser-Rastburg B (2006): The Galileo single frequency ionospheric correction algorithm. Paper presented at the Third European Space Weather Week, Brussels, Belgium, November 13–17.

Arbesser-Rastburg B, Jakowski N (2007): Effects on satellite navigation. In: Bothmer V, Daglis IA (eds): Space weather – physics and effects. Springer, Berlin Heidelberg New York: 383-402.

ARINC Engineering Services (2005): NAVSTAR GPS space segment / user segment L5 interfaces. Interface specification, IS-GPS-705, IRN-705-003.
Available at www.arinc.com/gps.

ARINC Engineering Services (2006a): NAVSTAR GPS space segment / navigation user interfaces. Interface specification, IS-GPS-200, revision D, IRN-200D-001. Available at www.arinc.com/gps.

ARINC Engineering Services (2006b): NAVSTAR GPS space segment / user segment L1C interfaces. Interface specification, Draft IS-GPS-800.
Available at www.navcen.uscg.gov/gps/modernization.

Arnold K (1970): Methoden der Satellitengeodäsie. Akademie, Berlin.

Ashby N (1987): Relativistic effects in the Global Positioning System. In: Relativistic effects in geodesy, Proceedings of the International Association of Geodesy (IAG) Symposia of the XIX General Assembly of the IUGG, Vancouver, Canada, August 10–22, vol 1: 41–50.

Ashby N (2001): Relativistic effects on SV clocks due to orbit changes, and due to earth's oblateness. In: Proceedings of the Annual Precise Time and Time Interval (PTTI)

Systems and Applications Meeting (33rd), Long Beach, California, November 27–29: 509–523.

Ashby N (2003): Relativity in the Global Positioning System. Living Reviews in Relativity, 6(1), at www.livingreviews.org/Articles/Volume6/2003-1ashby/.

Ashjaee J (1993): An analysis of Y-code tracking techniques and associated technologies. Geodetical Info Magazine, 7(7): 26–30.

Ashjaee J, Lorenz R (1992): Precision GPS surveying after Y-code. In: Proceedings of ION GPS-92, Fifth International Technical Meeting of the Satellite Division of the Institute of Navigation, Albuquerque, New Mexico, September 16–18: 657–659.

Ashkenazi V (2006): Geodesy and satellite navigation. Inside GNSS, 1(3): 44–49.

Ashkenazi V, Moore T, Ffoulkes-Jones G, Whalley S, Aquino M (1990): High precision GPS positioning by fiducial techniques. In: Bock Y, Leppard N (eds): Global Positioning System: an overview. Springer, New York Berlin Heidelberg Tokyo: 195–202 [Mueller II (ed): IAG Symposia Proceedings, vol 102].

Averin SV (2006): GLONASS system: present day and prospective status and performance. Presented at the European Navigation Conference GNSS-2006, Manchester, Great Britain, May 7–10.

Barker BC, Betz JW, Clark JE, Correia JT, Gillis JT, Lazar S, Rehborn KA, Straton JR (2000): Overview of the GPS M code signal. In: Proceedings of the 2000 National Technical Meeting of the Institute of Navigation, Anaheim, California, January 26–28: 542–549.

Bar-Sever Y, Young L, Stocklin F, Heffernan P, Rush J (2004): NASA's global differential GPS system and the TDRSS augmentation service for satellites. In: Proceedings of the 2nd ESA Workshop on Satellite Navigation User Equipment Technologies, NAVITEC 2004, ESTEC, Noordwijk, The Netherlands, December 8–10.

Bartone C, Graas F van (1998): Airport pseudolites for local area augmentation. In: Proceedings of IEEE PLANS, Publication 98CH36153, Palm Springs, California, April 20–23: 479–486.

Bauer M (2003): Vermessung und Ortung mit Satelliten – GPS und andere satellitengestützte Navigationssysteme, 5th edition. Wichmann, Karlsruhe.

Bedrich S (2005): Precise time facility (PTF) for Galileo IOV. Presented at the Workshop on Time and Frequency Services with Galileo, Herrsching, Germany, December 5–6.

Bennet JM (1965): Triangular factors of modified matrices. Numerische Mathematik, 7: 217–221.

Berg A van den, Dieleman P (2002): A practical interpretation of performance requirements for a global navigation satellite system. What does the user really need? In: Proceedings of the European Navigation Conference GNSS 2002, Copenhagen, Denmark, May 27–30.

Betz JW (2000): Design and Performance of code tracking for the GPS M code signal. In: Proceedings of ION GPS 2000, 13th International Technical Meeting of the Satellite Division of the Institute of Navigation, Salt Lake City, Utah, September 19–22: 2140–2150.

Betz JW (2002): Binary offset carrier modulations for radionavigation. Navigation, 48(4): 227–246.

Betz JW (2006): Free-space propagation loss. In: Kaplan ED, Hegarty CJ (eds): Understanding GPS – principles and applications, 2nd edition. Artech House, Norwood: 669–673.

References

Beutler G (1991): Himmelsmechanik I. Mitteilungen der Satelliten-Beobachtungsstation Zimmerwald, Bern, vol 25.

Beutler G (1992): Himmelsmechanik II. Mitteilungen der Satelliten-Beobachtungsstation Zimmerwald, Bern, vol 28.

Beutler G (1996): GPS satellite orbits. In: Kleusberg A, Teunissen PJG (eds): GPS for geodesy. Springer, Berlin Heidelberg New York Tokyo: 37–101 [Bhattacharji S, Friedman GM, Neugebauer HJ, Seilacher A (eds): Lecture Notes in Earth Sciences, vol 60].

Beutler G, Gurtner W, Bauersima I, Rothacher M (1986): Efficient computation of the inverse of the covariance matrix of simultaneous GPS carrier phase difference observations. Manuscripta Geodaetica, 11: 249–255.

Beutler G, Bauersima I, Gurtner W, Rothacher M (1987): Correlations between simultaneous GPS double difference carrier phase observations in the multistation mode: implementation considerations and first experiences. Manuscripta Geodaetica, 12: 40–44.

Beutler G, Gurtner W, Rothacher M, Wild U, Frei E (1990): Relative static positioning with the Global Positioning System: basic technical considerations. In: Bock Y, Leppard N (eds): Global Positioning System: an overview. Springer, New York Berlin Heidelberg Tokyo: 1–23 [Mueller II (ed): IAG Symposia Proceedings, vol 102].

Bevis M, Businger S, Herring TA, Rocken C, Anthes RA, Ware RH (1992): GPS meteorology: remote sensing of atmospheric water vapor using the Global Positioning System. Journal of Geophysical Research 97(D14): 15787–15801.

Bian S, Jin J, Fang Z (2005): The Beidou satellite positioning system and its positioning accuracy. Navigation, 52(3): 123–129.

Biancale R, Balmino G, Lemoine JM, Marty JC, Moynot B, Barlier F, Exertier P, Laurain O, Gegout P, Schwintzer P, Reigber C, Bode A, König R, Massmann FH, Raimondo JC, Schmidt R, Zhu SY (2000): A new global earth's gravity field model from satellite orbit perturbations: GRIM5-S1. Geophysical Research Letters, 27: 3611–3614.

Blomenhofer H, Ehret W, Su H, Blomenhofer E (2005): Sensitivity analysis of the Galileo integrity performance dependent on the ground sensor station network. In: Proceedings of ION GNSS 2005, 18th International Technical Meeting of the Satellite Division of the Institute of Navigation, Long Beach, California, September 13–16: 1361–1373.

Bock Y (1996): Reference systems. In: Kleusberg A, Teunissen PJG (eds): GPS for geodesy. Springer, Berlin Heidelberg New York Tokyo: 3–36 [Bhattacharji S, Friedman GM, Neugebauer HJ, Seilacher A (eds): Lecture Notes in Earth Sciences, vol 60].

Borge TK, Forssell B (1994): A new real-time ambiguity resolution strategy based on polynomial identification. In: Proceedings of the International Symposium on Kinematic Systems in Geodesy, Geomatics and Navigation, Banff, Canada, August 30 through September 2: 233–240.

Borre K, Akos DM, Bertelsen N, Rinder P, Jensen SH (2007): A software-defined GPS and Galileo receiver, a single-frequency approach – applied and numerical harmonic analysis. Birkhäuser, Boston Basel Berlin.

Boucher C, Altamimi Z (2001): ITRS, PZ-90 and WGS 84: current realizations and the related transformation parameters. Journal of Geodesy, 75(11): 613–619.

Branets V, Mikhailov M, Stishov Y, Klyushnikov S, Filatchenkov S, Mikhailov N, Pospelov S, Vasilyev M (1999): "Soyuz"–"Mir" orbital flight GPS/GLONASS experiment. In: Proceedings of ION GPS-99, 12th International Technical Meeting of the Satellite Division of the Institute of Navigation, Nashville, Tennessee, September 14–17: 2303–2311.

Breuer B, Campbell J, Müller A (1993): GPS-Meß- und Auswerteverfahren unter operationellen GPS-Bedingungen. Journal for Satellite-Based Positioning, Navigation and Communication, 2(3): 82–90.

Brigham EO (1988): The fast Fourier transform and its applications. Prentice-Hall, Englewood Cliffs.

Bronstein IN, Semendjajew KA, Musiol G, Mühlig H (2005): Taschenbuch der Mathematik, 6th edition. Deutsch, Frankfurt.

Brouwer D, Clemence GM (1961): Methods of celestial mechanics. Academic Press, New York.

Brown RG, Hwang PYC (1997): Introduction to random signals and applied Kalman filtering. With Matlab excercises and solutions, 3rd edition. John Wiley, New York Chichester Brisbane Toronto Singapore Weinheim.

Brunner FK, Gu M (1991): An improved model for the dual frequency ionospheric correction of GPS observations. Manuscripta Geodaetica, 16: 205–214.

Brunner FK, Welsch WM (1993): Effect of the troposphere on GPS measurements. GPS World, 4(1): 42–51.

Butsch F (2002): A growing concern: radiofrequency interference and GPS. GPS World, 13(10): 40–50.

Campbell J, Görres B, Siemes M, Wirsch J, Becker M (2004): Zur Genauigkeit der GPS Antennenkalibrierung auf der Grundlage von Labormessungen und deren Vergleich mit anderen Verfahren. Allgemeine Vermessungsnachrichten, 1: 2–11.

Cannon ME, Lachapelle G (1993): GPS – theory and applications. Lecture Notes for a seminar on GPS given at Graz in spring 1993.

Cannon ME, Sun H, Owen T, Meindl M (1994): Assessment of a non-dedicated GPS receiver system for precise airborne attitude determination. In: Proceedings of ION GPS-94, 7th International Technical Meeting of the Satellite Division of the Institute of Navigation, Salt Lake City, Utah, September 20–23, part 1: 645–654.

Capitaine N, Gambis D, McCarthy DD, Petit G, Ray J, Richter B, Rothacher M, Standish M, Vondrak J (eds) (2002): Proceedings of the IERS Workshop on the Implementation of the New IAU Resolutions. IERS Technical Note no. 29, Verlag des Bundesamtes für Kartographie und Geodäsie, Frankfurt/Main. Available at www.iers.org.

Centre National d'Etudes Spatiales (2007): CNES programmes, DORIS Web site. Available at www.cnes.fr/web/1513-doris.php.

CGSIC (1995): Summary record of the 26th meeting of the Civil GPS Service Interface Committee (CGSIC), Palm Springs, California, September 11–12.

CGSIC (1996): Summary report of the 27th meeting of the Civil GPS Service Interface Committee (CGSIC), Falls Church, Virginia, March 19–21.

Chen D (1994): Development of a fast ambiguity search filtering (FASF) method for GPS carrier phase ambiguity resolution. Reports of the Department of Geomatics Engineering of the University of Calgary, vol 20071.

Chen D, Lachapelle G (1994): A comparison of the FASF and least-squares search algorithms for ambiguity resolution on the fly. In: Proceedings of the International Symposium on Kinematic Systems in Geodesy, Geomatics and Navigation, Banff, Canada, August 30 through September 2: 241–253.

Chin M (1991): CIGNET report. GPS Bulletin, 4(2): 5–11.

Colcy JN, Hall G, Steinhäuser R (1995): Euteltracs: the European mobile satellite service. Electronics & Communication Engineering Journal, 7(2): 81–88.

Collins JP, Langley RB (1999): Possible weighting schemes for GPS carrier phase observations in the presence of multipath. Contract Report No. DAAH04-96-C-0086/TCN 98151 for the United States Army Corps of Engineers Topographic Engineering Center. Available at http://gge.unb.ca/Personnel/Langley/Langley.html.

Colombo OL, Bhapkar UV, Evans AG (1999): Inertial-aided cycle-slip detection/correction for precise, long-baseline kinematic GPS. In: Proceedings of ION GPS-99, 12th International Technical Meeting of the Satellite Division of the Institute of Navigation, Nashville, Tennessee, September 14–17: 1915–1921.

Conley R (2000): Life after selective availability. Newsletter of the Institute of Navigation, 10(1): 3–4.

Conley R, Lavrakas JW (1999): The world after selective availability. In: Proceedings of ION GPS-99, 12th International Technical Meeting of the Satellite Division of the Institute of Navigation, Nashville, Tennessee, September 14–17: 1353–1361.

Conley R, Cosentino R, Hegarty CJ, Kaplan ED, Leva JL, Uijt de Haag M, Dyke K van (2006): Performance of stand-alone GPS. In: Kaplan ED, Hegarty CJ (eds): Understanding GPS: principles and applications, 2nd edition. Artech House, Boston London: 301–378.

Coordination Scientific Information Center (2002): Global navigation satellite system – GLONASS – interface control document, version 5.0, Moscow. Available at www.glonass-ianc.rsa.ru.

Corrigan TM, Hartranft JF, Levy LJ, Parker KE, Pritchett JE, Pue AJ, Pullen S, Thompson T (1999): GPS risk assessment study. Final Report. VS-99-007, M8A01. Applied Physics Laboratory, The Johns Hopkins University, Maryland.

Cospas-Sarsat Secretariat (1999): Introduction to the COSPAS-SARSAT system, issue 5, rev. 1. C/S G.003. Available at www.cospas-sarsat.org.

Cospas-Sarsat Secretariat (2006a): Phase-out plan for 121.5/243 MHz satellite alerting services, issue 1, rev. 5. C/S R.010. Available at www.cospas-sarsat.org.

Cospas-Sarsat Secretariat (2006b): COSPAS-SARSAT 406 MHz MEOSAR implementation plan, issue 1 – revision 2. C/S R.012. Available at www.cospas-sarsat.org.

Cospas-Sarsat Secretariat (2006c): COSPAS-SARSAT system data, no. 32. Available at www.cospas-sarsat.org.

Counselman CC, Gourevitch SA (1981): Miniature interferometer terminals for earth surveying: ambiguity and multipath with the Global Positioning System. IEEE Transactions on Geoscience and Remote Sensing, GE–19(4): 244–252.

Creel T, Dorsey AJ, Mendicki PJ, Little J, Mach RG, Renfro BA (2006): New, improved GPS – the legacy accuracy improvement initiative. GPS World, 17(3): 20–31.

Dai L, Han S, Wang J, Rizos C (2001): A study on GPS/GLONASS multiple reference station techniques for precise real-time carrier phase-based positioning. In: Proceedings of ION GPS 2001, 14th International Technical Meeting of the Satellite Division of the Institute of Navigation, Salt Lake City, Utah, September 11–14: 392–403.

Daxinger W, Stirling R (1995): Kombinierte Ausgleichung von terrestrischen und GPS-Messungen. Österreichische Zeitschrift für Vermessung und Geoinformation, 83: 48–55.

DeCleene B (2000): Defining pseudorange integrity – overbounding. In: Proceedings of ION GPS 2000, 13th International Technical Meeting of the Satellite Division of the Institute of Navigation, Salt Lake City, Utah, September 19–22: 1916–1924.

Defense Science Board (2005): The future of the Global Positioning System. Office of the Under Secretary of Defense for Acquisition, Technology, and Logistics, Washington DC.

Deines SD (1992): Missing relativity terms in GPS. Navigation, 39(1): 111–131.

Delikaraoglou D, Lahaye F (1990): Optimization of GPS theory, techniques and operational systems: progress and prospects. In: Bock Y, Leppard N (eds): Global Positioning System: an overview. Springer, New York Berlin Heidelberg Tokyo: 218–239 [Mueller II (ed): IAG Symposia Proceedings, vol 102].

DeLoach SR, Remondi BW (1991): Decimeter positioning for dredging and hydrographic surveying. In: Proceedings of the First International Symposium on Real Time Differential Applications of the Global Positioning System. TÜV Rheinland, Köln, vol 1: 258–263.

Deo MN, Zhang K, Roberts C, Talbot NC (2003): An investigation of GPS precise point positioning methods. Paper presented at SatNav 2003, 6th International Symposium on Satellite Navigation Technology Including Mobile Positioning & Location Services, Melbourne, Australia, July 22–25.

Department of Defense (1995): Global Positioning System standard positioning service – signal specification, 2nd edition.

Department of Defense (2000): Standard practice for system safety. MIL-STD-882D.

Department of Defense (2001): Global Positioning System standard positioning service performance standard. Available from the US Assistant for GPS, Positioning and Navigation, Defense Pentagon, Washington DC.

Department of Defense, Department of Homeland Security, Department of Transportation (2005): Federal radionavigation plan. US National Technical Information Service, Springfield, Virginia, DOT-VNTSC-RITA-05-12/DoD-4650.5.

Department of the Air Force (2001): Approved lexicon of signal abbreviations. Memorandum for record SMC/CZ All. Headquarters Space and Missile Systems Center (AFMC), Los Angeles, California.

Dierendonck AJ van (1996): GPS receivers. In: Parkinson BW, Spilker JJ (eds): Global Positioning System: theory and applications. American Institute of Aeronautics and Astronautics, Washington DC, vol 1: 329–406.

Dierendonck AJ van, Braasch MS (1997): Evaluation of GNSS receiver correlation processing techniques for multipath and noise mitigation. In: Proceedings of the 1997 National Technical Meeting of the Institute of Navigation, Santa Monica, California: 207–215.

Dierendonck AJ van, Hegarty C, Scales W, Ericson S (2000): Signal specification for the future GPS civil signal at L5. Presented at the IAIN World Congress, San Diego, California, June 27.

Diggelen F van (1998): GPS accuracy: lies, damn lies, and statistics. GPS World, 9(1): 41–45.

Diggelen F van (2007): GNSS accuracy: lies, damn lies, and statistics. GPS World, 18(1): 26–32.
Dixon CS (2003): GNSS local component integrity concepts. Journal of Global Positioning Systems, 2(2): 126–134.
Dixon K (2005a): Satellite positioning systems: efficiencies, performance and trends. European Journal of Navigation, 3(1): 58–63.
Dixon K (2005b): StarFire: A global SBAS for sub-decimeter precise point positioning. In: Proceedings of ION GPS 2006, 19th International Technical Meeting of the Satellite Division of the Institute of Navigation, Fort Worth, Texas, September 26–29: 2286–2296.
Dixon K (2005c): Satellite positioning systems: efficiencies, performance and trends. European Journal of Navigation, 3(1): 58–63.
Dixon K (2007): Satellite navigation. Hydro International, 11(1). Available at www.hydro-international.com/issues/articles/id728-Satellite_Navigation.html.
Dixon RC (1984): Spread spectrum systems, 2nd edition, 3rd print. Wiley, New York.
Dobrosavljevic Z, Spicer JJ (2004): Sub-carrier and chip waveform shaping for generation of signals with controlled out-of-band spectral content. In: Proceedings of the 2nd ESA Workshop on Satellite Navigation User Equipment Technologies, NAVITEC 2004, ESTEC, Noordwijk, The Netherlands, December 8–10.
Dorsey AJ, Marquis WA, Fyfe PM, Kaplan ED, Wiederholt F (2006): GPS system segments. In: Kaplan ED, Hegarty CJ (eds): Understanding GPS – principles and applications, 2nd edition. Artech House, Norwood: 67–112.
Dvorkin V, Karutin S (2006): GLONASS: current status and perspectives. Presented at the Third ALLSAT Open Conference, Hannover, Germany, June 22.
Eissfeller B (1993): Stand der GPS-Empfänger-Technologie. In: Institute of Geodesy (ed): Proceedings of the Geodetic Seminar on Global Positioning System im praktischen Einsatz der Landes- und Ingenieurvermessung, Munich, Germany, May 12–14. Schriftenreihe der Universität der Bundeswehr München, vol 45: 29–55.
Ellum C, El-Sheimy N (2005): Combining GPS and photogrammetric measurements in a single adjustment. In: Proceedings of the 7th Conference on Optical 3D Measurement Techniques, Vienna, Austria, October 3–5: 339–348.
El-Mowafy A (1994): Kinematic attitude determination from GPS. Reports of the Department of Geomatics Engineering of the University of Calgary, vol 20074.
El-Sheimy N (2000): An expert knowledge GPS/INS system for mobile mapping and GIS applications. In: Proceedings of the 2000 National Technical Meeting of the Institute of Navigation, Anaheim, California, January 26–28: 816–824.
Erickson C (1992a): Investigations of C/A code and carrier measurements and techniques for rapid static GPS surveys. Reports of the Department of Geomatics Engineering of the University of Calgary, vol 20044.
Erickson C (1992b): An analysis of ambiguity resolution techniques for rapid static GPS surveys using single frequency data. In: Proceedings of ION GPS-92, 5th International Technical Meeting of the Satellite Division of the Institute of Navigation, Albuquerque, New Mexico, September 16–18: 453–462.
Essen L, Froome KD (1951): The refractive indices and dielectric constants of air and its principal constituents at 24 000 Mc/s. In: Proceedings of Physical Society, vol 64(B): 862–875.

Euler H-J, Goad CC (1991): On optimal filtering of GPS dual frequency observations without using orbit information. Bulletin Géodésique, 65: 130–143.
Euler H-J, Landau H (1992): Fast GPS ambiguity resolution on-the-fly for real-time applications. In: Proceedings of the Sixth International Geodetic Symposium on Satellite Positioning, Columbus, Ohio, March 17–20, vol 2: 650–659.
Euler H-J, Schaffrin B (1990): On a measure for the discernibility between different ambiguity solutions in the static-kinematic GPS-mode. In: Schwarz KP, Lachapelle G (eds): Kinematic systems in geodesy, surveying, and remote sensing. Springer, New York Berlin Heidelberg Tokyo: 285–295 [Mueller II (ed): IAG Symposia Proceedings, vol 107].
Euler H-J, Sauermann K, Becker M (1990): Rapid ambiguity fixing in small scale networks. In: Proceedings of the Second International Symposium on Precise Positioning with the Global Positioning System, Ottawa, Canada, September 3–7: 508–523.
European Commission (1999): Galileo, involving Europe in a new generation of satellite navigation services. COM 54 final, Brussels.
European Commission (2000): Cost benefit analysis results for Galileo. Commission staff working paper, Brussels.
European Commission (2003a): World wide radionavigation system – evaluation of the Galileo performance against maritime GNSS requirements. Submitted to the International Maritime Organization, Subcommittee on Safety of Navigation, NAV 49/14.
European Commission (2003b): Space: a new European frontier for an expanding union. An action plan for implementing the European space policy. White paper. Available at http://europa.eu.int.
European Commission, European Space Agency (2002): Galileo mission high level definition, 3rd issue.
European Council (1994): Council resolution of 19 December 1994 on the European contribution to the development of a global navigation satellite system (GNSS). Official Journal C 379, 31/12/1994 P.
European Council (2004): Council regulation (EC) No 1321/2004 of 12 July on the establishment of structures for the management of the European satellite radio-navigation programmes. Available at http://europa.eu.int.
European Parliament, European Council (2004): Directive 2004/52/EC of the European Parliament and of the Council of 29 April 2004 on the interoperability of electronic road toll systems in the Community. OJ L 166.
European Space Agency (1997): The Hipparcos and Tycho catalogues. ESA Publications Division, Noordwijk, The Netherlands, SP-1200, 17 volumes.
European Space Agency (1999): Gravity field and steady-state ocean circulation mission. Reports for mission selection. The four candidate earth explorer core missions. ESA Publications Division, Noordwijk, The Netherlands, SP-1233(1). Available at http://esamultimedia.esa.int/docs/goce_sp1233_1.pdf.
European Space Agency (2004): European cooperation for space standardization – glossary of terms. ESA Publications Division, Noordwijk, The Netherlands, ECSS P-001B.
European Space Agency (2005a): EGNOS fact sheet – 12: the EGNOS signal explained. Available at www.egnos-pro.esa.int/Publications/fact.html.
European Space Agency (2005b): EGNOS news, 5(2). Available at www.egnos-pro.esa.int/newsletter.

European Space Agency (2005c): Galileo – the European programme for global navigation services, 2nd edition. ESA Publications Division, Noordwijk, The Netherlands. Available at http://ec.europa.eu/dgs/energy_transport/galileo/documents/brochure_en.htm.

European Space Agency, Galileo Joint Undertaking (2006): Galileo open service. Signal in space interface control document (OS SIS ICD). Draft 0. 23/05/06. Available at www.galileoju.com.

Falcone M (2006): Galileo overall architecture. Course on Galileo held after the 3rd ESA Workshop on Satellite Navigation User Equipment Technologies, NAVITEC 2006, ESTEC, Noordwijk, The Netherlands, December 14.

Falcone M, Erhard P, Hein GW (2006a): Galileo. In: Kaplan ED, Hegarty CJ (eds): Understanding GPS – principles and applications, 2nd edition. Artech House, Norwood: 559–594.

Falcone M, Lucas R, Burger T, Hein GW (2006b): The European Galileo programme. In: Ventura-Traveset J, Flament D (eds): The European EGNOS project. ESA Publications Division, Noordwijk, The Netherlands, SP-1303: 435–455.

Feairheller S, Clark R (2006): Other satellite navigation systems. In: Kaplan ED, Hegarty CJ (eds): Understanding GPS – principles and applications, 2nd edition. Artech House, Norwood: 595–634.

Federal Aviation Administration (1999): Specification – performance type one, local area augmentation system, ground facility. US Department of Transportation, FAA-E-2937.

Feng S, Ochieng WY (2006): An efficient worst user location algorithm for the generation of the Galileo integrity flag. In: Proceedings of ION GPS 2005, 18th International Technical Meeting of the Satellite Division of the Institute of Navigation, Long Beach, California, September 13–16: 2374–2384.

Feng Y, Rizos C (2005): Three carrier approaches for future global, regional and local GNSS positioning services: concepts and performance perspectives. In: Proceedings of ION GNSS 2005, 18th International Technical Meeting of the Satellite Division of the Institute of Navigation, Long Beach, California, September 13–16: 2277–2287.

Fenton PC, Townsend BR (1994): NovAtel Communications Ltd. – what's new? In: Proceedings of the International Symposium on Kinematic Systems in Geodesy, Geomatics and Navigation, Banff, Canada, August 30 through September 2: 25–29.

Fernández-Plazaola U, Martín-Guerrero TM, Entrambasaguas-Muñoz JT, Martín-Neira M (2004): The null method applied to GNSS three-carrier phase ambiguity resolution. Journal of Geodesy, 78: 96–102.

Flament P (2006): The Lisbon objectives & status of the Galileo programme. In: Proceedings of the Workshop on Galileo for Small and Medium Enterprises, Progeny, Brussels, April 5–6.

Fliegel HF, Feess WA, Layton WC, Rhodus NW (1985): The GPS radiation force model. In: Proceedings of the First International Symposium on Precise Positioning with the Global Positioning System, Rockville, Maryland, April 15–19, vol 1: 113–119.

Flury J, Rummel R (2005): Future satellite gravimetry for geodesy. Earth, Moon, and Planets, 94: 13–29.

Fontana RD, Cheung W, Novak PM, Stansell TA (2001): The new L2 civil signal. In: Proceedings of ION GPS 2001, 14th International Technical Meeting of the Satellite

Division of the Institute of Navigation, Salt Lake City, Utah, September 11–14: 617–631.

Forden G (2004): The military capabilities and implications of China's indigenous satellite-based navigation system. Science and Global Security, 12: 219–250.

Forssell B, Martín-Neira M, Harris RA (1997): Carrier phase ambiguity resolution in GNSS-2. In: Proceedings of ION GPS-97, 10th International Technical Meeting of the Satellite Division of the Institute of Navigation, Kansas City, Montana, September 16–19: 1727–1736.

Fotopoulos G (2005): Calibration of geoid error models via a combined adjustment of ellipsoidal, orthometric and gravimetric geoid height data. Journal of Geodesy, 79: 111–123.

Frei E (1991): GPS – fast ambiguity resolution approach "FARA": theory and application. Paper presented at XX General Assembly of the IUGG, IAG-Symposium GM 1/4, Vienna, August 11–24.

Frei E, Beutler G (1990): Rapid static positioning based on the fast ambiguity resolution approach "FARA": theory and first results. Manuscripta Geodaetica, 15(4): 325–356.

Frei E, Schubernigg M (1992): GPS surveying techniques using the "fast ambiguity resolution approach (FARA)". Paper presented at the 34th Australian Surveyors Congress and the 18th National Surveying Conference at Cairns, Australia, May 23–29.

Fu Z, Hornbostel A, Konovaltsev A (2001): Suppression of multipath and jamming signals by digital beamformer for GNSS/Galileo applications. In: Proceedings of the 1st ESA Workshop on Satellite Navigation User Equipment Technologies, NAVITEC 2001, ESTEC, Noordwijk, The Netherlands, December 10–12.

Galileo Joint Undertaking (2003): Galileo, GOC – pre-selection phase, Annex 2: overview of the Galileo system and services. GJU/03/699/JT.

Gao Y (2006): What is precise point positioning (PPP), and what are its requirements, advantages and challenges? In: Lachapelle G, Petovello M (eds): GNSS Solutions: Precise point positioning and its challenges, aided-GNSS and signal tracking. Inside GNSS, 1(8): 16–18.

Gao Y, Chen K (2004): Performance analysis of precise point positioning using real-time orbit and clock products. Journal of Global Positioning Systems, 3(1–2): 95–100.

Gao Y, Shen K (2001): Improving ambiguity convergence in carrier phase-based precise point positioning. In: Proceedings of ION GPS 2001, 14th International Technical Meeting of the Satellite Division of the Institute of Navigation, Salt Lake City, Utah, September 11–14: 1532–1539.

García-Fernández M (2004): Contributions to the 3D ionospheric sounding with GPS data. PhD dissertation, Universitat Politècnica de Catalunya, Spain.
Available at www.tdx.cbuc.es.

Garin L, Rousseau J (1997): Enhanced strobe correlator multipath rejection for code and carrier. In: Proceedings of ION GPS-97, 10th International Technical Meeting of the Satellite Division of the Institute of Navigation, Kansas City, Montana, September 16–19: 559–568.

Gassner G, Brunner FK (2003): Monitoring eines Rutschhanges mit GPS-Messungen. Vermessung, Photogrammetrie, Kulturtechnik, 101(4): 166–171.

Geiger A (1988): Einfluss und Bestimmung der Variabilität des Phasenzentrums von GPS-Antennen. Eidgenössische Technische Hochschule Zürich, Institute of Geodesy and Photogrammetry, Mitteilungen vol 43.

Gendt G, Reigber C, Dick G (1999): GPS meteorology – IGS contribution and GFZ activities for operational water vapor monitoring. In: Proceedings of the Fifth International Seminar "GPS in Central Europe", Reports on Geodesy, Warsaw University of Technology, 5(46): 53–62.

Gerdan GP (1995): A comparison of four methods of weighting double difference pseudo range measurements. Trans Tasman Surveyor, 1(1): 60–66.

Gianniou M (1996): Genauigkeitssteigerung bei kurzzeit-statischen und kinematischen Satellitenmessungen bis hin zur Echtzeitanwendung. Deutsche Geodätische Kommission bei der Bayerischen Akademie der Wissenschaften, Reihe C, vol 458.

Gibbons G (2006): GNSS trilogy – our story thus far. Inside GNSS, 1(1): 25–31 and 67.

Gibbons G, Fenton P, Garin L, Hatch R, Kawazoe T, Keegan R, Knight J, Kohli S, Rowitch D, Sheynblat L, Stratton A, Studenny J, Turetzky G, Weill L (2006): BOC or MBOC? The common GPS/Galileo civil signal design: a manufacturers dialog, part 2. Inside GNSS, 1(6): 28–43.

Giraud J, Busquet C, Bauer F, Flament D (2005): Pulsed interference and Galileo sensor stations (GSS). In: Proceedings of ION GPS 2005, 18th International Technical Meeting of the Satellite Division of the Institute of Navigation, Long Beach, California, September 13–16: 914–925.

Gomi J (2004): Quasi-Zenith Satellite System (QZSS) program overview. In: Proceedings of the Munich Satellite Navigation Summit, Munich, Germany, March 23–25.

Görres B (1996): Bestimmung von Höhenänderungen in regionalen Netzen mit dem Global Positioning System. Deutsche Geodätische Kommission bei der Bayerischen Akademie der Wissenschaften, Reihe C, vol 461.

Görres B, Campbell J, Becker M, Siemes M (2006): Absolute calibration of GPS antennas: laboratory results and comparison with field tests and robot techniques. GPS Solutions, 10(2): 136–145.

Graas F van, Braasch MS (1991): GPS interferometric attitude and heading determination: initial flight test results. Navigation, 38(4): 297–316.

Graas F van, Braasch MS (1992): Real-time attitude and heading using GPS. GPS World, 3(3): 32–39.

Grafarend EW, Schwarze V (1991): Relativistic GPS positioning. In: Caputo M, Sansò F (eds): Proceedings of the geodetic day in honor of Antonio Marussi. Accademia Nazionale dei Lincei, Rome. Atti dei Convegni Lincei, vol 91: 53–66.

Grafarend EW, Shan J (2002): GPS solutions: closed forms, critical and special configurations of P4P. GPS Solutions, 5(3): 29–41.

Grewal MS, Andrews AP (2001): Kalman filtering. Theory and practice using MATLAB, 2nd edition. Wiley, New York Chichester Weinheim Brisbane Singapore Toronto.

Guier WH, Weiffenbach GC (1997): Genesis of satellite navigation. Johns Hopkins APL Technical Digest, 18(2): 178–181.

Gurtner W (1995): The role of permanent GPS stations in IGS and other networks. In: Proceedings of the Third International Seminar on GPS in Central Europe, Penc, Hungary, May 9–11: 221–239.

Gurtner W, Estey L (2006): RINEX: the receiver independent exchange format version 3.00. Available at http://igscb.jpl.nasa.gov/igscb/data/format/.

Gurtner W, Mader G (1990): Receiver independent exchange format version 2. GPS Bulletin, 3(3): 1–8.

Habrich H (1999): Geodetic applications of the global navigation satellite system (GLONASS) and of GLONASS / GPS combinations. PhD dissertation, University of Berne, Switzerland.

Habrich H, Beutler G, Gurtner W, Rothacher M (1999): Double difference ambiguity resolution for GLONASS/GPS carrier phase. In: Proceedings of ION GPS-99, 12th International Technical Meeting of the Satellite Division of the Institute of Navigation, Nashville, Tennessee, September 14–17: 1609–1618.

Hahn J (2005): Galileo time concept. Presented at the Workshop on Time and Frequency Services with Galileo, Herrsching, Germany, December 5–6.

Hammesfahr J, Dreher A, Hornbostel A, Fu Z (2001): Assessment of use of C-band frequencies for navigation. Deutsches Zentrum für Luft- und Raumfahrt, Institute of Communications and Navigation, DLR-GUST-003, version 3.0.

Han S, Rizos C (1996): Integrated methods for instantaneous ambiguity resolution using new generation GPS receivers. In: Proceedings of IEEE Position Location and Navigation Symposium PLANS'96, Atlanta, Georgia, April 22–26: 254–261.

Han S, Rizos C (1997): Comparing GPS ambiguity resolution techniques. GPS World, 8(10): 54–61.

Han S, Dai L, Rizos C (1999): A new data processing strategy for combined GPS/GLONASS carrier phase-based positioning. In: Proceedings of ION GPS-99, 12th International Technical Meeting of the Satellite Division of the Institute of Navigation, Nashville, Tennessee, September 14–17: 1619–1627.

Hartinger H, Brunner FK (1999): Variances of GPS phase observations: the SIGMA-ε model. GPS Solutions, 2(4): 35–43.

Hartung J, Elpelt B, Klösener KH (2005): Statistik. Lehr- und Handbuch der angewandten Statistik, 14th edition. Oldenbourg, München Wien.

Hatch R (1990): Instantaneous ambiguity resolution. In: Schwarz KP, Lachapelle G (eds): Kinematic systems in geodesy, surveying, and remote sensing. Springer, New York Berlin Heidelberg Tokyo: 299–308 [Mueller II (ed): IAG Symposia Proceedings, vol 107].

Hatch R (1991): Ambiguity resolution while moving – experimental results. In: Proceedings of ION GPS-91, 4th International Technical Meeting of the Satellite Division of the Institute of Navigation, Albuquerque, New Mexico, September 11–13: 707–713.

Hatch R, Euler H-J (1994): Comparison of several AROF kinematic techniques. In: Proceedings of ION GPS-94, 7th International Technical Meeting of the Satellite Division of the Institute of Navigation, Salt Lake City, Utah, September 20–23, part 1: 363–370.

Hatch R, Jung J, Enge P, Pervan B (2000): Civilian GPS: the benefits of three frequencies. GPS Solutions, 3(4): 1–9.

Hein GW (1990a): Bestimmung orthometrischer Höhen durch GPS und Schweredaten. Schriftenreihe der Universität der Bundeswehr München, vol 38-1: 291–300.

Hein GW (1990b): Kinematic differential GPS positioning: applications in airborne photogrammetry and gravimetry. In: Crosilla F, Mussio L (eds): Il sistema di posizionamento globale satellitare GPS. International Centre for Mechanical Sciences, Collana di Geodesia e Cartografia, Udine, Italy: 139–173.

Hein GW (1995): Comparison of different on-the-fly ambiguity resolution techniques. In: Proceedings of ION GPS-95, 8th International Technical Meeting of the Satellite

Division of the Institute of Navigation, Palm Springs, California, September 12–15, part 2: 1137–1144.

Hein GW, Issler JL (2001): Signal architecture & signal structure in satellite navigation. In: Course books for the summer school in Alpbach, Austria, July 17–26.

Hein GW, Pany T (2002): Architecture and signal design of the European satellite navigation system Galileo – status December 2002. Journal of Global Positioning Systems, 1(2): 73–84.

Hein GW, Godet J, Issler JL, Martin JC, Erhard P, Lucas-Rodriguez R, Pratt T (2002): Status of Galileo frequency and signal design. In: Proceedings of ION GPS 2002, 15th International Technical Meeting of the Satellite Division of the Institute of Navigation, Portland, Oregon, September 24–27: 266–277.

Hein GW, Avila-Rodríguez JA, Wallner S, Betz JW, Hegarty CJ, Rushanan JJ, Kraay AL, Pratt AR, Lenahan S, Owen J, Issler JL, Stansell TA (2006a): MBOC: the new optimized spreading modulation. Recommended for Galileo L1 OS and GPS L1C. Inside GNSS, 1(4): 57–66.

Hein GW, Avila-Rodríguez JA, Wallner S (2006b): The DaVinci Galileo code and others. Inside GNSS, 1(6): 62–75.

Hein GW, Avila-Rodriguez JA, Wallner S, Eissfeller B, Pany T, Hartl P (2007): Envisioning a future GNSS system of systems. Part 1. Inside GNSS, 2(1): 58–67.

Heinrichs G, Löhnert E, Mundle H (2004): GATE – the German Galileo test & development environment for receivers and user applications. In: Proceedings of the 2nd ESA Workshop on Satellite Navigation User Equipment Technologies, NAVITEC 2004, ESTEC, Noordwijk, The Netherlands, December 8–10.

Hernández-Pajares M, Juan JM, Sanz J, Orús R, García-Rodríguez A, Colombo OL (2004): Wide area real time kinematics with Galileo and GPS signals. In: Proceedings of ION GNSS 2004, 17th International Technical Meeting of the Satellite Division of the Institute of Navigation, Long Beach, California, September 21–24: 2541–2554.

Herring TA (1992): Modeling atmospheric delays in the analysis of space geodetic data. In: Munck JC de, Spoelstra TAT (eds): Refraction of transatmospheric signals in geodesy. Netherlands Geodetic Commission, Delft, new series, vol 36: 157–164.

Hewitson S, Wang J (2006): GNSS receiver autonomous integrity monitoring (RAIM) performance analysis. GPS Solutions, 10(3): 155–170.

Hilla S, Jackson M (2000): The GPS toolbox. GPS Solutions, 3(4): 71–74.

Hofmann-Wellenhof B, Moritz H (2006): Physical geodesy, 2nd edition. Springer, Wien New York.

Hofmann-Wellenhof B, Kienast G, Lichtenegger H (1994): GPS in der Praxis. Springer, Wien New York.

Hofmann-Wellenhof B, Lichtenegger H, Collins J (2001): GPS – theory and practice, 5th edition. Springer, Wien New York.

Hofmann-Wellenhof B, Legat K, Wieser M (2003): Navigation – principles of positioning and guidance. Springer, Wien New York.

Hollreiser M, Sleewaegen JM, Wilde W de, Falcone M, Wilms F (2005): Galileo test user segment – first achievements and application. GPS World, 16(7): 23–29.

Holmes JK (1982): Coherent spread spectrum systems, reprint 1990. Krieger, Malabar.

Hopfield HS (1969): Two-quartic tropospheric refractivity profile for correcting satellite data. Journal of Geophysical Research, 74(18): 4487–4499.

Hothem L (2006): The GPS modernization program and policy update. Paper presented at the XXIII International FIG Congress, Munich, Germany, October 8–13.

Huddle JR, Brown RG (1997): Multisensor navigation systems. In: Kayton M, Fried WR (eds): Avionics navigation systems, 2nd edition. Wiley, New York Chichester Weinheim Brisbane Singapore Toronto: 55–98.

Hudnut KW, Titus B (2004): GPS L1 civil signal modernization (L1C). The Interagency GPS Executive Board. Available at www.navcen.uscg.gov/gps.

Ilk KH, Flury J, Rummel R, Schwintzer P, Bosch W, Haas C, Schröter J, Stammer D, Zahel W, Miller H, Dietrich R, Huybrechts P, Schmeling H, Wolf D, Götze HJ, Riegger J, Bardossy A, Güntner A, Gruber T (2005): Mass transport and mass distribution in the earth system – contribution of the new generation of satellite gravity and altimetry missions to geosciences, 2nd edition. GOCE-Projektbüro Deutschland, Technische Universität München, GeoForschungsZentrum Potsdam (eds).

Institute of Electrical and Electronics Engineers (1997): IEEE standard definitions of terms for radio wave propagation. IEEE Std 211-1997.

Institute of Navigation (1997): Recommended test procedures for GPS Receivers, ION STD 101, revision C. Alexandria, Virginia.

International Association of Geodesy (1995): New geoids in the world. International Association of Geodesy, Bulletin d'Information 77, IGES Bulletin 4.

International Electrotechnical Commission (2005): IEC 61508 – Functional safety of electrical/electronic/programmable electronic safety-related systems.
Available at www.iec.ch.

International Telecommunication Union (2004): International Telecommunication Union radio regulations. Available at www.itu.int.

Irsigler M, Hein GW, Eissfeller B, Schmitz-Peiffer A, Kaiser M, Hornbostel A, Hartl P (2002): Aspects of C-band satellite navigation: signal propagation and satellite signal tracking. In: Proceedings of the European Navigation Conference GNSS 2002, Copenhagen, Denmark, May 27–30.

Issler JL, Lestarquit L, Grondin M (2001): Missions and radionavigation payloads. In: Course books for the summer school in Alpbach, Austria, July 17–26.

Jakowski N (1996): TEC monitoring by using satellite positioning systems. In: Kohl H, Rüster R, Schlegel K (eds): Modern ionosphere science. EGS, Katlenburg-Lindau, ProduServ GmbH Verlagsservice, Berlin: 371–390.

Jakowski N (2001): Space based atmosphere and ionosphere sounding. In: Course books for the summer school in Alpbach, Austria, July 17–26.

Janes HW, Langley RB, Newby SP (1991): Analysis of tropospheric delay prediction models: comparisons with ray-tracing and implications for GPS relative positioning. Bulletin Géodésique, 65: 151–161.

Japan Aerospace Exploration Agency (2007): Quasi-Zenith Satellite System navigation service, interface specification for QZSS (IS-QZSS). Draft version 0.1.

Jekeli C (2001): Inertial navigation systems with geodetic applications. Walter de Gruyter, Berlin.

Jin XX, Jong CD de (1996): Relationship between satellite elevation and precision of GPS code observations. Journal of Navigation, 49: 253–265.

Jong CD de, Lachapelle G, Skone S, Elema IA (2002): Hydrography. DUP Blue Print, Delft University Press.

Jong K de (2000): Selective availability turned off. Geoinformatics, 3(5): 14–15.

Jonge P de, Tiberius C (1995): Integer ambiguity estimation with the Lambda method. In: Beutler G, Hein GW, Melbourne WG, Seeber G (eds): GPS trends in precise terrestrial, airborne, and spaceborne applications. Springer, New York Berlin Heidelberg Tokyo: 280–284 [Mueller II (ed): IAG Symposia Proceedings, vol 115].

Jonkman NF (1998): Integer GPS ambiguity estimation without the receiver-satellite geometry. Delft Geodetic Computing Centre, LGR Series, vol 18.

Joos G (1956): Lehrbuch der Theoretischen Physik, 9th edition. Akademische Verlagsgesellschaft Geest & Portig K-G, Leipzig.

Joosten P, Tiberius C (2000): Fixing the ambiguities – are you sure they're right? GPS World, 11(5): 46–51.

Joosten P, Teunissen PJG, Jonkman N (1999): GNSS three carrier phase ambiguity resolution using the LAMBDA-method. In: Proceedings of GNSS'99, 3rd European Symposium on Global Navigation Satellite Systems, Genova, Italy, October 5–8, part 1: 367–372.

Julien O (2005): Design of Galileo L1F receiver tracking loops. PhD dissertation, Department of Geomatics Engineering, University of Calgary, Canada. Available at www.geomatics.ucalgary.ca/links/GradTheses.html.

Julien O, Cannon ME, Alves P, Lachapelle G (2004a): Triple frequency ambiguity resolution using GPS/Galileo. European Journal of Navigation, 2(2): 51–57.

Julien O, Macabiau C, Cannon EM, Lachapelle G (2004b): New unambiguous BOC(N,N) tracking technique. In: Proceedings of the 2nd ESA Workshop on Satellite Navigation User Equipment Technologies, NAVITEC 2004, ESTEC, Noordwijk, The Netherlands, December 8–10.

Jung J, Enge P, Pervan B (2000): Optimization of cascade integer resolution with three civil GPS frequencies. In: Proceedings of ION GPS 2000, 13th International Technical Meeting of the Satellite Division of the Institute of Navigation, Salt Lake City, Utah, September 19–22: 2191–2200.

Kalman RE (1960): A new approach to linear filtering and prediction problems. Journal of Basic Engineering, 82(1): 35–45.

Kaplan ED (2006): Introduction. In: Kaplan ED, Hegarty CJ (eds): Understanding GPS – principles and applications, 2nd edition. Artech House, Norwood.

Kaula WM (1966): Theory of satellite geodesy. Blaisdell, Toronto.

Kee C (1996): Wide area differential GPS. In: Parkinson BW, Spilker JJ (eds): Global Positioning System: theory and applications. American Institute of Aeronautics and Astronautics, Washington DC, vol 2: 81–115.

Keegan R (1990): P-code aided Global Positioning System receiver. US Patent Office, Patent no. 4,972,431.

Kelly JT (2006): PPS versus SPS – why military applications require military GPS. GPS World, 17(1): 28–35.

Kibe SV (2006): Indian SATNAV programme – challenges and opportunities. In: Proceedings of the First Meeting of the International Committee on Global Navigation Satellite Systems, The United Nations Office for Outer Space Affairs, Vienna, November 1–2.

Kim D, Langley RB (1999): An optimized least-squares technique for improving ambiguity resolution and computational efficiency. In: Proceedings of ION GPS-99, 12th International Technical Meeting of the Satellite Division of the Institute of Navigation, Nashville, Tennessee, September 14–17: 1579–1588.

Kim D, Langley RB (2000): GPS ambiguity resolution and validation: methodologies, trends and issues. Paper presented at the 7th GNSS Workshop – International Symposium on GPS/GNSS, Seoul, Korea, November 30 – December 2. Available at http://gauss.gge.unb.ca/papers.pdf/gnss2000.kim.pdf.

Kim E, Walter T, Powell JD (2006): Optimizing WAAS accuracy/stability for a single frequency receiver. In: Proceedings of ION GPS 2006, 19th International Technical Meeting of the Satellite Division of the Institute of Navigation, Fort Worth, Texas, September 26–29: 962–970.

Kim J, Sukkarieh S (2005): 6DoF SLAM aided GNSS/INS navigation in GNSS denied and unknown environments. Journal of Global Positioning Systems, 4(1–2): 120–128.

King RW, Masters EG, Rizos C, Stolz A, Collins J (1987): Surveying with Global Positioning System. Dümmler, Bonn.

Kleusberg A (1990): A review of kinematic and static GPS surveying procedures. In: Proceedings of the Second International Symposium on Precise Positioning with the Global Positioning System, Ottawa, Canada, September 3–7: 1102–1113.

Kleusberg A (1994): Die direkte Lösung des räumlichen Hyperbelschnitts. Zeitschrift für Vermessungswesen, 119(4): 188–192.

Klimov V, Revnivykh S, Kossenko V, Dvorkin V, Tyulyakov A, Eltsova O (2005): Status and development of GLONASS. Presented at the European Navigation Conference GNSS-2005, Munich, Germany, July 19–22.

Klobuchar J (1986): Design and characteristics of the GPS ionospheric time-delay algorithm for single-frequency users. In: Proceedings of PLANS'86 – Position Location and Navigation Symposium, Las Vegas, Nevada, November 4–7: 280–286.

Knight D (1994): A new method of instantaneous ambiguity resolution. In: Proceedings of ION GPS-94, 7th International Technical Meeting of the Satellite Division of the Institute of Navigation, Salt Lake City, Utah, September 20–23, part 1: 707–716.

Koch K-R (1987): Parameter estimation and hypothesis testing in linear models. Springer, Berlin Heidelberg New York London Paris Tokyo.

Kouba J, Héroux P (2001): Precise point positioning using IGS orbit products. GPS Solutions, 5(2): 12–28.

Kozai Y (1959): On the effects of the sun and the moon upon the motion of a close earth satellite. Smithsonian Astrophysical Observatory, Special Report, vol 22.

Kreher J (2004): Galileo signals: RF characteristics. Working paper of the Navigation Systems Panel (NSP), ICAO NSP/WGW: WP/36. Montreal, Canada, October 12–22.

Kreyszig E (2006): Advanced engineering mathematics, 9th edition. Wiley, Hoboken.

Kunysz W (2000): A novel GPS survey antenna. In: Proceedings of the 2000 National Technical Meeting of the Institute of Navigation, Anaheim, California, January 26–28: 698–705.

Kuusniemi H (2005): User-level reliability and quality monitoring in satellite-based personal navigation. PhD dissertation, Institute of Digital and Computer Systems, Tampere University of Technology, Finland.

Lachapelle G (1990): GPS observables and error sources for kinematic positioning. In: Schwarz KP, Lachapelle G (eds): Kinematic systems in geodesy, surveying, and remote sensing. Springer, New York Berlin Heidelberg Tokyo: 17–26 [Mueller II (ed): IAG Symposia Proceedings, vol 107].

Lachapelle G (1998): Hydrography. Lecture notes of the Department of Geomatics Engineering of the University of Calgary, Canada, No. 10016.

Lachapelle G (2003): Advanced GPS theory and applications. ENGO 625 Lecture Notes, University of Calgary.

Lachapelle G (2004): GNSS indoor location technologies. Journal of Global Positioning Systems, 3(1–2): 2–11.

Lachapelle G, Sun H, Cannon ME, Lu G (1994): Precise aircraft-to-aircraft positioning using a multiple receiver configuration. Canadian Aeronautics and Space Journal, 40(2): 74–78.

Landau H (1988): Zur Nutzung des Global Positioning Systems in Geodäsie und Geodynamik: Modellbildung, Software-Entwicklung und Analyse. Schriftenreihe der Universität der Bundeswehr München, vol 36.

Landau H, Euler H-J (1992): On-the-fly ambiguity resolution for precise differential positioning. In: Proceedings of ION GPS-92, Fifth International Technical Meeting of the Satellite Division of the Institute of Navigation, Albuquerque, New Mexico, September 16–18: 607–613.

Landau H, Ordóñez JMF (1992): A new algorithm for attitude determination with GPS. In: Proceedings of the Sixth International Geodetic Symposium on Satellite Positioning, Columbus, Ohio, March 17–20, vol 2: 1036–1038.

Landau H, Vollath U, Chen X (2002): Virtual reference station systems. Journal of Global Positioning Systems, 1(2): 137–143.

Landau H, Vollath U, Chen X (2004): Benefits of modernized GPS/Galileo to RTK positioning. In: Proceedings of the 2004 International Symposium on GNSS/GPS, Sydney, Australia, December 6–8: 92–103.

Langley RB (1997): GPS receiver system noise. GPS World, 8(6): 40–45.

Langley RB (1998): GPS receivers and the observables. In: Teunissen PJG, Kleusberg (eds): GPS for geodesy, 2nd edition. Springer, Berlin Heidelberg New York: 151–185.

Łapucha D, Barker R, Zwaan H (2004): Comparison of the two alternate methods of wide area carrier phase positioning. In: Proceedings of ION GNSS 2004, 17th International Technical Meeting of the Satellite Division of the Institute of Navigation, Long Beach, California, September 21–24: 1864–1871.

Leick A (2004): GPS satellite surveying, 3rd edition. Wiley, Hoboken.

Leitinger R (1996): Tomography. In: Kohl H, Rüster R, Schlegel K (eds): Modern ionosphere science. EGS, Katlenburg-Lindau, ProduServ GmbH Verlagsservice, Berlin: 346–370.

Leitinger R, Zhang M-L, Radicella SM (2005): An improved bottomside for the ionospheric electron density model NeQuik. Annals of Geophysics, 48(3): 525–534.

Lemoine FG, Kenyon SC, Factor JK, Trimmer RG, Pavlis NK, Chinn DS, Cox CM, Klosko SM, Luthcke SB, Torrence HM, Wang YM, Williamson RG, Pavlis EC, Rapp RH, Olson TR (1998): The development of the joint NASA GSFC and the National Imagery and Mapping Agency (NIMA) geopotential model EGM96. NASA Technical Paper NASA/TP-1998-206861, Goddard Space Flight Center, Greenbelt.

Levanon N (1999): Instant active positioning with one LEO satellite. Navigation 46(2): 87–95.

Li X (2004): Integration of GPS, accelerometer and optical fiber sensors for structural deformation monitoring. In: Proceedings of ION GNSS 2004, 17th International Technical Meeting of the Satellite Division of the Institute of Navigation, Long Beach, California, September 21–24: 211–224.

Li Z, Schwarz KP, El-Mowafy A (1993): GPS multipath detection and reduction using spectral technique. Paper presented at the General Meeting of the IAG at Beijing, P.R. China, August 8–13.

Lichten SM, Neilan RE (1990): Global networks for GPS orbit determination. In: Proceedings of the Second International Symposium on Precise Positioning with the Global Positioning System, Ottawa, Canada, September 3–7: 164–178.

Lichtenegger H (1991): Über die Auswirkung von Koordinatenänderungen in der Referenzstation bei relativen Positionierungen mittels GPS. Österreichische Zeitschrift für Vermessungswesen und Photogrammetrie, 79(1): 49–52.

Lichtenegger H (1995): Eine direkte Lösung des räumlichen Bogenschnitts. Österreichische Zeitschrift für Vermessung und Geoinformation, 83(4): 224–226.

Lichtenegger H (1998): DGPS fundamentals. Reports on Geodesy, Warsaw University of Technology, 11(41): 7–19.

Liu J, Shi C, Xia L, Liu H (2006): Development update – navigation and positioning in China. Inside GNSS 1(6): 46–50.

Logsdon T (1992): The NAVSTAR Global Positioning System. Van Nostrand, New York.

Lu G (1995): Development of a GPS multi-antenna system for attitude determination. Reports of the Department of Geomatics Engineering of the University of Calgary, vol 20073.

Lu G, Cannon ME (1994): Attitude determination using a multi-antenna GPS system for hydrographic applications. Marine Geodesy, 17: 237–250.

Lu G, Cannon ME, Lachapelle G, Kielland P (1993): Attitude determination in a survey launch using multi-antenna GPS technology. In: Proceedings of the National Technical Meeting of the Institute of Navigation, San Francisco, California, January 20–22: 251–259.

MacGougan G, Normark PL, Ståhlberg C (2005): Satellite navigation evolution – the software GNSS receiver. GPS World, 16(1): 48–52.

Mader GL (1990): Ambiguity function techniques for GPS phase initialization and kinematic solutions. In: Proceedings of the Second International Symposium on Precise Positioning with the Global Positioning System, Ottawa, Canada, September 3–7: 1233–1247.

Mader GL (1999): GPS antenna calibration at the National Geodetic Survey. GPS Solutions, 3(1): 50–58.

Malys S, Slater J (1994): Maintenance and enhancement of the World Geodetic System 1984. In: Proceedings of ION GPS-94, 7th International Technical Meeting of the Satellite Division of the Institute of Navigation, Salt Lake City, Utah, September 20–23, part 1: 17–24.

Mansfeld W (2004): Satellitenortung und Navigation. Grundlagen und Anwendung globaler Satellitennavigationssysteme, 2nd edition. Vieweg, Wiesbaden.

Martín-Neira M, Buck C (2005): A tsunami early-warning system – the PARIS concept. ESA Bulletin 124: 50–55.

Martín-Neira M, Toledo M, Pelaez A (1995): The null space method for GPS integer ambiguity resolution. In: Proceedings of DSNS'95, Bergen, Norway, April 24–28: paper no. 31.

Martín-Neira M, Lucas R, Garcia A, Tossaint M, Amarillo F (2003): The development of high precision applications with Galileo. In: CD-ROM-Proceedings of the European Navigation Conference GNSS 2003, Graz, Austria, April 22–25.

Martín-Neira M, Buck C, Gleason S, Unwin M, Caparrini M, Farrés E, Germain O, Ruffini G, Soulat F (2005): Tsunami detection using the PARIS concept. Progress in Electromagnetics Research Symposium, Hangzhou, China, August 22–26.

Mathur AR, Torán-Marti F, Ventura-Traveset J (2006): SISNeT user interface document, issue 4, revision 1. GNSS-1 Project Office.

McCarthy DD, Petit G (eds) (2004): IERS conventions (2003). IERS Technical Note no. 32, Verlag des Bundesamtes für Kartographie und Geodäsie, Frankfurt/Main. Available at www.iers.org.

Meindl M, Hugentobler U (2004): Exploiting EGNOS RIMS data for the determination of GEO precise orbits. In: Proceedings of the GNSS Final Presentation Day, ESTEC, The Netherlands, December 7.

Melchior P (1978): The tides of the planet earth. Pergamon, Oxford New York Toronto Sydney Paris Frankfurt.

Menge F, Seeber G, Völksen C, Wübbena G, Schmitz M (1998): Results of absolute field calibration of GPS antenna PCV. In: Proceedings of ION GPS-98, 11th International Technical Meeting of the Satellite Division of the Institute of Navigation, Nashville, Tennessee, September 15–18: 31–38.

Merrigan MJ, Swift ER, Wong RF, Saffel JT (2002): A refinement to the World Geodetic System 1984 reference frame. In: Proceedings of ION GPS 2002, 15th International Technical Meeting of the Satellite Division of the Institute of Navigation, Portland, Oregon, September 24–27: 1519–1529.

Mervart L (1995): Ambiguity resolution techniques in geodetic and geodynamic applications of the Global Positioning System. PhD dissertation, University of Berne, Switzerland.

Mervart L (1999): Experience with SINEX format and proposals for its further development. In: Proceedings of the Fifth International Seminar "GPS in Central Europe", Reports on Geodesy, Warsaw University of Technology, 5(46): 103–110.

Mikhail EM (1976): Observations and least squares. IEP, New York.

Mikhail EM, Gracie G (1981): Analysis and adjustment of survey measurements. Van Nostrand, New York.

Misra P, Enge P (2006): Global Positioning System: signals, measurements, and performance, 2nd edition. Ganga-Jamuna, Lincoln.

Misra PN, Abbot RI, Gaposchkin EM (1996): Integrated use of GPS and GLONASS: transformation between WGS 84 and PZ-90. In: Proceedings of ION GPS-96, 9th International Technical Meeting of the Satellite Division of the Institute of Navigation, Kansas City, Missouri, September 17–20: 307–314.

Moelker D (1997): Multiple antennas for advanced GNSS multipath mitigation and multipath direction finding. In: Proceedings of ION GPS-97, 10th International Technical Meeting of the Satellite Division of the Institute of Navigation, Kansas City, Montana, September 16–19: 541–550.

Montenbruck O (1984): Grundlagen der Ephemeridenrechnung. Sterne und Weltraum Vehrenberg, München.

Montenbruck O, Günther C, Graf S, Garcia-Fernandez M, Furthner J, Kuhlen H (2006): GIOVE-A initial signal analysis. GPS Solutions, 10(2): 146–153.

Moritz H (1980): Advanced physical geodesy. Wichmann, Karlsruhe.

Moritz H, Hofmann-Wellenhof B (1993): Geometry, relativity, geodesy. Wichmann, Karlsruhe.

Moritz H, Mueller II (1988): Earth rotation – theory and observations. Ungar, New York.
Mueller II (1991): International GPS Geodynamics Service. GPS Bulletin, 4(1): 7–16.
Mueller KT, Biester M, Loomis P (1994): Performance comparison of candidate US Coast Guard WADGPS network architectures. In: Proceedings of the 1994 National Technical Meeting of the Institute of Navigation, San Diego, California, January 24–26: 833–841.
Mueller TM (1994): Wide area differential GPS. GPS World, 5(6): 36–44.
National Imagery and Mapping Agency (2004): Department of Defense World Geodetic System 1984 – its definition and relationship with local geodetic systems, 3rd edition. NIMA Technical Report TR 8350.2, Bethesda, Maryland.
Available at http://earth-info.nga.mil/GanG/publications/index.html.
Nava B, Radicella SM, Leitinger R, Coïsson P (2005): Slant TEC data ingestion in the modified NeQuick ionosphere electron density model. Paper presented at the XXVIII General Assembly of the International Union of Radio Science, New Delhi, India, October 23–29.
Nayak RA, Cannon ME, Wilson C, Zhang G (2000): Analysis of multiple GPS antennas for multipath mitigation in vehicular navigation. In: Proceedings of the 2000 National Technical Meeting of the Institute of Navigation, Anaheim, California, January 26–28: 284–293.
Nee RDJ van (1992): Multipath effects on GPS code phase measurements. Navigation, 39(2): 177–190.
Neilan RE, Moore A (1999): International GPS service tutorial. Paper presented at the International Symposium on GPS, Tsukuba, Japan, October 18–22.
Niell AE (1996): Global mapping functions for the atmosphere delay at radio wavelengths. Journal of Geophysical Research, 101(B2): 3227–3246.
Ober PB (2003): Integrity prediction and monitoring of navigation systems. European Journal of Navigation, 1(1): 13–20.
Oppenheim AV, Schafer RW, Buck JR (1999): Discrete-time signal processing, 2nd edition. Prentice Hall, London.
Otsuka Y, Ogawa T, Saito A, Tsugawa T, Fukao S, Miyazaki S (2002): A new technique for mapping of total electron content using GPS network in Japan. Earth Planets Space, 54: 63–70.
Owen J (1995): GLONASS: Russian's equivalent navigation system. In: Public Release Version (1996): NAVSTAR GPS user equipment introduction, Annex A.
Pace S, Frost GP, Lachow I, Frelinger D, Fossum D, Wassem D, Pinto MM (1995): The Global Positioning System – assessing national policies. Research and Development (RAND) Corporation. Available at www.rand.org/pubs/monograph_reports/MR614.
Parkinson BW (1996): GPS error analysis. In: Parkinson BW, Spilker JJ (eds): Global Positioning System: theory and applications. American Institute of Aeronautics and Astronautics, Washington DC, vol 1: 469–483.
Perović G (2005): Least squares. University of Belgrade, Faculty of Civil Engineering. Translated from Serbian by S. Ninković.
Petovello MG, Cannon ME, Lachapelle G, Wang J, Wilson CKH, Salychev OS, Voronov VV (2001): Development and testing of a real-time GPS/INS reference system for autonomous automobile navigation. In: Proceedings of ION GPS 2001, 14th International Technical Meeting of the Satellite Division of the Institute of Navigation, Salt Lake City, Utah, September 11–14: 2634–2641.

Phelts RE, Enge P (2000): The multipath invariance approach for code multipath mitigation. In: Proceedings of ION GPS 2000, 13th International Technical Meeting of the Satellite Division of the Institute of Navigation, Salte Lake City, Utah, September 19–22: 2376–2384.

Philippov V, Sutiagin I, Ashjaee J (1999): Measured characteristics of dual depth dual frequency choke ring for multipath rejection in GPS receivers. In: Proceedings of ION GPS-99, 12th International Technical Meeting of the Satellite Division of the Institute of Navigation, Nashville, Tennessee, September 14–17: 793–796.

Polischuk GM, Kozlov VI, Ilitchov VV, Kozlov AG, Bartenev VA, Kossenko VE, Anphimov NA, Revnivykh SG, Pisarev SB, Tyulyakov AE, Shebshaevitch BV, Basevitch AB, Vorokhovsky YL (2002): The global navigation satellite system GLONASS: development and usage in the 21st century. In: Proceedings of the 34th Annual Precise Time and Time Interval (PTTI) Systems and Applications Meeting, Reston, Virginia, December 3–5: 151–160.

Prasad R, Ruggieri M (2005): Applied satellite navigation using GPS, Galileo, and augmentation systems. Artech House, Boston London.

Radio Regulatory Department (2006): Submission of the updated information of Compass system to the Fourth Resolution 609 (WRC-03) Consultation Meeting. Ministry of Information Industry, The People's Republic of China, RG/036/2006.

Radio Technical Commission for Aeronautical Services (2006): Minimum operational performance standards for Global Positioning System / wide area augmentation system airborne equipment. DO-229D, Special Committee no. 159, Washington DC.

Radio Technical Commission for Maritime Services (2001): RTCM recommended standards for differential GNSS (global navigation satellite systems), version 2.3. RTCM paper 136-2001/SC 104-STD, Radio Technical Commission for Maritime Services, Special Committee no. 104, Washington DC.

Ray JK, Cannon ME, Fenton PC (1999): Mitigation of static carrier-phase multipath effects using multiple closely spaced antennas. Navigation, 46(3): 193–201.

Remondi BW (1984): Using the Global Positioning System (GPS) phase observable for relative geodesy: modeling, processing, and results. University of Texas at Austin, Center for Space Research.

Remondi BW (1990a): Pseudo-kinematic GPS results using the ambiguity function method. National Information Center, Rockville, Maryland, NOAA Technical Memorandum NOS NGS-52.

Remondi BW (1990b): Recent advances in pseudo-kinematic GPS. In: Proceedings of the Second International Symposium on Precise Positioning with the Global Positioning System, Ottawa, Canada, September 3–7: 1114–1137.

Remondi BW (1991): NGS second generation ASCII and binary orbit formats and associated interpolation studies. Paper presented at the XX General Assembly of the IUGG at Vienna, Austria, August 11–24.

Remondi BW (2004): Computing satellite velocity using the broadcast ephemeris. GPS Solutions 8(3): 181–183.

Retscher G (2002): Accuracy performance of virtual reference station (VRS) networks. Journal of Global Positioning Systems, 1(1): 40–47.

Revnivykh SG (2004): Developments of the GLONASS system and GLONASS service. Presented at the UN/US GNSS International Meeting, Vienna, Austria, December 13–17.

Revnivykh SG (2006a): GLONASS status update. Presented at the UN/Zambia/ESA Regional Workshop on the Applications of Global Navigation Satellite System Technologies in Sub-Saharan Africa, Lusaka, Zambia, June 26–30.

Revnivykh SG (2006b): GLONASS status update. Presented at the First Meeting of the International Committee on Global Navigation Satellite Systems, Vienna, Austria, November 1–2.

Revnivykh S, Polischuk G, Kozlov V, Klimov V, Anfimov N, Bartenev V, Kossenko V, Urlichich Y, Ivanov N, Tylyakov A (2003): Status and development of GLONASS. In: CD-ROM-Proceedings of the European Navigation Conference GNSS-2003, Graz, Austria, April 22–25, Session A1.

Richardus P, Adler RK (1972): Map projections for geodesists, cartographers and geographers. North-Holland, Amsterdam London.

Rizos C, Han S (1995): A new method for constructing multi-satellite ambiguity combinations for improved ambiguity resolution. In: Proceedings of ION GPS-95, 8th International Technical Meeting of the Satellite Division of the Institute of Navigation, Palm Springs, California, September 12–15, part 2: 1145–1153.

Rosmorduc V, Benveniste J, Lauret O, Milagro M, Picot N (2006): Radar altimetry tutorial. [Benveniste J, Picot N (eds)]. Available at www.altimetry.info.

Roßbach U (2001): Positioning and navigation using the Russian satellite system GLONASS. Schriftenreihe der Universität der Bundeswehr München, vol 71.

Roßbach U, Habrich H, Zarraoa N (1996): Transformation parameters between PZ-90 and WGS84. In: Proceedings of ION GPS-96, 9th International Technical Meeting of the Satellite Division of the Institute of Navigation, Kansas City, Missouri, September 17–20: 279–285.

Rothacher M (2001a): Principles of operation and basic observation model. In: Course books for the summer school in Alpbach, Austria, July 17–26.

Rothacher M (2001b): Comparison of absolute and relative antenna phase center variations. GPS Solutions, 4(4): 55–60.

Rothacher M, Schaer S, Mervart L, Beutler G (1995): Determination of antenna phase center variations using GPS data. In: Gendt G, Dick G (eds): Proceedings of the IGS Workshop on Special Topics and New Directions, Potsdam, Germany, May 15–18, part 2: 205–220.

Roturier B, Chatre E, Ventura-Traveset J (2001): The SBAS integrity concept standardised by ICAO – application to EGNOS. In: CD-ROM-Proceedings of the European Navigation Conference GNSS 2001, Seville, Spain, May 8–11.

Rührnößl H, Brunner FK, Rothacher M (1998): Modellierung der troposphärischen Korrektur für Deformationsmessungen mit GPS im alpinen Raum. Allgemeine Vermessungsnachrichten 105(1): 14–20.

Saastamoinen J (1973): Contribution to the theory of atmospheric refraction. Bulletin Géodésique, 107: 13–34.

Sage A, Pande A (2005): A-GPS. The past, the present, and the future. Presented at the European Navigation Conference GNSS 2005, Munich, Germany, July 19–22.

Sandhoo K, Turner D, Shaw M (2000): Modernization of the Global Positioning System. In: Proceedings of ION GPS 2000, 13th International Technical Meeting of the Satellite Division of the Institute of Navigation, Salt Lake City, Utah, September 19–22: 2175–2183.

References

Satirapod C, Kriengkraiwasin S (2006): Performance of single-frequency GPS precise point positioning.
Available at www.gisdevelopment.net/technology/gps/ma06_19pf.htm.

Sauer K, Vollath U, Amarillo F (2004): Three and four carriers for reliable ambiguity resolution. In: CD-ROM-Proceedings of the European Navigation Conference GNSS 2004, Rotterdam, May 16–19.

Schaefer M, Thomsen S, Niemeier W (2000): A multi sensor system with cm-accuracy for the determination of dumping surfaces. In: Proceedings of ION GPS 2000, 13th International Technical Meeting of the Satellite Division of the Institute of Navigation, Salt Lake City, Utah, September 19–22: 28–34.

Schaer S (1997): How to use CODE's global ionosphere maps. Astronomical Institute, University of Berne. Available at http://www.aiub.unibe.ch.

Schaer S (1999): Mapping and predicting the earth's ionosphere using the Global Positioning System. Schweizerische Geodätische Kommission, Geodätisch-geophysikalische Arbeiten in der Schweiz, vol 59.

Schmitt G, Illner M, Jäger R (1991): Transformationsprobleme. Deutscher Verein für Vermessungswesen, special issue: GPS und Integration von GPS in bestehende geodätische Netze, vol 38: 125–142.

Schön S, Wieser A, Macheiner K (2005): Accurate tropospheric correction for local GPS monitoring networks with large height differences. In: Proceedings of ION GNSS 2005, 18th International Technical Meeting of the Satellite Division of the Institute of Navigation, Long Beach, California, September 13–16: 250–260.

Schwarz KP, El-Sheimy N, Liu Z (1994): Fixing GPS cycle slips by INS/GPS: methods and experience. In: Proceedings of the International Symposium on Kinematic Systems in Geodesy, Geomatics and Navigation, Banff, Canada, August 30 through September 2: 265–275.

Schupler BR, Clark TA (1991): How different antennas affect the GPS observable. GPS World, 2(10): 32–36.

Seeber G (2003): Satellite geodesy: foundations, methods, and applications, 2nd edition. Walter de Gruyter, Berlin New York.

Seidelmann PK (ed) (1992): Explanatory supplement to the astronomical almanac. University Science Books, Mill Valley.

Seidelmann PK, Fukushima T (1992): Why new time scales? Astronomy and Astrophysics, 265: 833–838.

Shaw ME (2005): Global Positioning System: a policy and modernization review. Presented at United Nations, International Committee on GNSS, December 1–2.

Shaw M, Sandhoo K, Turner D (2000): Modernization of the Global Positioning System. GPS World, 11(9): 36–44.

Shi C, Liu J (2006): GNSS status and developments in China. Presentation at the Civil Global Positioning System Service Interface Committee, 46th meeting, Fort Worth, Texas, September 26.

Singh A, Saraswati SK (2006): India heads for a regional navigation satellite system. Coordinates, a Monthly Magazine on Positioning, Navigation and Beyond, vol II, issue 11: 6–8.

Sjöberg LE (1997): On optimality and reliability for GPS base ambiguity resolution by combined phase and code observables. Zeitschrift für Vermessungswesen, 122(6): 270–275.

Sjöberg LE (1998): A new method for GPS phase base ambiguity resolution by combined phase and code observables. Survey Review, 34(268): 363–372.

Sjöberg LE (1999): Triple frequency GPS for precise positioning. In: Krumm F, Schwarze VS (eds): Quo vadis geodesia ...? Festschrift for Erik W. Grafarend on the occasion of his 60th birthday. Schriftenreihe der Institute des Studiengangs Geodäsie und Geoinformatik, Universität Stuttgart, Part 2, Report vol 1999.6-2: 467–471.

Sleewaegen JM, Wilde W de, Hollreiser M (2004): Galileo AltBOC receiver. In: Proceedings of the European Navigation Conference GNSS 2004, Rotterdam, The Netherlands, May 16–19.

Solar Influences Data Analysis Center (2007): Sunspot data. Available at http://sidc.oma.be/sunspot-data.

Spilker JJ (1996a): GPS signal structure and theoretical performance. In: Parkinson BW, Spilker JJ (eds): Global Positioning System: theory and applications. American Institute of Aeronautics and Astronautics, Washington DC, vol 1: 57–119.

Spilker JJ (1996b): Tropospheric effects on GPS. In: Parkinson BW, Spilker JJ (eds): Global Positioning System: theory and applications. American Institute of Aeronautics and Astronautics, Washington DC, vol 1: 517–546.

Spilker JJ (1996c): GPS navigation data. In: Parkinson BW, Spilker JJ (eds): Global Positioning System: theory and applications. American Institute of Aeronautics and Astronautics, Washington DC, vol 1: 121–176.

Spilker JJ, Natali FD (1996): Interference effects and mitigation techniques. In: Parkinson BW, Spilker JJ (eds): Global Positioning System: theory and applications. American Institute of Aeronautics and Astronautics, Washington DC, vol 1: 717–771.

Stansell T, Fenton P, Garin L, Hatch R, Knight J, Rowitch D, Sheynblat L, Stratton A, Studenny J, Weill L (2006): BOC or MBOC? The common GPS/Galileo civil signal design: a manufacturers dialog, part 1. Inside GNSS, 1(5): 30–37.

Stubbe P (1996): The ionosphere as a plasma laboratory. In: Kohl H, Rüster R, Schlegel K (eds): Modern ionosphere science. EGS, Katlenburg-Lindau, ProduServ GmbH Verlagsservice, Berlin: 274–321.

Su C-C (2001): Reinterpretation of the Michelson-Morley experiment based on the GPS Sagnac correction. Europhysics Letters, 56(2): 170–174.

Swann J (2006): Will GPS and Galileo have the same or interoperable reference systems? In: Lachapelle G, Petovello M (eds): GNSS Solutions: Reference systems, UTC leap second, and L2C receivers? Inside GNSS, 1(1): 20–24.

Swann J, Chatre E, Ludwig D (2003): Galileo: benefits for location based services. Journal of Global Positioning Systems, 1(2): 57–66.

Szabo DJ, Tubman AM (1994): Kinematic DGPS positioning strategies for multiple reference station coverage. In: Proceedings of the International Symposium on Kinematic Systems in Geodesy, Geomatics and Navigation, Banff, Canada, August 30 through September 2: 173–183.

Teunissen PJG (1993): Least squares estimation of the integer GPS ambiguities. Paper presented at the General Meeting of the IAG at Beijing, P.R. China, August 8–13.

Teunissen PJG (1994): A new method for fast carrier phase ambiguity estimation. In: Proceedings of PLANS'94 – Position Location and Navigation Symposium, Las Vegas, Nevada, April 11–15: 562–573.

Teunissen PJG (1995a): The least-squares ambiguity decorrelation adjustment: a method for fast GPS integer ambiguity estimation. Journal of Geodesy, 70: 65–82.

Teunissen PJG (1995b): The invertible GPS ambiguity transformations. Manuscripta Geodaetica, 20: 489–497.

Teunissen PJG (1996): GPS carrier phase ambiguity fixing concept. In: Kleusberg A, Teunissen PJG (eds): GPS for geodesy. Springer, Berlin Heidelberg New York Tokyo: 263–335 [Bhattacharji S, Friedman GM, Neugebauer HJ, Seilacher A (eds): Lecture Notes in Earth Sciences, vol 60].

Teunissen PJG (1999a): An optimality property of the integer least-squares estimator. Journal of Geodesy, 73: 587–593.

Teunissen PJG (1999b): A theorem on maximizing the probability of correct integer estimation. Artificial Satellites – Journal of Planetary Geodesy, 34(1): 3–9.

Teunissen PJG (2003): Integer aperture GNSS ambiguity resolution. Artificial Satellites – Journal of Planetary Geodesy, 38(3): 79–88.

Teunissen PJG (2004): Penalized GNSS ambiguity resolution. Journal of Geodesy, 78: 235–244.

Teunissen PJG, Verhagen S (2004): On the foundation of the popular ratio test for GNSS ambiguity resolution. In: Proceedings of ION GNSS 2004, 17th International Technical Meeting of the Satellite Division of the Institute of Navigation, Long Beach, California, September 21–24: 2529–2540.

Teunissen PJG, Jonge PJ de, Tiberius CCJM (1994): On the spectrum of the GPS DD-ambiguities. In: Proceedings of ION GPS-94, 7th International Technical Meeting of the Satellite Division of the Institute of Navigation, Salt Lake City, Utah, September 20–23, part 1: 115–124.

Teunissen PJG, Jonge PJ de, Tiberius CCJM (1995): A new way to fix carrier-phase ambiguities. GPS World, 6(4): 58–61.

Teunissen PJG, Jonkman NF, Tiberius CCJM (1998): Weighting GPS dual frequency observations: bearing the cross of cross-correlation. GPS Solutions, 2(2): 28–37.

Tiberius CCJM (1998): Recursive data processing for kinematic GPS surveying. Netherlands Geodetic Commission, Publications on Geodesy, vol 45.

Tossaint M, Samson, Toran F, Ventura-Traveset J, Sanz J, Hernandez-Pajares M, Juan JM, Tadjine A, Delgado I (2006): The Stanford-ESA integrity diagram: focusing on SBAS integrity. In: Ventura-Traveset J, Flament D (eds): The European EGNOS project. ESA Publications Division, Noordwijk, The Netherlands, SP-1303: 55–67.

Townsend BR, Fenton PC, Dierendonck AJ van, Nee DJR van (1995): Performance evaluation of the multipath estimating delay lock loop. Navigation, 42(3): 503–514.

Townsend B, Wiebe J, Jakab A (2000): Results and analysis of using the MEDLL receiver as a multipath meter. In: Proceedings of the 2000 National Technical Meeting of the Institute of Navigation, Anaheim, California, January 26–28: 73–79.

Tranquilla JM, Carr JP (1990/91): GPS multipath field observations at land and water sites. Navigation, 37(4): 393–414.

Tsui JBY (2005): Fundamentals of Global Positioning System receivers: a software approach, 2nd edition. Wiley, Hoboken.

Tziavos IN, Barzaghi R (eds) (2002): International Geoid Service, Bulletin no. 13, Special Issue. Proceedings of EGS 2001 – G7 Session "Regional and local gravity field approximation", Nice, France, March 25–30, 2001.

United Nations (2004): Report of the action team on global navigation satellite systems (GNSS) – Follow up to the Third United Nations Conference on the Exploration and Peaceful Uses of Outer Space (UNISPACE III), Vienna, Austria, July 19–30, 1999.

United States of America, European Community (2004): Agreement on the promotion, provision and use of Galileo and GPS satellite-based navigation systems and related applications. Available at http://ec.europa.eu/dgs/energy_transport/galileo.

US Army Corps of Engineers (2003): NAVSTAR Global Positioning System surveying. Engineer Manual no. 1110-1-1003, Department of the Army, Washington DC. Available at www.usace.army.mil/publications/eng-manulas/em1110-1-1003.

Uttam B, Amos DH, Covino JM, Morris P (1997): Terrestrial radio-navigation systems. In: Kayton M, Fried WR (eds): Avionics navigation systems, 2nd edition. Wiley, New York Chichester Weinheim Brisbane Singapore Toronto: 99–177.

Verhagen S (2004): Integer ambiguity validation: an open problem? GPS Solutions, 8(1): 36–43.

Verhagen S (2005): The GNSS integer ambiguities: estimation and validation. Netherlands Geodetic Commission, Delft, Publications on Geodesy ('Yellow Series'), vol 58.

Verhagen S, Joosten P (2004): Analysis of integer ambiguity resolution algorithms. In: CD-ROM-Proceedings of the European Navigation Conference GNSS 2004, Rotterdam, May 16–19.

Vermeer M (1997): The precision of geodetic GPS and one way to improve it. Journal of Geodesy, 71(4): 240–245.

Vollath U, Birnbach S, Landau H, Fraile-Ordoñez JM, Martín-Neira M (1999): Analysis of three-carrier ambiguity resolution technique for precise relative positioning in GNSS-2. Navigation, 46(1): 13–23.

Vollath U, Buecherl A, Landau H, Pagels C, Wagner B (2000): Multi-base RTK positioning using virtual reference stations. In: Proceedings of ION GPS 2000, 13th International Technical Meeting of the Satellite Division of the Institute of Navigation, Salt Lake City, Utah, September 19–22: 123–131.

Volpe National Transportation Systems Center (2001): Vulnerability assessment of the transportation infrastructure relying on the Global Positioning System. Final Report. Available at www.navcen.uscg.gov.

Wakker KF, Ambrosius AC, Leenman H, Noomen R (1987): Navigation and orbit computation aspects of the ESA NAVSAT system concept. Acta Astronautica 15(4): 195–208.

Walsh D (1992): Real time ambiguity resolution while on the move. In: Proceedings of ION GPS-92, Fifth International Technical Meeting of the Satellite Division of the Institute of Navigation, Albuquerque, New Mexico, September 16–18: 473–481.

Wang J (1999): Modelling and quality control for precise GPS and GLONASS satellite positioning. PhD dissertation, Curtin University of Technology, Australia.

Wanninger L (1997): Real-time differential GPS error modelling in regional reference station networks. In: Brunner FK (ed): Advances in positioning and reference frames. Springer, New York Berlin Heidelberg Tokyo: 86–92 [Mueller II (ed): IAG Symposia Proceedings, vol 118].

Wanninger L (1999): The performance of virtual reference stations in active geodetic GPS-networks under solar maximum conditions. In: Proceedings of ION GPS-99, 12th International Technical Meeting of the Satellite Division of the Institute of Navigation, Nashville, Tennessee, September 14–17: 1419–1427.

Wanninger L (2002): Virtual reference stations for centimeter-level kinematic positioning. In: Proceedings of ION GPS 2002, 15th International Technical Meeting of the Satellite Division of the Institute of Navigation, Portland, Oregon, September 24–27: 1400–1407.

Ward PW, Betz JW, Hegarty CJ (2006) Satellite signal acquisition, tracking, and data demodulation. In: Kaplan ED, Hegarty CJ (eds): Understanding GPS: principles and applications, 2nd edition. Artech House, Boston London: 153–241.

Weber R, Fragner E (1999): Combined GLONASS orbits. In: Slater JA, Noll CE, Gowey KT (eds): International GLONASS experiment (IGEX-98), Workshop Proceedings, IGS Central Bureau, Pasadena, California: 233–246.

Wei J, Xu D, Deng J, Huang P (2004): Synchronization for "Beidou" satellite terrestrial improvement radio navigation system. In: Proceedings of the 2004 International Conference on Intelligent Mechatronics and Automation, Chengdu, China, August.

Wei M, Schwarz KP (1995a): Analysis of GPS-derived acceleration from airborne tests. Paper presented at the IAG Symposium G4, XXI General Assembly of IUGG, Boulder, Colorado, July 2–14.

Wei M, Schwarz KP (1995b): Fast ambiguity resolution using an integer nonlinear programming method. In: Proceedings of ION GPS-95, 8th International Technical Meeting of the Satellite Division of the Institute of Navigation, Palm Springs, California, September 12–15, part 2: 1101–1110.

Wells DE, Beck N, Delikaraoglou D, Kleusberg A, Krakiwsky EJ, Lachapelle G, Langley RB, Nakiboglu M, Schwarz KP, Tranquilla JM, Vanicek P (1987): Guide to GPS positioning. Canadian GPS Associates, Fredericton.

Werner W, Winkel J (2003): TCAR and MCAR options with Galileo and GPS. In: Proceedings of ION GPS/GNSS 2003, Portland, Oregon, September 9–11: 790–800.

Wiederholt LF (2006): Stability measures for frequency sources. In: Kaplan ED, Hegarty CJ (eds): Understanding GPS – principles and applications, 2nd edition. Artech House, Norwood: 665–668.

Wielen R, Schwan H, Dettbarn C, Lenhardt H, Jahreiß H, Jährling R (1999): Sixth catalogue of fundamental stars (FK6), part I: basic fundamental stars with direct solutions. Veröffentlichungen Astronomisches Rechen-Institut, Heidelberg, vol 35, Braun, Karlsruhe.

Wieser A (2001): Robust and fuzzy techniques for parameter estimation and quality assessment in GPS. PhD dissertation. In: Brunner FK (ed): Ingenieurgeodäsie – TU Graz. Shaker, Aachen.

Wieser A (2007a): How important is GNSS observation weighting? In: Lachapelle G, Petovello M (eds): GNSS Solutions: Weighting GNSS observations and variations of GNSS/INS integration. Inside GNSS, 2(1): 26–33.

Wieser A (2007b): GPS based velocity estimation and its application to an odometer. In: Brunner FK (ed): Ingenieurgeodäsie – TU Graz. Shaker, Aachen.

Wieser A, Gaggl M, Hartinger H (2005): Improved positioning accuracy with high-sensitivity GNSS receivers and SNR aided integrity monitoring of pseudo-range observations. In: Proceedings of ION GNSS 2005, 18th International Technical Meeting of the Satellite Division of the Institute of Navigation, Long Beach, California, September 13–16: 1545–1554.

Wigger P, Hövel R vom (2002): Safety assessment – application of CENELEC standards – experience and outlook. Copenhagen Metro Inauguration Seminar, November 21–22.

Wilde W de, Sleewagen JM, Wassenhove K van, Wilms F (2004): A first-of-a-kind Galileo receiver breadboard to demonstrate Galileo tracking algorithms and performances. In: Proceedings of the 2nd ESA Workshop on Satellite Navigation User Equipment Technologies, NAVITEC 2004, ESTEC, Noordwijk, The Netherlands, December 8–10.

Wilde W de, Sleewaegen JM, Simsky A, Vandewiele C, Peeters E, Grauwen J, Boon F (2006): New fast signal acquisition unit for GPS/Galileo receivers. In: Proceedings of the European Navigation Conference GNSS 2006, Manchester, United Kingdom, May 7–10.

Wiley B, Craig D, Manning D, Novak J, Taylor R, Weingarth L (2006): NGA's role in GPS. In: Proceedings of ION GPS 2006, 19th International Technical Meeting of the Satellite Division of the Institute of Navigation, Fort Worth, Texas, September 26–29: 2111–2119.

Willis P, Boucher C (1990): High precision kinematic positioning using GPS at the IGN: recent results. In: Bock Y, Leppard N (eds): Global Positioning System: an overview. Springer, New York Berlin Heidelberg Tokyo: 340–350 [Mueller II (ed): IAG Symposia Proceedings, vol 102].

Witchayangkoon B (2000): Elements of GPS precise point positioning. PhD dissertation, University of Maine, Orono, Maine.
Available at www.spatial.maine.edu/SIEWEB/thesesdissert.htm.

Wolfe DE, Gutman SI (2000): Developing an operational, surface-based, GPS, water vapor observing system for NOAA: network design and results. Journal of Atmospheric and Oceanic Technology, 17(4): 426–440.

Wu JT, Wu SC, Haj GA, Bertiger WI, Lichten SM (1993): Effects of antenna orientation on GPS carrier phases. Manuscripta Geodaetica, 18: 91–98.

Wübbena G, Schmitz M, Menge F, Seeber G, Völksen C (1997): A new approach for field calibration of absolute GPS antenna phase center variations. Navigation, 44(2): 247–255.

Wübbena G, Schmitz M, Menge F, Böder V, Seeber G (2000): Automated absolute field calibration of GPS antennas in real-time. In: Proceedings of ION GPS 2000, 13th International Technical Meeting of the Satellite Division of the Institute of Navigation, Salte Lake City, Utah, September 19–22: 2512–2522.

Wübbena G, Bagge A, Schmitz M (2001): Network-based techniques for RTK applications. Paper presented at the GPS Symposium, GPS JIN 2001, GPS Society, Japan Institute of Navigation, November 14–16.

Wunderlich T (1992): Die gefährlichen Örter der Pseudostreckenortung. Habilitation Thesis, Technical University Hannover.

Xu G (2003): GPS – theory, algorithms and applications. Springer, Berlin Heidelberg New York.

Young R, McGraw GA (2003): Fault detection and exclusion using normalized solution separation and residual monitoring methods. Navigation, 50(3): 151–169.

Zhang W, Cannon ME, Julien O, Alves P (2003): Investigation of combined GPS/Galileo cascading ambiguity resolution schemes. In: Proceedings of ION GPS/GNSS 2003, Portland, Oregon, September 9–11: 2599–2610.

References

Zhu J (1993): Exact conversion of earth-centered, earth-fixed coordinates to geodetic coordinates. Journal of Guidance, Control, and Dynamics, 16(2): 389–391.

Zhu SY, Groten E (1988): Relativistic effects in GPS. In: Groten E, Strauß R (eds): GPS-techniques applied to geodesy and surveying. Springer, Berlin Heidelberg New York Tokyo: 41–46 [Bhattacharji S, Friedman GM, Neugebauer HJ, Seilacher A (eds): Lecture Notes in Earth Sciences, vol 19].

Zinoviev AE (2005): Using GLONASS in combined GNSS receivers: current status. In: Proceedings of ION GNSS 2005, 18th International Technical Meeting of the Satellite Division, Long Beach, California, September 13–16: 1046–1057.

Zogg JM (2006): Grundlagen der Satellitennavigation, User's guide. u-blox GPS-X-01006-B1. Available at http://telecom.tlab.ch/~zogg.

Zolesi B, Cander LR, Belehaki A, Tsagouri I, Pezzopane M, Pau S (2005): Geomagnetic indices forecasting and ionospheric nowcasting tools. Paper presented at the 2nd European Space Weather Week, Noordwijk, The Netherlands, November 14–18.

Zumberge JF, Heflin MB, Jefferson DC, Watkins MM, Webb FH (1997): Precise point positioning for the efficient and robust analysis of GPS data from large networks. Journal of Geophysical Research, 102(B3): 5005–5017.

Subject index

Numbers in boldface indicate headings and, thus, principal coverage of a topic.

Abel inversion 446
absorption 36, 63
accelerated motion 16, 249
accelerated reference frame 144
accelerometer 307, 452
accuracy measure **272**
accuracy of carrier phase 11
accuracy of code range 11
accuracy of pseudorange 9
ACF, *see* autocorrelation function
across-track component 32
active control network 457, **458**
active ranging system 55
A/D, *see* analog to digital
adjustment theory **238**
affine transformation 297, 300, 443
AFS, *see* atomic frequency standard
AGNSS, *see* assisted GNSS
aiding information 91, 93, 99, 103, 430
aircraft navigation 460
air drag 35
AL, *see* alarm/alert limit
alarm/alert limit 269, 270, 371, 382
albedo 39
aliasing 71, 88
Allan variance 68
almanac data **49**, 50, 85, 264, 338, 361, 414, 422, 429, 451
along-track component 32
AltBOC, *see* alternative BOC
alternative BOC 82, 101, 386, 390
altimetry 138, 345, 465
altitude 4, 6, 35, 92, 124, 140, 145, 322, 348, 358, 375, 377, 402, 404, 405, 423, 447, 470
ambiguity decorrelation 219, **227**
ambiguity function 188, 205, 218, **219**, 221, 222

ambiguity resolution 47, 112, 138, 168, 173, 187, **202**–206, 210, **213**, 214, **217**–219, **223**, 234, 236, 237, 329, 339, 384, 417, 428, 437, 444, 458
ambiguity search 205, 206, 219, **222**, **225**, 228, 231–234
ambiguity validation **236**
analog to digital 71, 85, 88, 90
angular momentum 14–17, 41
angular velocity 23, 42, 49, 57, 314, 345
anomaly 28–30, 32, 41, 45, 50, 51, 146, 147, 410, 411, 432
antenna area 64
antenna calibration 149, **150**, 152
antenna design **87**, 110, 158
antenna gain 64, 65, 73, 87, 158, 358
antenna ground plane 158
antenna height 149, 158
antenna offset 110, 433
antenna orientation 149, 151
antenna phase center 109, **148**–150, 152, 156, 159, 168, 169, 191
antenna swap 187, 206
antijamming 328, 332, 340
antipodal position 357
anti-spoofing 315, 319, **322**, 327, 334, 348, 456
apex 10, 263, 284
apogee 28, 409
a posteriori variance 223, 224, 240, 241
apparent sidereal time 23
Appleton layer 66
a priori variance 224, 239, 241
area-to-mass ratio 39
ARGOS **404**
argument of latitude 52, 349, 402, 432
argument of perigee 28, 50, 51, 412, 432
ARNS 61, 329, 331, 336, 357, 372, 383, 384

A-S, *see* anti-spoofing
ascending node 20, 28, 31, 41, 42, 349, 403, 412, 432
assistance system **429**, 463
assisted GNSS 429, 430, 464
astronomic position 1, 281
atmosphere 63–**65**, 67, 128, 129, 135, 140, 270, 292, 445, 446, 468
atmospheric effects **116**, 267, 411, 445
atmospheric pressure 129, 135, 167
atomic clock 6, 67, 88, 316, 349, 406, 411
atomic frequency standard 67, 68, 329, 356, 369, 375, 376, 383
atomic time 22, **23**, 315, 325, 347, 369, 410
attenuation 63–65, 73
attitude **10**–12, 234, 342, 350, 351, 377, **441**–444, 447, 452, 461, 465, 468
attraction 16, 35, 318, 468
augmentation system 6, 194, 414, 417, **420**, **421**, 424–425, **426**–**428**, 460
autocorrelation function 70, 71, 77, 81, 82, 92
auxiliary output chip 456
aviation application **460**
azimuthal projection 284, 285

Band-pass filter 69, 88, 89, 376
band-stop filter 69
bandwidth 70, 75, 76, 80, 81, 86, 98–101, 329–332, 355, 358, 359, 384, 385, 390
barometer 408, 415, 452
barycenter 13
barycentric dynamic time 22, 23
baseline error 27, 319
baseline length 188, 202, 203, 209, 210, 437, 439, 465
baseline solution 152, 196, 202, **257**, 258, 262
beat frequency 97, 98
beat phase 97, 98, 106–108, 194
Beidou 6, 310, **401**–403, **406**–**408**, 409, 414, 415, 426, 450
BER, *see* bit error rate
Bessel ellipsoid 293
best-fitting ellipsoid 293

bidirectional communication 404, 414, 415
Big Dipper 402
binary offset carrier 80–82, 94, 100, 101, 331, 335, 336, 361, 387–390, 413
binary phase-shifted key 74, 80–82, 94, 100, 101, 331, 332, 334, 357, 358, 386, 389, 413, 422
bit error rate 68, 73, 84, 103, 337, 338, 360, 392
Block I–III satellite 310, 323, 324, 326, 328, 331, 332, 335–337
BOC, *see* binary offset carrier
Boltzmann constant 86
boundary value problem 40, **42**, 43
BPF, *see* band-pass filter
BPSK, *see* binary phase-shifted key
broadcast ephemerides 49, **50**, 51, 170, 326, 337, 338, 361, 431, 432, 434

C/A-code 316, 319, 322, 330, 331, **332**–334, 336, 337, 347, 355, 357, **358**, 360, 362, 370, 400, 422, 431, 441, 448, 454–456
C/A-code pseudorange 319, 431, 454
calendar **24**, 125
calibration 138, 148–**150**, 151–153, 407, 409, 428, **455**
capacity **268**, 373, 374, 393, 397, 407, 460–462
carrier-aided code loop 100
carrier-aided tracking 93, 103
carrier frequency 68, **73**–75, 84, 96, 97, 104, 107, 128, 148, 149, 152, 327, 328, 330, 332, 334–336, 339, 354, 355, 357, 361, 382, 383, 387–390, 411, 413
carrier phase 11, 12, 48, 85, 93, 97, 98, 100, 102, 103, 107, 110, 111, 113–116, 119, 128, 149, 154–156, 159, **163**, 165–167, 172, 176, 189, 195, 197, 202, 203, **209**–211, 217, 218, 223, 235, **254**, 255, 329, 334, 384, 428, 433, 437, 441, 448, 449, 456, 458, 464, 465
carrier phase difference 114
carrier phase pseudoranges 113, 119

Subject index

carrier-to-noise power density ratio 86, 111, 332
carrier wave 84, 111, 116, 126, 316, 349
carrier wipe-off 98
Cartesian coordinates 22, 262, **277**, 278, 280, 293, 297, 299, 305, 345, 408
C-band 60, 65, 73, 376, 378, 380, 382, 383, 395, 421
CDMA, see code division multiple access
celestial body 22, 37
celestial ephemeris pole 16, 17, 20, 21
celestial reference system **16**, 17
central acceleration 37, 45
central force 36
central meridian 284, 286, 289
CEP, see celestial ephemeris pole
cesium 67, 316, 337, 349–351
CHAMP mission **140**, 292, 446
Chandler period 15
channel 74, 90, **91**, 92, 94, 96, 98, 103, 332, 334–336, 354, 356–358, 360, 374, 385–387, 389, 390, 392, 394, 407, 411, 426–429, 449, 454
channel multiplexing **91**
chip-by-chip time-multiplexing 331, 334
chip length 70, 75, 79–81, 84, 94, 100, 106, 110, 319, 332, 360, 386, 392
chipping rate 75, 80, 90, 101, 334, 337, 358, 360, 387, 389, 390
chi-square distribution 274
choke ring antenna 87, 158, 444
Cholesky factorization 224, 225
Cicada, see Tsikada
CIGNET, see cooperative international GPS network
CIO, see conventional international origin
circular error probable 275
circular frequency 57, 330
civil date 24, 25
Clarke ellipsoid 291, 293
climate monitoring 445
clock bias 4, 8, 49, 105, 109, 159, 161–165, 171, 172, 175, 189, 208, 252, 316, 319, 435
clock correction 49, 52, 162, 168, 169, 252, 328, 338, 352, 354, 361, 377, 379, 418, 435

clock drift 50, 51, 161, 166
clock error 4, 8, 9, 52, 105–107, 110, 147, 154, 166, 167, 170, 189, 221, 241, 253, 381, 411, 417, 423, 426, 445
clock frequency 51, 146
clock offset 50–52, 150, 161, 166, 202, 353, 380, 381, 444, 445
clock parameter 7, 48, 52, 205, 325, 326, 413, 414, 417
clock polynomial 147
C/N_0 86, 92, 332
code correlation 93, 101, 105
code data **209**–211, 224
code division multiple access 76, 84, 329, 355, 363, 384, 412
code generation 78, 333, 388
codeless technique 66, **101**, 102, 200, 309, 334, 456
code modulation **80**, 94, 413
code phase 94, 113
code pseudorange **11**, 67, **105**, 108, 109, **113**–116, 119, 120, 127, 155, 161, 166, 167, 171, 197, 319, 416, 431, 435, 436, 448, 454, 456
code pseudorange smoothing **113**
code range 11, 12, 48, 85, 90, 93, 102, 103, 110, 111, 128, 155, **161**, 163, 165, 166, **170**–174, 183, 195, 197, 198, 204, 209, 212, 217, 222, **252**–255, 262, 417, 433, 437, 449, 456, 458
code range noise 198
code tracking filter 100
cofactor matrix 223, 227, 229, 239, 240, 242–247, 249, 263–265, 269
coherent integration time 95, 96, 385, 429
coherent tracking 93, 103
cold start 91, 391, 454
commercial service **370**, 380, 381, 417
common point 259, 295–301
communication link 7, 55, 104, 325, 404, 414, 415, 429, 447
Compass 6, 307, **401**–403, 406, 450, 452, 467
confidence level 267, 273–276
conformal coordinates 289
conformal Lambert projection 284
conformal mapping 283, 284
conformal transformation 21

conical projection 284
continuity 69, **268**–270, 370, 371, 376, 379, 400, 410, 421, 427, 460, 461
control network 300, 433, **457**, **458**, 464
control segment 5, 7, 47, 50, 310, **324**–326, 342, **351**–**354**, 369, 378, 379, 470
control segment modernization **325**, **354**
conventional celestial reference system 16
conventional international origin 16, 17, 20, 21
conventional terrestrial reference system 16
convolutional encoding 103, 104, 338, 392, 393, 422
cooperative international GPS network 47
coordinate frame 13, 22, 277, 301, 345, 400
coordinate system 13–**15**, 17, 18, 21, 30, 31, 151, 228, 265, 269, 277, 280, 281, 290, 293, **302**, 303, 305, **313**, 342, **345**, 361, **369**, 400, 401, 406, 410
coordinate transformation **277**, 293, 400, 401
correlation function 69, 71–73, 94, 99, 159
correlation matrix 180, 181
correlation of double-differences **180**
correlation of single-differences **179**
correlation of triple-differences **181**
correlation spacing 71, 72, 96, 99, 100, 159
correlation technique 55, 456
COSPAS 372, 381, **405**, 406
Costas loop 98
covariance matrix 178–180, 205, 223, 224, 227, 228, 237, 239, 241, 244, 245, 249, 261, 423
covariance propagation 181, 240, 245, 258, 265
Cramer–Rao lower bound 101
crosscorrelation 69, 77–80, 92, 94, 95, 101, 183, 331, 332, 334–336, 355, 359
crustal deformation 465
cryosphere 292
crystal clock 3
cycle slip 113–115, **194**–202, 205, 222, 261

cylindrical projection 284

Data combination **111**, 159, 191, 213, 339
data exchange **193**, 431
data message 7, 55, 74–76, 83, 84, 90, 92, 97, 100, 104, 271, 316, 324, 330, 331, 334, 337–**339**, 371, 390, 392, 393, 415, 422, 427
data processing 140, 158, 159, 166, **193**, 431
data sampling rate 114
data transfer 437, **447**, 455
data transformation **277**
date conversion **25**
datum 2, 21, 22, 291–**293**, 295, 299, 300, 302, 457
datum transformation 21, 22, **293**, 295, 302
decomposition 228, 229, 235, 237
delay lock loop 97, 99–101, 159
DEM 407, 408, 415
demodulation 76, 96, 103, 104, 390, 391, 393, 394, 429
denial of accuracy **319**
density function 236, 272, 273, 275
design matrix 164, 174, 223, 226, 234, 238, 254, 261–263, 265, 295, 296, 306, 435
DGNSS 169, **170**–**172**, 415–417, 419, 420, 436, 447, 458, 468
DGPS, see differential GPS
differential correction 415, **416**–420, 426–429, 436, 447, 448, 450
differential GPS 407, 418, 419, 436, 437, 448
differential positioning 11, **169**–171, 417, **436**
differential system 415, **417**, 418, 420, 436
diffraction 61
diffuse reflection 61
digital signal processor 85, **90**, 92
dilution of precision 9, 110, 203, 222, 236, **262**–267, 283, 376, 400, 435, 440, 451
dipole antenna 149
discriminator function **71**–73, 82, 98–100, 335

Subject index

dispersion matrix 239
dispersive medium 62, 66, 118
distress alerting satellite system 406
disturbing potential 35–38, 43, 44, 307
dithering 319
DLL, *see* delay lock loop
DOP, *see* dilution of precision
Doppler, integrated 108, 115, 196, 198
Doppler data **108**, **165**, 166
Doppler effect 40, **143**
Doppler equation 166
Doppler frequency shift **59**, 446
Doppler shift 3, 9, 59, 89, 91–93, 97, 98, 100, 108, 109, 111, 114, 115, 165, 404, 411, 440
DORIS **403**, 446
double-difference 109, 149, 168, 174, **175**–177, **180**, 181, 184–188, 196, 197, 199, 202, 203, 206, 214–219, 222–227, 233–236, 255, 257, 259
downconversion 88–90, 96
downlink 55, 372, 383, 397, 402, 406, 414
dry atmosphere 129
dry component 65, 129–132, 135, 136
dry refractivity 130, 133
DSP, *see* digital signal processor
dual-frequency code 166, 210
dual-frequency data 109, 111, 206, 224, 225, 233, 318
dual-frequency phase 138, 196, **206**, 224
dual-frequency receiver 5, 48, 261, 318, 417, 418, 423, 433, 438–440, 454, 458, 465
dual-use system 310, 343, 354, 362, 402, 407, 456
dwell time 94
dynamic time 22, **23**, 24, 28

Early-minus-late power discriminator 100
earth-centered, earth-fixed 22, 431
earth rotation 21–23, 52, 147, 432, 435, 459, 465
earth's gravitational constant 145, 314, 345
earth's gravity field 44
earth's potential 36, 45
earth's rotation axis 15, 16, 23

earth's rotation vector 13, 14, 147
eccentric anomaly 28, 30, 32, 41, 146, 147, 432
eccentricity 28–30, 41, 50, 51, 140, 146, 278, 280, 375, 411, 412
ECEF, *see* earth-centered, earth-fixed
ECEF coordinates 161, 191, 278, 280, 345, 431
eclipse factor 39
ecliptic 13, 19
EGNOS 6, 365, 370, 424, **425**, 426, 454
electric field 56, 63, 145
electric forces 56
electric induction 56
electromagnetic spectrum **60**, 89
electromagnetic wave 56, **57**, 59, 61–63, 65–67, 73, 87, 101, 108, 116
electron density 65, 66, 118, 119, 123–126, 138, 140
electronic toll system 394, 395
elevation angle 63, 67, 110, 132, 134, 137, 159, 168, 411, 442, 447
elevation mask 323, 375, 380
ellipsoidal coordinates 22, 262, **277**, 278, 281, **283**, 289, 293, 297, 299, 305, 435, 451
ellipsoidal height 278, 279, 290, 293, 296, 297, 299–301, **305**, 408, 457
emission time 105, 106
emitted frequency 59, 67, 108, 146, 440
emitter 58, 59, 63, 143
ephemerides 3, 5, 7, 27, **49–52**, 55, 91, 104, 110, 111, 165, 170, 241, 283, 316, 319, 325, 326, 334, 337–339, 351, 352, 354, 355, 361, 362, 377, 379, 381, 390, 391, 393, 414, 417, 418, 421–423, 426, 429–434, 436, 448, 449, 454, 459
equatorial plane 13, 22, 45, 302, 345, 374, 414
equatorial system 31–33, 40, 43, 265
equipotential surface 291, 292
Euler angles 441
Eurofix 428
European Tripartite Group 425
EutelTRACS **415**

Fading 63, 96

Faraday rotation 56
FDMA, *see* frequency division multiple access
FEC, *see* forward error correction
Federal Aviation Administration 424, 427, 460
Fermat's principle 62, 119
fiducial points 293
filter 50, 69, 71, 83, 88, 89, 92, 98–103, 159, 199, 206, 219, **225**, 226, **244**, 246–248, 271, 326, 357, 376, 433, 451, 464
filtering 69, 71, 88, 158, 159, 167, 199, **238**, 241, 244, 247, 248, 250, 433
fixed ambiguity 214, 215
flattening 314, 345, 406, 410
flex-power 331, 335
FLL, *see* frequency lock loop
float ambiguity 204, 214–216, 237
FOC, *see* full operational capability
forward error correction 103, 338, 339, 392
Fourier transform 69, 70
fractional phase 107, 194
frame synchronization 103, 422
free space loss 73
frequency allocation 60
frequency diversity 104, 339, 394
frequency division multiple access 84, 354–357, 363, 428
frequency domain 68, 69, 82, 83, 94
frequency drift 51, 67, 68, 107, 440
frequency lock loop 98, 99
frequency offset 67, 92–94, 97, 98, 107, 329, 384
frequency response roll-off 93
frequency shift 40, **59**, 146, 147, 446
frequency spectrum 44, 60, 69, 70, 74–76, 80, 89, 96, 100
frequency stability 107, 316
frequency standard **67**, 85, 88, 92, 316, 328, 329, 353, 356, 369, 376, 411, 465
Friis transmission formula 64
full operational capability 310, 322, 324, 329, 336, 340, 342–344, 355, 368, 394, 395, 424–426

fundamental frequency 316, 319, 329, 383, 412

GAGAN 424, **426**
gain matrix 243, 246, 247, 249, 250
gain pattern 87, 158
Galilei transformation 142
Galileo **365**
Galileo Joint Undertaking 367, 368
Galileo operating company 368
Galileo services 368, **370**, 371, 384, 395, 457
Galileo terrestrial reference frame 17, 369, 400
Gamma-function 274
GATE 376
Gaussian distribution 86, 92, 95, 239, 245, 267, 270, 274, 381, 382, 423
Gaussian equations 35, 43
Gauss–Krüger coordinates 287, 289, 293
Gauss–Krüger projection 284, 287
Gauss–Markov model 239
GBAS, *see* ground-based augmentation
GDGPS 418
general relativity 13, 23, 68, **144**, 146
geocentric angle 37
geocentric distance 30, 36, 41, 42, 51, 52, 145
geocentric position 3, 37–39, 147, 161
geocentric system 13, 16, 40, 277, 278, 295, 306, 313, 345
geodetic reference system 280, 291, 410
geographic information system 465, 467
geoid 290, **291**, 292, 300–302, 457
geoidal height 290, 293, 297, 299–302, 451, 457
geomagnetic pole 123
GEO-ranging 421
GEOSTAR 402, **414**, 415
geostationary satellite 402, 405, 407, 409, 415, 418, 421, 424–426
GIOVE 375, 376, 394
GIS, *see* geographic information system
GJU, *see* Galileo Joint Undertaking
global coordinates **280**, 293
global ellipsoid 291
global frame 277
GLONASS **341**

GLONASS modernization 342
GLONASS services 347
GOC 368, 370
Gold code 78–80, 94, 332, 333
GPS 309
GPS III 324, 336, **340**
GPS modernization **310**, 311, 325, 326, 340
GPS services 315
GPS time 315, 337, 401, 410
GPS to Galileo time offset 369, 401
GPS week 25, 49–51, 314, 315, 338
GRACE 292, 446
gradiometry 465
Gram–Schmidt orthogonalization 257
gravitational constant 27, 28, 145, 314, 345, 432
gravitational field 144–146, 292
gravitational force 14
gravitational potential 68, 144
gravitational resonance 375
gravity 13, 39, 44, 45, 140, 292, 307
Greenwich hour angle 22
Greenwich meridian 13, 16, 22, 302, 345
Greenwich sidereal time 13, 17, 20, 22, 31, 49
ground-based augmentation 6, **426**, 427
ground control 352, 378, 379, 403
ground plane 87, 158, 444
ground wave 62
group delay 67, 119, 328, 338, 356
group velocity 66, **116**, 117
GRS-80 ellipsoid 280, 287, 289, 291, 293, 410
GTRF 17, 277, 369, 380, 400, 410

Hadamard variance 68
hand-over word 334, 337, 360
hazardously misleading information 270
heading 10, 441, 444
health status 337, 338, 360
height difference 291, **301**, 302, **305**, 466
height transformation **290**
Helmert transformation 293, 297, 400
high-accuracy signal 347, 348, 354, 355, 357, **360**, 362

high-sensitivity GNSS 464
HMI, *see* hazardously misleading information
Hopfield model **130–132**, 134
horizontal accuracy 169, 188, 347, 348, 428, 437, 439, 456
horizontal position 265, 266, 320, 431, 439
hot start 91
hour angle 22
HOW, *see* hand-over word
HPL 269, 270
Huygens–Fresnel's principle 61
hydrogen maser 67, 316, 346, 369, 376, 377
hydrosphere 292
hydrostatic atmosphere 129
hydrostatic component 65

ICAO 341, 343, 365, 372, 395, 427, 460
IERS, *see* International Earth Rotation Service
IGS, *see* International GNSS Service for Geodynamics
IMO 266, 365, 372, 395
inclinometer 307
indirect effect 35, 38, 39
inertial navigation system 11, 99, 202, 324, **452**, 468
inertial system 13, 14, 142, 452
inertial time 23, 27
information service 7, 48, 50, 315, **458**, 459, 461
initial integer number 107, 194
initialization 165, **187**, 206, 213, 247, 333, 335, 359, 360, 433, 438, 443
initial operational capability 310, 324, 340, 424, 425
initial value problem **40**
in-orbit validation 367, 368, 374–376, 394
INS, *see* inertial navigation system
instantaneous ambiguity resolution 187, 210
instantaneous navigation 12
instantaneous position 13, 15, 28, 165, 432
integer ambiguity 107, 114, 163, 189, 203–205, 214–217, 225, 236, 237
integrated water vapor 140

integration time 83, 95, 96, 99, 100, 385, 428, 429
integrity flag 271, 379, 381, 382
integrity monitoring 270, 271, 371, 414, 425, 427
intelligent transportation systems 467
interchannel biases 91, 241
interference signal 76, 83, 87
interferometry 16
interleaving 103, 392, 393
intermediate frequency 85, 88–90, 92, 96
intermodulation product 389, 390
international atomic time 22, 24, 315, 347, 369, 410
International Earth Rotation Service 16, 21, 24, 28, 194, 314, 345, 346, 369
International GNSS Service for Geodynamics 7, 48, 52, 139, 148, 166, 169, 194, 321, 346, 458, 459
International Telecommunication Union 60, 61, 68, 70, 83, 84, 124–126, 329, 330, 358, 375, 383, 384, 402
IOC, *see* initial operational capability
ionosphere-free combination 127, 128, 152, 166, 167, 209, 261, 329, 465
ionosphere map 138, 139
ionosphere term 127, 128, 197, 207, 209, 211
ionospheric aberration 465
ionospheric coefficients 122
ionospheric delay 66, 73, 339, 423, 428, 448
ionospheric effect 113, 128, 210, 324, 456
ionospheric path delay 121
ionospheric point 120–123, 139, 423
ionospheric refraction 66, 109, **118**, 119, 121–123, 126, 138, 140, 155, 197, 206, 208, 316, 331, 465
ionospheric residual 113, 122, 138, 197–201
ionospheric scintillation 63, 66
ionospheric tomography **138**
IOV, *see* in-orbit validation
IRNSS **414**, 450
ITRF 16, 17, 22, 48, 314, 345, 346, 369, 380, 400, 401, 410, 458

ITU, *see* International Telecommunication Union

Jamming **83**, 104, 269, 319, 332, 335, 372
Japanese geodetic system 410
JD, *see* Julian date
JPO 309, 315, 327
Julian date 24, 25, 315

Kalman filter 50, 159, 225, 226, **244**, 246–248, 271, 326, 433, 464
Kennelly–Heaviside layer 65
Kepler ellipse 32, 40, 43–45, 50
Keplerian elements 35, 125, 361
Keplerian orbit 28–30, 33, **40**
Keplerian parameters 32, 35, 40–44, 49, 50, 391
Kepler's equation 29
Kepler's law 3
kinematic application 168, 172, 204, 206, 210, 234, 253, 428
kinematic initialization **187**, 206
kinematic method 187, 438, 439
kinematic mode 12, 167, 440
kinematic point positioning 163, 165, 171, 172
kinematic relative positioning **185**, 438
kinetic energy 30, 144
Klobuchar model 122, 123
Krassovsky ellipsoid 406

LAAS **427**
LAD 417
LADGNSS 172, 173
Lagrange equations 35, 43
Lagrange interpolation 52, 53
LAMBDA method 234, 236, 237
Lambert projection 284
LBS, *see* location-based services
L1C-code **336**, 339
L2C-code 322, 324, 331, **334**
L5C-code 331, **335**
leap second 23, 24, 346, 347, 350, 361, 369

least-squares adjustment 122, 138, 151, 167, 187, 191, 204, 211, 212, 214, 215, 223, 224, 227, 228, 230, 233, 234, 236–**241**, 242, 244, 246, 254, **261**, 295, 296, 298, 301, 444
least-squares collocation 191, 307
Legendre function 36, 139
Legendre polynomial 36
level of crosscorrelation 79, 80, 95
LFSR, *see* linear feedback shift register
linear combination 111, **112**, 113, 127, 164, 174, 184, 191, 207
linear error probable 274
linear feedback shift register 78, 79, 332–335, 358–360, 385, 386, 390
linear frequency 57
linearization 223, 234, 238, **250**, 254, 255, 257, 261, 295, 443
line of position 10
line of sight 1, 10, 59, 62, 63, 67, 72, 73, 121, 131, 132, 149, 159, 169, 262, **282**, 423, 446, 451, 465
line-of-sight signal 73, 465
line-of-sight wave 62, 63
local-area augmentation 427
local-area DGNSS **172**
local-area differential 417
local coordinate frame 277, 301
local coordinate system 265, 269
local datum 299, 300
local ellipsoid 290, 291, 293, 299
local-level coordinates **280**, 281, 293, 303, 304, 442, 443
local-level frame 10, 435, 441–443, 452
local oscillator 88, 94
location-based services 460, **463**, 464
LOCSTAR **414**, 415
Loran-C 428, 447
Lorentz contraction **143**
Lorentz transformation **141**, 142, 143
loss of lock 66, 83, 107, 155, 438
Lunar laser ranging 16

Magnetic field 35, 56, 66, 125, 140, 465
Manchester coding 81
mapping **135**, **464**

map projection 283
maritime application 374, 420, 448, 460, **461**
maser 67, 316, 346, 369, 376, 377
mask angle 264, 265, 317, 375, 380
master control station 7, **324**–326, 411, 414
Maxwell's equations 56
MBOC, *see* multiplexed binary offset carrier
M-code 322, 324, 328, 331, **335**, 336, 340, 367
meaconing **83**, 372
mean angular velocity 42
mean anomaly 28–30, 45, 50, 51, 432
mean motion 20, 28, 51, 432
mean radial spherical error 276
mean sidereal time 22, 23
measurement-domain approach 416, 417, 436
Mercator projection **284**, 286
message decoding 103
meteorological data 129, 137, 140, 193, 194, 325
meteorology 123, 454, 465
modulation method **73**, 74, 390
moment of inertia 23
monitor station 7, 47, 50, 172, 314, 324, **325**, 326, 342, 458, 468
moon's node 20
MSAS 6, 424, **425**, 454
multipath 62, **72**, 73, 87, 92, 99, 109–111, 151, **154**–160, 169–172, 191, 195, 197, 200, 203, 204, 210, 213, 217, 235, 241, 267, 270, 271, 317, 381, 382, 390, 421, 423, 428, 432, 436, 440, 444, 464
multiple carrier ambiguity resolution **213**
multiplexed binary offset carrier 82, 83, 331, 336, 386, 388, 389
multipoint **258**, **261**, 262

NAD-27, *see* North American Datum 1927
narrow correlation spacing 72, 159
narrow correlator 82, 99, 217
narrow lane 112
national DGNSS 419, 458

National Marine Electronics Association 447, **450**, 451
navigation accuracy 5, 319
navigation message 76, 85, 96, 103–105, 122, 123, 125, 147, 161, 193, 194, 252, 315, 319, 325, 326, 328–330, 332, 334, 335, **337**, 338, 346, 349–351, 355, 357, **360**–362, 369, 370, 372, 374, 376, 381, 382, 384, 385, 387, **390**–394, 400, 401, 410, 411, 413, 414, 429, 453
navigation receiver 309, 343
navigation service 101, 328, 363, 366, 370, 374, 409, 410
NAVSTAR 5, 309, 459
NDGPS 419
near/far interference problem 428
NeQuick model 123–126, 371
network adjustment **257**
neutral atmosphere 65, 128
Newtonian mechanics 13, 27, 142
NMEA format **450**
noise level 86, 90, 98, 100, 107, 112, 155, 197, 198, 210
noise power 86–88, 96, 111, 168, 332
noise vector 249
normal distribution 178, 269, 270, 273, 274, 423
North American Datum 1927 291
NTRIP 449
null steering 87, 88
numerical eccentricity 28, 29, 278, 280
nutation 15, 17, **19**, 20, 23
Nyquist theorem 90

Oblateness term 36, 37, 44
obliquity 19, 131, 423
observables **105**
observation equation 38, 88, 109, 163, 166, 183, 185, 188, 204, 225, 226, 239, 244, 246, 249, 256, 257, 262, 307, 401, 408, 440
observation window 465
occultation 2
oceanic tides 38
oceanography 468

omnidirectional antenna 87
OmniSTAR 417
OmniTRACS **415**
one-dimensional transformation **300**, 301
one-way ranging system 55, 105, 397, 402–404
on-the-fly 172, 187, 206, **217**–219, 417, 438, 444
open service **370**, 402
orbital coordinate system 31
orbital error 27, 109, 170, 171, 203, 204, 320, 400, 428
orbital plane 28, 30–32, 35, 39, 41, 42, 52, 264, 322, 323, 348, 357, 374, 375, 402, 411, 414, 431, 432, 470
orbit determination 37, **39**, 40, 42, 47, 48, 352, 377, 379, 403, 404, 409
orbit dissemination **47**, 338
orbit improvement **43**, 293
orbit representation **30**
orthometric height 290, 293, 301
oscillations of axes **15**
osculating ellipse 33
OTF, *see* on-the-fly
outage 7, 261, 268, 323, 458
out-of-band 70, 71, 83, 84, 88, 90, 357, 376

Parceval's theorem 69
partial water vapor pressure 65
passive control network **457**, 458
path delay 121, 128, 130–133, 137, 140, 167
P-code 101, 102, 316, 318, 319, 322, 327, 330, **333**–335, 337, 338, 347, 348, 355, 357, 358, **360**, 362, 448, 455, 456
PE-90 17, 277, 345, 346, 361, 401
perigee 28, 30, 42, 43, 45, 50, 51, 409, 412, 432
perturbed orbit **43**
phase advance 67, 119
phase ambiguity 163, 171, 195, 208, 428
phase angle 57
phase center 109, **148**–150, 152, 156, 158, 159, 168, 169, 191, 444
phase combination 157, **178**, 195–197, 199, 208, 209, 255, 329, 384, 437

Subject index

phase equation 58, 106, 174, 175, 208, 212, 226
phase lock loop 97–99
phase model 107, 196, 207
phase modulation 73, 74, 390
phase pseudorange 11, 67, **106**, 107, 109, **111**–113, 116, 119, 120, 126, 149, 152, 165, 171, 189, 417
phase refractive index 118, 119
phase shift 74, 89, 96, 98, 99, 156, 157
phase velocity 62, 66, 67, 116, 117
phase wind-up 145, 168
phase wrap-up 145
pilot signal **83**, 99, 332, 334–336, 385, 387, 389, 390, 429
PL, *see* protection level
plane coordinates 191, 277, **283**, 293, 296, 297, 299
plasma frequency 122
plate tectonics 16, 465
PLL, *see* phase lock loop
point positioning 11, 12, 109, 110, 145, **161**–**166**, 169, 171, 172, 177, 183, 184, 190, **252**–**254**, 262, 265, 315, 415, 417, 418, 420, **431**, 433
Polaris 402
polarization 56, 87, 158, 328, 356
polar motion 15–17, **20**, 21, 23, 167
pole coordinates 20, 21
position accuracy 6, 11, 160, 169, 203, 214, 318, 373, 374, 400, 402, 403, 405, 411, 414, 416, 418–420, 422, 426, 427, 429, 431–433, 436, 439, 440, 451, 454, 460, 461
position determination **8**, 331, 390, 400, 403–405, 407, 409, 415, 420, 429–**431**, 445, 456, 464, 465
position error 27, 241, 269, 270, 275, 319, 381
postglacial rebound 465
postprocessing **12**, 174, 193, 404, 420, 433, 437, 447, 456, 458, 464, 468
power spectral density **68**–71, 75, 76, 80–83, 330, 357, 359, 385
PPS, *see* precise positioning service
PRARE **403**, 446

precession 15, 17, **18**, 19
precise ephemerides 5, 7, 27, 49, **52**, 418, 459
precise point positioning **166**, 417, 418, 433
precise positioning service 315, 316, **318**, 319, 330, **348**, 456
pressure 35, 36, **38**, 39, 51, 62, 65, 129, 135, 140, 167, 445
presurvey planning 465
prime vertical 277, 278, 286, 408
PRN, *see* pseudorandom noise
probability level 110, 272, 275, 317, 318, 320, 431, 445
propagation direction 56, 62
propagation effect **61**, 68, 73
propagation time 55, 328
protection level 269–271, 382, 423
PRS 372, 382, 385, 387, 389, 392
PSD, *see* power spectral density
pseudokinematic relative positioning **187**, 188, 437
pseudolite 374, **427**, 428
pseudorandom noise 75, 77–80, 82, 90, 105, 316, 320, **329**–334, 336, **357**, 358, **384**, 386, 389, 403, 413, 414, 421, 424
pseudorange 4, 6–9, **11**, 67, **105**–116, 119, 120, 125–127, 149, 152, 155, 161, 163, 165–167, 169–171, 189, 197, 270, 316, 319, 320, 325, 362, 416, 417, 431–436, 448, 454, 456, 458
pseudorange correction 169–171, 416, 448, 458
public-private partnership 368
public regulated service 367, **372**, 389, 392
PVS 431
PVT 85

QASPR 415
quadrature phase-shifted key 74
quality control 239
quasi-codeless technique 101, 102, 200, 456
quasi-inertial geocentric system 16
quasi-inertial terrestrial dynamic time 23

QZSS 337, **409–411**, 412–414, 450

Radial component 32
radial orbital error 170, 171
radial relative velocity 59
radial velocity 9, 59, 106, 108, 165, 440, 446
radio frequency 80, 85, **87–89**, 98, 331, 355
radio link 325, 354
radiometer 137
radionavigation plan **452**, 453
radio section **88**
Radio Technical Commission for Maritime Services 99–101, 346, 401, 417, 419, 427, **447**–449
rail application **461**
RAIM, *see* receiver autonomous integrity monitoring
range rate 9, 10, 40, 90, 108, 165, 169, 170, 172, 397, 403, 404, 416
range rate correction 169, 170, 172, 416
ranging code 55, 68, 74–**77**, 80, 83, 84, 91, 92, 94–97, 100, 101, 111, 119, 329, 330, 332, 334, 336, 355–358, 360, 371, 376, 382, 384–387, 389, 390, 392, 400, 413, 414, 421, 422, 427, 430, 453, 455, 456
rank deficiency 238
rapid static 155, 437, 438
rate-aided code loop 100
Rayleigh criterion 61
Rayleigh equation 117, 118
reacquisition 83, 91, 454
real-time accuracy 355
real-time application 464
real-time computation 437
real-time kinematic 172, 173, 188, 213, 419, 437–439, 448, 449, 454, 458, 464
real-time positioning 113, 169, 419, 420
receiver autonomous integrity monitoring 270–272, 370, 371
receiver calibration **455**
receiver clock 4, 8, 105, 107, 110, 144, **147**, 159, 162–164, 166, 167, 170–172, 175, 202, 221, 241, 252–254, 263, 319, 401, 435, 444, 445

receiver independent exchange (format) 48, 193, 194, 401, 447, **449**
receiver types 7, 108, 449
reconnaissance 467
recursive least-squares adjustment 212, **241**, 244, 246
redundancy 115, 174, 211, 212, 215, 238, 261, 270–272, 338, 339, 392, 400, 421, 432, 449, 451, 461
reference carrier 96, 106
reference ellipse 43, 45
reference frame 10, 16, 17, 21, 22, 47–49, 138, 139, 143, 144, 147, 277, 314, 345, 369, 400, 406, 457, 463
reference frequency 67
reference satellite 180, 184, 259
reference signal 105
reference site 5, 258, 259
reference system 13, **16**, 17, 142, 144, 309, **313**, 342, **345**, 361, **369**, 400, 401, **406, 410**
reflection 61, 109, 110, 154–156
refraction 61, 62, 66, 109, **118**, 119, 121–123, 126, **128**, 129, 138, 140, 155, 166, 170, 171, 188, 190, 197, 206, 208, 282, 305, 316, 331, 445, 465
refractive index 62, 63, 65, 67, 117–119, 128
refractivity 128–130, 133, 135
relative accuracy 267
relative baseline error 27, 319
relative positioning **11**, 12, 145, 147, 155, 171, 172, **173**, 174, **183**, **185**, **187**, 188, 227, **255**, 256, 262, 265, 433, **437**–439, 441, 458, 464
relativistic effect 13, 17, 35, **39**, **141**, **144**, 145, 147, 154, 162, 329, 356, 383, 412, 413, 445
relativity 13, 23, 68, 92, **141**–**147**, 169, 435
reliability 47, 103, 172, 203, 213, **272**, 340, 355, 366, 376, 420, 452, 460, 467, 468
remote sensing 445, 446, 468
repeatable accuracy 267
resonance effect 45
retroreflector 323, 351, 377
returnlink 381, 406
RF, *see* radio frequency

RF front-end 85, 87, 89, 98, 355
right ascension 28, 45, 49–51, 402, 412
RINEX, *see* receiver independent exchange (format)
RINEX format 194, **449**
RNSS 61, 329, 357, 383, 384
road application **463**
rotation angle 18, 229, 295, 298–301, 346, 401
rotation matrix 17, 20, 21, 31, 32, 265, 293–295, 297, 306, 435, 442, 443
roving receiver 165, 186, 187, 193, 415, 436
RTCM, *see* Radio Technical Commission for Maritime Services
RTCM format 419, **447**, 448
RTK, *see* real-time kinematic
rubidium 67, 316, 351, 376, 411

SA, *see* selective availability
Saastamoinen model **135**, 136, 137
safety integrity level 462
safety of life 335, 336, 355, **371**, 372, 385, 387, 390–392, 422, 424, 450, 458
safety-of-life service 336, **371**
Sagnac effect 147, 167
SAIF 413, 414
sampling rate 80, 90, 114
SARSAT 372, 381, **405**, 406
satellite antenna 84, 169, 328, 356, 433
satellite availability 6
satellite clock 9, 39, 40, 48–52, 105, 110, 111, **145**, 146, 161–163, 166, 169, 170, 175, 183, 189, 241, 252, 271, 316, 319, 320, 338, 342, 350, 352–354, 361, 379–381, 390, 391, 413, 414, 417, 421, 423, 433, 435, 436, 459
satellite diversity 104, 393
satellite ephemerides 3, 55, 165, 283, 316, 325, 381, 422, 426, 459
satellite geometry 6, 9, 155, 203, 217, 241, 263, 266, 381, 438
satellite health status 337, 360
satellite laser ranging 16, 47, 293, 314, 346, 380

satellite multiplexing **84**
satellite-only service 370, 453
satellite orbit 6, 12, **27**, 28, 39, 144, **145**, 146, 169, 190, 262, 292, 314, 325, 326, 337, 352, 360, 374, 377, 381, 404, 411, 421
satellite signal **55**
satellite visibility 7, 264, 340, 459
S-band 325, 327, 377, 380, 383, 403, 414, 418
SBAS, *see* space-based augmentation system
scale factor 22, 150, 227, 228, 262, 289, 293–295, 297–299, 301, 306, 346
scattering 61, 155, 202
scintillation 63, 66, 92, 99
SD, *see* selective denial
sea level 135, 137, 291, 292, 404, 465
search and rescue 351, 370, **372**, 392, 405, 406
search technique 187, 205, 206, **214**, 218, **222**, 224, 225, 234
second-order Doppler effect **143**
selective availability 311, 315, 318, **319**–322, 324, 343, 347, 348, 415, 456
selective denial 319, 321
semikinematic 438
semimajor axis 28, 36, 41, 44, 50, 51, 146, 231, 276, 313, 314, 345, 375, 406, 410, 412, 432
semiminor axis 30, 302
sensitivity 83, 88, 94, 95, 268, 464
sequential adjustment 206
service availability 370
session 138, 169, 183, 193, 194, 258, 353, 437–439, 465
Shannon theorem 90
shift vector 262, 293, 295, 297, 300, 306, 346, 442
sidereal day 28, 45, 151
sidereal time 13, 17, **20**, **22**, 23, 31, 49
signal amplification 90
signal components 67, 92, 316, 330, 382, 383, 385, 387, 389, 390
signal design **68**, 83, 84, 87, 382, 405
signal in space 269, 316, 380, 425, 430, 453
signal lock 12, 187, 195, 438

signal multiplexing **74**, 82
signal processing 68, **84**, 89, 90, 103, 195, 200
signal squaring 96, 99–101
signal structure 5, **68**, 316, 322, **327**, 347, 354, 366, **382**, **411**, 469
signal task force 366, 382
signal-to-noise ratio **86**, 87, 90, 96, 99, 101, 104, 111, 159, 168, 183, 195, 337, 433, 449, 456
SIL, *see* safety integrity level
similarity transformation 21, 262, 293, 296, 297, 299–301, 306, 443, 457
sine-cardinal function 70, 75
single baseline 183, 213, 234, 258, 259
single-difference 150, 171, **174**–177, **179**, 180, 182–186, 202, 219, 221, 257–259
single frequency 5, 111, 116, 124, 152, 169, **205**, 261, 370, 371, 394, 423, 438, 439, 445, 464, 465
single-frequency receiver 5, 111, 261, 370, 394, 423, 439, 464, 465
single-sided bandwidth 99, 100
SISA 381, 382
SISE 381, 382
SISNeT 425
SkyFix 417, 418
sky plot 283, 431
sky wave 62
SLR, *see* satellite laser ranging
smoothed code pseudorange 114, 116, 416
smoothing 11, 100, 111, **113**, 115, 116, 159, **250**, 433, 455
S/N, *see* signal-to-noise ratio
snapshot algorithm 271
SNAS **426**
Snell's law 62
software-based receiver 85, 470
SoL, *see* safety of life
solar activity 63, 66, 124, 465
solar panel 7, 39, 350, 351, 377
solar radiation 35, 36, **38**, 39, 51, 66
solar time 23
sovereignty 365, 461
space-based augmentation 6, 194, 409, **421**, 424–426

space-based augmentation system 6, 194, **421**–427, 449, 450
spaceborne receiver 263, 331, 356, 357, 465
space segment 5, 6, **322**–**324**, **348**, **350**, 369, 373, **374**, 376, 411, 414, 418, 421, 425, 453, 470
space segment modernization **324**, **350**
space-time coordinates 141, 142
special relativity **141**–144
spectral spreading **75**–77, 100, 335, 388
spectrum 44, **60**, 69, 70, 74–76, 80, 81, 83, 84, 86, 89, 96, 100, 322, 332, 380, 384, 389, 390, 469
speed of light 4, 39, 58, 62, 67, 100, 105, 117, 123, 141, 147, 161, 163, 176, 440
spherical error probable 276
spherical harmonics 44, 138, 292
split-spectrum signal 81, 322
spoofing **83**, 315, 319, **322**, 327, 334, 348, 372, 456
spot beam signal 332
spreading code 74, 76, 80, 84, 337, 390
spread spectrum 75, 76, 83, 84, 384
SPS, *see* standard positioning service
squaring technique 100, 101
standard atmosphere 135
standard deviation 214, 217, 223–225, 237, 266, 273, 292, 382
standard ellipse 231, 232, 275
standard ellipsoid 276
standard epoch 16–19, 24, 315
standardized variable 273
standard meridian 284
standard parallel 284
standard positioning service 315, **316**–319, 330, 343, **347**
Stanford diagram 270
StarFire 418
StarFix 417, 418
state-space approach 417
state vector 159, 225, 226, 244–250
static initialization **187**, 206
static point positioning 162–164, 184
static relative positioning **183**, 185, 437
static survey 155, 168, 183, 438
stereographic projection 284
stochastic variable 272–274
stop-and-go technique 438

Subject index

stratosphere 65, 128
strobe correlator 159
Student's distribution 223
submodulation **80**
sunspot 66, 67, 121, 124, 465
surface of position 10
surveillance mode 436
surveying **1**, 4, 9, **12**, 40, 109, 187, 262, 309, 400, 437, 460, **464**, 465, 467
synchronization 55, 102, 103, 324, 328, 335–338, 342, 351, 354, 356, 361, 379, 391–393, 407, 411, 422, 428, 445
system integration **450**–452

TACAN 329
TAI, *see* international atomic time
Taylor series 106, 147, 238, 245, 251
TCAR, *see* three-carrier ambiguity resolution
TDMA, *see* time division multiple access
TDT, *see* terrestrial dynamic time
TEC, *see* total electron content
telemetry word 337
temperature 62, 65, 86, 129, 131, 135–137, 140, 291, 350, 358, 445, 455
terrestrial data 277, 293, **302**
terrestrial dynamic time 22–24
terrestrial reference frame 16, 17, 21, 22, 47, 49, 314, 369, 400
tetrahedron 9, 263
thermal noise 86, 159
thermal radiation 39
three-body problem 37
three-carrier ambiguity resolution 205, 210, 213, 339, 374
three-dimensional transformation **293**, 294
tidal attraction 16, 35
tidal deformation 15, 23, 38
tidal effect 16, **37**, 51
tidal potential 14, 15, 38, 45
tide gauge 292
tiered code 79, 83, 386–388
time conversion **23**–25
time-delayed indirect signal, *see* multipath
time dilation **142**, 143
time division multiple access 84, 365

time of arrival 406, 465
time of transmission 55
time system 14, 20, **22**–24, 28, 55, 105, 106, 250, **315**, 337, **346**, 361, **369**, 401
time tags 174, 283
time to alarm/alert 269, 371, 414, 419, 461, 462
time to first fix 103, 104, 391, 393, 429
time transfer 309, 317, 318, 379, **445**, 468
TOA, *see* time of arrival
tomography **138**, 445, 446
torque 14
total electron content 66, 120–**122**, 124–**126**, 138–140
tracking loop 85, 91, **96**, 97, 99–103, 168, 430, 452
tracking network 7, **47**, 48
trajectory 266, 438, 452
Transit 4, 5, 8, 108, 309, 313, 314, 403
transition matrix 245, 247, 249
transmission loss 6, 63–65, 85
transmission time 108, 309, 337, 393, 409
transmitter 59, 62, 64, 68, 149, 168, 331, 349, 376, 404, 419, 427, 428, 462
transverse Mercator projection **284**, 286
travel time 105, 106, 460, 463
triangulation 1, **2**, 290, 457
trilateration 1, **2**
triple-difference 174, **177**, 178, **181**–183, 185, 186, 188, 196, 199, 201, 221, 225, 239, 257, 258
troposphere 62, 65, 110, 111, 128, 130, 131, **139**, 140, 154, 169, 191, 203, 252, 317, 436, 446
troposphere sounding **139**
tropospheric delay 65, 110, 128, 134, 135, 138, 166, 167
tropospheric path 128, 130–133, 140
tropospheric refraction 109, **128**, 129, 140, 166, 188, 190
tropospheric refractive index 65
true anomaly 28, 30, 432
Tsikada 5, 341, 403
TTA, *see* time to alarm/alert
TTFF, *see* time to first fix
Twin-Star 402, 415
two-dimensional transformation **297**, 298
two-sided bandwidth 70

two-way ranging system 55, 397, 406, 414

UERE, *see* user equivalent range error
UERE computation 110, 111, 241, 436
undulation 290
unit sphere 9, 28, 263
universal time 22–25, 122–125, 139, 315, 321, 369, 406
universal transverse Mercator 284, **289**
updated state vector 249, 250
update rate 326, 377, 454
uplink 55, 60, 349, 353, 374, 376, 378, 383, 397, 403, 404, 411, 414, 426
upload 7, 50, 125, 324–326, 337, 351
user differential range error 423
user equivalent range error 110, 111, 178, 241, 266, 267, 381, 436
user segment 7, 369, 373, 376, 407, 422, 431, **453**, 456, 457, 470
UT, *see* universal time
UTC 22–24, 315, 338, 346, 347, 351, 354, 361, 369, 371, 379
UTM, *see* universal transverse Mercator

Variance-covariance matrix 205, 223, 224, 237, 244
velocity determination **9**, 108, **440**
velocity vector 10, 30, 32, 33, 40–42, 45, 49, 50, 52, 146, 165, 404, 440
vernal equinox 1, 13, 16, 18, 19, 22
vertical delay 123, 417, 423
very long baseline interferometry 16, 47, 293, 314
virtual reference station 173, **188**–191, 203, 417, 458
visibility 7, 49, 262, 264, 283, 340, 349, 422, 437, 452, 459
Viterbi 393

VLBI, *see* very long baseline interferometry
VPL 269, 270
VRS 173, **188**–191, 458
vulnerability 319, 372

WAAS, *see* wide-area augmentation system
WAD, *see* wide-area differential
warm start 91, 454
WARTK 439
water vapor 62, 65, 129, 131, 135, 137, 140, 445
wavelength 11, 58, 61, 99, 101–103, 107, 110, 112, 113, 116, 117, 126, 163, 177, 197, 207, 211, 217, 219, 316, 329, 347, 357, 384
wave propagation **61**–**63**, 117
W-code 101, 102, 322, 334
weight matrix 227, 236, 239, 257, 260, 262, 263
wet component 65, 129, 131, 132, 134, 136, 140, 167
WGS-84 17, 47, 277, 280, 291, 295, 299, 313, 314, 346, 400, 401, 410, 432, 434
wide-area augmentation system 6, **424**, 426, 454
wide-area differential 407, 417, 421, 423, 424
wide lane 112, 217, 218

Yaw 10, 441, 442
Y-code 101, 102, 316, 318, 322, 327, 334, 455, 456

Zenith delay 131, 135, 137, 138, 140, 169
zero baseline 455
Z-tracking 102

SpringerGeosciences

Bernhard Hofmann-Wellenhof,
Helmut Moritz

Physical Geodesy

Second, corrected edition.
2006. XVII, 403 pages. 111 figures.
Softcover **EUR 59,–**
(Recommended retail price)
Net-price subject to local VAT.
ISBN 978-3-211-33544-4

"Physical Geodesy" by Heiskanen and Moritz, published in 1967, has for a long time been considered as the standard introduction to its field. The enormous progress since then, however, required a complete reworking. While basic material could be retained other parts required a complete update. This concerns, above all, the adaptation to the fact that the geometry can now be precisely determined by methods such as GPS, and that new satellite methods, combined with terrestrial methods, also make a detailed determination of the earth's gravitational field a possibility and a necessity. Highlights include: emphasis on global integration of geometry and gravity, a simplified approach to Molodensky's theory without integral equations, and a general combination of all geodetic data by least-squares collocation. In the second edition minor mistakes have been corrected.

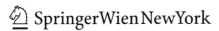

P.O. Box 89, Sachsenplatz 4–6, 1201 Vienna, Austria, Fax +43.1.330 24 26, books@springer.at, **springer.at**
Haberstraße 7, 69126 Heidelberg, Germany, Fax +49.6221.345-4229, SDC-bookorder@springer-sbm.com, springer.com
P.O. Box 2485, Secaucus, NJ 07096-2485, USA, Fax +1.201.348-4505, service@springer-ny.com, springer.com
Prices are subject to change without notice. All errors and omissions excepted.

Springer and the Environment

WE AT SPRINGER FIRMLY BELIEVE THAT AN INTERnational science publisher has a special obligation to the environment, and our corporate policies consistently reflect this conviction.

WE ALSO EXPECT OUR BUSINESS PARTNERS – PRINTERS, paper mills, packaging manufacturers, etc. – to commit themselves to using environmentally friendly materials and production processes.

THE PAPER IN THIS BOOK IS MADE FROM NO-CHLORINE pulp and is acid free, in conformance with international standards for paper permanency.